TEXT BOOK
OF
CRUSTACEA

TEXT BOOK OF CRUSTACEA

By

Dr. Amita Saxena

Associate Professor

College of Fisheries

G.B. Pant University of Agriculture & Technology

Pantnagar (Uttranchal)

D P H

DISCOVERY PUBLISHING HOUSE

NEW DELHI-110002

First Published-2005

ISBN 81-8356-016-4

Published by

DISCOVERY PUBLISHING HOUSE
4831/24, Ansari Road, Prahlad Street,
Darya Ganj, New Delhi-110002 (India)
Phone: 23279245 • Fax: 91-11-23253475
E-mail:dphtemp@indiatimes.com

Printed at:

Tarun Offset Printers,
Maujpur, Delhi-53

PREFACE

This Crustacea group has an enormous potential as for as fishery sciences are concerned. In this book mainly aquatic crustaceans are dealt which are ecologically important in food chain and there are few which could grow to the big size are well cultured in India and abroad. Before going for their cultural aspects one should know the structure their modifications and functioning of various parts. Lnowledge of functional anatomy leads to better culture practices as well as enhancement of production of various aquatic organisms. The food and feeding, digestion, absorption, respiration, Circulation of blood, haemolymph, acid base balance, Excretion Osmoregulation, Hormones, Reproductive mechanism fertilization and various larval stages are described in details. This book will be good for students, farmers and scientists also.

I am grateful to my Hon'ble Vice Chancellor dean students, scientists, colleagues and respected sisters who have provided me literature, help and expert opinions to compile this work. I am also thankful to Publisher for Publication. I am undebtful to God who induce strength and intellectual ideas to do such kind of work.

—Author

Contents

①

INTRODUCTION—CRUSTACEA

Arthropods, the most are of aquatic habitat and in mode of respiration, whose second and third somites may be jointed bear antennae and their forth somite of hair of mandible, mainly dealt here.

FACTORS IN THE EVALUATION OF THE CRUSTACEA

1. *Specialization of limbs :* The crustacea are essentially aquatic arthropods that fact along makes it a possible that if then the same appendages should combine the function of locomotion by swimming feeding by gathering particles from water respiration by exposing a thinly course surface to the medium and the receipt the sensory stimuli.
2. *Shortening of body :* Two other factor added to a perhaps consequent upon the specialization of limbs have taken part in bringing about the great variety of organisation which exists in crustacea one is shortening of the body as the efficiency of limbs increased by specialization.

   Then occurs a lessening of their number and finally the reduction.
3. *Development of carapace :* The special factor is the development from the linear part of the head of a stand fold the carapace by which two Important anterior region of the body is overhung and protected and the setting up the surrounding water of currents for purposes of respiration and feeding facilitated.

CLASSIFICATION

1. S. Class Branchiopoda

Free crustaceans with compound eye usually a carapace at least four pair of trunk limbs which are in most cases broad lobed and fringed on the inner edges with bristles.

1. *Order-Crustraca :* Its carapace stalked eye antenna of hair long size but that not biramus, trunk, limbs, numerous and all alike caudal somite unjointed.

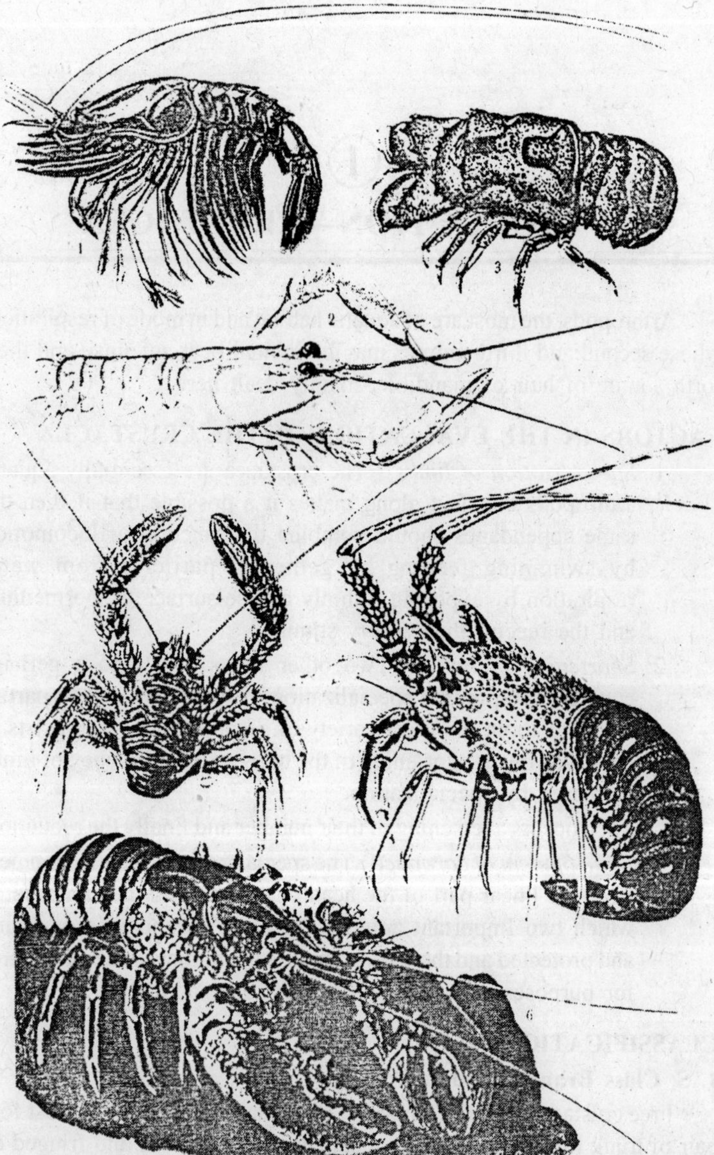

Fig. 1.1.

2. *Order-Lepostraca :* Fossils order represented by *lepodocaris.*
3. *Order-Notostraca :* Carapace a broad shield, above trunk, compound eye sessile and close together, trunk limbs numerous thick pair of little ring considerably from the rest multidenticulate caudal rami.
4. *Diplostraca :* Compressed, carapace enclosing trunk and limbs compound eye sessile. Antenna large and biramus.

2. Sub Class

Ostracoda : Few crustaceans with a bivalve shell and abductor muscle and not more than two recognizable pairs of trunk limbs.

3. Sub Class

Copepoda : Free or parisitic crustaceans without compound eye or carapace typically size pairs of thoracic limbs of with the 1st is always branched and the else are often uniramus.

4. Sub Class

Branchiura : Crustacea temporily parasitic on fish with compound eye and suctorial mouth carapace like expansion of head. Unsegmented limbs.

5. Sub Class

Cirripedia : Most of part paired hermaphrodite crustaceans, without compound eye in the adult with carapace mandibular palp and typically size pairs of thoracic limb. It is diveded into following orders:

Order

1. *Thoracica :* Cirripedia with alimentary canal. Size pairs of thorasic limbs, no abdominal somites.
2. *Acrothoracica :* Separate sexes, have an alimentary canal less than 6 pairs of thoracic limbs,
3. *Apoda :* Hermaphrodite, no mantle, no thoracic limbs and no anus.

Order

Ascothoracica : Parasitic with alimentary canal and 6 pairs of thoracic appendages.

Sub Class

Malacostraca : Crustaceanse with compound eye, untypically stalked, typically a carapace covering, the thorax of 8 somites and abdominal of 6 somites.

Order

Leptatraca : Have 7 abdominal somites large carapace, that fee fed to any thoracic somite.

Order

Hoplocarida : Shallow carapace fused to 3 thoraxic somites.

Order-Syncarida : No carapace.

Orders-Eucarida : Carapace fused to the thorasic segment No ostegites.

General Structure

Cuticle : The cuticle of a crustaceans is safety for the jointed legs, usually stout selender to the size of the animal, but is thinner and flexible in many parasite genera, it is often strengthened by calcification.

Somites : These are embryological indications that the body should be regarded as containing, besides the somites an anterior presegmental region to which the eyes belong the somite never all remain distinct in the adults.

Tagmatization

Head : The head always contain besides the region of the eyes and the embryonic flat somite, the somites of 5 pairs of appendages –2, the antennules and (2) antennae, pleural and 3-the mandibles, rostrum and maxilllae.

Thorax and abdomen : The head, though it varies in extent. They decide boundary between thorax and abdomen it is sometimes difficult to fix. The somite which bears the genital opening is always the last somite of the true thorax.

Carapace : A structure very commonly found in crustacean In the shell or carapace, a dorsal fold of skin arising from the border of the half and tending for a greater or less distance over the trunk. The carapace may fuse with the dorsal side of some or all of the thoracic somites in front of the carapace is continuous with the dorsal plate which represent the terge of the head the cervical groove.

The nervous system of crustacean exhibit a very complete such of stages from the ideal arthopod condition.

Sense organ : Sense organs are well developed. Eyes are of two kinds, the compound eye, of which a pair is usually present except in the copepoda the median eye are sessile or stalked.

Digestive system : The alimentary canal is with very rare exceptions straight at its anterior end. The midgut usually bulge from its anterior end.

Digestion : Digestion is extracellular, therefore frequently the seat of mechanical process and sometimes of chemical action by juices secreted by the rnidgut and intestine.

In the hindegut the faeces are passed to the anus

Excretion : The principal organ of the extretion are two pairs of gland known as the antennary and maxillary glands"

Respiration : In many of the smaller crustacean and mostly in the copepoda the respiration takes place through the general surface of body in forms with stouter which are more or less bulky bodies. This is supplemented or replaced by the use of special organ upon which the cuticle remain pithed. The gills are common respiratory organ.

Vascular system : The vasculai system is seen in its most proportionate condition in the Brachiopoda the heart runs the whole length of the trunk situated above the gut. Blood, sinus and pericardium present.

Reproductive system : It is usual with animals that are free and active the sexes are separate as in the great majority of the crustaceans.

The spermatozoa are very varied in form and other complex structure usually but not always they are immobile. They are transferred to the female in spermatophores form.

The ova usually have much yolk and meroblastic centrolecithal cleavage, but some time are less yolky? Undergoes total cleavage gastrulation may be invagination.

Larval Forms

Typically the crustaceans hatches as a replicas larva of a minute creature from egg, shaped with the broad end in front, unsegmented, the median eye is the only organ of vision, fore, mid and hindgut can be recognised in the alimentry canal in the branchiopoda larvae. Other larval forms zoea, metazoea, cypris, megalopa etc. are also present in various curstaceans. The nauplius so transformed gradually into the adult.

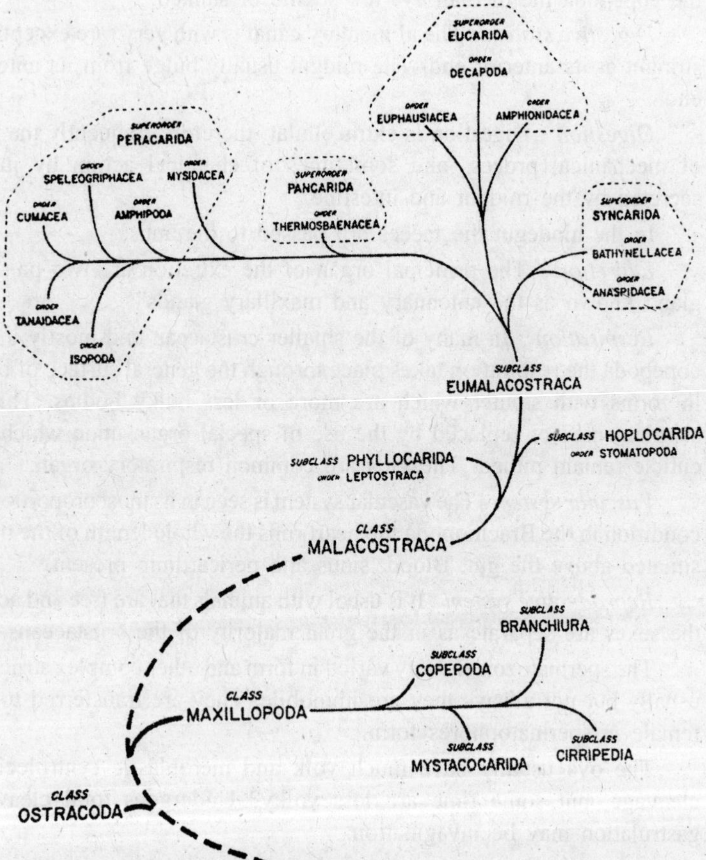

Fig. 1.2 : A visual representation of the Bowman and Abele classification of crustacea. This is not intended to indicate phylogenetic relationships and should not be so interpreted. The dashed line at the base emphasizes uncertainties concerning the origins of the five classes and their relationships to each other.

②

TAXONOMY OF CRUSTACEA

Approximately 30,000 species make up this Subphylum. Most are aquatic; of these, the majority are marine but some are found in fresh water. Members of the Subphylum include lobsters, crabs, crayfish, shrimp, copepods, barnacles, and several other groups of organisms. All have two pairs of antennae, a pair of mandibles, a pair of compound eyes (usually on stalks), and two pair of maxillae on their head, followed by a pair of appendages on each body segment (crustacean bodies usually are made up of head, thorax, and abdomen, although the segments composing these tagmata differ among different Classes). The appendages are primitively branched (biramus), and although this condition is modified in many species, adults always have at least some biramus appendages. Crustaceans respire via gills. Like other arthropods, all have a hard but flexible exoskeleton.

Most crustaceans are free-living, but some are sessile and a few are even parasitic. Most use their maxillae and mandibles to take in food. The walking legs, including specialized chelipods, may be used to help capture prey. Some crustaceans filter tiny plankton or even bacteria from the water; others are active predators; while still others scavenge nutrients from detritus.

Most crustaceans are dioecious. The actual mechanisms by which fertilization is achieved, vary greatly. Some crustaceans hatch young that are like miniature adults; others go through a larval stage called a nauplius.

Many species, including lobsters, crayfish, barnacles, and crabs are important to human economies, some very much so. Others, such as krill, are at the base of extremely important marine food chains. Still others are crucial in recycling nutrients trapped in the bodies of dead organisms.

- Class *Remipedia*
- Class *Cephalocarida*
- Class *Branchiopoda* (fairy shrimp, water fleas, etc.)

- Class *Maxillopoda* (ostracods, copepods, barnacles)
- Class *Malacostraca* (isopods, amphipods, krill, crabs, shrimp, etc.)

REMIPEDIA

These primitive crustaceans, which were discovered in submerged caves only in 1980, have a long trunk made up of many segments, each bearing a pair of biramus swimming appendages. Not much is known of their biology.

Cephalocarida

These tiny crustaceans make a living by feeding on benthic marine detritus. Only nine species are known. The second pair of maxillae of cephalocarids closely resembles the appendages of the thorax. In this and other characteristics cephalocarids resemble what we imagine primitive crustaceans might have been like. Cephalocarids also lack abdominal appendages. They have small compound eyes that are buried in the exoskeleton, rather than being raised on stalks as in most other crustaceans. As in branchiopods and malacostracans, cephalocarids feed by generating currents with their thoracic appendages. These currents bring in food particles, which are trapped and passed anteriorally along a ventral groove leading to the mouthparts.

Branchiopoda

(Fairy shrimp, water fleas, etc.)

The four orders that make up this class include brine shrimp, tadpole shrimp, water fleas, and clam shrimp. Most live in fresh water or in small, salty lakes or ponds. Most are small (under 2 cm). The number of segments and appendages possessed by an individual varies considerably among groups, but in most forms the number of segments is large and tagmosis is minimal. About 800 species are known; most feed by filtering small organisms and organic particles from the water. Currents that move water for feeding are generated by thoracic appendages, and food particles are passed anteriorally along a ventral groove leading to the mouthparts.

Maxillopoda

(Ostracods, copepods, barnacles)

Maxillipods include barnacles, copepods, mystacocarids, tantulocarids, branchiurans, ostracods, and related groups. Most species are small. Most feed by means of their maxillae (rather than filter feeding using thoracic appendages to move water); barnacles, however, are an exception. Barnacles feed with thoracic appendages, but in a way that is unique among crustaceans. Other characteristics of maxillipods including a basic plan of 5 head and 10 trunk segments (6 thoracic and usually 4 abdominal), followed by a terminal telson. The abdominal segments typically lack appendages; appendages elsewhere on the body are usually biramus.

Barnacles grow to encrust solid structures placed in marine environments, including docks and pilings and also the bottoms of ships.

Malacostraca

With over 20,000 species, Malacostraca is by far the largest Class of Crustaceans. Its members, which are primarily marine but also occupy some freshwater and terrestrial habitats, are extremely diverse. They include isopods (sowbugs), amphipods, euphausiids (krill), beachhoppers, mantis shrimp, and a very large order, the Decapoda, that contains many kinds of shrimp, crabs, and crayfish. Malacostracans are characterized by a trunk with eight thoracic and six abdominal somites, each bearing a pair of biramus appendages.

As in branchiopods and cephalocaridans, primitive malacostracans feed by filtering food from water; currents that move water for feeding are generated by thoracic appendages. Food particles are passed anteriorally along a ventral groove leading to the mouthparts. Other malacostracans are predatory, capturing prey by a variety of means and tearing it into edible pieces with their appendages. Yet others scavenge organic material by scraping it from the substrate, from living fish in the case of cleaner shrimp, or even from fallen leaves and other plant material in the case of terrestrial isopods. A few species are even parasitic, living on other crustaceans or on fishes.

The Class Malacostraca contains a number of species of considerable economic significance. These include edible lobsters, shrimp, crayfish and crabs. Equally or more significant are the many malacostracans which, as larvae or adults, contribute to plankton and as such are at the base of an immensely important marine food chain.

Species included in the Animal Diversity Web:

Order Decapoda

Paralithodes carpischatica

Macrocheira kaempferi (Giant Japanese Spider Crab)-t-

Order Isopoda

Order Stomatopoda

Gonodactylus curacaoensis (Dark Mantis)

Recognition

Diverse! All non-insect, non-arachnid aquatic arthropods. Most have 2 pairs of antennae, and many pairs of segmented appendages, typically with gills for gas exchange. The appendages are basically biramus (protopod, endopod, exopod). The head has 5 pairs of appendages: first antennae (antennules), second antennae (or just antennae), mandibles, first maxillae, second maxillae.

The trunk is primitively composed of many similar segments each with similar biramus appendages, more advanced groups have uniramus appendages, fewer segments that are more specialized or fused into distinct regions (thorax and abdomen), part of the thorax may be fused to the head (cephalothorax).

The trunk appendages are often highly modified (maxillipeds, gnathopods, pereiopods, pleopods, and uropods). Malacostracan pereiopods may be chelate and have 7 segments (coxa, basis, ischium, merus, carpus, propodus, and dactylus), the exopodite is often loss.

Terminal telson is usually not counted as a segment.

Life Cycle

Most crustaceans are dioecious with copulation, but parthenogenesis is common and hermaphroditism sometimes occurs. Marine forms often have a series of planktonic larval forms (nauplius, zoea, megalops, etc.). Freshwater species tend to have more direct development and to brood the eggs. Females may produce a single or multiple broods per season or alternate between sexual and non-sexual (parthenogenetic) reproduction. Molting may continue after the adult stage is reached. Resistant eggs are common. Non-emergent, only resistant stages may disperse over land. Cyclomorphosis occurs in some species.

Trophic Roles

Free-swimming and filter-feeding by way of setae is primitive in Crustacea and still widespread. Others are benthic or intermittent swimming filter-feeders. Others crawl or burrow and feed as herbivores, scavengers or carnivores. A few are parasitic.

Communities

Often rather simple, at least by insect standards, with few species, though some in great abundance. Crustaceans occupy nearly all oxygenated aquatic habitats, including intermittent ponds, saline ponds, hot springs and deep lakes. The species are separated by micro-habitat, habits, body size, food size, seasonality, or time of activity. They are the dominant macroinvertebrates in the open water of large lakes and reservoirs, but shallow, heavily vegetated lentic water has the most species. Some species have considerable diurnal vertical migrations. Many species occur in interstitial areas, subterranean waters (trilobites), and groundwater.

Economic Importance

Important primary and secondary consumers in many aquatic systems, critically so in lentic waters. Some species support direct fisheries and aquaculture for human food or fish bait, important fish and wildlife foods, some may help control aquatic vegetation, many are good environmental indicators for metals, acid rain, global warming, etc. Parasites may destroy hatchery fish, some serve as intermediate hosts to other parasites, some may out-compete fish for food or prey on the fish directly, some damage underwater wood, dense populations may clog filters, burrowing crayfish may denude farmland and destroy earthen dikes and dams.

Zoogeography

Most major groups occur worldwide, species of small body size or those with resistant eggs may also be globally distributed. In general, most groups have a few species that are widespread and many species that are endemic to single stream systems or to caves, some of these are endangered. Many groundwater species evolved in deep refuge from surface fluctuations, some of these survived under the continental ice sheets.

Collecting

Plankton nets and routine benthic samples will get many of the “important” species. Many require very fine nets. Some plankton species require large nets worked very deep or on dark nights. Crayfishes may be better trapped or picked up at night by using headlamps, interstitial and ground-water species may be filtered from pumped water or taken in springs. Lighted bottle traps take many species in abundance, even very small ones.

Preservation

Small crustaceans can be adequately fixed directly in 70-80% ethanol. Larger specimens are best fixed in neutral 5% formalin and transferred to 70% ethanol within a few days. Kahle's fluid quickly decalcifies crustacean exoskeletons, but preserves the rest very well. Isopropyl alcohol is of no use with crustaceans as they become explosively brittle. Briefly boiling the specimens in water or 70% ethanol is often a suitable substitute for formalin. Most color patterns are lost during storage so notes should be taken while the specimens are fresh.

Taxonomy

Fairly well studied at the order and family level. Higher level relationships are more difficult. Some species are well known, but many are known only from the original description, and some remain undescribed. Juvenile instars and the "wrong" sex can be very difficult or impossible to identify. Many species identifications require carefully prepared slide mounts.

Higher classification of freshwater Crustacea

- Subclass Branchiopoda-phyllopods, First antennae reduced.
  - Order Notostraca-tadpole shrimps
  - Order Anostraca-fairy shrimps
  - Order Conchostraca-clam shrimps
  - Order Cladocera-water fleas
- Subclass Ostracoda-seed or mussel shrimps, no compound eyes, carapace calcified
  - Order Podocopa-2nd antennae with endopodite length greater or equal to the exopodite
- Subclass Copepoda-copepods, nauplius eye, but no compound eyes, no carapace
  - Order Harpacticoida-1st antennae with 5-9 segments, benthic and interstitial
  - Order Cyclopoida-1st antennae with 10-16 segments, planktonic and benthic
  - Order Calanoida-1st antennae with 16-26 segments, planktonic
  - Order Siphonostomatoida-specialized parasites
- Subclass Branchiura-sessile compound eyes, shieldlike carapace
  - Order Arguloida-fish lice
- Subclass Malacostraca-8 segmented thorax, 6 segmented abdomen
  - Superorder Syncarida-freshwater, fossil marine

- Order Anaspidacea-South America and Australia only.
- Order Bathynellacea-minute, reduced blind interstitial forms

- Superorder Pancarida-eggs carried in a dorsal brood pouch
  - Order Thermosbanacea-small, subterranean forms
- Superorder Peracarida-eggs carried in marsupium under the thorax, no zoea
  - Order Mysidacea-opossum shrimps
  - Order Isopoda-isopods or sowbugs
  - Order Amphipoda-amphipods or scuds
- Superorder Eucarida-eggs carried under the abdomen, zoea larvae
  - Order Decapoda-3 pairs of maxillipeds and 5 pairs of legs on thorax
    - Suborder Natantia- shrimp
    - Suborder Reptantia- crayfish

Artificial key to higher taxa of Crustacea (Thermosbaenacea and Bathynellacea omitted)

la. Eyes stalked 2

Ib. Eyes sessile or absent 5

2a. Carapace absent .... Anostraca

2b. Carapace present 3

3a. Carapace free posteriorly, all legs biramus .... Mysidacea

3b. Carapace fused, some legs uniramus .... Decapoda 4

4a. First two pair of pereiopods chelate.................. Natantia

4b. First three pair of pereiopods chelate................. Reptantia

5a. Carapace absent 6

5b. Carapace present 11

6a. Abdominal appendages lacking .... Copepoda 7

6b. Abdomen with pleopods 10

7a. Highly modified parasites .... Siphonostomatoida

7b. Not highly modified for parasitism, rarely parasitic 8

8a. Abdomen not much narrower than thorax .... Harpacticoida

8b. Abdomen much narrower than thorax 9

9a. First antennae almost as long or longer than body .... Calanoida

9b. First antennae not reaching beyond the cephalothorax .... Cyclopoida

10a. Body dorsal-ventrally compressed .... Isopoda

10b. Body laterally compressed .... Amphipoda

11a. Carapace univalved 12

11b. Carapace bivalved 13

12a. Thorax with 4 pairs of appendages .... Branchiura

12b. Thorax with 40 or more pairs of appendages .... Notostraca

13a. Carapace not enclosing the head .... Cladocera

13b. Carapace enclosing the entire body 14

14a. Thorax with 10 or more pairs of appendages .... Conchostraca

14b. Thorax with 3 or fewer pairs of appendages .... Ostracoda

Some changes from Pennak

- Eubranchiopoda and Oligobranchiopoda are abandoned as unnatural taxa.
- Eucopepoda is discarded, Branchiura is elevated to subclass and the copepod suborders are elevated to orders.
- The more specialized parasitic copepods, Calioida and Lemaeopodoida are combined into the order Siphonostomatoida. *Lernaea* is now placed in the Cyclopoida.
- The genus *Hyalella* is generally placed in its own family the Hyalellidae rather than the Talitridae.

Synopsis of Neoartic Orders

- Anostraca-fairy shrimp, many pairs of thoracic appendages, no carapace, compound eyes stalked, second antennae of males modified as claspers, filter feeders, swim up side down, 180 species, none are marine, characteristic of temporary pools or extreme environments, *Artemia* in saline lakes.
- Notostraca-tadpole shrimps, sessile eyes, antennae reduced, depressed body, large shield-like univalved carapace, many pairs

of thoracic appendages, about 11 widely distributed species, all freshwater.

- Conchostraca-clam shrimps, completely enclosed within a bivalved carapace, sessile eyes, second antennae used for swimming, many trunk appendages, 180 species, all freshwater.
- Cladocera-water fleas, bivalved carapace covers only the trunk, large and sessile eyes, large second antennae used for swimming, only 4-6 trunk appendages, body segments fused, egg brooders, 450 species, mostly freshwater, planktonic and littoral.
- Podocopa-seed shrimps, entire body enclosed within a bivalved calcified carapace, compound eyes lacking, both antennae well developed and used for swimming, three pairs of thoracic appendages, body segments fused, abdomen reduced, mostly found in vegetation, 5000 species, marine and freshwater.
- Harpacticoida-copepods with 5-9 segments in the first antennae, abdomen not much narrower than the thorax, 2800 species, mostly marine and benthic.
- Cyclopoida-copepods with the first antennae with 10-16 segments and not reaching much beyond the cephalothorax, abdomen (urosome) much narrower than the thorax (metasome), 450 species, planktonic and benthic.
- Calanoida-copepods with the first antennae with 16-26 segments and about as long or longer than the body, abdomen much narrower than the thorax, 2300 species, mostly marine and planktonic.
- Siphonostomatoida-copepods that are highly modified for parasitic existence, 1430 species, marine and freshwater.
- Arguloida-fish lice, body flattened, carapace covers most of the body, eyes sessile, mouthparts suctorial, first maxillae forming suckers, both pair of antennae reduced, abdomen short and bilobed, four pair of thoracic appendages, 150 species worldwide.
- Thermosbaenacea-reduced carapace fused only to the first thoracic segment and used as a dorsal brood pouch in the females, pleopods reduced or absent, 9 species, thermal water, saline wells, ground-water.
- Bathynellacea-small (0.5-3 mm), elongate, blind, all 8 thoracic segments are free, pleopods reduced or absent, 100 species, interstitial and ground water.

- Mysidacea-opossum shrimps, carapace extends over most of the thorax but not fused with the last 4 segments, eyes stalked, 1 or 2 pairs of maxillipeds, 6 or 7 pairs of biramus swimming legs, 5 pairs of pleopods the fourth is very long in males, 780 species, all but 42 are marine.
- Isopoda- isopods or sowbugs, body flattened, no carapace, eyes sessile, first thoracic segment fused with the head, 1 pair of maxillipeds, 1 pair of gnathopods, 6 pairs of pereiopods, 5 pairs of flattened pleopods, 0-5 abdominal segments fused with the telson, 1 pair of uropods, over 4,000 species, mostly marine, also terrestrial, freshwater, many triglobits.
- Amphipoda- scuds, body laterally compressed, no carapace, eyes sessile, first thoracic segmented fused with the head, 1 pair maxillipeds, 2 pair gnathopods, 5 pair pereiopods, 3 pair of narrow pleopods, 3 pairs of uropods, 6000 species, mostly marine, also freshwater, some terrestrial, many triglobits.
- Decapoda- crayfishes, shrimps, crabs, eyes stalked, carapace fused with all 8 thoracic segments, 3 pairs or maxillipeds, 5 pairs of pereiopods (decapod), 5 pairs of pleopods, 1 pair of uropods, 10,000 species, mostly marine, also freshwater and terrestrial.

CLASSIFICATION OF MALACOSTRACA

Subclass	Malacostraca
Superorder	Leptostraca
Order	Nebaliacea
Superorder	Haplocarida
Order	Stomatopoda
Superorder	Syncarida
Orders	Anaspidacea, Bathynellacea, Stagocaridacea
Superorder	Pancarida
Orders	Thermosbaenacea
Superorder	Peracarida
Orders	Spelaegriphacea, Mysidacea, Cumacea, Tanaidacea, Isopoda, Amphipoda
Superorder	Eucardia
Orders	Euphausiacea, Decapoda.

Table 2.1 : L Classification of Crustaceans

S.No.	Sub-Class	Order	Super family	Family	Example
1.	Cephalocarida				i. *Lightiella acuninate*
					ii. *Hutchiasoniella macrocartha*
					iii. *Lightiella incisaella*
					iv. *Lightiella serendipita*
					v. Lightiella sendersiella
2.	Mystacocarida				
3.	Branchiura				*Angulers*
4.	Brachiopoda	Anostraca		Branchinetidae	*Branchinecta gigus*
				Artemidae	*Artenia salina*
				Chirocephalidae	*Chirocophalopsis bundyi*
				Branchipodidae	*Branchipus stagnalis*
				Streptocephalididce	*Streptocephalus arcitus*
		Lepostraca (extinct)			
		Notostraca		Triopsidae	*Triops cancriformes*
				Lepidurusidae	*Lepidurus arcticus*
		Conchostraca		Lyncidae	*Lynceius branchyurus*
				Limnadridae	*Limnadia lenticularis*
				Leptestheridae	*Leptestheria dehalacaensis*
				Cyzicidar	*Ceanestheriella*
				Cyclestheriidae	*Cyclestheria*

S.No. Sub-Class	Order	Super family	Family	Example
	Cladocera			
			Leptodoridae	*Leptodora pindtii*
		Sidoidea	Sididae	*Jiphanoroma brachyurum*
			Horopededidae	*Horopedium gibberum*
		Anomopeda	Daphniidae	*Daphnia pulex*
			Bosminidae	*Bosmina longirostris*
			Macrothricidae	*Itigocrysotus*
			Chydoridae	*Chydorus hydroallona*
		Polyphemoidea		*Polyphemus pediculus*
5. Ostracoda	Myodocopa			
	Cladocopa			
	Podocopa			
	Platycopa			
6. Copepoda	Calanoida			
			Calinaeidae	*Calinus*
			Centropeogidae	*Centropage kroyer*
			Temoridae	*Temora brigicornias*
			Diaptomidae	*Diaptomus oregoneasis*
			Pontellidae	*Pentella meadi*
	Harpacticoida			
	Cyclopoida	Suborder-Gnathostoma		*-Cyclops, Merocyclops*
		Suborder- Poecinostoma		*-Bomolochus*
		Suborder-Siphonostoma		*-Xenocoeliems brumpti*
	Monstrilloida		Monstrillidar	*Monstrilla helgolandica*
	Calioida			*Caligus rapax, Pandarus*
	Notodelphoida			*Salmincola salmonea*

S.No.	Sub-Class	Order	Super family	Family	Example
7.	Cirripedia	Ascothoracica		Synagogidae	*Synagoga*
				Lauridae	*Laura*
				Dendrogasteridae	*Dendrogaster*
		Thoracica			*(Barnacles)*
			Suborder-Lepadomorpha	Scalpellidae	*Scalpellium*
				Lepadidae	*Lepas fasciailaris*
			Suborder-Verrucomorpha		*Veruco stmoemia*
			Suborder-Ballanomorpha	Chthamalidae	*Catophragmus*
				Balanidae	*Balanus*
		Acrothoracica	Suborder-Apygophora	Trypeteridae	*Baccinum*
			Suborder-Pygophora	Lithoglyptidae	*Lithoglyptus*
				Cryptophialidae	*Cryptophialus*
		Rhizocephala			
			Suborder-Kentrogonidar	Sacashnidae	*Sacculina*
			Suborder-Akentrogonida		*Dupnorlrs*
		Apoda			*Protolepas*
8.	Malacostraca		Suborder-Phyllocarida		
		Leptostraca		Neballidae	*Paranebahiaca*
			Superoder-Haplocarida		
		Stomatopoda		Sqnallidae	*Squill mantis*

S.No. Sub-Class	Order	Super family	Family	Example
		Superorder-Synacarida		
	Anaspidacea			*Anaspides tasmaniae*
	Stygocaridacea			*Parastygocaris*
	Bathynellacea			*Bathynella magna*
		Superorder-*Eucarida*		
	Euphansiacea		Euphausiidae	*Euphosia superba*
	Decapoda			
		Suborder-Natantia	Pencidea, Caridea Stenopidea	
		Suborder-Reptantia	Macrura, Anonura, Brachyura	

SUBCLASS MALCOSTRACA

In this there are nineteen segments, five cephalic, eight thoracic and six abdominal, exceptionally seven. Carapace shows variation in different species, and some it is vestigial. Thoracic appendages are mostly for walking, and abdominal appendages for swimming. Except in a few species abdominal furca is absent. The last pleopod is a uropod. Male genital opening is on the sixth, and female on the eighth trunk segment.

SUPERORDER LEPTOSTRACA

They are primitive Malacostraca with only one order Nebaliaceae, marine in habit. The size usually varies from 4mm to 40mm. They show a combination of Entomostracan and malacostracan characters. Abdomen has seven segments and a telson which ends in a caudal furca. Uropods are absent. A bivalved carapace covers the thorax and part of abdomen. A movable rostrum is present. Compound eyes are present, in some reduced or absent .The last abdominal segment is with out appendages. Thoracic appendages are flattened and expanded like the foliaceous limb and also possess epipodite. First four abdominal appendages are biramus, next two uniramus. Antennary and maxillary gland are present.

The sexes are separate : Eggs are carried by the mother between the thoracic legs. Development is direct. After two or three molts the young leaves the mother, which resembles the adults.

Nebaliacea are mostly bottom dwellers from coastal areas to depths below 2000 m among stones and vegetables. They are filter feeders, but the filtering mechanism does not appear to be well developed. *Nebalia longicornis* has been obtained from Krusadi Island , India. Nebaliopsis typical is a bathypelagic form found at 1000 m depth.

SUPER ORDER HOPLOCARIDA

Body is elongated with a dorsoventral flattening. Carapace is fused to the first two thoracic segment, and partly covers the two following segments. The last four segments are uncovered. Respiratory chamber is not formed by the carapace as in decapods. Anteriorly the carapace has a small movably articulated rostral piece. The antennules have three flagellae. Anterior five pairs of thoracic legs have no exopodites , but have epipodites, and are directed forwards as maxillipeds ending in claws, the second legs developed into massive raptorial arms for capturing prey. Posterior three legs are small and biramus having three segments on portopodites, pleopods are biramus used for swimming, and large branching gills for respiration are found on Exopodites. In males first pair of pleopods are modified copulatory organs. Uropods are forked and form a tail fan along with the broad telson, which may possess ribs and thorn-like projections.

The sexes are separate : Eggs are carried among the last three pairs of chelate maxillipeds, or laid in the burrow of the female. In development two types of larvae are found, the Erichthus-type and Alima-type. The stomato-pods are called mantis-shrimps and are of somewhat large size, from 20 mm to more than 30 cm. They are marine in habit and bottom -dwellers along the coasts. Some can tolerate brackish-water. They are predators on crustaceans, molluscs and fishes. *Squilla raphides* and *S. wood-masoni* are found on Indian coasts. *Squilla mantis, Lysiosquilla maculata* and *Oratosquilla oratoria* are fished commercially for food.

SUPERORDER SYNCARIDA

These crustaceans form some of the primitive Malacostraca of small size form 0.5mm to 5cm. Body is elongated and cylindrical. The eight thoracic segments are free and similar to each Other without any specialization. Carapace is absent. The head is fused to the first thoracic Segment or may be free. The thoracic limbs are non-chelate and most of them are biramus. Mandible has no lacinia mobilis. Brood pouch is absent.

In Anaspidacea head is fused to the first thoracic segment. Stalked or sessile compound eyes are present. Appendages are present on thorax and abdomen. Plate-like epipodite on thoracic legs assist in respiration. Statocysts without statoliths are found on the first segments of the antennules. First two pairs of pleopods are modified into copulatory organs as in decapods. Uropods are present.

In stygocaridacea head is fused to the first thoracic segment. Eyes are absent. The first thoracic leg is a maxilliped. Pereopods are present, pleopods are Absent except the copulatory organs and uropods.

In Bathynellacea head is not fused to the first thoracic segments. There are no eyes. Pereopods are present, the last appears to be Copulatory. Pleopds are lacking or reduced. Uropods are present. The sexes are separate. Eggs are deposited on other objects. Development is direct, but all parts are not present in juveniles, and adult structures are developed by molting.

In Bathyn-llacea larval stage is found with antennae and mouthparts. Development is not known in Stygocaridacea, but the youngest animal found has all adult limbs, indicatig direct development.

Syncarida are fresh- water dwellers in mountain lakes, caves and springs. They swim and crawl on vegetation and mud and even burrow. Many are inerstitial in habit. They are supposed to be detritus filters feeders, eating microscopic organisms. *Anaspides* is predaceous. *Anaspides tasmaniae* is found in mountain lakes of Tasmania at a height of 2000 to 4000 feet. Micraspides calmani is another example. Parabathynella Malaya is found in streams in Assam. Bathynellanatans lives at a temperature of 47°C to 50°C in Europe, Japan and other countries. Stygocaris is found in chile.

SUPERORDER PANCARIDA

Small crustaceans, 1 to 4 mm long coming in a single order Thermosbaenacea. Carapace is fused to the first thoracic segment and may extend above the next two segments. Thorax and Abdomen are not demarcated by constriction. In females a dorsal brood pouch is formed by the posterior extension of the carapace and also a branchial chamber at the sides. Mandible has lacina mobilis. Eyes are absent. The first thoracic limbs are maxillipeds having epipodites and by their action water is drawn into the branchial chamber, and exchange of oxygen takes place through the blood lacunae of the thin carapace. There are five biramus pereopods behind the maxillipeds without epipodites . The last two segments have no limbs, but the eight thoracic segments have a pair of flexible *panes* on its sternite. There are two pairs of uniramus pleopods on the first two abdominal somites , and a biramus uropods on the sixth.

The sexes are separate : Development is direct and takes place within the brood pouch, but how the eggs enter the pouch is not known. The young resembles the adult in *Thermosbaena*, and in *Monodella argentarii* few limbs are absent in the young.

Algae, diatoms, bacteria, detritus and vegetable particles are found in the gut, but the method of feeding is not known. Vegetables may be broken by the rasping action of the mandibles. They are interstitial fauna found in fresh and brackish waters, and in hot springs. *Themosbaena mirabilis* lives in hot springs in Tunisia of 34% at 98°F-118°F, lower temperature of 95°F being harmful to it. *Monodella argentarii* lives in complete darkness in fresh water in a Tuscan cave. *M. stygicola* lives in brackish water in South Italian cave. *M. relicta* is found in the thermomineral springs on the shores of the Dead Sea.

Superorder Peracarida : Small crustaceans, the head fused to the first thoracic segment. If a carapace is present it fuses only with the anterior four thoracic segments leaving the posterior four segments free.

The protopodite of antenna has three segments. Mandible has a lacina mobilis. Stalked or sessile eyes present. In adult females there is a brood pouch or *marsupium* under the thorax, formed by flat plates or oostegites extending from coxae of thoracic legs.

Spelaeogryphacea has only species *Spelaeogryphus lepidops* of about 7 to 9mm long, got from a stream in Cape Town, S. Africa. Carapace is short and is fused to the first thoracic somite. First thoracic leg is a maxilliped having a cup-shaped respiratory organ at its base. Vascular branchial gills are found on fifth, sixth and seventh thoracic legs. Four abdominal legs are swimming. Fifth one is vestigial and sixth is a uropod forming a tail-fan with the telson. Eyes are absent. Brood pouch is present under the thorax of female. The animal is a detritus feeder, but there is no filter feeding mechanism.

The Mysidacea , commonly known as opposum shrimps , are small ones, size varying from 1mm to 3cm. Head fuses with not more than three thoracic segments forming a cephalothorax. The carapace covers the thoracic segments except the last one or two, and forms lateral branchial chambers. Compound eyes are stalked and movable. The first thoracic leg is a Maxilliped, sometimes second and third also. The exopodite of antenna is a scale-like *Squame*. Respiratory epipodite gills are found on thoracic legs. Pleopods may be redused in females. Uropods have statocyst, and form tail-fan. Mysidacea are marine, some live in brackish water and rarely in fresh water. Some are burrowing and some frequent moist sand. Most are filter feeders, some are predaceous. *Mesopodopsis orienalis* in an Indian species found in marine, brackish and fresh water area.

Cumacea are very small in size, from 0.5 to 28 .6 mm lengths. Carapace is very much bulged out forming a brachial cavity, and the abdomen is narrow. Carapace fused with anterior 3 to 4 or 5 to 6 thoracic segments. Anteriorly carapace extends as 'pseudorostrum'. Compound eyes are sessile, often fusing to form a median one. Anterior three thoracic appendages are maxillipeds. The first maxillipeds have epipodites, which extend into the branchial chamber and bear a row of gills. Females have no pleopods , but uropods are narrow and do not from a tail fan. Cumacea are mostly marine, a few live in brackish or fresh water. They live partly buried in sand or mud. Filter-feeding mechanism is present in cumaceans, but many are predaceous. *Bodotria choprai, Eocuma travancoricum* and *Gigacuma halei* are found on the Kerala cost of India.

Tanaidacea are very small ones, size usually within 3mm. Bigger ones are seen more as an exception. Body has a slight dorsoventral flattening. Carapace fused with first two thoracic segments and forms a branchial chamber on either side. The first thoracic limbs is a maxilliped, which has respiratory epipodites? The second one is very big provided with a powerful chela. Eyes are on movable stalks, pleopods may be absent in females.

Isopods are slender and long. Tanaidacea are marine, but some have taken to brackish-water and freshwater habitats. They usually shelter under leaves and other object, and some make tubes or make burrows. They are filter feeder on *microorganisms*, and can also pick up food by the chelae. Adult males do not feed. *Tanais philetaerus* is found on Kerala and Bombay coast. Isopods are small in size, from 1mm to 35mm. Some are parasitic, on fishes, crustaceans etc., and have undergone degeneration of body structures. Some isopods are terrestrial in habit. Body is dorsoventrally flattened, with rare exceptions. Head is fused to the first thoracic segments. Carapace is absent. Compound eyes are sessile, or on short movable stalks. Body segments project laterally over the legs. First thoracic leg is a maxilliped, sometimes seconds also. Pleopods are foliaceous , and due to blood supply serve respiration also. Uropods are fan-shaped or styliform. Heart is posteriorly placed near the branchiae. Isopods are marine, mostly found in intertidal region, but also inhabit brackish and fresh water areas. Some are terrestrial, found in moist places, and some are parasites on fishes, prawns etc. Isopods are not filter feeders. They feed on detritus, decayed plants, carrion, and also prey upon small animals. *Cirolana pleonastica* attacks fishes in Chilka Lake and eat away the flesh. Several specie a of *Sphaeroma* and *Limnoria* bore into timber. *Stegoalpheon kempi* is a parasite in the branchial cavity of *Alphaeus crassimanus* in Indian seas. *Hemiporcellio carinatus* is a terrestrial Isopod living under stones and dried weeds in Chilka Lake area, and *Ennurensis hispidus* is another found in sandy areas at Mandapam and Pamban, India. Amphipoda are also small crustaceans, size from 5mm to 14cm. Body is laterally compressed, without a carapace. Head is fused to the first or even second thoracic somite. The first thoracic leg is maxilliped, the base of each fused to form a kind of labium. Some of the thoracic legs have epipodites for respiration. Compound eyes are not stalked. Isopods are styliform. Heart is anteriorly placed corresponding to the position of gills. Amphipods are mostly marine, some are interstitial. They feed bybrousing on sand, carrion etc., while some are predaceous. Some feed on micro-organism by filtering. *Melitu zeylanica* is found in the brackish waters of Kerala. *Corophium trioenyx* is abundant in Indian estuaries. *Cyamis boopis* is an ectoparasite on humpback whales. *Chelura terebrans* is a woodborer.

Sexes are separate in Peracarida, the young developing in the brood pouch of the female. Some Isopod parities like Cryptoniscidae and Crmothoidea are protrandric hermaphrodites. Development is direct, the young resembling the adult. In some the young hatched does not posses all the appendages. In Isopods parthenogenesis is met with in *Trichnooniscus*. Juvenile stages are not known in *Spelaeogriphus*.

SUPERORDER EUCARIDA

In Eucarida are included the largest and the most specialized crustaceans. The carapace is fused to all the thoracic segments forming a cephalothorax. Compound eyes have movable peduncles. Mandibles do not have lacinia moblis. Brood sac is absent. There are well developed respiratory branchiae. Heart is short and saccular situater in the thorax. The digestive gland, hepatopancreas, is large and ramified . Different types of larvae are hatched from the egg.

In Euphausiacea the specimens come to 1 to 2 inches in length. Carapace is fused to all thoracic segments, but does not extend very much to the sides, and so the gills are exposed. Only one series ofbrachiae, the podobranchiae, develop on the coxae of thoracic legs. No thoracic leg is specialised as maxillipeds. These legs have setae on them, which are arranged for filter feeding.

Pleopods are biramus , used for swimming . The first two pleopods in the male are modified as copulatory organ. Uropods form a narrow tail-fan along with an elongated telson. There is no statocyst. Compound eyes are present which may have, in some, a constriction by which one part gives a dorsal and the other part a side vision. Occasionally a nauplius eye may persist under the rostrum of the adult.

Sexes are separate. Eggs are carried by the female on pleopods. In some cases they are laid in water. A *nauplius* hatches from the egg, which has no carapace. It molts and forms a metanauplius having a carapace, which gets an elongated abdomen and becomes a *Calyptopsis* stage (protozoea) possessing paired immovable eyes. Later, pleopods develop and eyes become movable and the stage is called furcilla. This passes into a *cryptopia* stage (post-larva). Euphausiacea are entirely marine, living in the open ocean. They are permanent swimmers never resting on the bottom, and are world -wide in distribution. Most of them are luminescent possessing photophores on eye - stalks, bases of second and seventh thoracic limbs, and on the first four abdominal segments ventrally. *Thysanopoda cornuta*, found in the Gulf of Alaska is a giant form measuring nearly four inches. In the Antarctic seas *Euphausia krohni* and *E. superba* are food of whales. Larvae and post-larvae of Nyctiphanes sp. Of 9.9 mm size appeared swarming near Trivandram coast in 1946 (Kurian, 1953).

The decapoda contain the biggest crustacean like prawn, lobsters and crabs. The head and thorax are fused to form a cephalothorax. The carapace is fused to all the thoracic somites, and also extend to the sides covering the gills and forming a branchial chamber. The exopodite of the second maxillae is a well-developed scale. The first three thoracic appendages are modifided as maxilliped, and the rest are pereiopods, some of which may be chelate . The pleopods are biramus. There series of branchiae are usually present, the podobranchiae on the coxae of legs, the arthrobranchiae on the arthroidal membranes, and pleuobranchiae on the sides of thoracic wall. A statocyst is usually present on the proximal segment of the antennae.

Sexes are separate. In most decapods females carry eggs between the pleopods. In Penaeidae eggs are laid in the open sea where they develop. Different types of larvae are met with. There is also direct development in most freshwater decapods without any free zoea stage. Decapods are mostly marine, but many can tolerate brackish water and some are fresh water dwellers. They rest and move on the substratum, but swim temporarily for food capture or migration. Decapoda is divided into two suborders, *Natantia and Reptantia.*

In Natantia body is laterally compressed, the carapace has well-developed rostrum; second antenna has a large scale-like exopodite. Thoracic legs are slender, some of the anterior ones chelate. These are swimming crustaceans, the prawns and shrimps. There are three sections in Natantia, the Penaeidea, Caridea and Stenopidea. In Caridea gill filaments are thin plate-like, arranged in two rows on a central axis like the leaves of a book, termed *phyllobranchiate*, in stenopidea gills are *trichobranchiate*, the gill filaments broken into finger like bits, appearing like a bristles of a bottele brush, and in penaeidea gills are *dendrobranchiate*, the gills filament broken into very fine tiny filaments. Petasma is present only in penaeidea. The edible marine prawns of the genera *Penaeus, Metapenaeus*, and the fresh water genus *Macrobrachium* come under this sub-order. *Periclimenes brevicarpalis* is found associated with a giant anemone, *Stoichactis giganteum* in the sea along the tuticorin coast .India. This prawn can enter the entire cavity of the anemone and get trapped when the latter contracts, and come out when the anemone expands and open the cavity. *Caridina laevis* is a small fresh water prawn fonds in the ponds and wells in India. *Stenopus hispidus* has the habit of cleaning fishes living among corals. *Hippolysmata ensirostris* is a commercial prawn of India (Kurian and Sebastian, 1982).

In reptantia the body is dorsoverrtrally flattened, and the rostrum and scale like expedite of antenna are reduced or absent. Abdomen is well developed or reduced very much. Pleopods also undergo reduction in size, not very much used for locomotion. Pereiopods are very strong, usually without exopodites, the first one being very strong and chelate. These are mostly crawlers. These are three sections under Reptantia, the *Macrura, Anomura* and *Branchura*. Sometimes *Macrura* is split into two sections. The *Palinura* and *Astacura*.

In Macura the abdomen is well developed and held in an extended position, telson and uropods forming an expanded tail fan. The carapace is depressed with a strong lateral keel, or is cylindrical. The rostrum is small or absent. The first four pereiopods are cylindrical. In *Palinura* they possess subchelae, rarely chelae, while in Astacura the first three are chelate. Lobsters and crayfishes come under this section. *Panulirus ornatus* and *Thenes orientalis* are Indian lobsters, and *Penilurus sewelli*, is a deep sea form. *Homarus americanus* is an American lobster attaining a length of 60cm and a weight of 22kg. Astacidae are mostly freshwater cray fishes.

In Anamura the abdomen is well developed, but is often bent ventrally, flexed towards the thorax. Sometimes the abdomen is asymmetrical. The carapace is not fused to the epistome, the base of the labrum. The last thoracic stomite is free, and the appendage of the segment is reduced. The pereopods is chelate. Antennal scale is redused . Uropods are present. Hermit crabs and sand crabs belong to this section. Hermit crabs have undergone asymmetry when adult, but there larvae known as *glaucathoe* are symmetrical. *Birgus latro* is a giant crab, 32cm long has become terrestrial and in able to climb coconut trees to eat the soft kernal of tender fruits. They are found on some uninhabited islands of the Andaman group. They visit the sea for egg laying. The Paguridae or the hermit crabs are asymmetrical and they introduce their abdomen into empty molluscan shells, the anterior thoracic region and appendages remaining outside. *Clibinarius olivaceous*, *Diogenes avarus* and *Coenobita cavipes* are hermit crabs found in the Chika Lake. The sand crab dig into the sand and take in food by filtering the water through the feather-like setae on the second antennae. *Emerita asiatica* and *Albunea syminta* are found along the Indian beaches. *Petrolisthes ohshimae*, of Porcellanidae, found at Tuticorin coast creeps around the giant anemone, *Stioichactis giganteum*, for protecton from enemies (Nair and Mahadevan, 1965).

Brachyura contains the true crabs. The carapace is very large and aneriorly fused to the epistome. There is no rostrum. Abdomen is short and thin and firmly folded under the cephalothorax, and is not used for locomotion. The first pereiopods is unusually large and chelate. There are no uropods. The third to fifth pleopods are absent in males. Crabs mostly run sideways. *Potunus sanguinolentus, P. pelagicus* and *Charybdis cruciata* are some of the edible Indian crabs. *Uca annulepis* is a small crab found in the estuaries of India, always waving one of the chelipods, which is bigger than the other. *Pseudocarcinus gigas* is a giant crab of Australia whose carapace is 40cm wide. The giant spider crab of Japan, *Macrocheira kaempferi* has a leg span of 3.5m.

Suborder : Natantia

The most primitive decapods are the SHRIMP LIKE MACRURANS (sub-order Natantia), with about 2000 species which inhabit all possible environments between the deep sea and fresh water, but which are mostly marine. The body is usually laterally compressed, and a tail fan is present on the relatively long abdomen. Despite this, the shrimps spend most of their time on the bottom, walking abound on their five pairs of walking legs or using them to dig in the sediment. The long first antennae, which are held close together, form a tube through which fresh water is brought to the body for respiration, as in the genus *Solenocera* from the Mediterranean. A similar structure is used by the crabs *Emerita* and *Albunea*, while *Corystes* uses its second antennae for the same purpose.

A shrimp which is caught commercially along North America's eastern coast and in the Caribbean Sea is *Penaeus setifer* (L-up to 20 cm). Its abdomen is sold as "crab tails" in preserved form all over the world. The "crab tails" of South America, Australia, and the Mediterranean region usually are also *Penaeus* species. In Italy, for example, it is *Penaeus trisulcatus*. Many genera of shrimps live almost exclusively in the deep sea, such as *Pandalus borealis* in the North Atlantic; others, conversely, are bound to the coastal zones, among them the genus *Hippolyte* with almost 200 species. The species name of *Hippolyte varians* ("changeable") honors this creature, since it takes up the background color of the green or brown alga on which it happens to be sitting. This takes place through the action of various color cells in the epidermis and is, in part, controlled through the eyes. This can be shown when the animal is kept in dim light conditions, for the red and green tones change to a pale blue. *Hippolyte varians* is also pale blue during the night. Many species live communally with sea anemones, and in this case are completely transparent, apart from a few points of color.

Family : Alpheidae

The snapping shrimps (family Alpheidae), live in coral heads or the shore regions of tropical seas. Their first pair of walking legs is especially strong, and is provided with a pincer. Unilaterally enlarged pincers-that is, larger left-or right-hand pincers-are differentiated from the smaller one on the other side. Members of the genus *Alpheus* are able to shoot a sharp stream of water with the snapping pincer, accompanying this with a loud report. This is used as a defensive mechanism, but also aids in prey capture, since it has a shock effect. In the snapping shrimps, which often live as pairs in holes they dig, beautiful examples of closely cooperating communities can be demonstrated. Such a symbiosis can be found between the shrimps and fish. Here the fish, which has a greater flight distance, indicates the presence of danger which the shrimp is not yet able to perceive, while the shrimp provides the fish with a refuge in its hole.

The Common Shrimp

The common shrimp (*Crangon crangon*; family Crangonidae) is also a member of the Natantia. This species inhabits the Wattenzee in large numbers, despite that body of water's great variations in temperature and salt content. Such a variable environment can be inhabited only by specially adapted forms. The common shrimp is especially well equipped for this. The total annual tonnage of this shrimp captured comprises up to five percent of the Wattenzee's total fisheries catch. It is used as food for animals or humans, often as a delicacy. Only small shrimps, under 5 cm in length, are used as animal food. The common shrimp feeds at night on other small crustaceans, worms, and mollusks; during the day, it hides in the soft mud. Its life expectancy is three years; during this time, a single female can produce up to 20,000 offspring. Although the common shrimp is also found in the Baltic, the common baltic shrimp (*Palaemon Squilla*) belongs to another family, Palaemonidea. This species, however, occurs in much smaller numbers and much less regularly than the common shrimp.

Many palaemonids inhabit fresh water., especially in the tropics; some even inhabit submerged caves. Such cave dwelling shrimps are present in the Cargo lakes in the Balkans. In these, vision is reduced to the point of total absence of the eyes, and the sense of touch has developed into the main means of determining changes in the surroundings, something which can be summerised from the exceedingly long antennae.

The richness of their genetic background enabled the palaemonids to evolve in several directions; they thus provide good examples of special developments in behavior and adaptation to the environment. Numerous species have become symbiotic in sponges, sea anemones, corals, and even fish. This can lead to a very delicate "cleaner symbiosis"; in this, the Caribbean shrimp *Periclimenes petersoni* attracts fish by waving its antennae from. "its" sea anemone. It is then allowed to remove external parasites from the fish's scales, opercula, and other parts of the body which are difficult to reach, and may even enter the fish's mouth to clean up the remains of its meal.

Suborder : Reptantia

The *crawlfish* (suborder Reptantia: "the crawlers") can easily be Suborder: Reptantia distinguished from the shrimps (Natantia: "the swimmers") by their dorso-ventrally flattened bodies and the usually powerfully developed pincers on the first pair of walking legs. The abdominal legs are no longer adapted for swimming, but are used by the female as the place to attach her eggs. This group includes the spiny lobsters (family Palinuridae), recognizable by their antennae, which are longer than the body, and the spiny processes of the carapace. Large gripping pincers are lacking. The commercial use of spiny lobsters in all parts of the world is attributable to their meaty abdomens.

The European Spiny Lobster

The *European spiny lobster* (*Palinurus vulgaris*; Bl. up to 45 cm, weight up to 8kg; lives on the rocky coasts of the Atlantic and the Mediterranean. It is a valuable food animal. It feeds at night on snails, mollusks, and dead animals; it must therefore be caught with pots or bait. The size and weight given above are reached by lobsters in good fishing areas, such as the coast of Brittany; they are then about ten to fifteen years old. The cape spiny lobster (*Jasus lalandis*) is a popular item on the menu in South Africa and Australia; the same is true of other *Palinurus* species on the coasts of North and South America. Not only gourmets but also scientists are interested in these animals: the American species *Palinurus* argus has seasonal migrations; in this, hundreds of animals walk one after the other for more than 100 km along the bottom of the sea, and are thus able to discover suitable areas with sufficient food. It is possible that the sounds the animal produces, which were discovered only a few years ago, aid in regulating these migrations.

In the related scyllarids (family Scyllaridae), the *slipper lobsters*, the second antennae are very much compressed and covered with spine-like processes which aid in defense; these give the animals a body form which cannot be confused with any other creature. Members of the genera *Scyllarus* and *Scyllarides* live on rocky shores and frequently on coral reefs.

The *true macrurans* (suborder Astacura) are best characterized by the european lobster (*Homarus, Gammarus*; and the american lobster (*Homarus americanus*) typical representatives of the lobsters (family Homaridae), and by the *european crayfish* (*Astacus astacus*), a member of the crayfish family (Astacidae).

The lobsters inhabit rocky coasts, feeding on mollusks and dead animals during their nightly excursions, after which they always return to their home holes. This fidelity to one area is a result of their excellent ability to orient, something which they have in common with nearly all the Higher Crustacea, and of their learning ability. Seasonal wanderings have also been shown to occur in lobsters. The famous Heligoland lobster became reduced in number and size because too many were caught, and special protective measures had to be taken to save it. The females first become sexually mature at six years of age, when they are about 25 cm in length, and lay about 8000 eggs; females 40 cm in length can produce 30,000 eggs. Old lobsters molt only every second year.

Some lobsters are deep-sea inhabitants; this indicates that these are, evolutionarily speaking, extremely old. In fact, the lobster like crustaceans have been well represented in fossil form since the Jurassic. One lobster species inhabiting deep water is the *king lobster* or norway lobster (*Nephrops norvegicus*; which is usually found along the coasts of Britany and Norway, but has also been caught in the Mediterranean. It is recognizable by its narrow pincers.

The *crayfish* (family Astacidae) are restricted in distribution to the northern hemisphere. In the southern hemisphere the ecological niche they occupy is taken over by the *parasticids* (family Parasticidae) in South America and Madagascar (but not in Africa), and in Australia by the *Austroastacids*(family Austroastacidae).

The European crayfish (*Astacus astacus*; L, ♂♂ up to 25 cm, ♀♀ up to 18 cm) also hunts for its food at night, feeding on both plant and animal material. Male crayfish can live twenty years. The crayfish lives only in very clean water, where it hides in a hole in the bank; therefore, its population has decreased drastically because of increasing water pollution. In addition, a fungus disease termed "crab plague" (infection by the fungus *Aphanomyccs astaci*) almost wiped out the whole European crayfish population in 1870, and the present-day population is still threatened by it. The *american crayfish* (*Orconectes limosus*; Bl. about 10 cm), introduced into Europe in 1890, is immune to the disease and can also survive in dirty water. It does not dig holes, and can also be seen feeding during the day. Although it cannot be compared in size with the European crayfish, it is now considered the most common inland crustacean used for human consumption.

Intermediate Forms

The *intermediate forms* (suborder Anomura) combine, in their body Intermediate forms structure, the macrurans with the brachrurans. Members of this group are the Indo-Pacific mole crabs and the marine and terrestrial hermit crabs, together with the lithodid crabs which are not true brachrurans. In addition, the western Atlantic molecrabs, curious creatures which are completely pincerless and spend their lives buried in the sand, are also included.

The mole crab (*Thalassina anomala*) lives, buried like a mole, in the mangrove swamps of the Indo-Pacific. Here it sieves food particles out of the mud, piling the mud it has sorted in huge hills which can reach a height 2 m and a width of 10 m. Since it feeds completely underground, it must continually enlarge its tunnels. As a result, it can cause great damage to rice paddies near the shore.

Marine Hermit Crabs

Many visitors to the coast know the crabs which hide their soft Marine hermit crabs abdomens in snail shells and more rarely in the shells of other mollusks. These are the *marine hermit crabs* (family Paguridae. Since a snail shell is usually spirally wound in a right-hand direction, the abdomen of the crab is usually also bent from the last larval stage (glaucothoe) onward. The last pair of walking legs, together with an area of wart-covered skin, hold the shell tight. The soft abdomen takes part in respiration and thus serves as a type of "gill." The asymmetrical body form of the hermit crab is also recognizable from the pincers, one of which is .much larger than the other; the large pincer acts as a type of lid, closing the entrance to the snail shell. Since the crab grows with each molt, the snail shell it inhabits soon becomes too small. To obtain a new home, it must, therefore, investigate objects in the surroundings to determine whether they would be suitable and the right size. The senses of sight and touch determine the weight of the object as well as its size and "fit."

An interesting feature of the marine hermit crabs is the great variety of associations they have with coelenterates which are attached to the shell in which the hermit crab lives. Sponges can also grow over the hermit crab's "house." There are a large number of mollusks whose shells are used by the hermit crab; from the whelk of the North Sea coasts to the straight tusk shell, which is not spirally wound, and the flat shells protect themselves.

REFERENCES

1. Bigchow, A.S. (1968). Classification and Phlogeny Pergaman Press
2. Chandiasekaran, S.V. (1989). Fishery Bulletin Seattle, Washington.
3. Gardiner, Mary, S. (1968). Biology of Invertebrates, Academic Press, London.
4. Hyman Henrietta Liffie (1967), the Invertebrates, Printice Press, New Delhi.
5. Jordon. L.E. (1967). Invertebrate Zoology, S. Chand and Co., Meerut.
6. Kaestner Alfred. Invertebrate Zoology, Vol.-3 Interscience Publication; 2nd ed., 1986.
7. Kotpal L.R. (1993-94). A Text Book of Arthropoda Rastogi Publication Meerut.
8. Kurian. C.V and Sebastian. V.O. Prawns and Prawn Fisheries of India. H.P.C. (India), 2nd ed, 1982.
9. Rakshnal. R. (1970), Invertebrates, Academic Press, New York.
10. Snodgras E.R. (1965). A Text Book of Arthropod Anatomy Aayner Publication Company, New Delhi.
11. Zwuch Y.A. (1970). Encylopedia (Animal Life) Reinhold Company, New York.

3

BODY PLAN AND APPENDAGES

Of the great array of species constituting the Order Decapoda, the vast majority are marine. In the United States only the Astacidae (crayfishes), about 12 species of Palaemonidae (fresh-water prawns and river shrimps), and four species of Atyidae are found in fresh waters. These representatives, totaling about 320 species, are characteristic and common inhabitants of a wide variety of environments, including most types of running waters, shallows of lakes, ponds, sloughs, swamps, underground waters, and even wet meadows where there is no open water. They are absent from the greater portion of the Rocky Mountain region, although one species occurring in Pacific slope drainages has migrated as far eastward as the Great Basin and the headwaters of the Snake, Yellow-stone, and Missouri rivers.

To this list should be added at least one grapsoid crab, *Platychirograpsus typicus* Rathbun, which has become well established in the Hillsboro River, Florida. Possibly this river crab occurs also in Gulf coast streams of other states since it is common along the east coast of Mexico.

General characteristics. Crayfish, crawlfish, crawdads, or crabs, as they are locally and variously known, are all more or less cylindrical, and the body and appendages are strongly sclerotized. The compound eyes are large, stalked, and movable. Total length, not including the antennae, ranges from 10 to 150 mm. The six abdominal segments are all distinct, but the head and thoracic segments are fused to form a large cephalothorax. A carapace covers the cephalothorax, and there is a large gill chamber on each side between the body and the downward extensions of the carapace. The cervical groove of the carapace roughly delimits the head and thoracic regions, and the areola is a narrow more or less hourglass-shaped median area of the carapace just posterior to the cervical groove. The anterior end of the carapace forms an elongated rostrum which bears a short spine (acumen) at its extremity.

The 19 pairs of serially homologous appendages are basically biramus but modified for a wide variety of functions. Detailed accounts of their morphology may be found in most zoology texts and they will be only briefly discussed here. Beginning at the anterior end, each of the first antennae (antennules) has two slender, segmented, sensory flagella. The dorsal surface of the basal segment of the first antenna bears the opening of the statocyst, an organ of balance. The second antenna has a flattened basal scale and a single long flagellum.

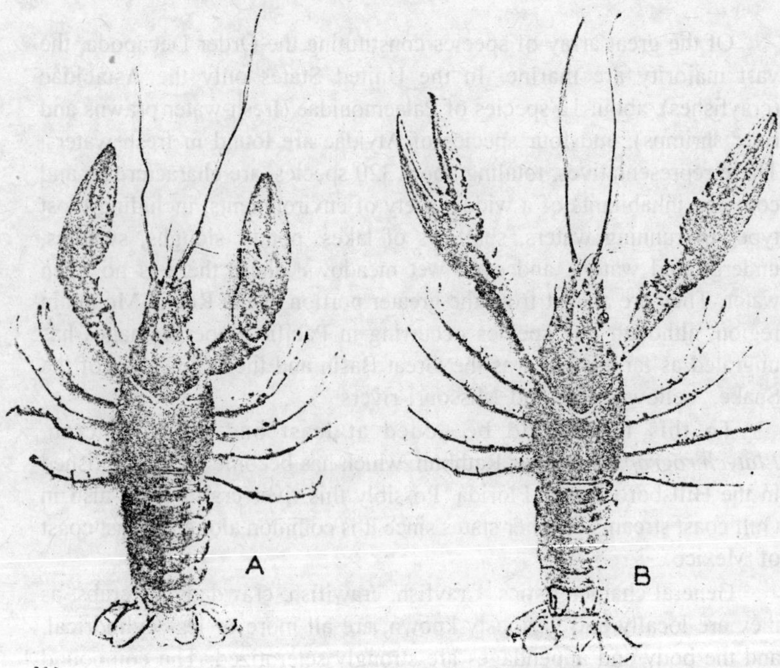

Fig. 3.1 : A, dorsal view of female *Orconectes*, × 0.7; B. dorsal view of male *Orconectes*, × 0.7.

The next five pairs of appendages greatly overlap one another and are used chiefly for handling and mincing the food. The heavy, crushing mandibles lie on either side of the mouth. The first maxillae, which follow, are small, flattened, and delicate. Each of the second maxillae has a long, flat extension projecting laterally beyond the point of attachment of this appendage. This is the "sailer"; its rapid beating creates a posterior-anterior current of water through the gill chamber, thereby ensuring a continuous supply of oxygenated water for the gills in the gill chamber. The first, second, and third maxillipeds are progressively more robust and clearly biramus. The third maxillipeds are used for cleaning the antennae.

The last five pairs of appendages of the cephalothorax are large walking legs, or pereiopods (hence the name "Decapoda"). Each consists of seven segments. Beginning at the base, these are as follows: coxa, basis, ischium, merus, carpus, propodus, and dactylus. Sometimes the suffix "podite" is used for each segment; thus; coxopodite, basopodite, and so on. The coxa and basis are always short and compact. The first three pereiopods are clawed, or chelate, the dactylus being the movable "finger" of the claw, and the propodus constituting the "hand" and immovable finger. The first leg, or chela, is greatly enlarged in the crayfish and is used for crushing the food and as an offensive and defensive weapon. It is of little use in locomotion, walking being chiefly a function of the last four pairs of legs. The second and third legs are also used in handling and mincing the food.

Near the base of the ischium of each walking leg and marked by a double crease is a "breaking joint." When a crayfish leg is seized or irritated, reflex action causes strong contraction of special muscles at the breaking joint, and the whole leg breaks off at this point. Consequently the collector must exercise care if he is to secure suitable specimens. Perfect specimens are uncommon since appendages are often lost as the result of crayfish fighting among themselves. Missing legs are regenerated in miniature at the subsequent molt but later become larger.

The first five abdominal segments bear pleopods, or swimmerets. In the female they are all more or less biramus and are used as places of attachment for the incubating eggs, but in the male the first two pairs are modified and used during copulation. The last abdominal segment bears a flat terminal telson and a pair of flat biramus uropods. The telson and two uropods collectively form the broad tail fan.

Mature crayfish exhibit such sexual dimorphism that there is no difficulty in distinguishing the two sexes. Males usually have the larger chelae and narrower abdomen. In the Cambarinae the third segment of some of the male pereiopods bears a small hook used during copulation. In the female the genital pores are on the basal segments of the third pereiopods, but in the male they are on the fifth pereiopods. The annulus ventralis, or seminal receptacle, is a grooved, elliptical, calcified area in the midline between the bases of the fourth and fifth pereiopods of the female. The first pleopods of the female are usually similar to the others, but in the male they have lost their biramus character and are heavily sclerotized, grooved, and specialized at the tip for sperm transfer.

The male second pleopods are also somewhat specialized, and both pairs are bent forward so that they lie against the median ventral surface of the body between the bases of the pereiopods.

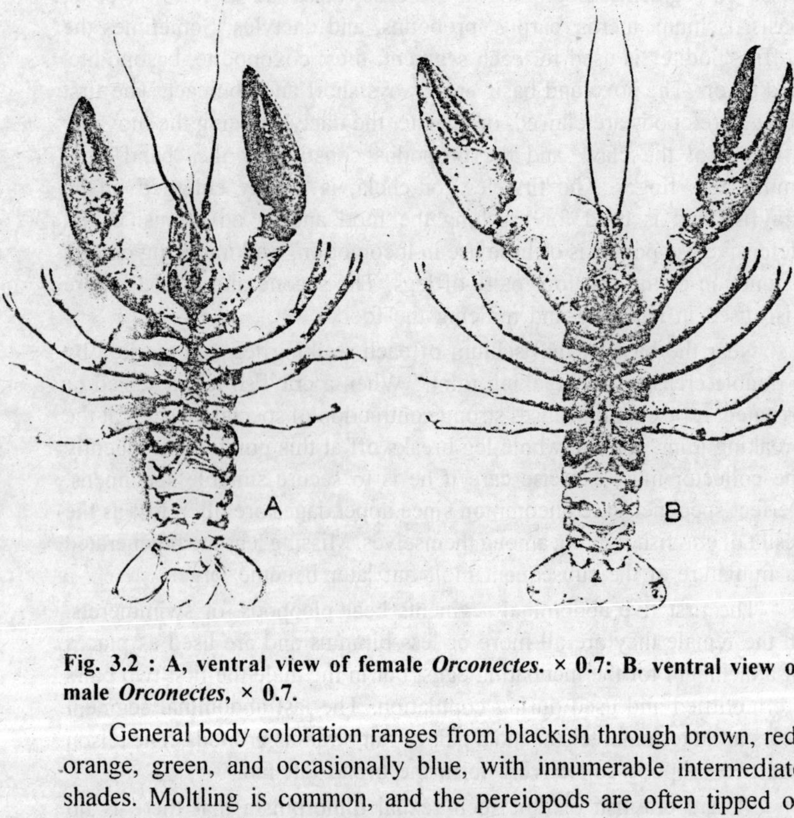

Fig. 3.2 : A, ventral view of female *Orconectes*. × 0.7: B. ventral view of male *Orconectes*, × 0.7.

General body coloration ranges from blackish through brown, red, orange, green, and occasionally blue, with innumerable intermediate shades. Moltling is common, and the pereiopods are often tipped or banded with brighter colours than the rest of the animal. Many species show a wide range of colour. *Orconectes immunis*, for example, may be bluish, blackish, brown, greenish, or red. In general, newly molted specimens are more brightly coloured than older specimens, especially since the latter usually have an accumulated coat of dirt, debris, and algae.

The coloration of a species often varies in accordance with the coloration of the substrate. In most crayfish such colour adaptations are decidedly limited and occur only over a period of weeks or months. In *Palaemonetes patudosus*, however, variation.

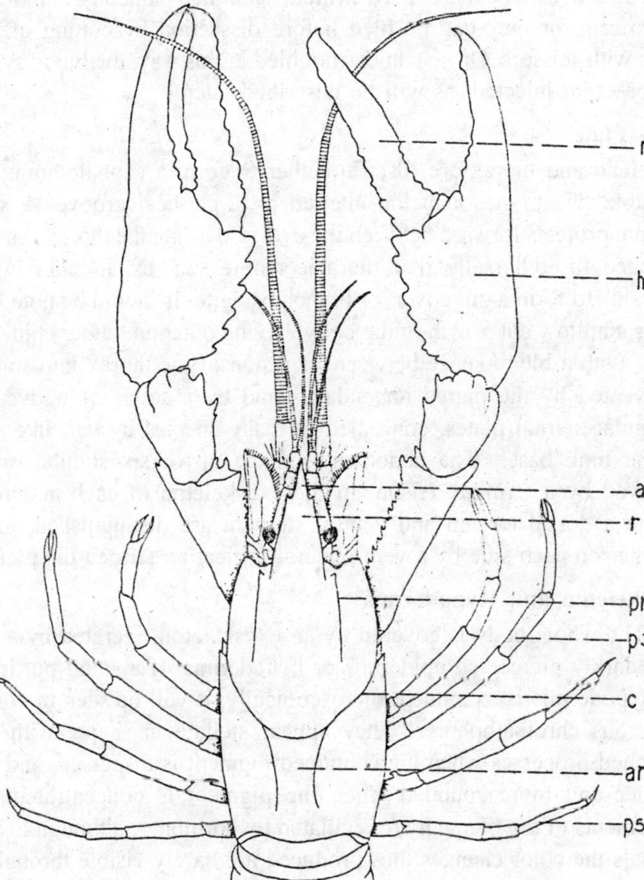

Fig. 3.3 : Structure of a typical crayfish, a, acumen: ar, areola; c, cervical groove: h, hand; mf. movable finger of chela; p3, third leg (pereiopod); p5, fifth leg (pereiopod); pr, postorbital ridge; r. rostrum; t. telson; u. uropod. (Modified from Hobbs, 1942a).

Animal can be narcotised with a saturated aqueous solution in chloroform, or they can be bled before dissection by cutting off, the limbs with scissors. On specimens not bled in this way, the blood system can be color-injected, as will be described later.

Body Plan

Head and thorax are fused together to form a cephalothorax, the probable line of fusion being marked by a cervical groove. A sharp rostrum projects forward between the eyes. Covering the thorax is a hard carapace, fused dorsally to all thoracic somites, and expanding down on each side to form a gill cover, or branchiostegite. It should be noted that this is a hollow outgrowth of the body, having outer and inner epidermal walls, with a blood space between. Division of the thorax into somites is revealed by the paired appendages and by a series of midventral, triangular sternal plates, concealed in many species by hair-like setae on the limb bases. The abdomen is made up of six somites and is followed by a terminal telson. In the exoskeleton of each abdominal somite a dorsal tergum and ventral sternum are distinguished, joined together on each side by a ventrolateral projection termed the pleuron.

Exoskeleton and Hypodermis

The whole body is covered by an exoskeleton, secreted by a very thin, darkly pigmented epidermis or hypodermis. If a small portion of the hypodermis is examined microscopically, it will be seen to contain numerous chromatophores. They appear stellate in shape, with fine, branched processes when the contained pigment is dispersed, and look smaller and more rounded when the pigment is concentrated. The movements of the pigment are regulated by hormones, although in adult animals the color changes thus produced are barely visible through the thick exoskeleton. The exoskeleton is hardened by calcium salts, except at the flexible joints and in the period just after a molt. Molting of adult crayfishes occurs, in at least some species, only twice a year. Many species are restricted to waters with a calcium content above a certain minimum.

Paired Appendages

The appendages vary much in appearance; but are thought nearly all to be derived from a single basic type. Refer to the figure given under *Heteromysis* will show the parts of this basic, biramus appendage. In the simplest case there is a protopodite, divided into two segments, the coxa and basis, from the latter of which arise a medial endopodite and a more

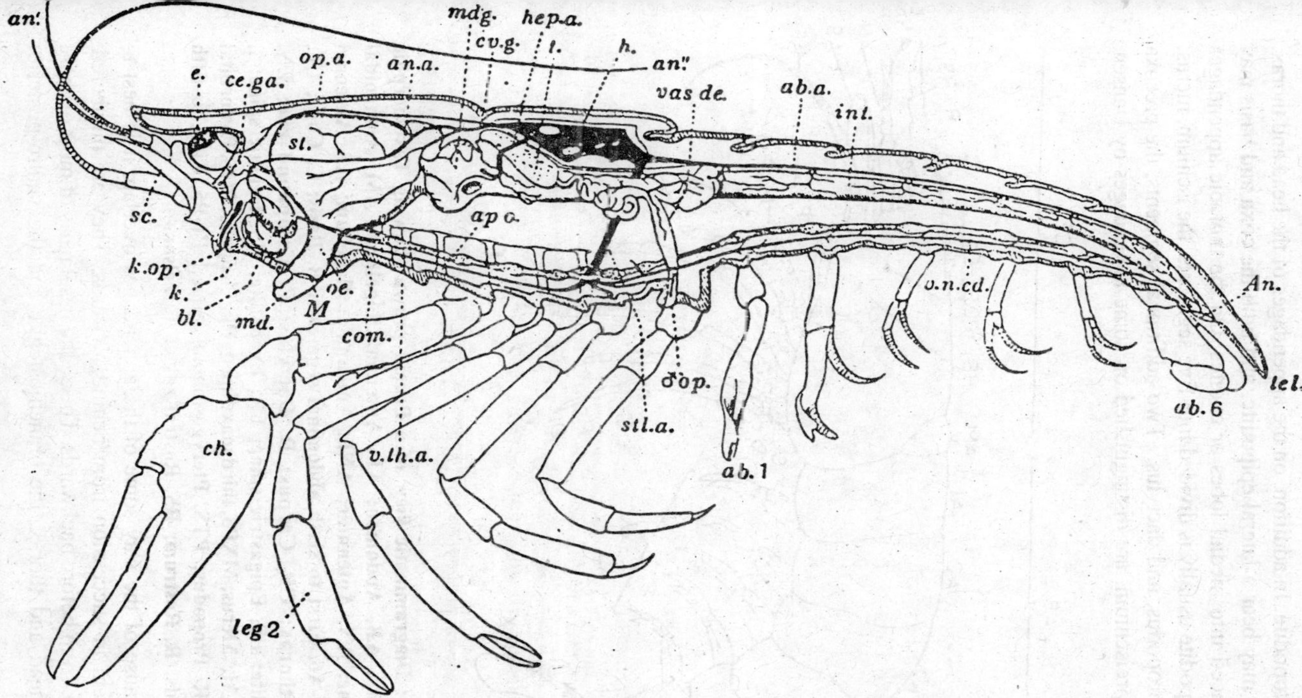

Fig. 3.4 : *Astacus fluviatilis*, dissected from the left-hand side. From Shipley and MacBride. an.' antennule; an." antenna; md. maible; M. mouth; sc. scale of antenna; An. anus; tel. telson; ♂ ♂ op. male genital opening; ch. chela; legli, first walking leg; ab. I, the first abdominal appendage, modified for copulation; ab. 6, the last abdominal appendage; oe. oesophagus; st. "stomach"; *mdg.* mesenteron; *cv.g.* cervical groove; int. intestine; ce.ga. cerebral ganglion; *com.* nerve collar; *v.n.cd.* ventral nerve cord; e. eye; h. heart; *stl.a.* stemal artery; *ab.a.* dorsal abdominal artery (posterior aorta); *v.th.a.* ventral thoracic artery; *op.a.* ophthalmic artery (anterior aorta); *an.a.* antennary artery; *hep.a.* hepatic artery; t. testes; *vas de.* vas deferens; *apo.* apodemes; k. kidney (green gland); *bl.* bladder; *k.op.* kidney opening. Ventral abdominal and left gastric arteries shown but not labelled.

lateral exopodite. In addition, on the appendages of the head and thorax, the coxa may bear a lateral epipoditc, and both the coxa and basis may be produced into medial lobes or endites. On the thoracic appendages the cndopodite usually is divided into five segments, the ischium, merus, carpus, propodus, and dactylus. Two additional segments, the precoxa and the preischiuin, are distinguished on some appendages by Hansen.

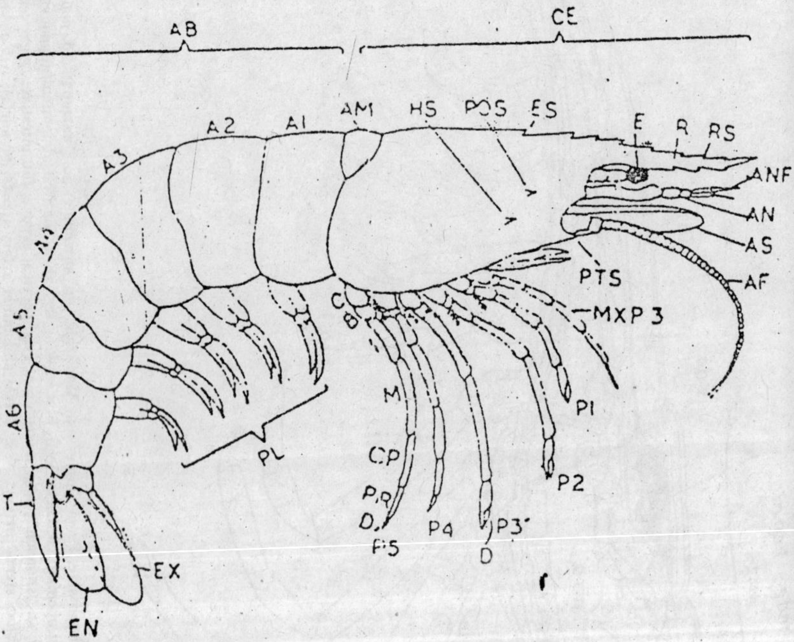

Fig. 3.5 : Diagrammatic view of a Penaid prawn showing Taxomorphic characters. AB, Abdomen; AF, Antenna! flagellum; AM, Arthroidal membrane; AN, Antennule; ANF, Antennular flagellum; AS, Antennal scale; Al-A6, First to sixth abdominal segments; B. Basis; C, Coxa; CE, Cephalothorax; CP, Carpus; D. Dactylus; E, Compound eye; EN, Endopodite; ES, Epigastric spine; EX, Exopodite; HS, Hepatic spine; I, Ischium; M,' Merus; MXP3, third maxilliped; PL, Pleopods; POS. Postorbital spine; PR, Propodus; PTS, Pterygostomian spine;' P1-P5, First to fifth pereiopods; R, Rostrum; RS, Rostral spine; T, Telson.

Because of the way some of the appendages overlap, it is best to study them in succession from behind forwards. They are numbered, however, from before backwards. To see the basal parts of most of those on the head and thorax, the branchiostegite and the ventro-posterior covering of the head should be removed from one side. This can be done by cutting with scissors from the dorsoposterior margin of the branchiostegite, along a line high on the side of the thorax, then curving ventrally a short distance in front of the cervical groove. The small portion of exoskeleton of the head thus included has to be carefully freed from underlying tissues with the aid of a sharp scalpel. So far as possible the appendages should be studied *in situ.* If it is necessary to remove an appendage for closer study, this is best done by cutting with a scalpel as close to the body as possible, while pulling the appendage with forceps held in the other hand. Afterwards the corresponding limb of the other side can be studied *in situ.* Observations of limb movements on living specimens ought also to be made.

The most posterior appendages are termed the uropods. On these the endopodites and exopodites are modified as flute, rounded blades. When expanded, the two uropods, together with the telson, form a wide tail fan, by means of which the animal can dart backwards. Farther forward are the pleopods or swimmer of the first five abdominal somites. The third, fourth, and fifth pleopods are simple, biramus, and best developed in the female, in which sex they carry the eggs; the attaching cement for this purpose is said to be secreted by so-called integumental glands in the exopodites and endopodites, and sometimes adjacent regions. In both sexes the pleopods beat so as to cause a current of water, but the functional significance of this is uncertain. The second pleopods resemble the succeeding three pairs in the female, but in the male crayfish are modified as stiff copulatory appendages. In the male *Homarus* the only modification of the second plcopods is a still extra process on each, medial to the endopodite. The first pleopods are much reduced in the female (in certain species sometimes absent), while in the male they again are modified as stiff copulatory appendages, (In crayfishes of the southern hemisphere the first pleopod are absent in both sexes.)

On the thorax the last five pairs of appendages are called pereiopods. They all are uniramus, exopodites being absent. Each of the first three pairs terminates in what is termed a chela, so constructed that the dactylus can bite against a prolongation of the propodus. The term chelipod is appropriate for these throe pairs of appendages but usually is restricted to the large first pair. This pair comes into use chiefly as weapons of defense or in predatory feeding, while the two smaller pairs are used for cleaning the body and for picking up small particles of food. The large first pereiopod, as compared with succeeding pairs, has undergone some torsion, bringing the dactylus into a medial rather than a lateral position. A clear concept of this torsion can be gained by determining the axis of flexure at each joint. Of the first two pereiopods and indicating these axes by means of pins appropriately inserted. Incidentally it will be found that the line of flexure differs at every joint, a feature which gives the limbs a wide range of mobility. The large chelipods further differ from the other pereiopods in that the basis and ischium are firmly fused together (though occasionally this condition is observed on the other pereiopods too). Perhaps associated with this fusion is the fact that it is especially the large chelipods which can be cast off, or autotomized, when the limb is wounded or firmly held. The break on such occasions always occurs at a specific fracture plane in the basi-ischium, possibly representing the plane of fusion between the two segments, though more probably lying just distal to this. On the remaining pereiopods, fracture planes also occur, just distal to the joint between basis and ischium, or, in Hansen's scheme, at the line of fusion between the ischium and preischium. Autotomy of these limbs, however, is relatively rare, and lobsters seem unable to cast them off without an assisting pull applied distally. After autotomy of any limb, loss of bipod is minimized by a diaphragm and valve which may be seen closing the end of the stump. Other features of interest to observe on the pereiopods include special setae and spines on the dactylus and propodus of the fifth or fourth and fifth legs, used by the female to clean the abdomen and pleopods before egg-laying; also, in males of certain species of crayfish, large spiny projections from the ischia of the third and/or sometimes an adjacent pair of pereiopods, used to lock under the legs of the female during mating. The male and female genital openings also should be noted, located on the medioposterior borders of the coxae of the fifth or of the third pair of pereiopods respectively.

APPENDAGES OF PRAWN

Antennule

1. They form the preoral, cephalic appendage.
2. The antennules are attached one on either side.
3. The protopodite consist of three segments—Precoxa, coxa and basis.
4. Proximally the precoxa bears a depression contains the opening of the opening of the statocyst on the dorsal side and a spring lobe called the stylocerite on its outer margin.
5. The middle pedomere or coxa is short and cylindrical.

6. The basis carries two long many jointed feelers.
7. The antennules bear sensory setae and are tactile.

Antenna

1. It is one on either side just below the antennules.
2. The peduncle or protopodite is greatly swollen due to the presence of excretory organ within which opens by a minute renal aperture on the inner margin of coxa.
3. The endopodite is represented by a many jointed feeder, while enopodite is in form of a broad leaf kike plate or scale.
4. The antennae are sensory, excretory and balancing in function.

Maxillubae

1. Small thin leaf like appendages.
2. Inner borders of coxa and basis are covered with painted spines and prejects inwards as jaws.
3. The endopodite forms a curved process bifurcated at the open.
4. Endopdite is absent
5. Mandible helps is manipulation of food.

Mandible

1. The are strong calcified bodies lying one on either side of mouth.
2. Entire mandible consists of coxa differentiated into two parts of apophysis and head.
3. Head forms two processes, the indan process bearing six dental plants and incisor process ending in three teeth.
4. Enopodite is absent.
5. Mandibles are masticatory in function.

Maxilla

1. Thin, leaf like mouth appendages.
2. The small coxa is particully divided while the large basis formes a bifuracated gnatho base internally.
3. The endopodite is small while the endopodite forms a large expanded fan shaped roller, the movement of which creates a water current passing over the gills.
4. The maxillae help in respiration and in manipulation of food.
5. *First Maxillipeds*

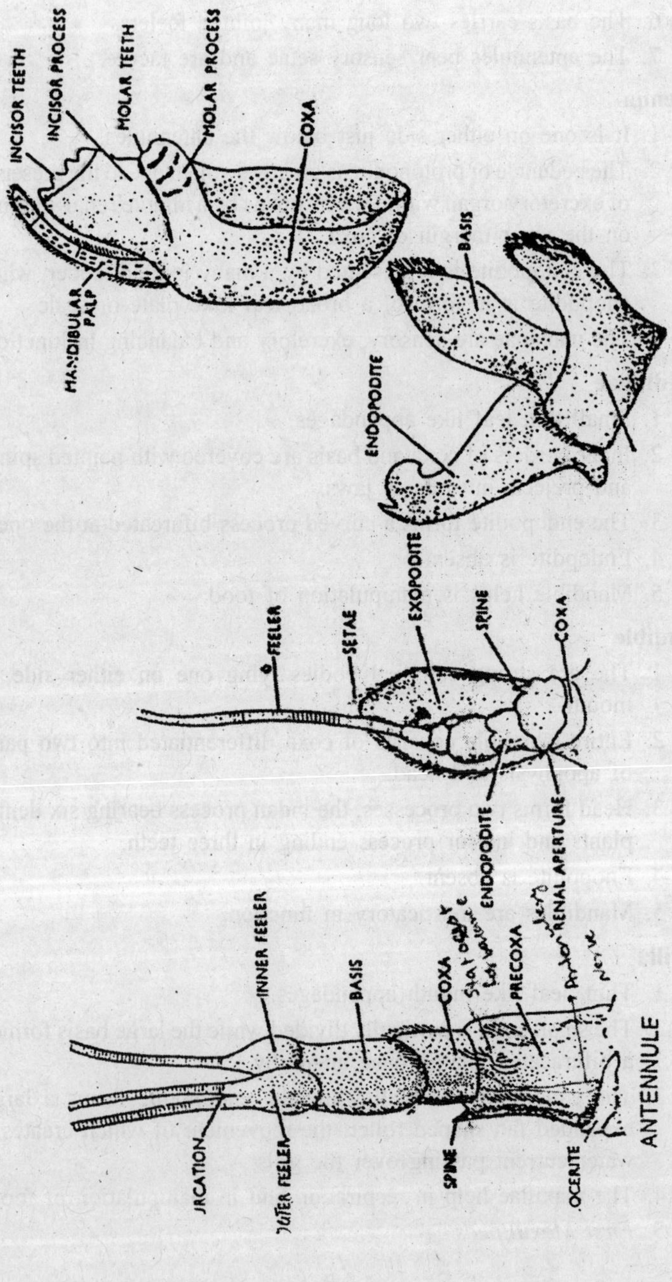

Fig. 3.6. Appendages of prown (Contd...)

1. They are thin and leaf like with the inner borders of coxa and basis forming endites or gnathobases.
2. The outer side of coxa bears a bilobed respiratory epipodite or primitive gill.

6. *Second Maxillipeds*
   1. The coxa bears an epipodite and a gill (Podobranch) on its outer margin.
   2. The basis carries a long slender and unjointed endopodite covered with setue along its distal half and a five segmented endopodite.
7. *Third Maxillipeds*
   1. They look leg like in appearance and have the some parts as second maxillipeds.
   2. These 3 pairs of maxillipeds take part in feeding and hold. The food in position while the mandible mastigates it.
8. *Walking legs*
   1. A typical walking leg like the 4th consists of a two jointed protopodite and a five jointed endopodite.
   2. They differ from manillipeds in their greater size and absence of enopodites and epipodites.
9. *Typical Abdominal Appendages*
   1. In a typical appendages, the protopodite consists of ring like coxa and a cylindrical basis.
   2. The basis bears two flattened leaf like rami and the exdopodite and the outer and larger enopodite.
10. *First Abdominal Appendages*
    1. In first pair the appendage intesna is absent and the endopodite is greatly reduced in size the rest of the structure is typical.
11. *Second Abdominal Appendages*
    1. In the second pair of male alone there is an additional rod like setue bearing process called the appendia masculina, in between the appendia intenna and the endopodite.
    2. The rest of the structure is typical.
12. *Uropods*
    1. The 6th pair of abdominal appendages, called uropods or tail feet, are large an lie one on either side of the telson forming tail fin, which enabbes the frown to take a backward

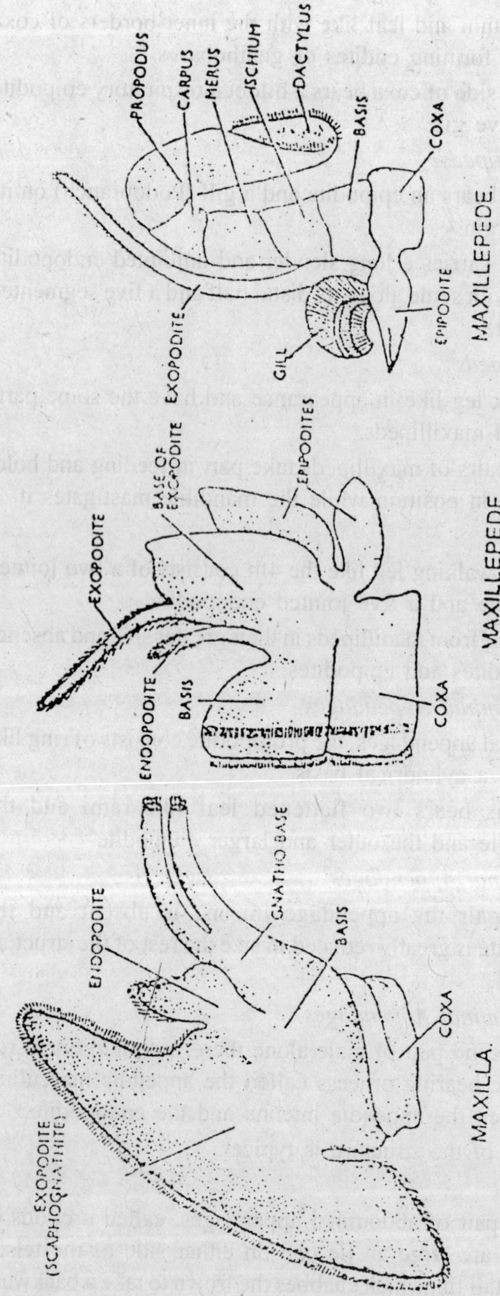

Fig. 3.7. Appendages of prown (Contd...)

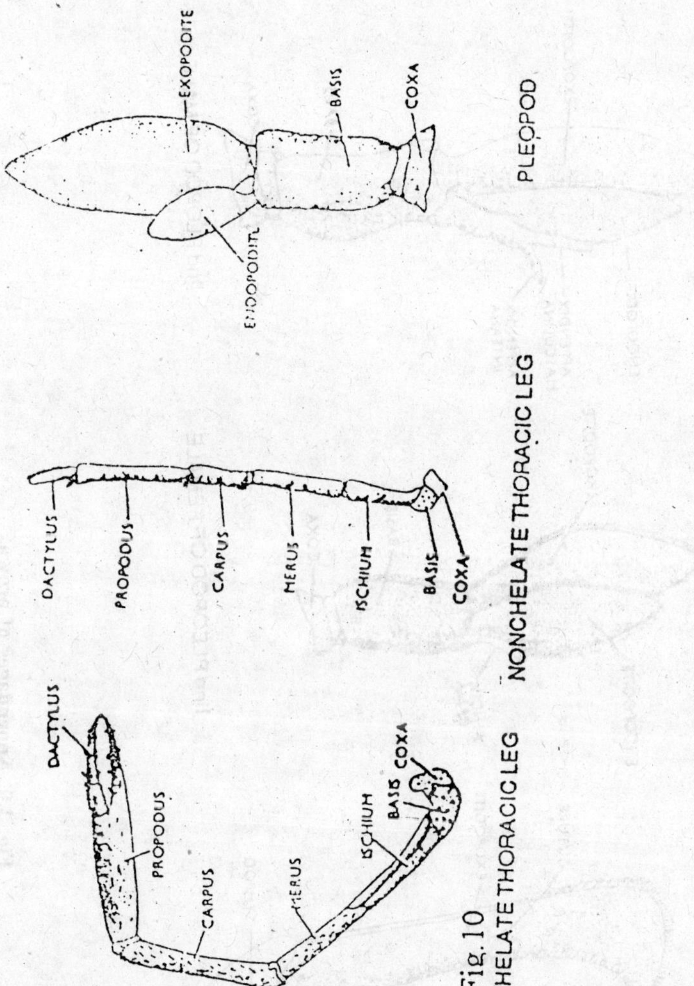

Fig. 3.8. Appendages of prown (Contd...)

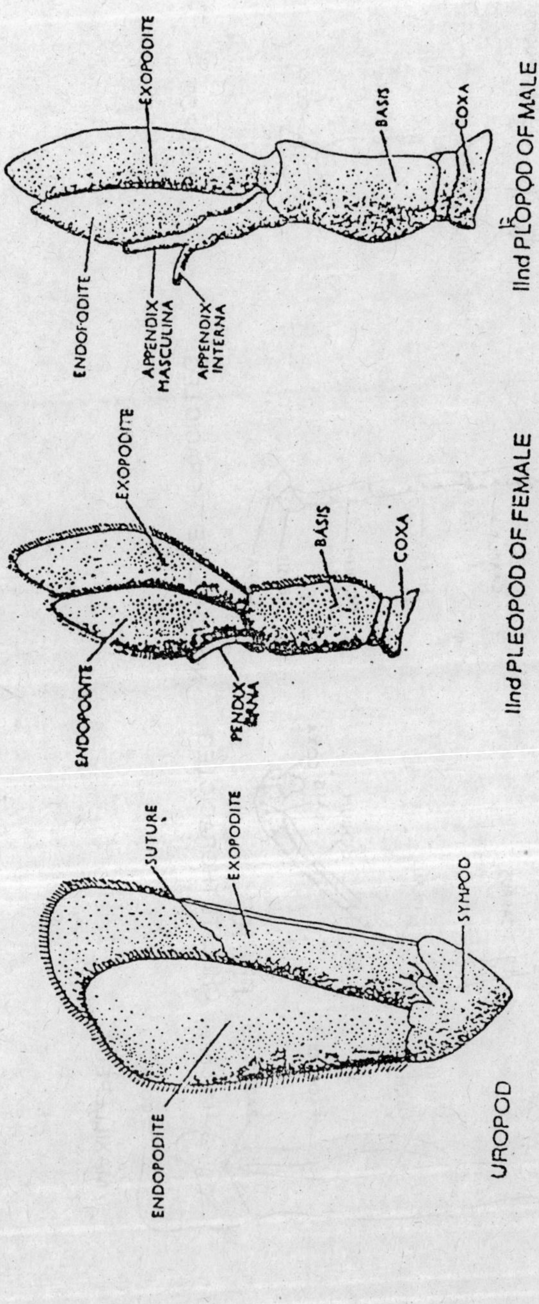

Fig. 3.9. Appendages of prown

spring in the water.

2. Their margins, except the outer border of the enopodite, are fringed with numerous setae.

Table 3.1 : How antennale differ from antenna.

Antennale	Antenna
1. They are lies below the bases of eye-stalked.	1. They lies below the antennules.
2. Antennae are larger than antennae.	2. They are smaller than antennae.
3. Antennules bears sensory setae and tactile in function.	3. Antenna are sensory encretory and balancing in function.
4. Squama absent	4. Scale and squama is present.
5. Excretory organs are absent protophodite is greatly	5. Due to excretory organ swollen.

CEPHALOTHORAX

The cephalothorax is completely fused dorsally and does not show any kind of segmentation.

Cephalon 5 (pairs)	Preoral	Somites 1-3	Antennule Antenna Mandible
	Postoral	Somites 4-5	1st Maxilla 2nd Maxilla
Thorax 8 (pairs)		Somites 6-8 Somites 9-13	Maxillipeds Pereiopods

The terga and pleura of the cephalothorax are fused to form the large carapace. This is a broad shield, the width of which is about one and half times of its length The cephalic region of the carapace is heavily calcified. This region is bounded behind and at the sides by very deep grooves. It is divided into areas by means of depressions and grooves and each area is named after the organ lying immediately below it. The cervical groove is most prominent transverse groove and is situated about mid-way along the carapace. The cervical groove separates the cephalic region from the thoracic region. This groove is seen as transverse median depression The cephalic portion of the carapace is divided into the facial and gastric regions. The facial region is separated from the rest of the carapace and is divided into median frontal region and two lateral orbits regions.

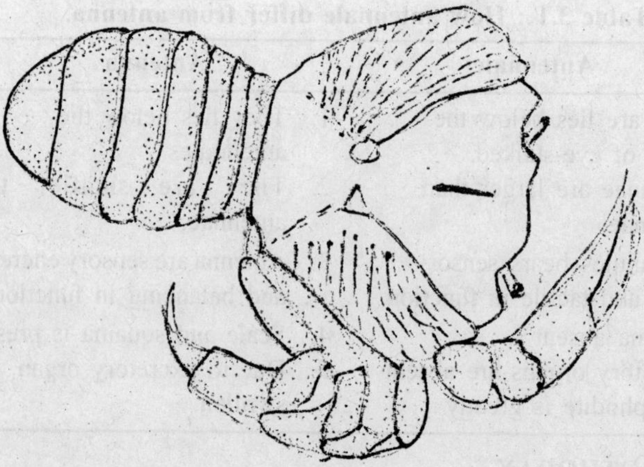

Fig 3.10.

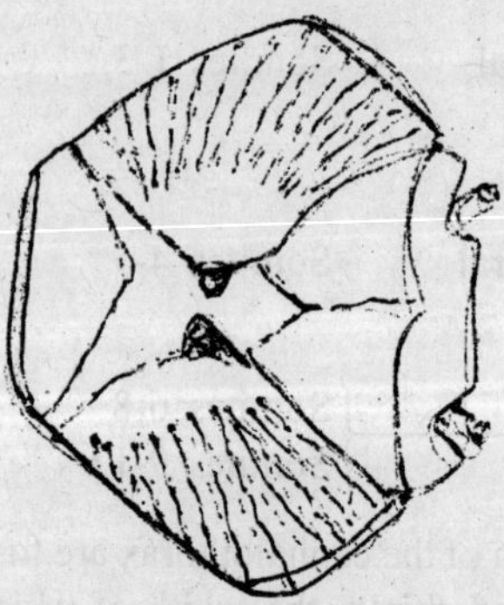

Fig 3.11.

The median triangular portion is the gastric region and it is bounded by cervical groove. The median portion is divided into two anterior protogastric, median anterior mesogastric region a pair of posterior matagastric and a median posterior urogastric region. The lateral portion is known as hepatic region. The thoracic region is the whole area posterior to the cervical groove. It is divided by two longitudinal grooves, branchiocardiac grooves into median cardiac and two lateral branchial region. The ventral inflected portion of the carapace is divided into two parts by well defined suture, which may be termed as the pleural groove. The carapace passes outward almost horizontally and then suddenly bends inward up to the bases of walking legs on either side of the membranous wall of the thorax which is called as 'branchie straggle as it encloses the branchial cavity.

The anterior border of carapaces situated in orbits. The median portion is a rostrum which consists of two lateral lobes. It is continued ventrally as a median plate which separates the two cavities in which are situated eye peduncles. The posterior border of the carapace is horizontal and is continuous behind with tergum of the first abdominal segment.

Pre-oral Cephalothoracic Sterna

The rostrum passes backwards ventrally as a rectangular plate. This plate is separated from the first sternum by well defined suture, which forms a septum between the articular cavities of the two optic peduncles.

The first sternum lies in the segment of the first antenna and separates the articular cavities of these appendages. From the ventral side of the first sternum arises the posterior and inner sides of the sockets of the antennules and the lateral prolongation extends as far as the bases of the second antennae.

The epistoma is a broad plate in front of the mouth, immediately behind the first sternum: It represents the united stems of second (antennary) and third mandibular somite. The labrum is soft fleshy lobe attached to the middle legion of the posterior border of the epistoma.

The sterna of the thoracic region are all fused together as a single triangular shaped plate situated between the bases of the paired post-oral cephalothoracic appendages. Transverse grooves are present which mark the division of this region into segments. The surface of the fused sterna is concave laterally in order to place the abdomen which is always in flexed condition. This concavity is well marked in the mates. On the sternum of the fourth thoracic segment are two small tubercles which fit into two concavities on, the abdomen and thus form an affective locking apparatus. The sternum of the third thoracic somite of the female bars a pair of large external genital openings. At the outer and posterior corners of each sternum are backwardly directed region which are called as episterna.

Epimera

Cephalic epimera are difficult to identify but they are represented by the regions between the outer portion of the articular cavities of the somite and the carapace. Epimera of the third (mandibular) and two last cephalic somites (maxillary) the represented by the membranous roof of the prebranchial chamber at each side. This is continuous thoracic epimera.

Thoracic epimera form the inner wall of the branchial chamber by continuous thin plates at each side and support the gills. The lower border of the ecimera commences immediately above the base of the thoracic appendages. They pass upward and inward and are continuous above with membranous roof of the branchial chamber. Thoracic epimera are fused to form a convex wall which is divided into segments by vertical suture. These sutures indicate the lines of secretion between the various somites of the thorax.

Abdomen

The abdomen is continuous with the posterior part of the cephalothorax and is connected by means of an arthrodial membrane so that it can allow considerable movement between the two regions. The abdomen is small and it is clearly applied to the sternal region of the cephalothorax. This region differs in two sexes being much broader in female than in male.

Female abdomen consists of six somites and a telson all of which are freely movable. There are four pairs of appendages, one pair being born on the second and rest three on the following somites respectively. The hindgut opens on the ventral side of the telson at the anus.

The male abdomen is slightly shorter, narrower than the female. There are only two pairs of appendages present, one pair being born on the first and other on the second somite. These appendages are modified to act as copulatory organ. In male the abdomen is very closely applied to the thorax because of the larger tubercles present it on the thoracic sternum.

Appendages

There are five pairs of appendages on the head and eight pairs on the thorax. There are four pairs of abdominal appendages in the female and only two pairs in the male. The appendages are as follows:

Cephalon → Somite →

1. 1st antenna (Antennule)
2. 2nd antenna
3. Mandible
4. 1st maxilla
5. 2nd maxilla

Thorax → Somite →

6. 1st maxilliped
7. 2nd maxilliped
8. 3rd maxilliped
9. 1st pereiopod
10. 2nd pereiopod
11. 3rd pereiopod
12. 4th pereiopod
13. 5th pereiopod

Abdomen → Somite →

	Female	Male
14.	Absent	1st Pleopod
15.	1st Pleopod	2nd Pleopod
16.	2nd Pleopod	Absent
17.	3rd Pleopod	Absent
18.	4th Pleopod	Absent
19.	Absent	Absent

Cephalothoracic Appendages

First antenna : The first antenna or antennule is generally situated in a deep depression on the ventral side of the cephalic sternum. The depression in which the first antenna lies is bounded by the rostrum from the front side and from behind by the lateral expansion of the first sternum. The outer boundary is formed by the inner edge of the second antenna and inner boundary by the median portion of the first sternum. The appendage consists of protopodite which is formed of three joints or places. The basal joint is the broadest one from which is given off on its inner side a two jointed portion. These three constitute the protopodite. From the end of distal segment of the protopodite two many jointed flagella arise, an inner endopodite which is quite small and an outer exopodite which is large and bars on its inner side many long setae.

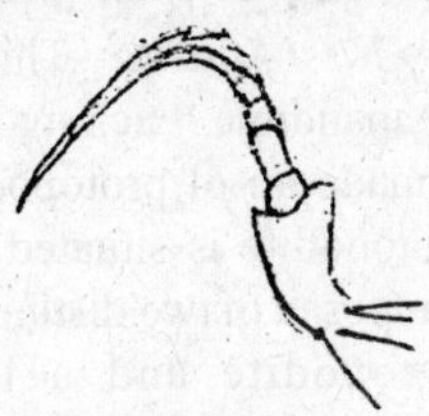

Fig. 3.12.

The small setae are the olfactory setae which lie on the ventral side of the exopodite. There is a longitudinal groove on the dorsal side of the basal segment of the protopodite where the auditory sac pens to the exterior.

The second antenna : The second antenna lies lateral to the antennules just near to the eye. This consists of basal large portion which is called as protopodite and partly it is fused to the carapace. At the distal end of this arises a long flagellum which consists of two long bass; segments at the lower side. The remaining part of the flagellum consists of a number of short rings. At the posterior side of the large basal segment is the operculum which covers the excretory opening The operculum represents as a exopodite and large basal portion as a basopodite and these together from the protopodite. The flagellum represents as an endopodite.

Fig. 3.13.

Mandible : It is a strongly calcified elongate structure which is divided into two parts, an inner one which covers the ventral region of the mouth and acts as 'jaw' and outer one, the apophysis. The apophysis of the mandible is a long bar to which are attached the tendons of mandibular muscles. At the outer end of the apophysis lie the tendon of the external aductor muscle and behind this lies the tendon of external abdutor muscle. The tendon of the internal abductor muscles is three attached to the posterior and inner side of the apophysis. The mandibulary palp is a little jointed structure which arises from he inner side of the apophysis. This structure is probably used for clearing the bitting edges of the mandible. The mandible is attached to the episterna by means of small projections below the palp.

Fig. 3.14.

First Maxilla : This lies immediately behind the mandible. Each maxilla is thin and small and is made up of protopodite and endopodite. The protopodite is situated on the inner side and is composed of two distinct parts, a narrow proximal coxopodite and a large basopodite. The endopodite arises from the outer side of the basopodite and consists of a broad proximal leaf

Fig. 3.15.

like region and a narrow distal region. The exopodite is absent. From distal end of both parts of the protopodite arise strong setae. The maxilla performs the function of passing feed to the mouth.

Second maxilla : The second maxilla is thin and foliaceous like the maxillulae. It consists of an inner protopodite, a median endopodite and outer scaphoganthite (exopodite). The protopodite is composed of coxopodite and basopodite both of which are bilobed. The two lobes of the coxopodite are slender and bilobed. The two lobes of the coxopodite are slender and long and are separated from each other. The basopodite is some that broader than coxopodite. The endopodite is a small unjointed process but the exopodite is large and forms an expansive broad plate of irregular shape which is called as scaphognathite. This lies in respiratory chamber and by means of its rapid movement it bales the water out of the branchial chamber.

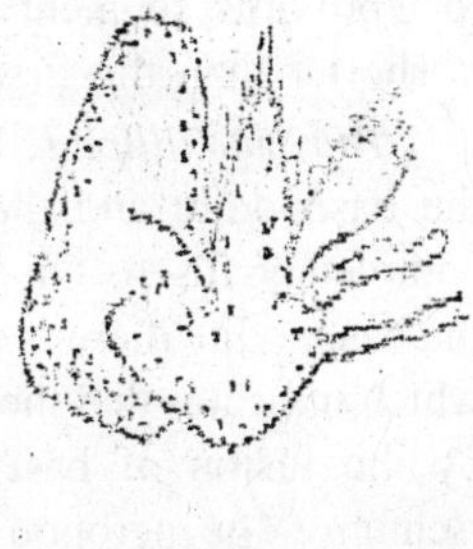

Fig. 3.16.

First maxilliped : In the first maxilliped the coxopodite and basopodite of the protopodite are flattened leaf-like and project on the inner side. The coxopodite is a long lamella having two rows of setae on its outer side. The endopodite is generally situated in between the protopodite and exopodite. It is unjointed, membranous and its distal portion is flattened. The exopodite is long slender and consists of long proximal segment. At the distal end of this there are many jointed flagella which project inward. From the outer side of the protopodite is given off a long flagellum (epipodite) which passes back into branchial chamber above the gills.

Fig. 3.17.

Second maxilliped : The protopodite in the second maxilliped is much reduced but still the coxopodite and basopodite are visible. The endopodite is comparatively larger than the endopodite of the first maxilliped and it is segmented. It arises from the basipodites and divided into five segments. The first segment is the ischipodite which is a small one. The second segment is the meropodite which is the

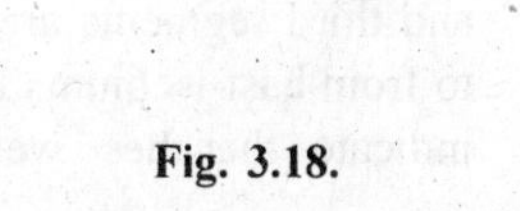

Fig. 3.18.

longest one. The remaining three distal segments are carpopodite, propodite and dactylopodite respectively. These three segments are small and they turn inward at right angles to the meropodite. The exopodite is again long and slender as in the previous appendage. It consists of long proximal segment and distally many jointed flagellum. The flagellum is shorter than the first maxilliped.

Third maxilliped : In the third maxilliped the basopodite and the meropodite are not completely fused. On the dorsal surface of the basopodite there is a long vertical groove which indicates that the basopodite is formed by the fusion of basis and ischium (basi-ischium). The meropodite is broad and square in shape and partly fused on the basi-ischium. To this are attached the remaining three segments, *i.e.* carpopodite, propodite and dactylopodite. They are rather hard and chitinous having pointed spines on them. The endopodite and exopodite are closely applied to each other. The endopodite is much flattened so as to form and effective operculum covering the complete mouth parts. The flagellum of this appendage passes back into branchial cavity from the front side of the anterior inhalant branchial aperture so that the size of the aperture is considerably reduced. The flagellum is richly provided with strong setae and it acts as a 'strainer'.

Fig. 3.19.

First pereiopod or chela : It is the largest appendage in the body. It consists of seven segments. A comparison with the third maxilliped show that the two proximal segments represent the protopodite and remaining five, endopodite. The exopodite is absent. The propodus is elongated distally, beyond its articulation with the dactylus so that this distal elongation comes opposite to the dactylus thus forming a pair of chela with sharp terminal claws or teeth. As the dactylus is freely movable it is called as a 'finger' and propodus being called as 'palm'. The second and third segments are the basis and ischium which are fused together to from basi-ischium. The basi-ischium has thin groove around it which indicates that these were originally two separate segments. This groove

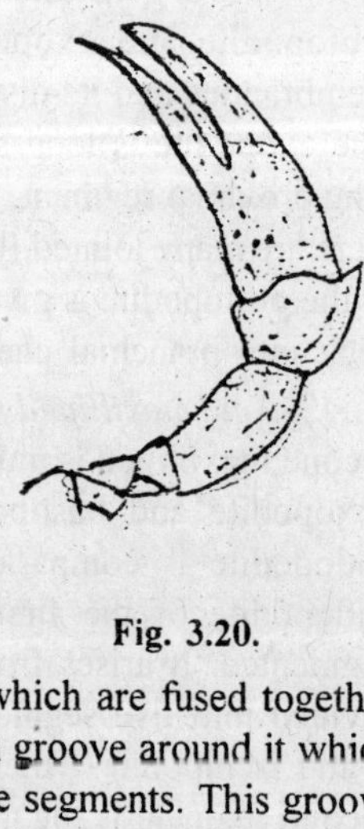

Fig. 3.20.

is known as fracture plans because it is at this point the animal fractures the limbs during the process of self amputation. The coxopodite is attached to the body by means of two hinges, one being dorsa and the other ventral. The motion of the coxopodite is in a horizontal plane moving backward and forward. The merus is well developed but has a very little movement. The carpus is short and has a pointed spinelike structure in the inner side. The two hinges of the carpus are situated dorsally and ventrally. The segment moves forwards and downwards. On the dorsal side of the basiischium and merus there are irregular groves. These are the lines of absorption. The whole of the appendage is called as chelate leg and is used to catch the prey and transfer it to the mouth.

Pereopods : These are the walking legs. The structure of the walking leg is very similar to that of the chela except that in the walking leg the propodite does not possess an out growth which is in conjunction with the dactylus. So the walking leg terminates in a single claw and is not in chelate form. The distal three segments are capable of moving and they flexed upon the proximal segment in a vertical plane.

Fig. 3.21.

Abdominal Appendages

Pleopods (Female) there are four pairs of appendage on the female abdomen. Each pair is situated on the second, third fourth and fifth somite respectively. Except the first pleopod all the three pars of pleopods are similar in structure. Each pleopod is attached to the abdomen by a basal piece which is known as the protopodite. From this is given off two places. An outer exopodite and an inner endopodite. In the first pair

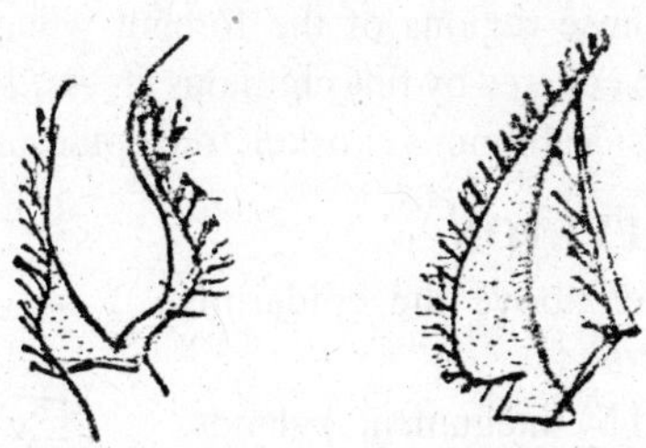

Fig. 3.22.

of pleopods the exopodite is cylindrical and it has rows of setae, whereas the endopodite is as long as the exopodite having transverse grooves on it. Many long setae arise from these transverse grooves. A remarkable change is noticed from the second pleopod. The exopodite here is modified into an elongate fan-like structure having many bundles of setae along its whole border. The endopodite is similar as in the first pleopod. These pleopods are mainly used for the attachments of the eggs.

Pleopods (Male) : The are two pairs of appendages on the male abdomen. They are situated on the first two somites. The first pair consists of two pieces a broad basal portion which is the protopodite and the elongate distal portion is the endopodite. The endopodite appears just like a tube. The second pair of appendage consists of two parts. The basal portion is the protopodite and attached to this is a vertical rod called as endopodite. The vertical rod-like portion on the second pleoped fits into the tube of the first pleopod.

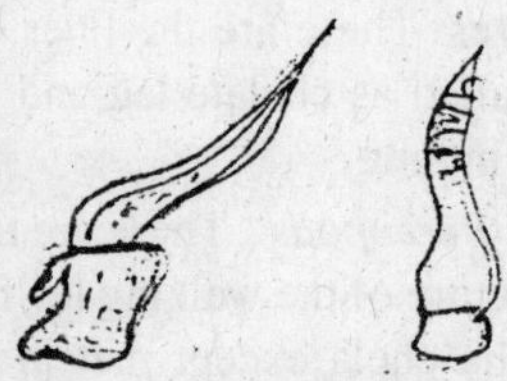

Fig. 3.23.

The outer surface of the body of the crab, *Baryteiphusa*, is covered by the chitinous exoskeleton which serves partly as a protective covering and also as a means of attachment far the muscles. The main portion of the exoskeleton is strongly calcified except between the movable somites of abdomen and also between the articulating segments of the appendages. The exoskeleton remains uncalcified in order to allow free movement. The outer portion of the ventral region of the abdomen is not so calcified. The outer walls, the floor and the roof of the branchial and prebranchial chamber are extremely thin. The chitinous lining of the foregut and hindgut which are continuous with the exoskeleton are uncalcified, except in those regions of the foregut where ossicles are present. The gills are also covered by fine chitinous layers. The integument mainly consists of 3 parts: a chitinous exoskeleton, epidermis and dermis.

CHITINOUS EXOSKELETON

The Chitinous layer above the epidermis consists of several layers

Cuticle : This is laid by integumental glands. The uppermost layer is the cuticle which is thin, hyaline and without any visible structure covering the whole of the chitinous exoskeleton.

Fig. 3.24.

Its continuity is interrupted at internals by the duct of cutaneous glands.

Pigment layer : This layer of the endocuticle is heavily calcified and is characterised, by the presence of a granular deposit of a melanin-like pigments. It is derived from the underlying epidermal cells and is found only where hardening by calcium impregnation has taken place. The horizontal laminae typical of the endocuticle as a whole are clearly evident in the pigmented layer.

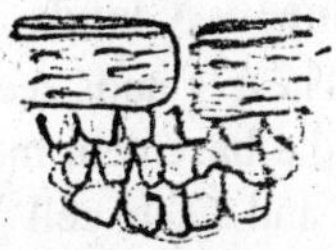

Fig. 3.25.

Calcified layer : This is the broadest layer of the cuticle in the fully formed exoskeleton. It is colourless and richly impregnated with calcarious salts. Like the pigment layer, it shows striations parallel to the surface, but the laminae are generally broader than in the pigments layer. This layer forms the greater part of the hard exoskeleton because new laminae are constantly being added to this region until the exoskeleton attains its maximum thickness.

Fig. 3.26.

Noncalcified layer : It is very thin and composed of delicate laminae parallel to the surface. This layer remains in a very soft condition and is not formed until the calcified layer is completed.

Fig. 3.27.

Epidermis

The epidermis consists of a single layer of columnar cells resting on a baserrent membrane. These cells give rise to the chitinous layers of the integument and vary in size both in different parts of the body and during the molting cycle. In dorsal integument of the hard crab the cells of the epidermis are columnar but in the soft crabs cells are comparatively much greater in length.

Dermis

The dermis lies below the basement membrane and varies in thickness, it consists of a network of connective tissue fibres and scattered cells. There is a layer of pigment calls close to the basement membrane. There are two kinds of cells present in connective tissue which contain reserve material. These are the 'Cellules Proteiques'. The former contains glycogen and are present in great numbers especially when the period of ecdysis approaches. The cutaneous glands are also embedded in the dermis.

Cutaneous Glands

These glands seem to be scattered throughout the connective tissue near the basement membrane. Each gland consists of a round duster of cells and each cell has a nucleus situated basally. Each cell of the globular mass is in contract with small cavity on its inner side. The central cavity receives the secretion from, various cells, and is connected with a long narrow duct enabling glandular secretion to pass to the exterior. This duct is intracellular in nature and is lined by mandible and in the walls of esophagus.

Immediately infront of the mouth there is a compact mass of cutaneous glands at each side which open on the surface of the epistoma. These glands have the same structure of the typical cutaneous gland. The cutaneous elands are also present on the endopodite of the maxilliped. On the endopodite of the pleopods of the female there are closely packed cutaneous glands. They secrete the cement like structure which attached the eggs to the endopodites of the abdominal appendages. The exact functions of the various cutaneous glands in different parts of the body are not clearly known. The gland of the epistoma and metastoma may produce a secretion which is poured on the food as it enters the mouth.

The ages of full grown male crabs (size is the carapace width in mm except in the case of *Paralithodes* which carapace length is given).

Species	Size	Age years	Source
Halicarcinus australis	18	1	Lucas and Hodgekin, 1970
Cyclograpsus punctatus	20	2	Broekhuysen 1941
Aratus pisoni	23	1.5	Warner, 1967
Rithropanopeus harisii	26	5	Turoboyski, 1973
Pachygrapsus crassipes	47	2.5	Hiatt, 1948
Curcinus maenas	65	4	Crothers, 1967
Callinectes sapidus	140	1-2	Tagatz, 1968
Paralithodes camtschatica	170	17	Hoopes and Karinen, 1972
Cancer magister	207	5	Butler, 1961
Cancer pagurus	250	15-20	Bennct, 1974

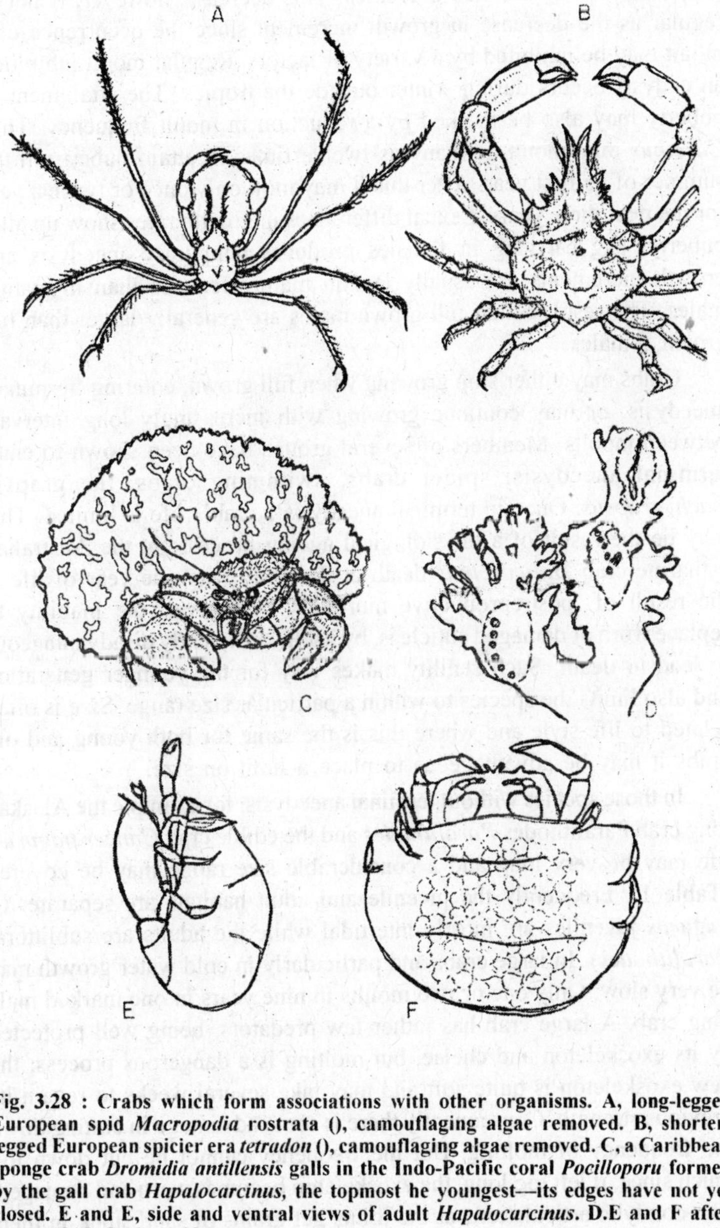

Fig. 3.28 : Crabs which form associations with other organisms. A, long-legged European spid *Macropodia* rostrata (), camouflaging algae removed. B, shorter-legged European spicier era *tetradon* (), camouflaging algae removed. C, a Caribbean sponge crab *Dromidia antillensis* galls in the Indo-Pacific coral *Pocilloporu* formed by the gall crab *Hapalocarcinus,* the topmost he youngest—its edges have not yet closed. E and E, side and ventral views of adult *Hapalocarcinus*. D.E and F after Potts. 1915. Scale = 10 mm.

In older crabs not only does the growth increment decrease, the frequency of molting also decreases. This decrease, however, is not as regular as the decrease in growth increment since the occurrence of a moult may be inhibited by a variety of factors. Regular moult inhibition an ecdysis occurs during winter outside the tropics. The attainment of puberty may also be marked by a reduction in moult frequency. Thus *Carcinus* may moult as many as twelve times to attain puberty in the summer of its first year. After this it may molt only once or twice a year for the next three years. Sexual differences in growth rates show up after puberty: egg carrying in females produces temporary anecdysis and growth increments are usually less in mature females than in mature males. The result is that full grown males are generally larger than full grown females.

Crabs may either stop growing when full grown, entering 'terminal' anecdysis, or may continue growing with increasingly long intervals between moults. Members of several groups have been shown to enter terminal anecdysis: spider crabs, swimming crabs, the grapsid *Pachygrapsus*. Once in terminal anecdysis a crab's life is limited. This may be the result of a physiological ageing process: in the Australian estuarine crab *Halicarcinus* death at the end of its one year of life is the result of 'postreproductive morbidity'. However, the inability to replace worn or damaged cuticle is, by itself, sufficiently disadvantageous to lead to death. Such senility makes way for the younger generation and also limits the species to within a particular size range. Size is often related to life-style and where this is the same for both young and old crabs it may be advantageous to place a limit on size.

In those species without terminal anecdysis, for example the Alaskan king crab Paralithodes *Paralithodes* and the edible crab *Cancer pagurus*, life may be very long and a considerable size range may be covered (Table 1). Frequently the juvenile and adult habitats are separate: *C. pagurus* juveniles are mostly intertidal while the adults are sublittoral *Paralithodes*). In large crabs and particularly in cold water growth may be very slow; Only one or two moults in nine years in one marked male king crab. A large crab has rather few predators, being well protected by its exoskeleton and chelae, but molting is a dangerous process; the new exoskeleton is quite soft and may take several weeks to regain its previous strength. Consequently there is some advantage in cutting down the frequency of molting. But the frequency cannot be cut down too much since, if left too long, the exoskeleton becomes worn and damaged. The condylic articulations at the joints get grains of sand stuck in them and no longer operate smoothly; numerous epifaunal animals such as barnacles and tube-worms settle on the skeleton and may become a heavy burden. Thus molting retains some value as a periodic 'spring cleaning'. Death in these large slow-growing animals probably eventually occurs as a result of the disadvantages of long intermolt periods.

Relative Growth

Growth of a part of the body (y) relative to another part, for instance carapace width, (x) is best expressed graphically and the relationship may take a number of different forms. In the simplest case there may be no change in the relative proportions of v and r and no change in shape with growth, the relationship is linear of the form $y = bx$ and growth is described as isometric. Alternatively there may be a constant difference between the growth rates of x and y producing a progressive change in shape with growth. This relationship has the form $y = bx^a$ where the exponent a is a measure of the difference in growth rates. The graph is an exponential curve arid the type of growth is described as allometric, positive when $a < 1$ (y increases in size relative to x) and negative when $a > 1$. Lastly the relative growth of x and y may change during the growth of the crab, again producing changes in shape.

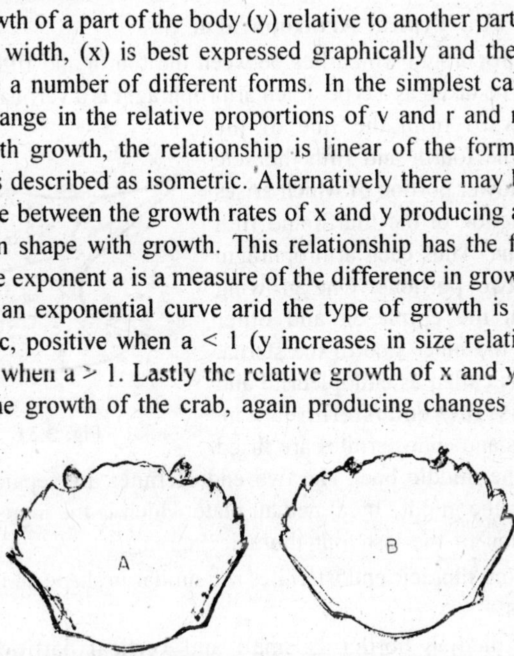

Fig. 3.29 : The change in shape with growth of the carapace of the shore crab *Carcinus maenas*. A is the carapace of a juvenile crab 13mm wide, B is the carapace of a mature crab 66mm wide.

The Endophragmal Skeleton

The endophragmal; skeleton is a system of calcified plates in the thorax which mainly for the attachment of the muscles of the proximal region of the appendages in the region of the body, It is also useful in supporting and protecting certain portions of the viscera. This system consists of number of inwardly projecting

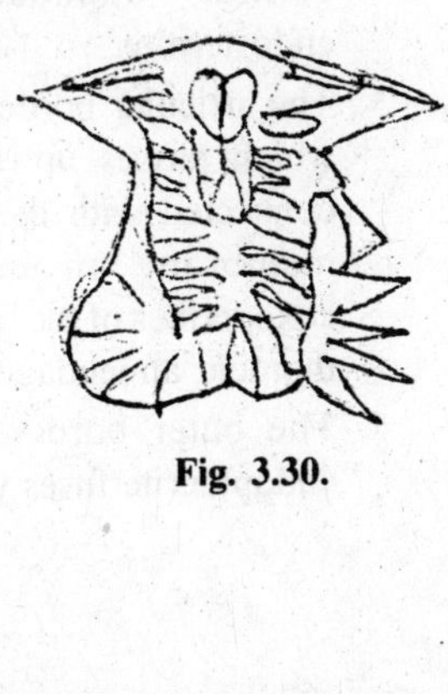

Fig. 3.30.

plates arranged transversely so as to divide the inner portion of the cephalothorax into series of irregular compartments.

Each portion is called an arthrophrgam. Each arthrophragm or plate arises at the junction of the two somites and is formed by inturning of the edges of the epimera and sternum of each segment. Thus each plate is double as it is duplicated throughout by neighbouring segments.

Description of a Typical Arthrophragm

The arthrophragm which lies between the fourth and fifth thoracic somites may be taken as a type. Each arthrophragm is a vertical partition growing inward from the line of the junction of the fourth and fifth thoracic epimera the lower portion of which arises from the junction of the fourth and fifth thoracic sterna. Thus each arthrophragm consists of two portions. One growing inward from the epimeron and other arising from the inner side of the sterna. The former is called as endopleurite and later one is the endosternite. The endopleurites and endosternites are fused together in the middle line. The two endosternites are separated from each other in the middle by a median plate which is the ingrowth from median portion of the last four thoracic Sterna.

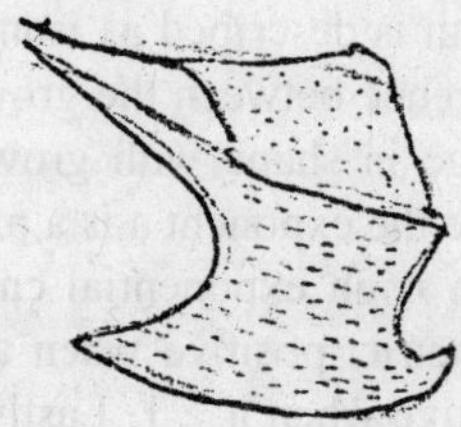

Fig. 3.31.

All the four thoracic endosternites are similar in shape and have five main borders:

1. The median border is small and vertical partition of the endosternite is in contact with the median plate.
2. The sternal border which is in contact with the sternum forms a ventral boundary of the endosternite.
3. The articular border is a portion which passes upward and it is connected with the arthropodial membrane in contact with coxopodites of the fourth and fifth thoracic appendages.
4. The outer border is portion where the lower end of the endopleurite fuses with the endosternite. The apodemal formen

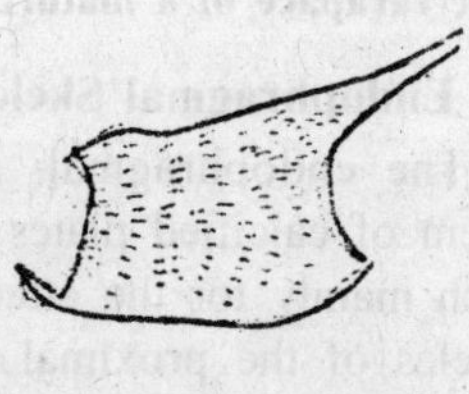

Fig. 3.32.

which is described in the *Cancer* is completely absent in *Barytelphusa.*

5. The dorsal border is a upward and outward portion from the mealan line forming almost a semicircle.

The endopleurite is rectangular in shape and four borders can be clearly seen.

1. The inner border is a portion which is in contact with the outer border of the endosternite. The inner border and outer border get fused at this region.
2. The articular border of the endopleurite is similar to that of endosternite. It is in contact with the arthropodial membrane connecting the coxopodite of the fourth and fifth thoracic appendages.
3. Epimeral border is the one which is in contact with the four and fifth thoracic epimera.

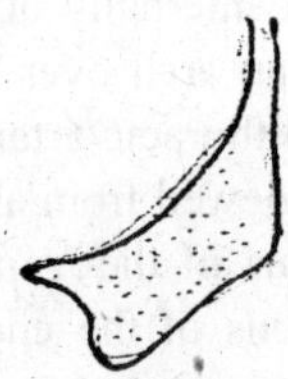

Fig. 3.33.

4. Dorsal border is the border free end of the endopleurite.

All the arthrophragms of the post-oral cephalothoracic region are built on the same plan as above. Each somite has one endosternite and one endopleurite at each side. The last five thoracic arthrophragms are very similar to the one described above.

The last thoracic arthrophragm differs considerably from the rest. It is not divided into endopleturite and endosternite. It is, therefore, said that this arthrophragm represents the fused endosternite and endopleurite.

The thoracic sternum is extremely convex antero-posteriorly. The epimeral wall also, instead of having fat surface as found in the animals of the group Macrura, is extremely convex on its outer face. Due to convex shape of the sternum and epidermal wall there is much complexity in the endophragmal skeleton in the last five thoracic somites.

The anterior post-oral endophragmal system consists of last three thoracic and two post-oral cephalic arthrophragms. The arthrophragms of this region differ much from those present in the posterior thoracic somites. All the arthrophragms of the anterior region have no median plate as inch. The median plate appears with the fourth thoracic arthrophragm. All the sterna of the post-oral region anterior to the fourth thoracic arthrophragm are fused together.

Endoskeleton

The exoskeleton not only is continued internally as linings for the fore and hindgut, but also projects internally as so-called apodcmes, secreted by inpocketings of the epidermis. Collectively, these apodemcs form the endoskeleton. Some of them take the form of tendons for the attachment of certain muscles. Others are in the form of hard, skeletal parts. Of these the most conspicuous are those in the floor of the thorax, which form a structure called the endophragmal shelf. This shelf is a latticework made up of curiously shaped apodemes, of which two pairs project internally at each junction between somites. The median pair forms an arch over the sternal sinus, in which the nerve cord and the ventral thoracic artery lie. Each member of the lateral pair is Y-shaped when viewed from above, with the stem of the Y directed outwards and the arms of the Y meeting the tops of successive endosternal arches. Functions of the endophragmal shelf are to protect the ventral nerve cord, to contribute one component in the joint for the articulation of each of the pereiopods, and to provide attachment for a number of muscles, particularly those of the appendages below, and the large flexor abdominis above.

Before each molt certain parts of the endoskelcton are resorbed, making the molt mechanically possible.

④

DIGESTIVE SYSTEM

INTRODUCTION

The digestive tract is divided into a foregut comprising the esophagus, cardiac stomach, and pyloric stomach; a midgut, comparing the hepatopancreas and part of the intestine; and a hindgut, comprising the rest of the intestine and the rectum. The internal epithelium of the foregut and of the hindgut is ectodermal and secrets a cuticular lining which is cast as each molt.

All parts of the digestive tract should first be studied *in situ* in the animal. To facilitate this study the body wall of the cephalothorax should be cut away on one side, if this has already been done care has to be taken to avoid breaking the delicate connections of the hepatopancreas and intestine behind the stomach. After a general study has been complete, the different portions of the digestive tract may be removed from the body for study of parts not otherwise visible. However, unless a second specimen is to be available, this should not be done until certain a parts of the nervous system have been studied. The stomach and the hindgut should be slit open midventrally, and their internal structure observed under water. A small part of the hepatopancreas should be mounted for study under the microscope.

From the mouth the short, vertical esophagus ascends to open into the floor of the cadiac stomach. The opening is guarded by a amedian anterior and a pair of lateral valves, all beset with setae pointing into the stomach. On the anterior wall of the cardiac stomach a large oval area is present ion each side; before each molt this area secretes a calcareous disk known as a gastroliths immediately after the molt the two gastroliths, being no longer protected by the overlying cuticular lining, are dissolved in the stomach. Probably the calcium salts thus released are absorbed and used for prompt hardening of certain essential structure such as the mandibles, tips mandibles, tips of the chelae, and teeth of a structure termed the gastric mill.

It is as the site of this gastric mill that cardiac stomach is chiefly important. The grinding elements of this mill are three teeth, one dorsal and two lateral, much more used for chewing the food than are the mandibles or other mouth parts. The mechanism for moving the teeth and other parts of the cadiac stomach is very complex, involving some thirteen pairs of muscles and a framework inside the stomach wall composed of over two dozen distinct plates, ossicles, and bars, small setae on many parts of the stomach lining point towards the region where the three teeth bite together.

In the cardiac stomach the food not only is ground up hurt also is digested by enzymes from the hepatopancreas. Only after the food has been ground into fine particles or is in solution can it pass into the next division of the stomach. Larger particles have to be rejected through the esophagus a constriction marks the boundary between the cardiac stomach and the much smaller pyloric stomach. Internal folds and ridges of the stomach lining, beset with numerous setae, from a complex system of filters and channels, through which small particles are admitted from the cardiac stomach into the pyloric stomach and thence either into the intestine or into the intestine or into the left or right hepatopancreas. Several main routes are present.

(1) a route which starts just behind the dorsal tooth and passes back as a dorsal groove of the pyloric stomach, termed the midgut filter, which presently divides into left and right channels oppening dorsally into the midgut.

(2) a pair of routes each of which starts below one of the lateral teeth, curves laterally around a medioventral projection termed the cardiopyloric valve, then joins with its fellow behind this to pass back medially between a pair of plates which forms a so-called muscular press, about midway between the upper and lower limits of the pyloric stomach; from here this route opens into the midgut.

(3) A route which starts behind the esophageal opening , as a ventral groove of the cardiac stomach; this groove soon forks and each division passes ventrolaterally around the cardio-pyloric valve, and enters on of the two so-called gland filter, which lie within a pair of prominent ventral swelling of the pyloric stomach; at this point materials in solution, or particles small enough to pass between the tiny setae in the gland fiters (spaced about 10 μ apart), pass on into channels leading to the opening of the left

and right division of the hepatopancreas; particles too large to pass through the gland fiters can pass dorsally by converging channels which join route 2 in front of its entrance into the midgut.

The walls of the pyloric stomach are stiffened internally by some nine plates and bars, complexly articulated with one anther. About half a dozen muscles serve to regulate the size of the various channels and to squeeze particles through some of the filters.

The region which appears to by the posterior part of the pyloric stomach is the beginning of the midgut. This lacks a cuticular lining and possesses a cuboidal rather then a columnar type of epithelium, its junction with pyloric stomach slants ventroanteriorly, into a very delicate and easily broken hepatopancreatic duct opens on each side, just dorsopostererior to the swelling of the pyloric stomach which houses the glands liters. Four valvular flaps, dorsal, lateral and ventral, project into the midgut from the pyloric stomach. A short caecum projects dorsoanteriorly from the midgut, pressed so closely against the pyloric stomach that it frequently is overlooked. In crayfishes, these usually can be seen, just posterior to this point, a slight transverse ridge which markes the end of the midgut and the beginning of the hindgut. The latter runs back as the intestine almost to the telson, where it enlarges slightly as the rectum. The anus opens on the ventral surface of the telson. The internal surface of the hindght is thrown into six longitudinal folds and is covered throughout with a cuticular lining. In the lobster the condition is rather different. Here the midgut includes the whole intestine, which runs back to the last abdominal somite, where it meets the hindgut modified as a relatively large rectum; a posterior midgut caecum is given off dorsally just before the junction within the rectum and runs back most of the length of the sixth abdominal somite. In larval lobsters the midgut is said to be short and hindgut long, as in crayfishes.

The very large hepatopancreas is divided into left and right divisions, although these are united under the pyloric stomach. The two division represent outgrowth from the midgut and are invested by a very thin membrane. When this membrane is broken, the tiny terminal tubules which make up the bulk of the organ will be seen to fray out if the specimen is under water. The total internal surface offered by these tubules is very large, since they are extremely numerous and the two divisions of the hepatopencreas extend nearly the length of the cepholothorax and into part of the abdomen. The color of the

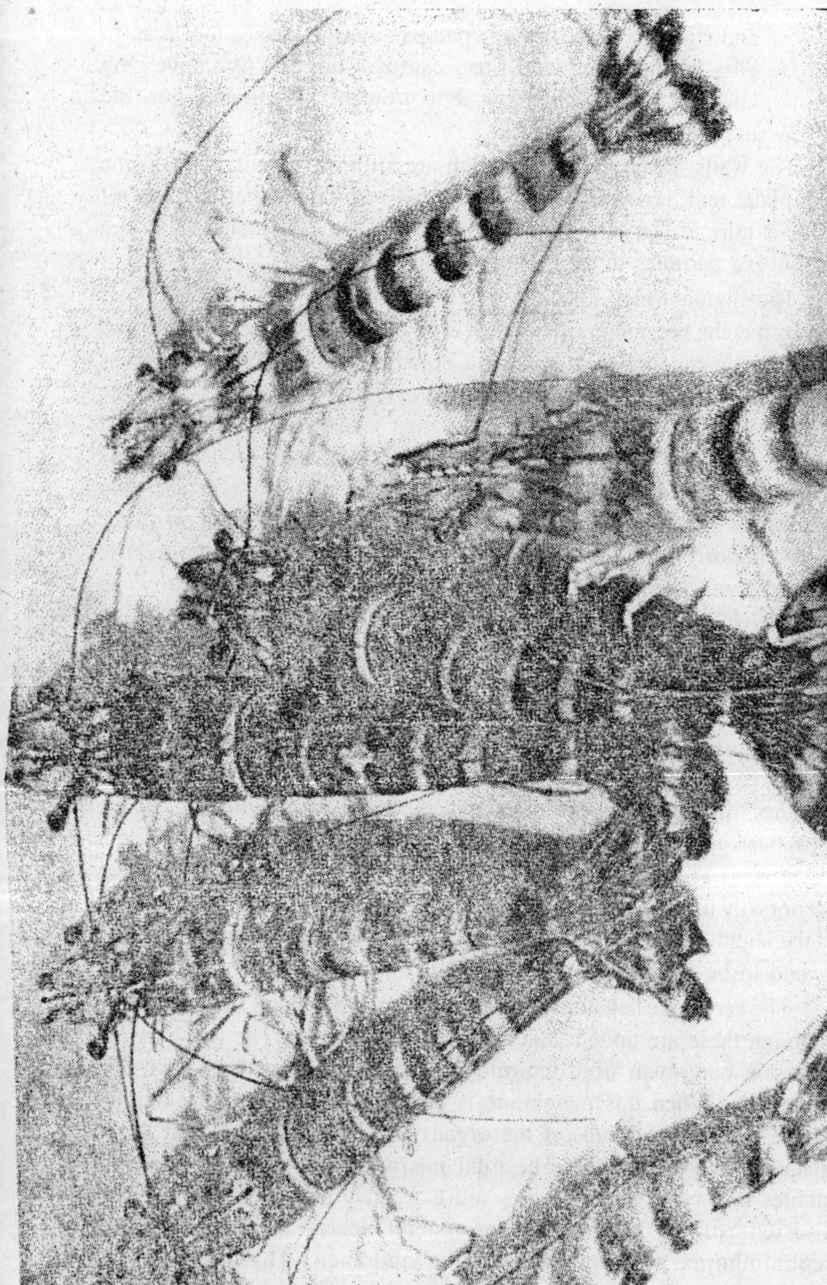

Fig. 4.1 : Tiger shrimp, *Penaeus monodon*.

hepatopancreas varies in fresh specimens from whitish, through yellowish, to various shades of green, depending on the season and the stage of the molt cycle.

Histologically the tubules and ductes of the hepatopancreas show four types of cells. These are

(1) Numerous columnar cells with a striated border, important in absorption and in storage of reserve food substances.

(2) Less abundant, vacuolated, often Pear-shaped cells, which secrete the digestive enzymes.

(3) Still less frequent, tiny cells near the basement membrane, which are thought to replace spent gland cells.

(4) Circular and longitudinal muscle fibers said to occur outside the basement membrane, but of such minute size as to be almost undetectable in ordinary sections.

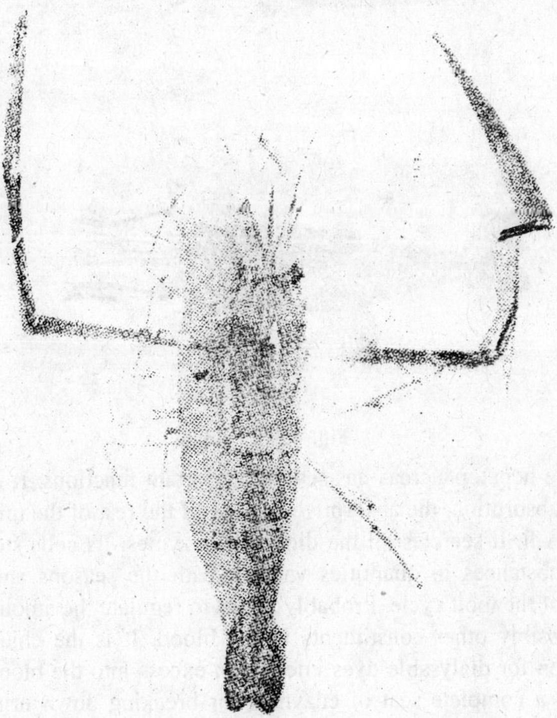

Fig. 4.2 : Male.

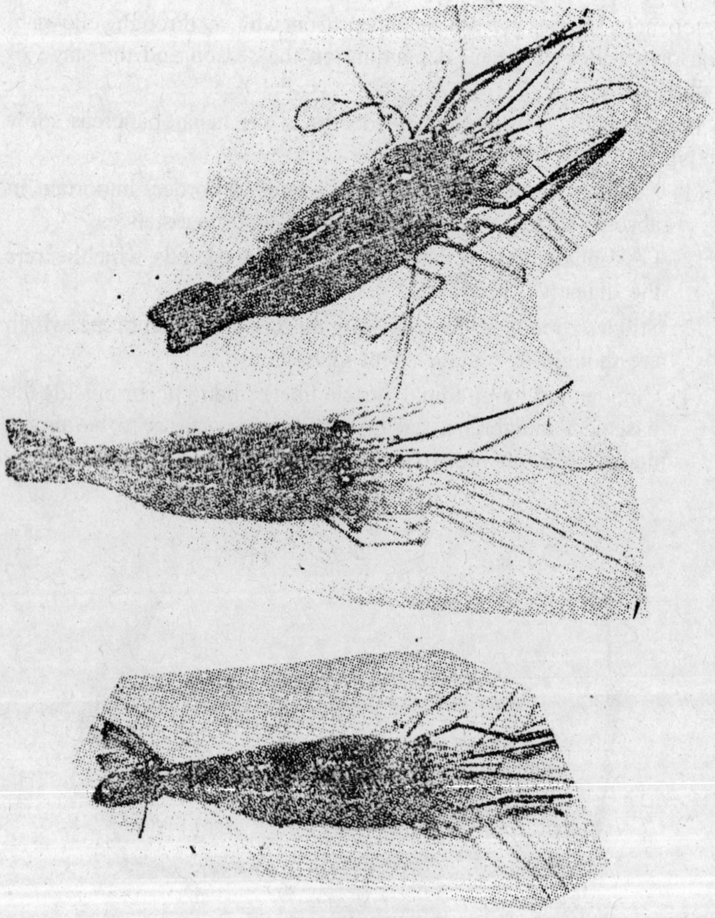

Fig. 4.3 : Female.

The hepatopancreas has several important functions. It is the chief site of absorption, the absorptive surface of the rest of the midgut being very small. It secretes all the digestive enzymes. Its cells store reserve food substances in quantities varying with the seasons and with the stages of the molt cycle. Probably it acts to regulate the amount of sugar and possibly other constituents in the blood. It is the chief organ of excretion for dialysable dyes injected in excess into the blood. Its cells contain a complete seat of enzymes for breaking down urine through uric acid and urea to ammonia.

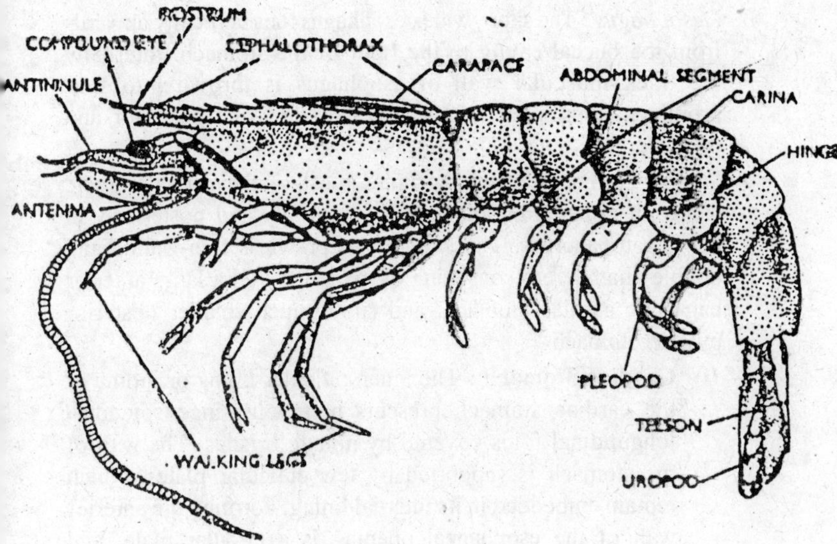

Fig. 4.4.

Digestive System of Cray Fishes

I. Alimentary Canal

The alimentary canal consists of three distinct regions:

(i) the foregut, comprising the mouth, buccal cavity, oesophaghs and stomach,

(ii) the midgut, and

(iii) the hindgut, the foregut and hindgut are lined internally by cuticle, called intema, which is shed with the exoskeleton when the animal molts. The mid-gut is lined internally by endoderm.

1. *Mouth :* This is a large and slit-like aperture lying midventrally below the anterior end of the head. It is bounded in front by the shield-like, fleshy labrum, laterally by the plate like incisor processes of madibles are behind by the biolobed labium.

2. *Buccal cavity :* The mouth leads into a short buccal region. It is antero-posteriorly compressed and has a thick cuticular lining which is irregularly folded. The molar processes of mandibles lie opposite each other in the buccal cavity to crush the food between them.

3. *Oesophagus :* The short, wide oesophagus runs vertically upwards from the buccal cavity to the floor of the stomach. Internally the thick muscular wall of esophagus is thrown into four prominent longitudinal folds, one anterior, one posterior and two lateral.
4. *Stomach :* The stomach occupies most of the cephalothracic cavity. It remains buried laterally, ventrally and posteriorly in the hepatopancreas. The stomach of prawn is thin-walled and double-chambered, consisting of two parts : (i) a large anterior bag- like cardiac stomach, and (ii) a much smaller posterior pyloric stomach
   (i) Cardiac Stomach : The inner cuticular lining or intima of the cardiac stomach presents numerous, inconspicuous, longitudinal folds covered by minute bristles. The wall of the stomach is supported by few cuticular plates, which remain embedded in its internal lining. Forming the anterior wall of the esophageal opening is a circular plate, and behind it on the roof of the stomach is a lanceolate plate. A large triangular plate is embedded in the middle of the floor of cardiac stomach. It is called the hastate plate, because it looks like the head of a spear. The upper surface of the hastate plate is armed with a thick growth of delicate setae and carries a distinct median ridge with gradually sloping sides. The posterior triangular part of the plate is depressed, fringed with setae along its edge and forms the anterior border of the cardio-pyloric aperture. A longitudinal cuticular supporting rod supports each lateral side of the hastate plate beneath. A narrow lateral groove runs along either lateral side of the hastate plate. The floor of each lateral groove is covered by a cuticular plate which resemble an open drain pipe and is called the groove plate. Each lateral groove is bounded on its inner side by the supporting rod and on the outer side by a long cuticular ridged plate. The inner border of each ridged plate is fringed all along with a row of delicate bristles, forming a comb- like structure, so that it is also named as a combed plate. The bristles bridge over the lateral groove and partially overlap the lateral margin of the hastate plate, where they constantly keep moving in a living prawn. The two combed plates are

unted anteriorly, thus completely enclosing the hastate plate except that their incurved posterior ends remain separated by is folded inwards to form a prominent lateral longitudinal fold. The two folds are very low anteriorly but gradually increase in height posteriorly and also bend inwards to form the sides of the cardio-pyloric aperture. These folds are also known as the guiding ridges because they guide the food towards the cardio-pyloric aperture.

The cardio-pyloric aperture is narrow, X-shaped and is guarded by four valves. The anterior valve by is formed by the depressed posterior part of the hastate plate, the posterior valve by a semi-lunar fold of the stomach wall, and the two lateral valves by the large flap-like posterior ends of the guinding ridges. This aperature leads into the pyloric stomach.

(ii) Pyloric Stomach : The pyloric stomach is a small and narrow chamber lying below the posterior end of the cardiac stomach. Its lateral walls are thick, muscular and prominently folded inwards, so that the cavity of the stomach is imperfectly divided into a big ventral chamber and a small dorsal chamber, continuous by a narrow vertical passage. The floor of the ventral median longitudinal ridge which divides it into two lateral compartments. The floor is covered by a filter plate. The filter plate has the from of an inverted half-open note-book and appears A-shaped in cross section. Each of the two rectangular surfaces of the filter plate bears a series of alternating longitudinal ridges and grooves. The ridges bear rows of bristles forming a felt-like covering over the grooves. The side of walls of the ventral chamber are also covered with closely-set bristles which, together with the filter plate, from an efficient strainer or filter. This pyloric filtering apparatus allows only liquid food to pass through it. The paired openings of the hepatopancreatic ducts lie behind the filtering apparatus, just below the junction of the dorsal chamber of pyloric stomach and mid-gut, these openings are guarded by a group of elongated setae arising from the posterior end of the median ridge of the filter plate.

The dorsal chamber gives out a small blind caecum dorsaily and then leads behind into the midgut. The junction of the two is guarded by one median dorsal and two lateral groups of elongated setae that project backwards into the midgut. These groups of setae strain the food entering the midgut and prevent its regurgitation into the dorsal chamber.

5. *Midgut :* The midgut with endodermal living is a long, narrow and straight tube running back along the median line, between the extensor and flexor muscles, upto the sixth abdominal segment. Its lumen is wide at the anterior end but reduced posteriorly due to the presence of longitudinal folds.

6. *Hindgut :* It has ectodermal living. It is the shortest portion of the alimentary canal, leading from the midgut to the auns. Its anterior swollen muscular part, called the intestinal bulb or rectum, bears many internal longitudinal folds. The terminal narrow, tubular part of the hidgut opens to the exterior through the anus, which is a sphinctered longitudinal slit-like opening situated on a raised papailla at the base of the telson.

II. Hepatopancreas

It is a large, bilobed dense and organs glandular mass, which lies below the gonad and nearly fills up the cephalothoracic cavity. It surrounds the stómach on its lateral, ventral and posterior sides. It consists of numerous bracnching tubules completely held together by connective tissue. The wall of the tubules consists of a single layer of columnar epithelium which is made up of: (i) grandular cells, (ii) ferment cells, (iii) hepatic cells with globules of fat, and (iv) replacing or basal cells. The epithelium rests on a basement membrane. The tubules rejoin to form larger and larger canals, finally forming the two large hepato-pancreatic ducts, which open into the ventral chamber of the pyloric stomach just behind the pyloric stomach just behind the pyloric filter plate.

The hepatopancreas combines in itself the function of pancreas, small intestine and liver of higher animals. Functioning as the pancreas, it secrets digestive enzymes, which can digest carbohydrates, proteins and fats; as the midgut, it absorbs the digested food material and as the liver it serves as an important storage organ for glycogen, fat and calcium. Some intracellular digestion also seems to take place in it.

Feeding and Digestion

The food of prawn consists of algae moss, weeds and small insects. Small particles of food are picked up by the chelate legs and are carried by them to the mouth. Near the mouth the second maxillipeds hold them and the mandibles tear and cut them. The gnathobases of a maxillipeds, maxillae and maxilluae transfer these smaller pieces of food to the mouth. In collecting large pieces of food the chelate legs are helped by the third maxillipeds. The food is passed into the buccalcavity by the incisor processes of mandibles and the labrum. In the buccal cavity the molar processes of the mandibles masticate the food. The masticated food is transferred to the cardiac stomach through the esophagus. In the cardiac stomach through the esophagus. In the cardiac stomach the food in ground further. Due to the sieve-like filetering mechanism present at the junction of cardiac and pyloric stomachs only finally ground particles can pass into the pyloric stomach where the food is filtered again by the pyloric filter before passing into the midgut.

The hepatopancreas produces protease, lipase and several carbohydrate hydrolyzing enzymes, which are poured into the pyloric stomach. From the pyloric stomach these enzymes pass forwarded so that chemical digestion also takes places in the cardiac stomach. Absorption of the products of digestion occurs through the lining of the midgut and hepatopancreas. The hepatopancreas have some cells which store excess of food as glycogen and fat, they also store calcium. Undigested particles pass on into the posterior end of the intestine, where they are gathered together into feaces and egested through and anus.

The alimentary canal consists of three distinct regions.

1. Foregut comprising the mouth, buccal cavity, oesophagus and stomach.
2. Midgut.
3. Hindgut.

Mouth is a large slit like apperture lying mid ventrally below the anterios and of the head. Mouth leads in a shart buccul cavity the short, wide oesophagus suns vertically upwards form the buccul cavity to the floor of stomach.

Stomach occupies most of the cephalothoraxic cavity and remains burried laterally anterior and posterior into the hepatopancreas.

Hepatopancreas is a large, bilobed, dense, orange in colour glandular mass, which easily fills up the cephalothoracic cavity. In surrounds the stomach from its lateral ventral and posterior sides.

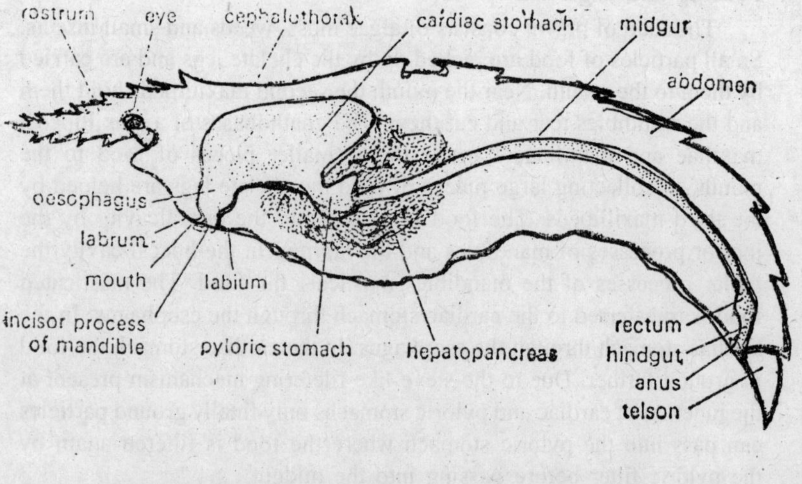

Fig. 4.5 : *Palaemon* : Alimentary canal in lateral view.

- It is the dorsal view of the floor of cardiac stomach.
- A large triangular plate is embedded in the middle of the floor of cardiac stomach it is called hastate plate.
- Circular plate is present in anterior side.
- Beneath the circular plate esophageal opening is present.

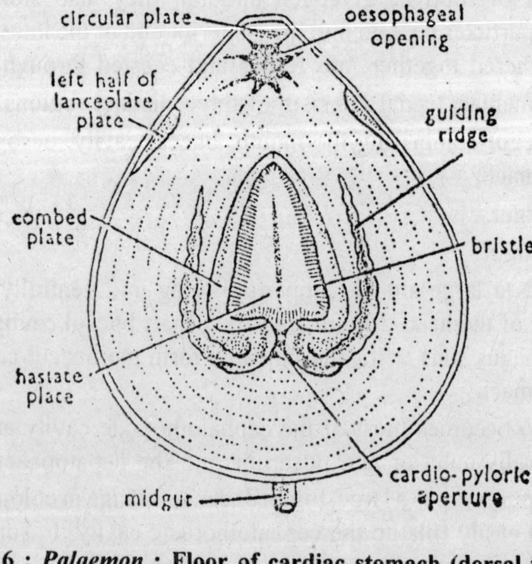

Fig. 4.6 : *Palaemon* : Floor of cardiac stomach (dorsal view).

- Outer border of the hastate plate is fringed all along with a row of delicate bristles forming a comb like structure. So that is also known as a combed plate.
- Lateral groove is present.

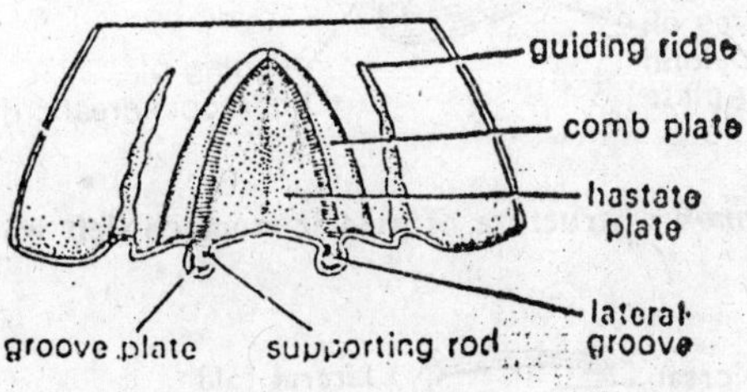

Fig. 4.7 : *Palaemon* : Floor of cardiac stomach cut across.

- Cardiac stomach open into pyloric stomach.
- Opening of hepatopancreas is situated on this region.

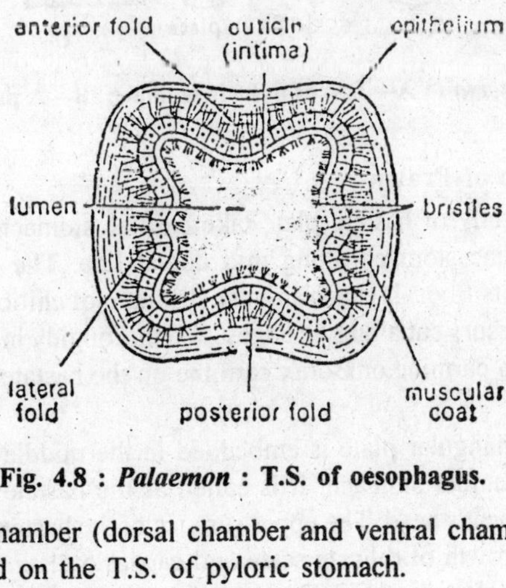

Fig. 4.8 : *Palaemon* : T.S. of oesophagus.

- Two chamber (dorsal chamber and ventral chamber) are seen situated on the T.S. of pyloric stomach.
- Pyloric filter plate is present.

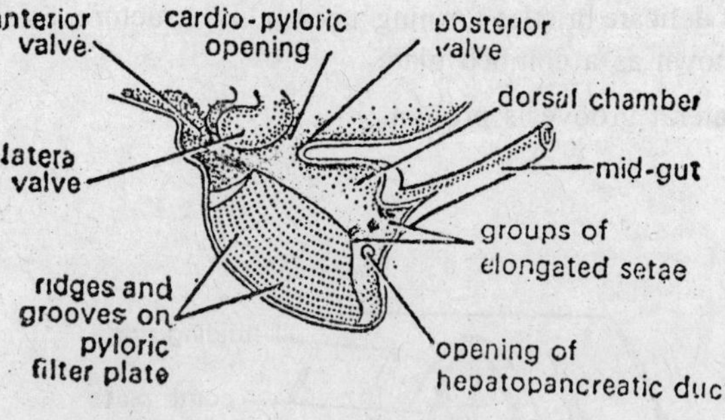

Fig. 4.9 : *Palaemon* : Structure of pyloric stomach (left wall removed).

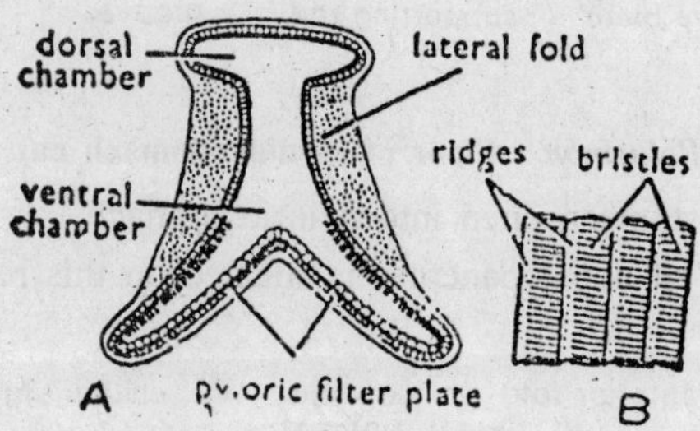

Fig. 4.10 : *Palaemon* : A—T.S. of pyloric stomach. B—A part of the pyloric filter

Hastate Plate of Prawn

For the study of hastate plate take cardiac stomach of prawn. Cut open the cardiac stomach along mid dorsal line. The hastate plate is present along its floor. It is a triangular structure of chitionus nature with the help of scissors cut along its edges. We thoroughly in water and stain in picro-indigo carmine or Borax carmine on the hastate plate and filter plates.

A large triangular plate is embedded in the middle of the floor of the cardiac chamber stomach. It is called as the hastate plate. Because it look like a spear's head. The upper surpace of the hastate plate is armed with a thick growth of dilicate setae and carries a distinct median ridge with gradually sloping sides. The posterior trianular part of the plate is depressed and fringed with setae along its posterior border and touches the cardio-pylosic aperture. Each lateral side of the hastate plate is supported and beneathed by a longitudinal cuticular supporting rods. A narrow lateral groove runs along either lateral border of the hastate plate. The cuticular plate covering floor of each lateral groove is bounded on its inner side by a long the supporting rods and on the outer side by a long cuticular ridges plate. The inner border of each ridges plate is fringed all along with a row of dilicate bristles forming a comb like structure. So that is also known as a combed plate.

The differences between pyloric and cardiac stomach.

Cardiac Stomach	Pyloric Stomach
1. Anterior part of the stomach is cardiac stomach	1. Posterior part of the stomach is pyloric stomach.
2. Esophageal opening is present	2. Hepatopancreas opening is present.
3. Hastate plate, circular plate, combed plate are present in this stomach.	3. Pyloric filter plate is present in pyloric stomach.
4. Oesophagous open into the cardiac stomach.	4. Pyloric stomach open into intestine.

Filter plate helps in digestion?

Filter plate is situated on the pyloric stomach. Food particles are filtered with the help of pyloric filter plate, when the food particulars (particles) are large and are not passes through filter plate, these particles again return into the cardiac stomach for churning.

Thus the filter plate help in digestion of food.

Digestion

The digestive secretion of the hepato-pancreas is paired through the hepato-pancreatic ducts in to the pyloric stomach whence it reaches the cardiac stomach and mixes with the food. The cardiac stomach expands and contracts in the living prawn, so that the food is churned and also digested by the action of the digestive ferments. As the food passes over the hastate plate, the moving bristles of the combed plates cut it in to smaller particles. The semi liquid and semi-digested food is filtered through the bristles of the combed plates into the lateral grooves below, whence it's carried into the lateral grooves below, whence it's carried into the ventral chamber of the pyloric stomach through the cardio-pyloric aperture. Here the digested and liquified food filters through the pyloric filtering apparatus to enter the hepatopancreas through the hepatopanoreatic ducts, and then become absorbed. The residual food, consisting of undigested coarser particles, ascends the dorsal pyloric chamber and from there enters the mid-gut, while the residual faecal matter passes an to the hind-gut and ultimately expelled through the anus, at the time of defecation.

Physiology of Digestion

Digestive tract consist of a fore-gut, lined with cuticle and derived from the stomodaeum; a midgut lined with endoderm; and a hind-gut.

The mid-gut surface is almost invariable amplified by one or more diverticula, which may serve as digestive glands or as sites for absorption.

The hind-gut may be long or short and is generally responsible for faeces formation, although some absorption may also occur in the hind gut.

Fore-gut

The crustacean fore-gut is quite variable as might be expected considering the difference in size and dietary habits found among crustacean.

On it's way into the fore-gut, food passes the mandibles. The mandibles fragments food for swallowing but do not adequately reduce it to particles that can be effectively digested. Filtering setae in the fore-gut are used to strain out larger particles, preventing their entry into the stomach.

Mid-gut

Most of the Crustacea have a short mid-gut, but the forms with a simple, tubular esophagus. It is often a simple tubular intestine. Digestion and absorption of food be are time consuming processes, and food must be kept in the mid-gut region long enough to complete them. Where there are no valves or other devices at the end of the mid-gut, same other system of delaying food passage is needed.

Diverticula and caeca are most effective for the strategic delay of food particles, and are often the most important regions of the mid-gut from a functional point of view. It is known that in some of the larger species enzymes are secreted and food is absorbed in the diverticula. Where the diverticula are extensively branched they are usually termed digestive gland or sometimes the term hepertopancrees is used. Tall cells serve for the absorption and storage of food, and perform enzymes secreting cells are scattered among them. At the base of the mucosal epithelium are small, basal cells that are thought to replace the absorptive and secretory cells as they deteriorate. While particles enter the digestive glands, as well as the caeca digestion appears to be wholly extracellular in most crustacean.

Hind-gut

Even when the hind-gut is long it remains relatively simple in form. A narrow tube with a chitinus lining it is typically equipped with intrinsic muscles used in defecation. In some instance the hid-gut has extrinsic dilator muscles used for anal pumping, undoubtedly a respiratory adaptation. Crustacea faeces are somewhat compacted and often mixed with mucus. Presumably absorb considerably large quantities of water from the hind-gut contents.

Parasitic Adaptation

Where food is taken in, from the host through the body surface of the parasite, the gut tends to deteriorate. Branchiura, the fish lice, have large caeca that are emerged with hast tissue fluids. Some of the suctorial isopods have a very extensively modified gut. The *entoniscids* have 3-chamber in the-gut after the esophagus. The anterior chamber is their walled and has many and the posterior chamber in heavily muscled. The middle comportment is equipped with a typhlosole-like, setae strainer. The anterior and posterior chambers contract rhythmically evidently pulling the ingested fluid over the straining apparatus. The detail of digestion, however, and not known.

The food and feeding habits of crustacea are as varied as their ways of life. Many of the decapods are omnivorus. This must no be taken as necessarily indicating that all in *grist* to the *gastric* mill, however, for while some, such as *Carcinus*, *Nephrops,* and *Homarus* are scavengers, favouring a mainly meat diet, others. Such as malacostracan, have a greater predilection for vegetation. True vegetarians are rare among the higher crustacea. Wood lice, it is true, will sometimes attack living plants; but both they and the terrestrial amphipods feed primarily on rotting vegetation. Individual choice is some times important even where there is apparently parting of opportunity to select the diet. For example, in a population of *Gammarus duebeni* studied by Hynes, some were found to have eaten only animal material, some only plant material, and some both.

The marine isopod *Limnavia* (the gribble) finds both food and shelter by tunnelling into wood piles and, where numerous, can cause severe damage to wooden harbour installations.

The herbivores par excellence of the crustacea are found amongst the cladocera and copepoda, where diet consists largely of unicellular algal, though many are not averse to supplementing thin with planktonic animals and even smaller members of their crustacean groups. It is well known that crabs will attack and eat any individuals which are incapacitated by multing, or damage, but many other forms also, such as *Asellus aquaticus* (copepoda), will attack their fellows, especially, if other forms of good are lacking. Crabs are often partial to decaying flesh, *Asellus* and *Neomysis* are more fastidious, eating their fellows only when alive or recently dead.

Method of Feeding

Decapods, such as *Nephrops,* seize their food with their chelal and pass it directly, or via the small chelalte on the fourth walking legs, to the third mandible which slice the food into chunks, which are swallowed and chewed into smaller particular by the chitinous teeth of the gastric mill. By contrast, the mandibles of the Natantia themselves masticate the food into fine particular, and the gastric mill of these forms in thus rendered unnecessary and is reduced.

Filter Feeding

The filtering of particular form the water is practised mainly by the smaller Species, Ostracoda, Copepoda, Cladocera, etc. and sessile forms such as the Cirripedia, but even in the malacostraca some species, *e.g., Callinanrra, Upogebia* and mysids, feed either wholly or in part by this means. The fact that particular are filtered form a stream of water must not be taken to indicate that the feeders are totally unselective in their diet except on the basis of particular size some filter feeders certainly do display the capacity to take one type of particular and reject other. For instance *calanus* takes the alga Lauderia in preference to *Chaetoceros*. *Evadne* is another selective feeder, taking *Ceratium furca* and *Peridinium* but ignoring diatoms and even other species of *Ceratium*.

The mode of filter feeding is variable; barnacles, such as *Balanus*, rhythmically sweep the water with their long hair. Such vigorous searching of the water for food is necessary in forms which can exist so far up the beach that they may be covered by water for only a few hours of the day but it would clearly be wasteful of energy in continuously submerged pelagic forms, indeed a specimen of *Concoderma* showed a more spread as a net and only periodically withdrawn. Occasionally, however, a single appendage flicked at high speed towards the mouth, presumably after being touched by particular is the water.

The trunk limbs of anostracon branchiopods, such as *Artemia* are used to createḋ a current of water in the gully between the bases of the legs. Fine setae on the appendages prevent the escape of any particular contained in this current and the food is ultimately gathered together and taken to the mouth along a groove on the ventral surface. Modifications and specializations of this general technique are found in *Daphnia* and other cladocerans. Here the third and found fourth trunk limbs combine the functions of producing the current and straining off particular of food, whilst bristles on the first and second limbs prevent excessively large particular form entering the system.

By contract with the branchiopods, copepods produce the feeling current by movement of the antennal, mandibles and maxillae, and filter off particular with maxillae for straining off particular but in the care of there forms the current is produced by the exopoditer of the thoracic legs. Filter feeding in mysids is frequently supplemented by the direct seizure of larger particulars by the mouthparts. In mature females of *Neomysis* the latter method predominater, possibly because the brood pouch interferes with the direction of the filter feeding current. A comparable effect has been observed in the larger euphausids, such as meganyctiphanes, where the smaller individuals are filter feed on and the larger ones primarily carnivorious.

The mud-shrimp *Upogebia pugettensis* which constructs burrows some feet long, is interesting in that its first two pairs of limbs are specialized both for carrying mud and for filter feeding through both functions of course, cannot be carried on simultaneously. When digging, the limbs act as a mud basket for removal of the shell; at other times they filter particulars form the constant stream of water (created by other limbs) which is caused to pass through the burrow.

Structure of the Gut

The gut crustacea is usually a straight tube, though in some cladoceram it may be coiled. It is divided into three main regions, the fore-gut (stomodaeum) mid-gut (merentron) and hind-gut (proctodaeum). The fare and hind-guts arise form ectodermal thickenings, and their linings secrete cuticle. The mid-gut lining is mesodermal and lacks cuticle.

In the decapods the fore-gut is sub-divided into an oesophagus cardiac stomach and pyloric stomach. Stomachs are also present in many other forms, though in same cases the mesenteron as well as the fore-gut may contribute. The structures are, therefore, not homologous with those of the decapods. The cardiac stomach of decapods is large and sac-shaped. It has a thick chitinous layer, which may be calcified in certain regions to form ossicles which are the masticatory apparatus of the gastric mill. Muscles move the ossicles to grind the food into fine particles before it is passed to the pyloric stomach.

The gastric mill tends to be especially well developed in those species which swallow large chunks of food, *e.g.*, most Reptantia. Apart from the malacostraca, only in the ostracoda it is apparent.

Food leaving the cardiac stomach is passed through the cardio-pyloric valve which permits the passage of only small particles.

Behind the pyloric stomach is the mid-gut and its lateral expansions, the hepatopancreas or mid-gut gland. The mid-gut itself is of variable length, being long in *Homarus* but short is crabs. The hepatopancreas is a mass of fine blendly ending tubules which unite in to a common hepatopancreatic duct opening into the anterior end of the mid-gut. A complex chitinous filter formed by the fore-gut protects the entrance of channels leading into the hepatopancreatic duct and this prevents all but the finest particular form passing up the duct to the absorptive cells of the hepatopancreatic gland.

In *Nephrops* an anterior dorsal caecum and a posterior diverticulum, which open at the junction of the mid-and hind-guts apparently provide additional absorptive surface. The hindgut is a short, short tube opening at the anus.

Cilia are of course absent in the crustacea, and so all movement of food in the gut is by peristalsis and anti-peristalsis. Peristaltic and anti-peristaltic waves are apparent in the mid-gut and are even more pronounced in the hind-gut where in *Nephrops* they may occur from three to five times a minute. Presumably such co-ordinated muscle movements are under the control of the extensive nerve plexus in the gut wall.

A curious feature of the muscles responsible for peristalsis is that whilst in the mid-gut the longitudinal muscles arrangement is present in the hind gut. Unlike the situation in vertebrates, the gut-muscles are all striated.

Digestion and Absorption

The evidence at present available suggests that the digestive enzymes are produced mainly, possibly even solely, by the however, can be passed forward into the cardiac stomach and imitiate digestion in this region of the fore-gut. The pH of the fluid in the cardiac stomach of *Nephrops,* is slightly acid which is suitable for the pH optimum of the proteolytic enzyme present. This enzyme will digest proteins such as fibrin, casein and peptone. The comparable enzyme responsible for caseinase activity in the crayfish *Orconectes Virilis* is found to have two pH maxima at pH 6 and pH 7.5 to 8.5. There does not appear to be any enzyme comparable with vertebrate chymotrypsin in *Orconectes,* since the specific substruate for Chymotryprin, BTEE (N-benzoyle tyrosine ethyl ether), is not hydrolyred by the gastric extract. Pepsin like enzymes also appear to be taking.

Abundant enzymes are commonly present, breaking down carbohydrates and related compounds. Thus the stomach fluid of *Astacus* contains cellobiase (cellobiose→glucose), alphagalactosidase (melibiose, maltose →galactose→glucose), saccharase (invertase) (Sucrose→glucose + fructose), beta- galactosidase (lactose→galactose-glucose), maltase (maltose→glucose), isomaltaso (dextrans and isomaltose→glucose) alpha-xylosidase (xylase→xylose), amylase(starch→maltose), β (beta)-1, 3-gluconase (glucon→glucose),beta 1 4 glucanase (cellulase) (cellulose →isomaltose), mannonase (Mannan→mannase), xylonase (xylan→ xylose) and a chitinase. Most of these enzymes, including cellulase and chitinase, are also present in extracts of the hepatopancreas. It is probable, therefore, that the enzymic activity is not due to bacterial activity. Naturally this is difficult to prove categorically, but at least seems fairly certain in the case of the chitinase, as bacteria plated from the gut do not show the ability to attack chitin. By contrast , the chitinase of porcellis scaber does seem to be produced by bacteria. The pH optima of the cellulase and chitinase are lower than that of the proteases already mentioned, being 4.0 to 4.5 and 3.0 to 4.0 respectively.

A cellulase is apparently absent in *Nephrops* and *Ligia oceanica* but, as might have been expected, occurs in extracts of the hepatopancreas (digestive tubules) of the wood-boring isopod *Limnoria lignorum.* Here again, no bacteria showing celluase activity can be found in the gut; So the enzyme is almost certainly produced by the gland itself.

Starch, glycogen, sucrose, maltose and lactose; but not inulin or raffinose, are digested by *Nephrops* hepatopancreas extracts. The halase occurs in the guts of the *woodlice Porcellio* and *Oniscus*. Alginic acid and pectin are not digested by *Astacus*; but there is no evidence to show whether more fully vegetarian forms such as *Maia* have the necessary enzymes or not.

In *Nephrops* there is a lipase which will split olive oil; Various esters, such as methyl acetate, amyl acetate, ethyl acetate and butyl acetate, are also digested. *Butyrin* from milk is converted to butyric acid. Young failed to demonstrate the presence of bile salts in *Nephrops*, but their presence in crayfish has since been demonstrated by Vonk.

Absorption

The fore-and hind-guts of crustacea are reported to act merely as semi-permeable membranes, and there is as yet no evidence that they can absorb material other than water. Thus, when glucose is introduced into the fore-gut and this region of the gut is *ligated*. The glucose is not taken up by either the fore-or hind-guts. This is in contrast with the insect gut (*Teherne*) where numerous material can be absorbed form the hind-gut.

The mid-gut and hepatopancreas are, therefore, the primary absorptive areas. Digestive products and fine particular are passed up the hepatopancreatic duct in to the gland. Digestion is almost wholly extra-cullar in arthropods, unlike the care in the majority of invertebrates. This has complicated the identification of the all actually responsible for absorption. In consequence there has been some controversy as to whether cells in the hepatopancreas combine secretory and absorptive functions, or are specialized for either one function or the other. As frequently happens in biological controversies, it has now been found that there was a measure of truth on both sides. Thus, the most recent study of the problem has shown that all cells in the hepatopancreas arise from embryonic cells at the end of the tubules and that each cell is at first absorptive, than secretory, than fibrillar. The production rate of cells is greater than the death rate, so that tubules elongate as the animal increase in size.

Oil droplets are absorbed in the mid-gut. Salt and water are actively taken up from the gut by forms whose blood can be maintained hypotonic to the medium. Such forms drink the medium either orally, and if the medium is dyed, the dye becomes concentrated in the gut as the salts and water are absorbed. Careful measurments of the osmotic pressure of the gut fluid in *Artemia* indicate that although it is less concentrated than the medium, it is considerably to the blood. Clearly, therefore the salts must be taken up from the gut fluid. The manner in which water is taken up is more obscure. Either than gut wall is able to transport water actively against the concentration gradient, as is shown to be possible in the insect rectum, or alternatively the osmotic concentration immediately in contact with the gut cells may perhaps be lowered below the blood concentration, so that the water can be taken up passively by osmosis, even though the bulk of fluid in the gut has a higher concentration than the blood.

Efficiency of Digestion

Studies on feeding rates and the efficiency of digestion have been confined almost solely to the copepods.

Many of the observations and calculations on the amount of water swept clear daily by a form such as *Calanus* are conflicting ranging from about 5 cc to over 1,000 cc per day. A recent study however, indicates that the average amount of food taken in daily during the summer months by female *Calanus helgolandicus* is equivalent to some 25% of the dry body weight. Naturally, when such high rates of food intake are occurring passage of food along the gut is very rapid. The gut of a starved copepod will be filled within one hour when offered food and will be emptied in one hour when food is withdrawn. Digestion therefore has to be very efficient if all the material taken in is to be utilized in the short time available. In fact utilization is not complete. When feeding on an abundant source of phytoplankton, *Calanus* may produce faecal pellets at the rate of one every five to seven minutes. There pellets are green and, perhaps not surprisingly, sometimes contain undamaged cells. It has even been possible to start a new phytoplankton culture form viable cells taken form the faeces. Nevertheless, the digestive processes are quite efficient, since 50-90% of the food taken in is assimilated. The precise level, however, tends to vary with the food organism offered, flagellates such as *Decrateria. Chromulina* and *Chlorella* being lass effectively digested than other forms.

Nutrition

Both rate of growth and egg-laying capacity have been used as indices to test the suitability of various diets; and the results indicates that, at least in the care of some omnivores, not all requirements are met by a single food type. For example *Gammarus* fed solely on animal food grow less rapidly than those fed on both animal and plant materials. Conversely larval *Palaemonetes* reared solely on plant material (*Nitzchia, Chlamydomonas, Nanochloris, Porphyridium* and *Pyramimonas*) do less well than those given animal food also in the form of *Artemia naupli.*

Starved copepods stop laying eggs. Egg-laying capacity is restored in the care of *Calanus* by feeding with diatoms such as chaetoceros, skeletonema or Rhizosolenia; but flagellates such as *Chlorella*, Hemiselmis and *Dicrateria* are apparently lacking in some essential food factor, since if fed alone to *Calanus* they do not restore egg-production in starved individuals. Indeed *Chlorella* appears to pass undigested through the gut. Growth and reproductive capacity of the harpacticoid *Tigriopus japonicus* have also been found to vary widely according to the species of algae with which it is fed.

Nutritional requirements vary at different times of the year. Poikilotherms do not require more calories during winter months to maintain the body temperature as homeotherms do; instead, they are less active, and so have lower basic metabolic rates. Growth and reproduction also tend to be slower at low temperatures, again economizing on food requirements during the winter.

Nevertheless, as was pointed out by Raymont, there is still an apparent discrepancy between the maintenance requirements of copepods for food and the amount actually present in the water during winter. This discrepancy may be more apparent than real; for over-wintering copepods show little reduction in the fat content (Stage V *Calanus*) during the winter months; which implies that the food intake is adequate to supply the needs of the animal without recourse to reserves.

Feeding Migrations

Many species make migrations form one place to another in connection with either feeding or reproduction. In the latter category are the land crabs which migrate ebmasses to the sea to reproduce, on which occasions they have been described as sounding like a regiment of cuirassiers on the march; as they bounce from rock to rock on their headlong descent to the water. The chinese wool-handed crab, *Eriocheir* makes an extensive feeding migration to fresh water, young crabs penetrating many tens or even hundreds of miles up the North European rivers. They return ultimately, however, to saline waters to breed.

The diurnal migration of many planktonic forms, especially copepods, is well known. Even such small forms as the copepod *Calanus* may migrate daily, from a depth of 400 metres during the day to near the surface at night, whilst migrations from 200 to 1,000 metres are made by many euphaurids, decapods, mysids and amphipods.

Finally, we must mention the extraordinary migration of the Americon spiny lobster *Panulirus* though its precise purpose remains in doubt. On these migration the animals move along the bottom parallel with the shore in chains of 3-30 individuals, the members composing each chain positioning themselves in single file, each with its cephalothorax lying immediately above the tail of the one in front. This crustacean 'Conga' lacks little but the rhythm.

The first good experimental demonstration of absorption of digest trients by the digestive gland was obtained by Speck and Urich (1970). They showed, in the' crayfish *Orconectes limosus,* that between 82 and 93% of ^{14}C-labeled nutrients were absorbed in the wall of the digestive gland, and 5% in the wall of the midgut. Few studies have been made on the nature of absorption of nutrients in Crustacea have shown that glycine and lysine transport by the midgut of *Penaeus marginatus* is an active, carrier-mediated process with a sodium requirement showed that glucose transport by the midgut of *Macrobrachium rosenbergii* was an active, sodium-dependent process, but the precise mechanism was unknown. No information is available on this process in the digestive gland, the major site for absorption glucose was rapidly metabolized after absorption, with only about 9% passing unchanged into the blood. The rest was converted mainly into maltose, galactose, trehalose, organic acids, and amino acids. Amino acids and palmitic acid were also rapidly metabolized after absorption. Such rapid metabolism would assist absorption by reducing the internal concentration of the nutrients. Found that glycerol tri[l-^{14}C] palmitate which had been incorporated into food pellets began leaving the digestive gland of *Nephrops norvegicus* only 4 hr after feeding, indicating a rapid translocation from digestive gland to the blood. By 24 hr about 75% of the absorbed tripalmitate had been metabolized.

The digestive and absorptive functions of the gland are well established, but the multitude of other functions attributed to it are not so clear, as *Van Weel (1974)* points out when objecting to the term "hepatopancreas" for this structure. The R cells obviously store lipids, but these cells would represent not more than 50% of the mass of the living organ which is only 4-10% of the body wet weight Thus, the neutral fat was less than 0.5% of the body weight in *Carcinus,* while glycogen did not exceed 0.02%. After 4 weeks starvation neutral fat levels fell to about one-third while those of glycogen fell to about one-half, but the actual amounts metabolized represent only about 0.3% of the body weight. Similarly, Schafer (1968) found that mainly proteins and lipids were metabolized by *Penaeus duorarum* during starvation. Found that digestive gland lipids were the main source of energy in

Orconectes limosus during the first 2 weeks of starvation, and thereafter proteins were metabolized could not find evidence of metabolism of digestive gland lipids during starvation in *Nephrops norvegicus,* but there was significant decrease in dry weight of other tissues. A similar decrease was observed in *Panulirus.* The digestive gland "reserves" would therefore seem to be of minor value in prolonged starvation.

The crustacean midgut also serves as an organ for accumulation and metabolism of calcium and copper and for the inactivation of a number of other potentially toxic metals. Calcium is stored as small concentrically structured granules. These granules appear to have the additional role of immobilizing Zinc and Lead Copper, which is toxic to Crustacea as free ions, is stored in proteinaceous granules in the shrimp *Crangon* and in a number of other non-decapod Malacostraca Cadmium will induce metallothionein formation. Presumably other, potentially toxic, heavy metals are detoxified in the same way. Although, the digestive gland has a well-defined role in storage, metabolism, and detoxification of a number of metals, its role as an excretory organ is unclear. Methylene blue was excreted by the digestive gland, but detoxified heavy metals do not seem to be excreted from this organ. Thus, its main role in this respect may be for storage excretion.

Although, the foregut is controlled by a complex neuromuscular system that has been researched in detail no information is available on nervous control of the digestive gland, but it seems likely that a number of its functions are under endocrine control. The effects of eyestalk ablation and eyestalk extracts on lipids, carbohydrates, oxidative and lipolytic enzymes in the digestive glands of *Scylla serrata* has been observe. The responses obtained suggest a wide range of control by neurosecretory hormones. Eyestalk hormone control of amylase synthesis and synthesis of other digestive enzymes is under a similar control. Nothing is known about control of cellular proliferation, absorption, or myofibril contraction, and it is possible that one or more of these functions are under endocrine control.

Assimilation Efficiencies

Digestibility may tie influenced by the morphology of food components and anatomy of the digestive tract as well as the more obvious features such as biochemical composition of the food and characteristics of digestive enzymes. Rate of ingestion can affect the assimilation efficiency. Assimilation efficiencies provide information about potential food value and, in some cases, about the process of digestion. Thus high assimilation efficiencies 95-98%) of bacteria, which were uniformly' labelled with ^{14}C in the shrimp *Metapenaeus bennettae,* indicate that bacteriolytic enzymes were present. Bacteriolytic enzymes should also lyse blue-green algae. Reports of reasonably high assimilation efficiencies of blue-green algae substantiate this *Daphnia longispina* assimilated 100% of the green algae *Elakatothrix and Ankistrodesmus,* which suggests that *Daphnia* possesses all the enzymes necessary to digest completely the complex polysaccharides in the cell walls of these algae.

An assimilation efficiency of 100% for the alga *Chlorella* eaten by *Daphnia pulex,* but as he did not use algae that were uniformly labeled with ^{14}C, his results apply only to the assimilation of compounds labeled during 1 hr of photosynthesis. The high assimilation efficiencies of these compounds in *Chlorella* and some blue-green algae also support the hypothesis that *Daphnia pulex* possesses enzymes that lyse these types of cells.

Table 4.1 : Recent References to Assimilation Efficiencies of Some Food Items in Crustaceans

Food	Species	Assimilation efficiencies(%)	References
Bacteria	*Hyalella* (Amphipoda)	60-89	Hargrave (1970)
	Metapenaeus (Decapoda)	80-98	Moriarty (1976)
Blue-green algae	*Hyalella*	5-15	Hargrave (1970)
	Thermocyclops (Copepoda)	35-58	Moriarty *et al.* (1973)
	Diaptomus (Copepoda)	30-73	Schindler (1971)
	Daphnia (Cladocera)	18-51	Schindler (1971)
	Metapenaeus	48-87	Moriarty (1976)
Green algae	*Hyalella*	55	Hargrave (1970)
	Diaptomus	13-50	Schindler (1971)
	Daphnia	10-100	Schindler (1971)
Diatoms	*Hyalella*	75	Hargrave (1970)
	Diaptomus	20	Schindler (1971)
	Daphnia	38	Schindler (1971)
	Calanus (Copepoda)	36-88	Conover (1966)
	Penaeus (Decapoda)	87	Condrey *et al.* (1972)
Fungi	*Palaemon* (Amphipoda)	65-77	Barlocher and Kendrick (1975)
Protein	*Palaemon* (Decapoda)	80-98	Forster and Gabbott (1971)
	Pandalus (Decapoda)	75-97	Forster and Gabbott (1971)
	Penaeus	62-92	Condrey *et al.* (1972)
Lipids	*Penaeus*	75-99	Condrey *et al.* (1972)

William Dall and David J.W. Mariasty (1982)

In most crustaceans, it is probable that any food which passes rapidly through the foregut is poorly digested, as was shown in *Metapenaeus bennettae.* Powell's (1974) description of the action of the gastric mill and circulation of fluid in the foregut of a decapod implies that mechanical and biochemical digestion proceed in the foregut and therefore, would be more complete when food remains there longer. The cell walls of bacteria and blue-green algae have similar biochemical compositions, but some blue-green algae are not easily digested, whereas bacteria are generally smaller than blue-green algae, it is likely that in crustaceans physical degradation of blue-green algae, especially of clumps of filaments, is as important as enzymatic lysis. Green algae vary considerably in size and biochemical composition, and thus assimilation efficiencies vary. Values for carbohydrate assimilation efficiency, although available in the literature, are not given here because they have no significant meaning. Unlike proteins and lipids, which are degraded as a class by a small group of proteinases and lipases, respectively, each type of carbohydrate is degraded by a specific enzyme or group of enzymes. Because crustaceans have carbohydrases in different amount and composition peritrophic membranes secreted within tubules of digestive gland it product gut from abrasive material in faeces, disintegrates after digestion. Defaecation occurred in hind gut in shrump longitudinal pad grasps the feed pellet in it peritonic membrane and rhythmically expellit. In lobsters and crabs hindgut longitudinal folds containing longitudinal muselebundles with an after layer of circular muscle, longitudinal ridges of rectal pads are also associated with it.

(a) Starch Similar Compounds

α-Amylase (endo-α-1,4-glucanase) activity has been demonstrated in all Crustacea that have been investigated. It has been isolated from *Hormarus americanus,* and in its acidic (pH 5.2), requirement for chloride, and molecular weight (41,000) it is similar to α-amylases from other organisms. α-Amylase has also been isolated and characterized from the isopod *Asellus aquaticus.*

For the complete digestion of starch and glycogen, at least two other type of enzymes are necessary, viz., either an oligo-α-1,6-glucosidase or a branching enzyme and a maltase (α-1,4-glucosidase), but they have no been isolated or characterized from crustaceans. α-1,4-Glucosidase activity has been found in all crustaceans.

(b) Cellulose

Cellulose (β-1,4-glucan) is hydrolyzed by a mixture of enzymes which together are loosely referred to as cellulase. The main components are endo-β-1,4-glucanase, exo-β-1,4-glucanase (C_1) and β-glucosidase (cellobiase). The early hypothesis that soluble cellulose (*e.g.*, carboxymethyl cellulose) could be degraded by a Cx activity and cellobiase only, whereas native (insoluble) cellulose required the prior attack of a C_1 component, is no longer tenable. Cellulose hydrolysis requires the synergistic action of the three types of enzymes. Cellobiase has been found in many animals which cannot or do not utilize cellulose, but this activity may well be due to a β-glucosidase with broad specificity. Cellulolytic activity has been found in many crustaceans, but as the substrate commonly used was carboxymethyl cellulose, this activity does not necessarily indicate that both exoglucanases and endoglucanases are present. Digestion of insoluble (crystalline or native) cellulose has been demonstrated in the isopod *Limnoria* spp. The amphipod *Gammarus pulex*, and possibly in the mysid *Mysis stenolepis* the assimilation efficiency of cellulose in the mysid was about 50% but as their data indicated an instantaneous efficiency of about 25%, it is likely that at least some of the absorption of ^{14}C was not the result of cellulose digestion.

Probably many Crustacea synthesize cellulases, although some cellulase activity may be due to microorganisms either in the gut or in the food prior to ingestion. Cellulases may have two functions in digestion: (1) to convert cellulose into glucose as an energy source; (2) to enable other digestive enzymes to penetrate a plant cell wall more detailed studies are needed on the occurrence, origin, and specificities of cellulolytic enzymes in animals that digest cellulose.

(c) β-1,3-Glucans

β-1,3-Glucans are storage or structural polymers in many algae, fungi, and protozoans and, therefore, are a potential source of energy for crustaceans that feed on algae or microorganisms. β-1,3-Glucanases (laminarinases) probably occur in all Crustacea and, together with cellulases amylases, and chitinases, are regarded by Piavaux as part of the general digestive carbohydrases.

(d) Chitinases

Chitinobases has been partly separated from digestive gland fluid in two species of *Cancer*; it has a molecular weight of 110,000-125,000. Because chitin is associated with other glycans and proteins, its digestion would require the concerted action of many different enzymes. In Crustacea, the cuticle contains at least some $CaCO_3$ and in heavily armoured species this may be a major component. Its effect on digestion is not known; the acidic pH of the digestive fluid in most Crustacea would help in removing small amounts of $CaCO_3$.

A problem that arises in considering chitin digestion is: If all Crustaceans contain a digestive chitinase, how does the chitinous peritrophic membrane avoid digestion? The removal of digestive fluid, together with a rise in pH and the association of other compounds such as mucopolysaccharides with the peritrophic membrane may inhibit residual chitinase activity.

(e) Other Polymers

In addition to the more common polysaccharides referred to above, algae and microorganisms contain many other less well characterized polymers. Glycosidases that hydrolyze some of these compounds do occur in crustaceans that feed on them, but little is known of their activities.

Glycosidases would have a function in degrading glycolipids and glycoproteins in addition to mucopolysaccharides and polysaccharides. De-oxyribonuclease and ribonuclease probably occur widely as digestive enzymes.

pH of crustaceans remains 5-7.pH in foregut fluid during feeding could fluctuate around neutrality and all digestive enzymes might be active enough at neutral pH for a efficient digestion.

In deep sea amphipods greater physiological activities are performed by micro organisation microflora compared to free living bacteria. A possible function of microbial activity in digestion in digestive tract is supply of vitamins and essential amino acids It suggests that they might be more widely distributed among invertebrates, and that the low molecular weight proteinases in crustaceans may be similar enzymes.

Table 4.2 : Comparison of Amino Acid Composition of Crustacean and Vertebrate Serine Proteinases

Amino acid	Crayfish trypsin[a]	Shrimp trypsin[b]	Crab collagenase[c]	Dogfish trypsin[d]	Bovine trypsin[e]
Lysine	5	5	2	5	14
Histidine	5	5	5	8	3
Arginine	2	3	5	7	2
Aspartic acid	30	30	29	24	22
Threonine	15	10	24	7	10
Serine	17	24	17	17	33
Glutamic acid	21	24	13	16	14
Protine	10	11	12	10	9
Glycine	28	28	28	28	25
Afanine	16	16	21	16	14
Half-cystine	6	8	8	12	12
Valine	18	18	20	17	17
Methionine	2	2	3	8	2
Isoleucine	15	14	17	14	15
Leucine	16	10	13	15	14
Tyrosine	11	10	8	12	10
Phenylalanine	7	6	8	1	3
Tryptophanine	3	3	2	5	4
Total amino acids	227	227	235	222	223

[a]Zwilling and Tomasek; 1970.
[b]Gates and Travis, 1969.
[c]Elsen *et al.*, 1973.
[d]Bradshaw *et al.*, 1970.
[e]Walsh and Neurath, 1964.

Willcan dall and David J.W. Moriarty (1982)

Peptidases : The presence of various peptidases has been reported, mainly based on studies with synthetic substrates and crude mixtures Kleine (1967b) has separated carboxypeptidase A, arylamidase, and a dipeplidase using gel filtration of foregut fluid and digestive gland extracts from *Astacus astacus* and *Orconectes limosus*. He found that the molecular weight of the carboxypeptidase from the digestive gland was greater than that from the foregut fluid. Carboxypeptidases A and B have been isolated and characterized from the digestive gland of *Penacus setiferus*. Some recent studies which indicate the presence of various peptidases are given in Table 4.3. The dipeptidases may be associated with the brush border of absorptive cells in the digestive gland, rather than being secreted into the foregut fluid.

Zymogens : Number of zymogens (inactive precursors) of crustacean proteolytic enzymes have been found, and some workers have argued that they do not occur. If this is so, a question to be answered is how do the secretory cells protect themselves from proteolytic activity? The collagenase from the crab *Uca* has aspartic acid at its amino terminus, in contrast to valine or isoleucine usually present in the related proteinases of vertebrates. Thus as the vertebrate mechanism for activation could not occur, Eisen *et al.*, (1973) suggest that a zymogen form does not exist.

Table 4.3 : Some Recent Reports of Peptidases from Decapoda.

Enzyme	Species	Location	References
Carboxy-peptidase A	*Penaeus setiferus*	d.g.	Gates and Travis (1973b)
	Astacus astacus	d.g., f.f.	Klene (1967b), Kleine and Ponyi (1967)
	Astacus astacus	f.f.	Sonneborn *et al.*, (1969)
	Cambarus affinis	d.g., f.f.	Kleine (1967b), Kleine and Ponyi (1967)
	Homarus americanus	d.g., f.f.	Brockerhoff *et al.*, (1970)
	Orconectes virilis	d.g. f.f.	DeVillez (1965)
Carboxy-peptidase B	*Penaeus setiferus*	d.g.	Gates and Travis (1973b)
	Astacus astacus	f.f.	Zwilling *et al.*, (1969)
	Astacus astacus	f.f.	Sonneborn *et al.*, (1969)
	Astacus leptodactylus	f.f.	Zwilling *et al.*, (1969)
	Orconectes virilis	f.f.	DeVillez (1965)
Arylamidase	*Astacus astacus*	d.g.	Kleine (1967b), Kleine and Ponyi (1967)
	Cambarus affinis	d.g.	Kleine (1967b), Kleine and Ponyi (1967)
	Macrobrachium lamarrei	d.g., f.f.	Murth (1977)
Dipeptidase	*Astacus astacus*	d.g.	Kleine (1967b)
	Cambarus affinis	d.g.	Kleine (1967b)
	Macrobrachium lamarrei	d.g., f.f.	Murthy (1977).

[a]d.g., Digestive gland; f.f., foregut fluid.

In the aquatic environment, crustaceans have exploited every type of niche and ecological diversity is paralleled by the diversity of food eaten. The micro crustaceans typically feed on micro algae, whereas larger crustaceans range from burrowing detritus feeders to active predators of mollusks and fish. The dietary diversity is mostly reflected in the diversity of external appendages and mouth parts, whereas, in general, the alimentary canal follows a simple plan which tends to be uniform for a given family. Most of our knowledge of digestive physiology is derived from the larger Decapods, although there has been appreciable research on the Amphipoda and Isopoda. Earlier knowledge of function was very fragmentary, but over the last 20 years experimental research on digestive physiology, using modern methods, has grown considerably. As a group adult barnacles are completely sessile; a range of parasitic species living on host body fluids or soft tissues seen within the Maxillipoda.

ESSENTIAL AND REQUIRED DIETARY COMPONENTS

1. The artificial diet must be in a form as acceptable to the animal as its natural diet. This includes its physical state (usually solids, such as pellets or suspended particles are preferred) as well as taste.
2. Proportions of the constituents, the ratios of the major and of minor constituents all appear to be critical.
3. Use of radioisotope-labelled precursors in xeric animals may show nonessentials due to synthesis by gut symbionts.
4. Nonessentiality of minor and trace dietary constituents cannot be assumed unless the test animal has passed through several generations, because of possible carryover from parents to egg.
5. Trace contaminants from supposedly pure chemicals may invalidate results.

The principal regions of the gut are foregut, midgut and hindgut.

THE FOREGUT

The proventriculus is always divided into and anterior part that usually serves as a crop in macrophagous feeders. The posterior end of this chamber constricts toward a gastric mill comprised principally of a median dorsal ossicles. Subsidiary ossicles may be present. The gastric mill leads into posterior part of the preventriculus, which is in turn divided into dorsal and ventral chambers. The dorsal chamber bears lateral grooves, leads into the midgut or directly into the hindgut. The ventral chamber contains the filter-press, which leads to the digestive gland. The floor of the anterior proventriculus bears a median groove and two ventrolateral grooves, with fringing dense setae. The ventrolateral grooves leads to the filter-press, these structures being present in most, if not all, Decapoda. Additional grooves may be present in some genera, and various setose plates are characteristic of the burrowing thalassinids. There is a complex system of muscle attachments over the surface of the proventriculus, particularly around the gastric mill.

Until recently, concepts of the decapoda foregut were derived almost entirely from morphology. It was inferred that secretion from the digestive gland entered the anterior proventriculus and it was assumed that the function of the ventral grooves was to transmit these secretions. After passing backward through the gastric mill, the food was sieved ventrally by the filter press, which permitted only fluids and the finest particles to enter the digestive gland for final digestion and absorption.

As food enters the anterior chamber (AC) of the proventriculus it is penetrated by fluid from the digestive gland that flows forward dorsolaterally in grooves (PCG) in the posterior chamber (PC). Trituration and further mixing with fluids occurs at the gastric mill ossicles (O). The food mass the continually being manipulated by the lateral plates of the anterior chamber and forced into the gastric mill. Eventually fluid passes from the food mass into the ventral grooves (VG) of the anterior chamber. Dense setae exclude larger particles above 1μm, and finally into the openings of the digestive gland (DG). Fluid from the digestive gland is pumped dorsally into the dorsolateral grooves, joined by fluid squeezed from the food mass in the posterior chamber. Some fluid is also pumped in and out of diverticulum of the midgut. The combined fluids then pass forwards to the anterior chamber. The circulation is driven by the pumping action of the filter-press and associated structures, probably aided by other pumps. The lateral grooves (LG) of the anterior chamber may also have roles, as yet undefined, in fluid circulation. The rate of flow along the channels is measured in seconds and the entire circulation time is probably only of the order of minutes. Presumably dissolved nutrients are absorbed, and the fluid, perhaps with the addition of more enzymes, is then returned to the general proventricular circulation. Alternatively, the fluid may circulate round the proventriculus a number of times before entering the tubules of the digestive gland, in which case tubular absorption and secretion could be expected to alternate.

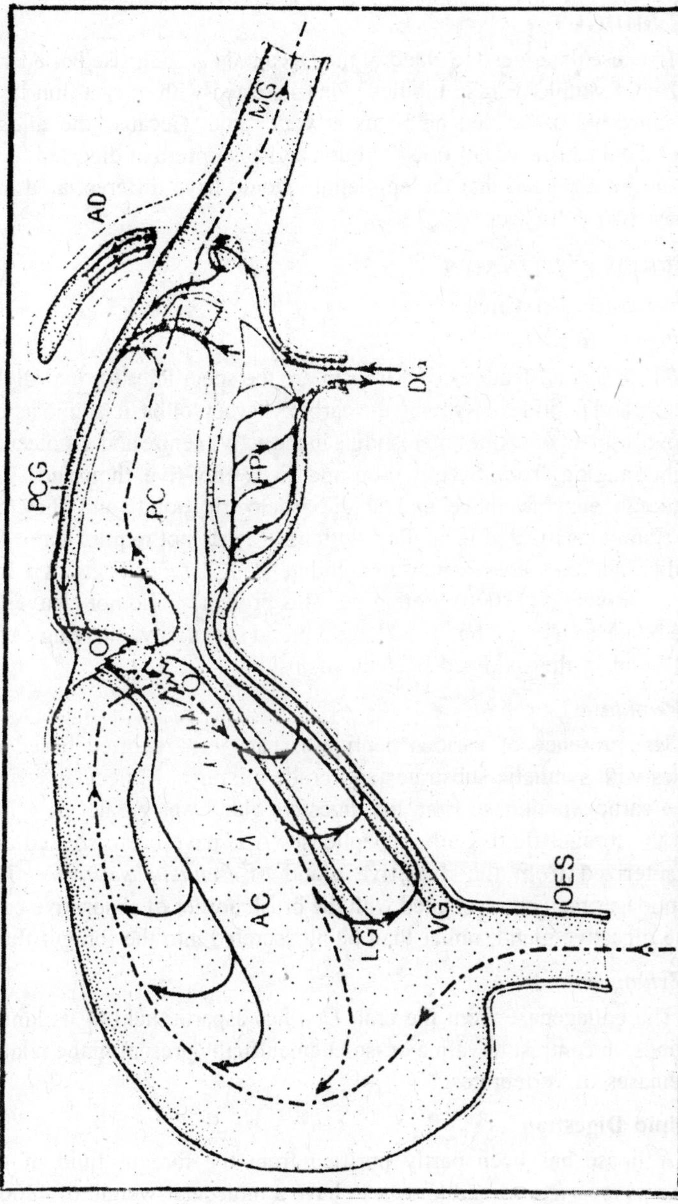

Fig. 4.11.

THE MIDGUT

Because the digestive gland of the Decapoda is comprised of a large number of simple, fragile tubules being invested with only a thin layer of connective tissue and autolysis is very rapid. Because the midgut serves the dual role of enzyme secretion and absorption of digested food, it could be expected that the epithelium would have differentiated into at least two cell types.

DIGESTIVE ENZYMES

1. Proteolitic Enzymes

(a) Proteinases

The essential features of a trypsin are the specificity for hydrolysis of esters and peptides involving the carboxyl group of basic amino acids, the involvement of serine and histidine in the active centre and a molecular weight ranging from twenty thousand to twenty-five thousand. The crustacean enzymes have and acidic isoelectric point, and they are irreversibly inactivated in acidic solutions, they do not require Ca^{++} for stability, and they are resistant to autodigestion. *Astacus* contains a low molecular weight (11000) proteinase. This proteinase did not cleave the same bonds as trypsin, but it did have a broad specificity, cleaving every third bond in the oxidized β chain of insulin.

(b) Peptidases

The presence of various peptidases has been reported based on studies with synthetic substrates and crude mixtures. Molecular weight of the carboxypeptidase from the digestive gland was greater than that from the foregut fluid. Carboxypeptidases A B have been isolated and characterized from the digestive gland of *Penaeus setiferus*. The dipeptidases may be associated with the brush border of absorptive cells in the digestive gland, rather than being secreted into the foregut fluid.

(c) Zymogens

The collagenase from the crab *Uca* has aspartic acid at its amino terminus, in contrast to valine or isoleucine usually present in the related proteinases of vertebrates.

2. Lipid Digestion

A lipase has been partly purified from the foregut fluid of the *Homarus americanus* and known to have a molecular weight of 43000, which is similar to that of hog pancreatic lipase; optimum activity occurred around pH 7. The action of lipase in lobster foregut fluid toward triolein is similar to that of vertebrate pancreatic lipase; in that it seems to specifically hydrolyze the Alpha- linkage. Triolein was cleaved mainly into Alpha, beta diacylglycerol and beta-monoacylglycerol. Small amounts of Alpha, Alpha-diacylglycerol and Alpha-monoacylglycerol were formed. The Crab *Cancer pagurus* synthesizes fatty acylsarcosyl-taurines and secretes these into its foregut fluid to aid in the digestion and absorption of lipids. A fatty acyltaurine has been found in *Astacus leptodactylus*, taurine or other amino acids occurs in *Homarus gammarus*. As celles of *steroids* and lecithins are readily formed with these emulsifying agents, thus promoting their absorption.

3. Carbohydrate Digestion

Hydrolysis of carbohydrates is a complex process involving more than one reaction mechanism, the two most common being transglycosylation and condensation. Some glycosidases have broad specificities, determined mainly by the nature of sugar contributing the reducing group (the aglycone) and the type of linkage (Alpha or Beta). Others have narrower specificities, determined also by groups attached to the aglycone. It does not necessarily follow, therefore, that the range of different glycosidases present in a mixture of digestive enzymes can be determined by incubating the mixture with various substrates. Thus many of different carbohydrases that have been reported as present in the digestive tracts of Crustacea may not exist. Digestive enzymes, especially those present in low quantities, may not be secreted by the animals, but may be derived from the food organisms themselves or from microorganisms growing in the gut. As there is such diversity in the composition of structural carbohydrates, secretion of extracellular enzymes by bacteria in the gut could be important in the lyses or digestion of algae, fungi, and other components of food eaten by herbivorous or detritivorous crustaceans.

THE HINDGUT

In the microcrustaceans such as copepods the hindgut is usually a simple, cuticle-lined tube, but in barnacles it is differentiated into an anterior much folded region which acts as a sphincter and a posterior expanded chamber with slit like anus. In lobsters and crabs, the hindgut has interior longitudinal folds containing longitudinal muscle bundles, with an outer layer of circular muscle. Tegumental glands are present, but these do not appear to open into the lumen of gut. The maximum complexity of the hindgut is seen in terrerestrial isopods. There is a small

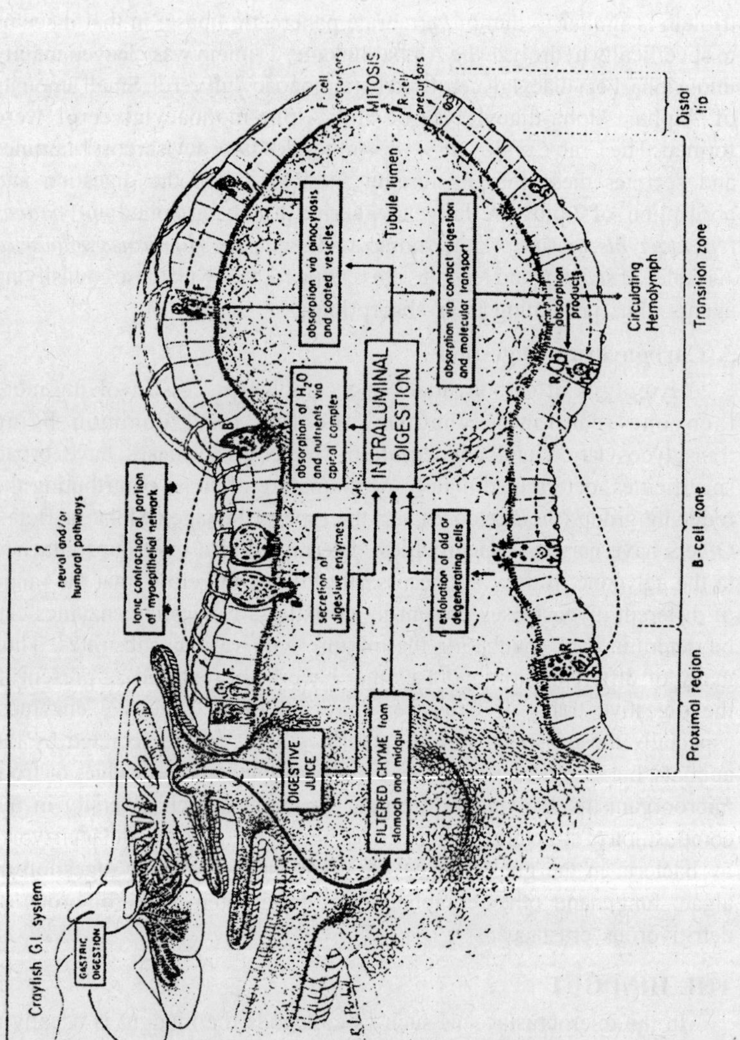

Fig. 4.12 : Scheme of digestive gland function showing the differentiation of cell types (see text), secretion of enzymes, and cellular absorption. Relative zone lengths are not drawn to scale. Proximal region and β-cell zone have been reduced 9 ×. (From Loizzi, 1971).

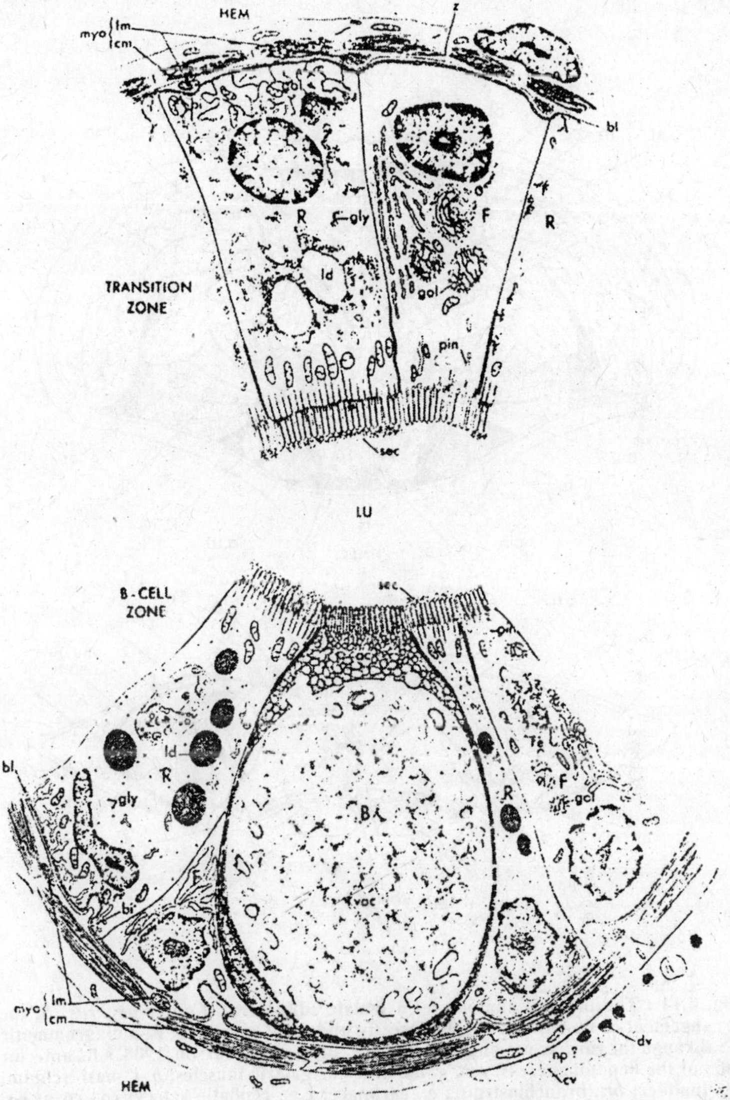

Fig. 4.13 : Schematic drawings of transition zone and β-cell zone in the digestive gland tubule showing ultrastructural details. β, F, R, cell types (see text : apical complex; bi, basal invaginations; bl, basal lamina; cm, circular muscle fiber: cv, clear vesicle; dv. dense-core vesicle; Fe, iron granule in supranuclear vacuole; gly, glycogen: gol, Golgi body; HEM, blood surrounding tubule; ld, lipid droplet, lm, longitudinal muscle fiber; Lu, lumen of tubule; mvo. myoepithelial network; np?, possible neurosecretory prucess; pin, pinocytic channels and vesicles; sec, surface enteric coat; vac, vacuole (From Lotzzi, 1971).

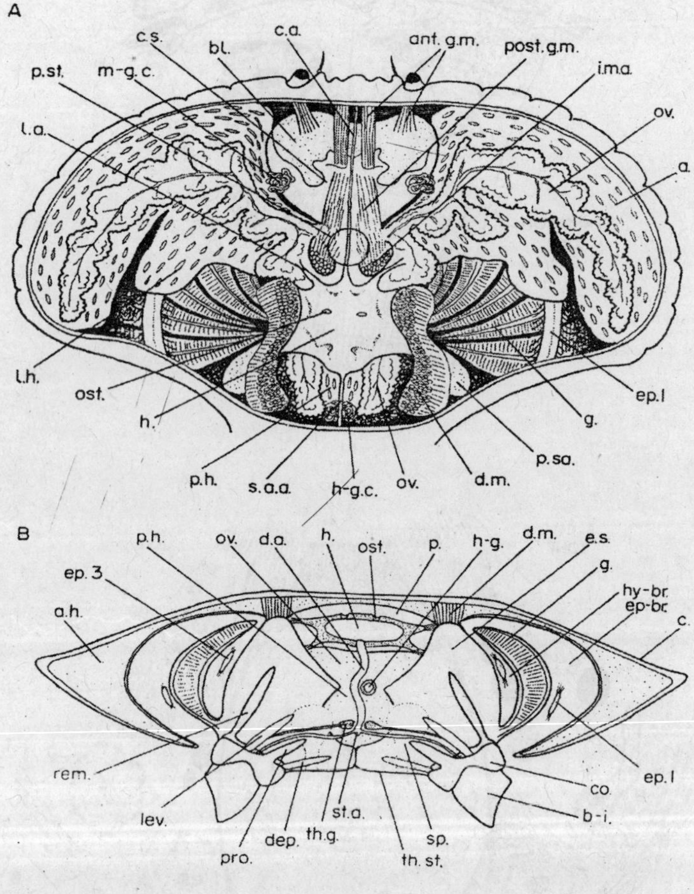

Fig. 4.14 : The internal anatomy of a female edible crab *Cancer pagurus*. A, the arrangement of tissues and organs directly under the carapace; B, a diagrammatic TS through the cardiac region, anterior view. Both after Pearson, 1908. a.h., anterior lobe of the hepatopancreas; *ant. g. m.*, anterior gastric muscles; *b.-i.*, basi-ischium; *bl.*, bladder; *br.*, branchiostegite; c., carapace; c.a., cephalic artery; co., coxa; c.s. cardiac stomach; d.a., descending artery; dep., apodeme of the depressor muscle of the basi-ischium; d.m., dorsal muscle; ep. l, epipod of the 1st maxilliped; ep. 3, epipod of the 3rd maxilliped; ep.-br., epibranchial space; e. s., endophryangeal skeleton; g., gill; h., heart; h.-g., hind-gut; h.-g. c., hind-gut caecum; hy.-br., hypobranchial space; i.m.a., interna Tadductor muscle of the mandible; l.a., lateral artery; lev., apodeme of the levator muscle of the basi-ischium; l. h., lateral lobe of the hepatopancreas; m.-g.c., mid-gut caecum; ost., ostium; ov., ovary; p., pericardium; p.h., posterior lobe of the hepatopancreas; p. so., pericardial sac; p.st., position of pyloric stomach; post. g.m., posterior gastric muscles; pro., apodeme of the promoter muscle of the coxa; rem., apodeme of the remotor muscle of the cox s.a.a. superior.

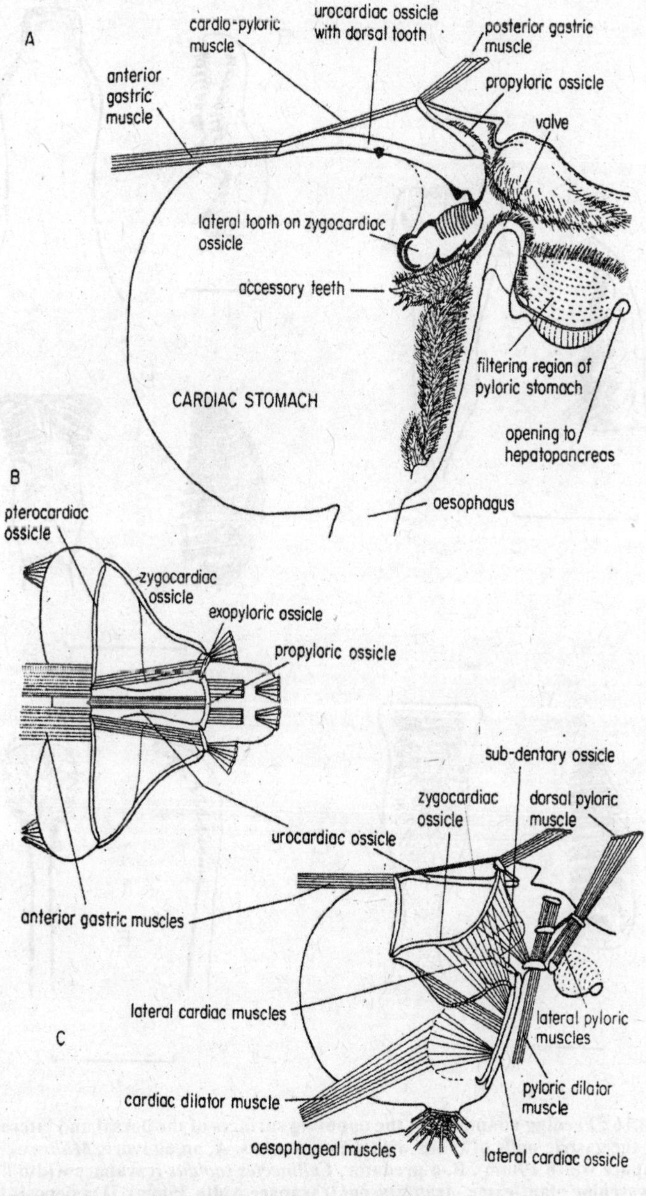

Fig. 4.15 : The stomach and gastric mill. A, the gastric mill seen by looking into the stomach from the side, the near half of the stomach has been removed. B, dorsal view and C, side view of an entire stomach.

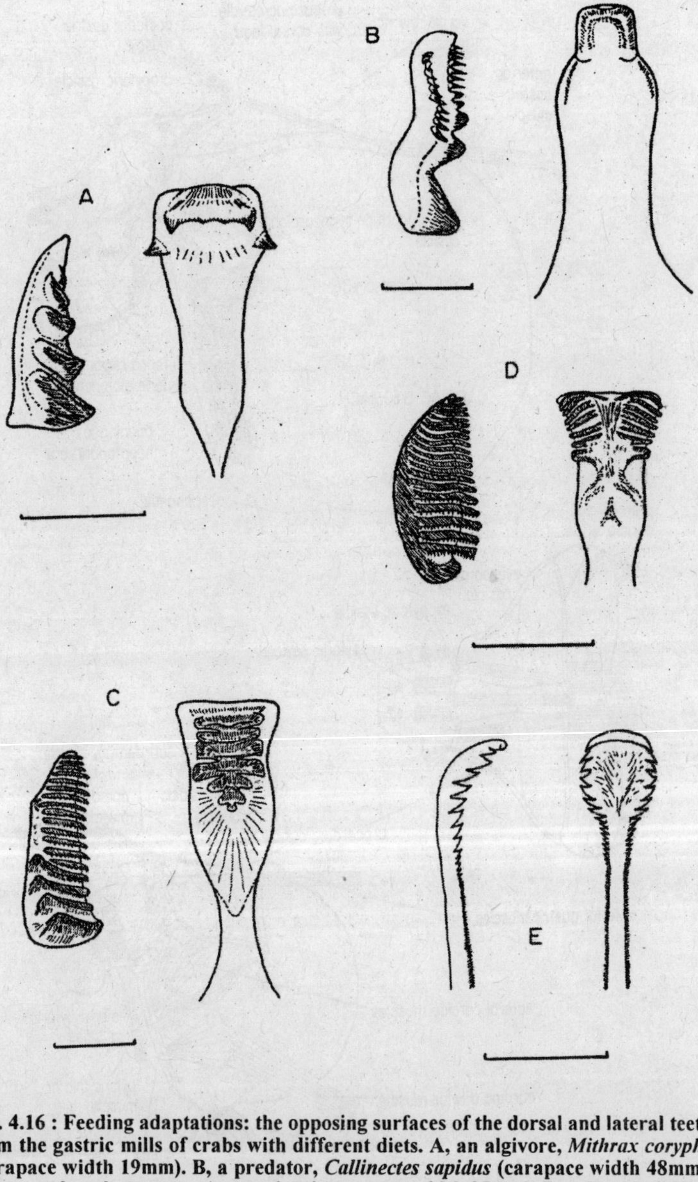

Fig. 4.16 : Feeding adaptations: the opposing surfaces of the dorsal and lateral teeth from the gastric mills of crabs with different diets. A, an algivore, *Mithrax coryphe* (carapace width 19mm). B, a predator, *Callinectes sapidus* (carapace width 48mm). C, a vascular plant-eater, *Aratus pisoni* (carapace width 25mm). D, a deposit-feeder, *Uca* spp. (carapace width 16mm). Anterior is downwards, only left lateral teeth are shown. E. spoon-tipped hairs from the 2nd maxilliped of *Uca* sp. collected from a sandy-mud habitat. Scale for A, B, C and D= 1 mm; for E=0-l mm.

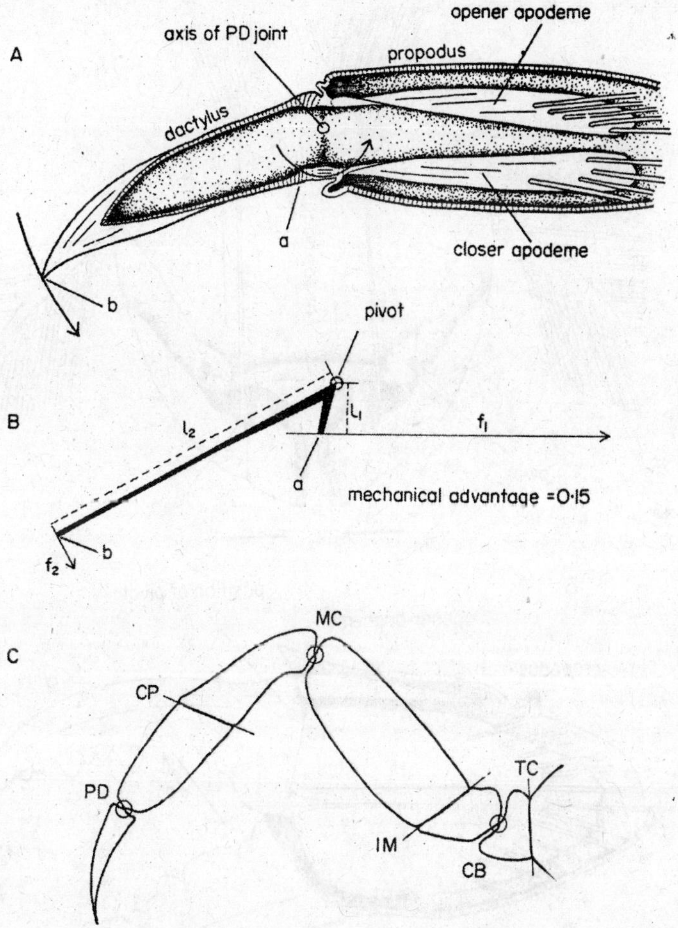

Fig. 4.17 : The lever system of a crab's leg. A, the dactylus and propodus dissected from the side with all but a few muscle fibres removed; the arrows show the movements of the dactylus at a and b which would result from closer muscle contraction. B, the lever system of the closer muscle in A above; force (f_1) is applied to the system at a and by the system (f_2) at b, the arrows indicate directions and relative magnitudes of the forces, da.shed lines indicate lever lengths (l_1 and l_2). C, a whole leg showing alternating planes of movement at successive joints; circles and lines represent the axes of joints: circles indicate axes perpendicular to the paper, lines indicate axes in the plane of the paper; joints are named according to the podomeres that they connect.

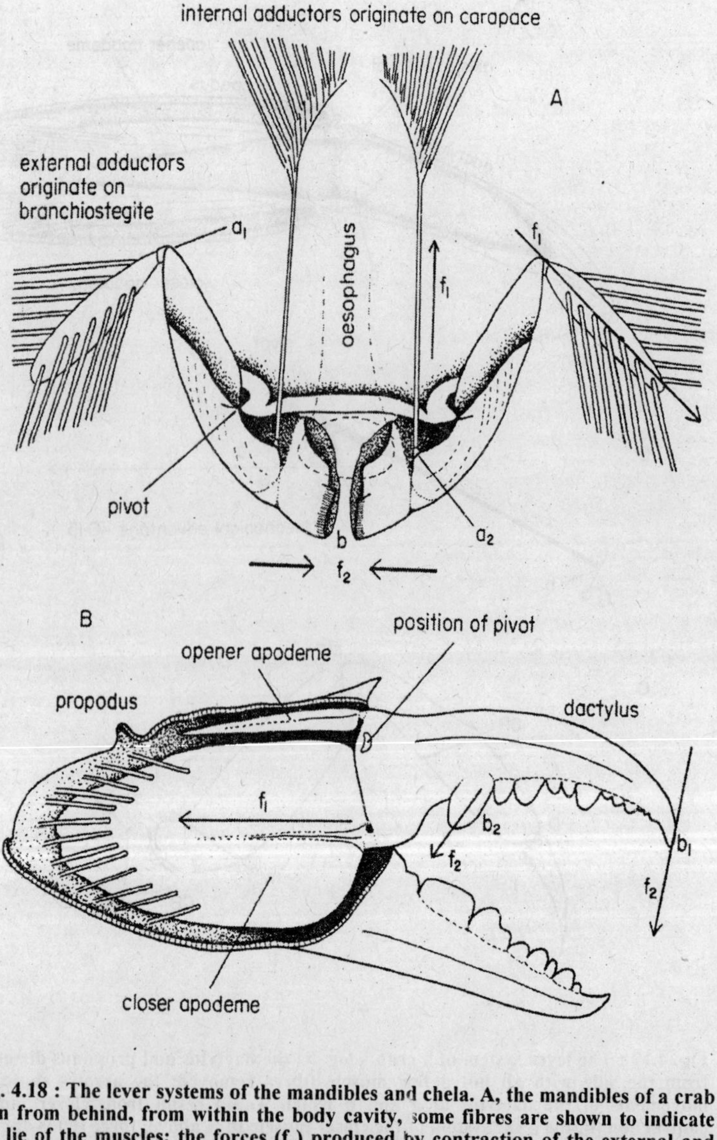

Fig. 4.18 : The lever systems of the mandibles and chela. A, the mandibles of a crab seen from behind, from within the body cavity, some fibres are shown to indicate the lie of the muscles; the forces (f_1) produced by contraction of the external and internal adductors are applied to the mandibles at a; and a_2 respectively, the resultant forces (f_2) are applied by the mandibles at b. B, a chela dissected from the side with all but a few of the muscle fibres removed; the force produced by contraction of the closer muscle (f_1) results in a force (f_2) applied at any point chosen by the crab (b_1, b_2) along the biting edge of the dactylus.

proventriculus and the midgut is limited to four major regions: the anterior region is largest and has a dorsal typhlosole; a shorter region follows, then a sphincter and a rectum. The isopod *Philoscia muscorum* lives on decaying leaf litter. The muscular hindgut may also pump water into the gut and press it into the mouth. Such hairs tend to be more numerous and finer than the grasping and tearing bristles of carnivores. Scraping algivores mainly gather fine filamentous green algae and diatoms but other algivores, such as the European wrinkled crab *Xantho incisus*, pluck that each lever length is measured as the shortest distance between the pivot and the line of action of the force. As the joint bends l_1 will increase slightly and then decrease, and f_2 will change its direction. The relationship between the forces and the lever lengths in a system with a frictionless pivot is $f_1 l_1 = f_2 l_2$, thus for a given f_1 the magnitude of f_2 is increased either by an increase in l_1 or a decrease in l_2. The factor by which the magnitude of the force is altered (the mechanical advantage) is f_2/f_1 and is equal to the ratio of the lever lengths, l_1/l_2. These simple leverage principles are in everyday use in such mechanisms as doorhandles and pliers in which large forces are produced by applying smaller forces further away from the pivot.

REFERENCES

1. David MC Farland, Animal Behaviour Oxford University Press, Oxford, 1981
2. H. H. Kornbuber Handbook of Crustacean Physiology, Academic Press, New York, 1979.
3. J.W., Moriarhy, Functional Aspects of Nutrition and Digestion William Dall and David, London, 1982.
4. Lindha H. Mantel, Internal Anatomy and Physiological Regulation, Academic Press, New York, 1983.

5

EXCRETION AND OSMOREGULATION

Basic plan of excretory system is an endsac or coelomic catcher closing the rammants of coelom an excretory canal a labyrinth, exit duct and large bladder in malacostracans the openings are on the sixth or maxillary segment.

Morphology and fine structure of the maxillary glands of *Artemia salina* were described by Tyson (1968, 1969). In these glands there is a central end sac surrounded by three coils of the excretory duct, ending in a terminal duct. Ultrafiltration is thought to occur across the wall of the end sac, which is held open against the flow of hemolymph by a "scaffold" consisting of tubules and their connective tissue attachments to the end sac (Tyson, 1968). Filtrate then travels through the efferent duct where it is modified and finally to the cuticle-lined terminal duct. Fine structure of cells supports the suggested function of each region. Cells in the end-sac epithelium are variable in height with branched, podocytic basal endings that rest on the basement membrane. These foot processes interdigitate, and cells are separated by a filtration slit membrane. Fine structure of the epithelial cells in the efferent tubule reveals prominent apical (luminal) microvilli and infoldings of the basal surface, often associated with mitochondria. Both adherent junctions and desomosomes are found on the apical surface of the cells. Cells in the terminal portion of the duct are smaller, with fewer infoldings of the membrane, fewer mitochondria, and a lack of apical specializations. Similar organization and cell types are found in the excretory organs of the barnacles *Semibalanus bafanoidos* (formerly *Balanus balanoides*) and *C. hamvri* (White and Walker, 1981).

Best studied have been the antennal glands of the crayfish. These consist of a coelomosac, labyrinth, tubule or nephridial canal, which may be regionally specialized, and an extensive bladder Riegel and Cook, 1975). Cells of the coelomosac resemble those of the end sac of *Artemia salina*. In addition to the podocytic processes, mitochondria, dense bodies, lysosomes, and vacuoles are present. Fine structure of the filtration surface has been investigated by Schaffner and Rodewald (1978). They found that the filtration barrier is provided by the slit diaphragms between adjacent foot processes of the epithelium. Proteins up to the size of hemocyanin and ferritin (MW about 450,000) cross the basement niembrane but do not penetrate through the slits into the urinary space. This situation contrasts with that in vertebrate glomeruli, where the primary barrier to filtration of proteins is thought to be the basement membrane.

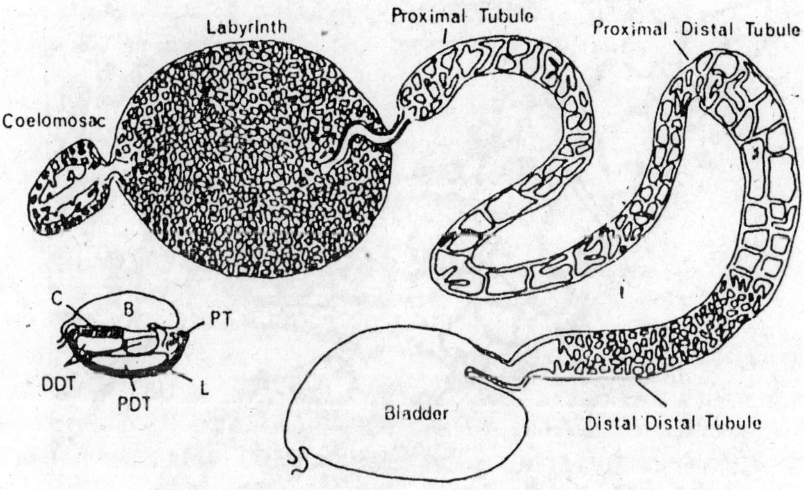

Fig. 5.1 : Diagram of the antennal gland of a crayfish showing the major regions of the gland in extended form and the relationships among the regions *in vivo*. (From Riegel, 1972; reprinted with permission).

Cells in the labyrinth and nephridial canal, or tubule, resemble those of the tubule in *A. salina*, with some regional variations in size and histochemical properties. In the proximal labyrinth, cells have a well-defined apical brush border with many mitochondria. The basal surface is extensively infolded, and long, narrow intercellular spaces are common. The distal portion of the labyrinth contains cells with active endocytolic vesicles, particularly at the apical surface, and less extensive basal infoldings. These cells may take up proteins. In the nephridial canal, cells lack an apical brush border, although extensive basal infoldings are again present with associated milochondria.

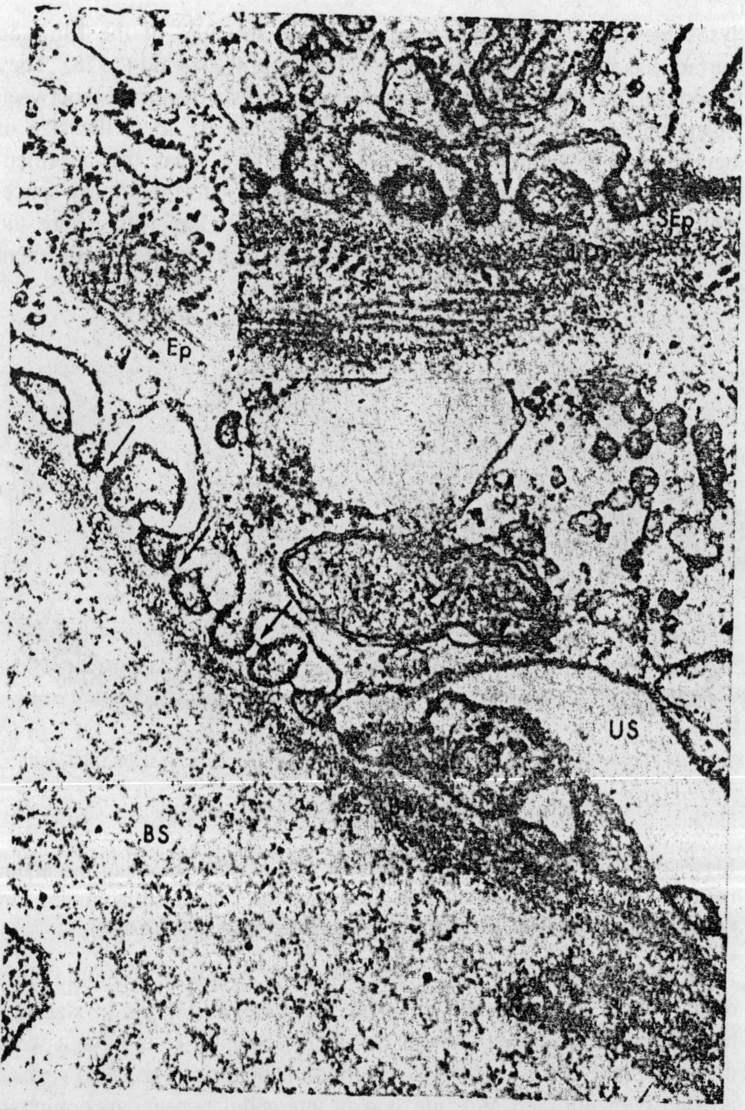

Fig. 5.2 : Coelomic sac wall in the crayfish, immersion-fixed with formaldehyde-glutaraldehyde. The two structural layers which comprise the wall, the fibrillar basement membrane (BM) and the epithelial cell layer (Ep) are evident. Slit diaphragms bridge the gap between adjacent foot processes of the epithelium (arrows). BS, blood sinus; US, urinary space. × 24,000 (Inset) The circular profiles of a slit diaphragm can be seen within a slit cut in cross section (arrow). An asterisk indicates collagen fibrils in a thickened area of the lamina densa (LD). SEp, subepithelial layer, × 29,500. (from Schaffner and Rodewald, 1978, Reprinted with permission of Academic Press).

The bladder, which may represent 2-4% of the body weight in crabs and crayfish, has recently been found to be the site of uptake and secretion of ions and water. Cells of the bladder have an apical brush border and glycocalyx, basal infoldings, and vacuoles that may be related to secretion of organic acids.

Structure of junctions between cells in the labyrinth and canal of *Orconcctes virilis* have been studied by freeze fracture. The sepatate junctions appear as parallel rows of particle's, continuous around the apex of each cell; they resemble junctions in several insect tissues. These junctions do not form permeability barriers between cells, since horseradish peroxidase passes between the cells via an extracellular route. There are also desmosome-like connections on the basal and lateral surfaces of the cells.

ROLE IN VOLUME REGULATION

In all crustaceans, the renal organ functions in volume regulation and in regulating concentration of nutrients, solutes, and divalent ions. In freshwater hyperregulators, the renal organ also functions in compensatory NaCl reabsorption.

Urinary volume in some crabs is inversely proportional to salinity of the external medium, but in others, need for renal compensation of osmotic water gain is lessened by reduction in osmotic water permeability across the body surface. In most estuarine species the upper limit of the excretory volume is established by the animal's ability to compensate for renal salt loss, since the urine produced is isosmotic to the blood. Although the antennal gland in these species appears to have no direct role in compensatory salt balance, alterations in amount of urine produced clearly play a role in regulation of internal volume and pressure. In a rigid-bodied animal, the presence of an escape route for water is vital for survival of the animal under conditions of increased water uptake such as occur in dilute media.

Several studies have been performed on rate of urine production as a function of external salinity. In *Cancer magister*, *Carcinus maenas, Pachygrapsus crassipes, Coniopsis cruentata, and Hemigrapsus nudus,* urine flow increases from 1 to 5% body weight per day in crabs acclimated to 100% sea-water to 11-60% body weight per day for crabs acclimated to dilute seawater. The rate of urine production doubles in *Callinectes sapidus* when the crabs are transferred from full seawater to 50% seawater. However, in some crabs, such as *Cancer productus* and *Rhithropanopeus harrisii,* such an increase does not occur.

Increased rate of urine flow could be produced by decreased osmolality of hemolymph or by changes in internal volume or pressure; all of these could result from exposure to dilute medium. Recent experiments by Holliday (1978a) in *C. magister* showed that in this crab the signal for the increased urine production is increased volume of hemolymph, since urine flow increased when isosmotic salt solution was injected, but not when osmolality of blood was reduced by removal of hemolymph and replacement of the same volume with distilled water. The study by Davenport (1972a, b, c) on *Pagurus bernhardus* mentioned above also implicates blood volume as a critical factor in determining the rate of urine production, and the importance of urine flow in reducing blood volume.

In a related study on *Carcinus mavnas*, found that hydrostatic pressure, as recorded from the foregut or leg sinus, increases when the animals are transferred from 100% to 50% sw. After 3-4 hr., pressure falls sharply, then rises again. This sequence is repealed several times over several hours. Associated with the abrupt decreases in pressure is the release of urine from the nephropores. Animals with blocked nephropores show a steady increase in pressure and eventually die. Both of these studies support the role of increased urine flow in reducing a water load. However, there are differences between the two animals. In *Cancer magister*, volume loading does not affect internal pressure in the antennal artery (which supplies the antennal gland) or the heart. Thus, Holliday feels that the response is truly a volume response, perhaps mediated by receptors similar to those described in the thorax of *Pagurus bernhardus* by Pilgrim (1974). Norfolk found, in *Carcinus maenas*, that production of urine increased within a few minutes of transfer to dilute medium, before internal volume or pressure could have increased significantly. In addition, on initial transfer to dilute seawater, the crab exhibited a 20-min lag period before increased urine production began, while a second exposure to dilute medium only resulted in a lag of 5 min. Thus, Norfolk postulated a chemical mediator that is activated or produced in response to changes in salinity acting on external receptors. This chemical mediator could increase production of urine before internal pressure became high, although release of urine would still be sporadic. A second exposure to dilute medium would find the mediator readily accessible and would account for the shorter lag time. This chemical system could be related to the system that causes a decrease in apparent water permeability in *C. maenas* exposed to dilute seawater also found that water permeability was adjusted before internal conditions were changed when *C. maenas* was exposed to dilute seawater. Thus in this animal, an early warning or anticipatory system could operate so that the animal changed its permeability, rate of urine flow, and perhaps its rate of sail uptake on direct response to the external medium, rather than depending on changes in the internal medium. A system such as that found in *Cancer magister* could play a secondary role in *Carcinus maenas.* In an estuarine animal exposed regularly to changes in salinity, the *C. maenas* system would enable the animal to undergo minimal variation in internal milieu and would lend to damp the effect of changes in salinity. Recent studies of *Callinectes sapidus* exposed to a fluctuating salinity in the laboratory have shown that there is a lag in change of osmolality when the animal goes from seawater to dilute seawater, but not in the opposite direction as in *C. maenas,* this lag could represent the time needed to mobilize chemical mediators.

MECHANISMS OF URINE FORMATION AND REABSORPTION OF SOLUTES

The mechanism of urine formation in crustaceans has been described as ultrafiltration, based on studies employing the markers inulin and ^{15}Cr EDTA (Riegel, 1972; Riegel and Cook, 1975). In decapods, the antennal artery supplies the antennal gland, and crabs can lie rendered anuric by severing this vessel (Holliday, 1978). II has also been suggested by Riegel that pressure-driven filtration can be enhanced by the presence of "formed bodies" in the coelomosac. These lysosomal-like bodies sequester solute and are hydrophilic; their presence in the coelomosac fluid enhances the rate of formation of the primary urine (Riegel, 1970a,b, 1977a,b).

Fluid entering the coelomosac, the primary urine, is isomotic to blood Modification by reabsorptive and secretory processes occurs during its passage through the labyrinth and renal tubule and in the bladder. Owing to technical difficulties in sampling the fluid from distinct regions, the mechanisms of modification are poorly known. Cicropuncture studies on the most successful preparation, the crayfish antennal gland, have yielded the results shown in Fig. 5.3.

Several authors have demonstrated glucose reabsorption and its inhibition by phlorizin. In *Pachygrapsus crassipes* (Gross, 1967) and *Cancer magister* (Holliday, 1978b) glucose is reabsorbed from perfusate introduced into the bladder. Glucose reabsorption in the antennal gland of the crayfish has also been observed, but the site of reabsorption has not been identified. The bladder of the crayfish does appear to be the site of amino acid reabsorption.

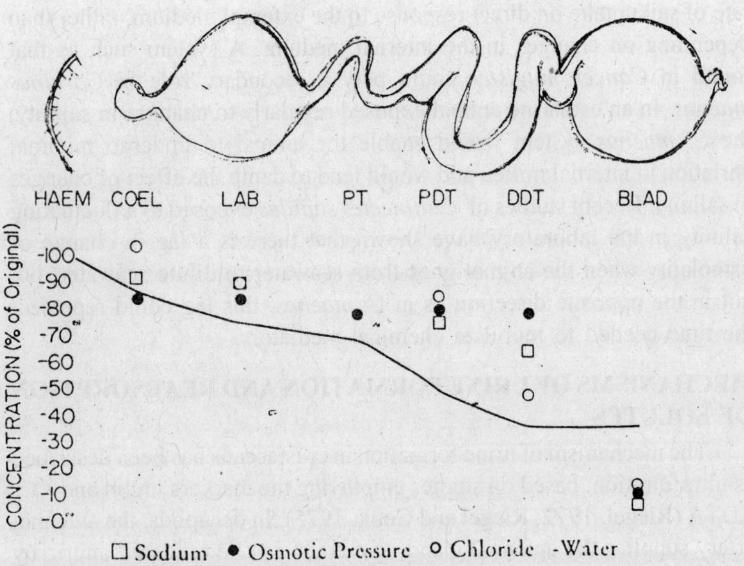

Fig. 5.3 : Changes in the Na, Cl, and H_2O content and osmotic pressure (OP) of urine in the antennal gland of the crayfish. Haem, hemolymph; coel, coelomosac; PT, proximal tubule; PDT proximal distal tubule; DDT, distal tubule; blad, bladder. Data in Riegei, 1963, 1965, 1968; Cook, 1973. (From Riegel and Cook, 1975, Reprinted with permission of Gustav Fischer Verlag.)

The role of the antennal gland in accomplishing ionic regulation is well established, and many of the recent experiments are discussed by Riegel and Cook (1975). Renal handling of ions is usually assessed by measuring concentration of the ion in the urine (U_{ion}) as compared to the concentration in the hemolymph (H_{ion}), with the assumption that, in a filtration-based system, U/H = 1 for any ion that has been filtered but neither reabsorbed nor secreted. However, in the absence of data on movement of a marker substance, such as inulin or radioactively labelled EDTA, there is no assurance that water has not been reabsorbed following the reabsorption of ions. In those cases where marker data are available, animals producing urine isosmotic to the blood usually have a U/H ratio close to 1 for Na and Cl and slightly above 1 for K, Mg, and Ca.

All crustaceans examined possess at least a limited capacity for reabsorption of NaCl, but in most estuarine hyperregulators this contribution to total ionic balance is minimal.

In *Callinectes sapidus* in fresh water, only 12% of the primary urine is reabsorbed; urinary sodium loss is 40% of the total sodium efflux. These authors also note that the renal organ of C. *sapidus* contributes little to the acid-base balance and excretes only 1-2% of the total NH_3. In hyporegulators the renal organ is also ineffective in compensatory NaCl regulation since urine concentrations of Na and Cl are the same as those in the blood. On the oilier hand, the renal organ of freshwater hyperregutators does effect net salt reabsorption through production of a dilute urine. This ability has been correlated with the presence of an elongated tubular segment which, in the crayfish, can be differentiated into a thinner proximal and a thicker distal region. Kamemoto and Tullis (1972) found that the distal region also contained the highest concentration of Na-K ATPase in the antennal glands of *procambarus clarkii* and *P. blandingi*.

The kinetics of NaCl transport have yet to be defined. Holliday (1980) has shown net flux of Na and Cl from the bladder to hemolymph in the isolated bladder of *Cancer magister*. Reabsorption of Na in this species is not inhibited by ouabain, but it is dependent upon aerobic metabolism. Kamemoto (1961) found that the bladder of crayfish is rich in acetylcholinesterase and that reabsorption of Na was inhibited by the cholinesterase inhibitor eserine. Na-K ATPase is also found in antennal glands of several terrestrial crabs (Towle, 1981). Thus, though the mechanisms tor Na uptake are still unknown, the uptake of salt may provide the driving force for reabsorption of water by the bladder (Holliday, 1980).

NaCl transport may be linked to Mg secretion. They note that with increasing salinity there is an increase in U/H ratios of Mg and a proportional decrease in U/H for Na. Mg secretion in the renal organ of *Palaemon serratus* is an active process, against an electrochemical gradient. Although there is reabsorption of fluid from the primary urine, it cannot completely account for the U/H concentration of Mg. These authors also found that Mg secretion is inhibited by ligation of eyestalks, independent of changes in rate of water reabsorption. However, Ablation of eyestalks had no effect on Mg concentration in blood or urine of *Cancer magister*, but he did concur that secretion is an active process that can be demonstrated both *in vivo* and in isolated bladder preparations.

The grapsid and ocypodid crabs increase the secretion of Mg greatly when they are in concentrated seawater and regulating hypoosmotically; the amount of Na in urine is correspondingly reduced. Yet, if secretion of Mg is coupled to reabsorption of 2 Na, osmotic balance necessitates reabsorption of water as well. Holliday (1980) found that the net reabsorption of Na from urine was apparently four times that of the net secretion of Mg. However, Mg secretion appears to be nonelectrogenic and quabain sensitive. He speculates that Mg transport from the blood across the serosal surface of the bladder cell may be Na dependent. As mentioned above, the excess Na influx is likely to function in reabsorption of Cl and water. Application of techniques for perfusion of the antennal artery and isolation of the bladder tissue should facilitate further examination of these mechanisms.

HORMONAL CONTROL OF PERMEABILITY TO WATER AND IONS

A. General Considerations

The presence of central control systems regulating permeability to water and ions has been assumed for many years. Early studies related changes in water permeability to the events immediately preceding and accompanying ecdysis; these will be discussed below. Regulation of permeability in intermolt animals has been studied both in the whole animal and on tissues likely to be involved in regulation; particularly gill, gut, and antennal gland. However, it is important to remember that gill and gut are multifunctional tissue's and that changes in permeability may be related to changes in gas exchange, acid-base balance, feeding, and drinking. Thus the control systems may be regulated by behavioral, ventilatory, or cardiovascular responses as well as by direct changes in the boundary tissues themselves. In view of these facts, proposed that the term "apparent water permeability" be used to describe the changes that can be measured in the intact animal when it is exposed to changes in salinity or treated with putative controlling factors.

With the exception of substances directly involved in molting and reproduction, hormones in crustaceans are produced in and released from neuroendocrine centers in the central nervous system, particularly in the eyestalk (X-organ-sinus gland complex), thoracic ganglionic mass (TGM) and brain.

In contrast to the extensive work done in annelids and molluscs, only one study has correlated changes in the appearance of neurosecretory cells with an osmoregulatory response in a crustacean. The medial neurosecretory cells in the protocerebrum of *Anemia salina* increase their secretory activity when the animals are kept in very dilute seawater, that is, under conditions where they are hyperosmotic to the medium. In concentrations greater than about 30% seawater the animals are hypoosmotic. When the animals are in this condition, the neurosecretory cells dimmish their rate of secretion, become filled with acidophilic granules, and appear to become inactivated. Osmotic receptors are thought to be located near those cells. The nature and function of the secretion are unknown and its role in the normal life of the animal is uncertain. Since the Anostraca are thought to have arisen in fresh water, this neurosecretory system may be an evolutionary remnant of the group's adaptation to that medium.

Other evidence for neuroendocrine control of permeability comes from studies on whole animals, where parts of the neuroendocrine system are removed and then replaced, and on isolated systems in which the tissue is treated with in extract of the neuroendocrine tissue. In only one case has a substance been found in or isolated from hemolymph that directly affects permeability to water or to ions in an intermolt animal. Thus the substances obtained from homogenates of neuroendocrine tissues and thought to be released by them, either into the hemolymph or in proximity to the tissue involved, should be referred to as "factors" rather than as hormones.

The principal organs for excretion are the antennary gland?anus or frequently referred to as the green glands. They are located, one on each side, anterior to the ventral portion of the cardiac stomach. In the crayfish each consists of a ventral, glandular part. connected posteriodorsally with a more dorsal bladder, which drains anteriorly into a duct leading ventrally to the opening already seen on the ventral surface of the antennal coxa. The glandular part is firm and shaped like a planoconvex lens, with the flattened surface directed upwards. This bladder is likely to be overlooked, since it usually is collapsed and its walls are very thin and membranous. The distended bladder may be considerably larger than the glandular part and pressed posteriorly against the stomach, hepatopancreas, and other organs.

The glandular part has a complex internal structure, difficult to make out completely even in sections. It begins blindly as a dorsoanterior coelomosac, having complexly infolded walls. This part opens anteriorly into the labyrinth, made up of a maze of channels occupying the peripheral parts of the organ. The labyrinth ultimately opens into a long and very much convoluted tubule, which occupies most of the central portion of the organ and leads finally to the opening into the bladder. If the entire gland is removed to a watch glass, the limits of its main regions usually can be recognized.

The cells of all three of these divisions of the glandular part of the nephron are secretory in appearance. Branches from the antennal and ventral thoracic arteries provide an unusually copious blood supply, their minute terminal branches discharging into tiny hemocoelic spaces which surround all the ramifications of the gland. The bladder is non-secretory and receives from the antennal artery a much more restricted arterial blood supply.

It is reported that no valves or sphincters occur between any of the internal parts of the gland. A valve is present only at the external opening. Here a soft operculum covers most of the external opening. However, the actual excretory pore is only a thin, crescentic slit near the anterior border of this operculum, the concavity of the crescent being directed forwards. Any attempt to inject or inflate the bladder will fail, unless the needle or tube is aimed accurately at the crescentic slit rather than at the operculum as a whole.

In the antennal gland of the lobster certain important differences must be noted. The coelomosac occupies most of the dorsal aspect of the glandular part, with the labyrinth surrounding it peripherally and extending under it vcnfcrally. A tubule is absent, probably correlated with the fact that the lobster secretes a scanty, isotonic urine, in contrast to the copious, hypotonic urine secreted by the fresh-water crayfish. From the anteriomedial portion of the labyrinth a white projection extends ventroposteriorly; this projection contains a number of parallel tubes which open into the duct from the bladder near its external orifice. There is no direct opening into the bladder, and thus the bladder can be filled only by the backing up of urine through its duct. This duct runs downward from the bladder in front of the glandular part of the organ, then turns and runs ventroposteriorly to the external opening, located in the center of the operculum.

The work of the antennal glands in excretion and osmoregulation is supplemented by the hepatopanereas and the gills. The roles of the hepatopancreas in excretion have already been mentioned. A passive exchange of ions and water molecules occurs between the blood in the gills and th surrounding water, and there is evidence also that the gills are able to take in certain ions against a concentration gradient. When dyes which are non-dialysable or of large molecular size are injected into the blood, they are accumulated by certain cells in the gills, as well as by the cells of thc coclomosacs in the antennal glands.

The principal excretory and organs in Crustacea are so called head glands associated with antennae and with the IInd maxillae. They are modified Coelomoducts.Which probably occurred in a more extended series in ancesftal Crustacea.

Hutchinsoneiella, a recently discovered primitive genus, has a of segmental organs which might be representative of the ancestral condition, both sets of head glands (antennal and maxillary) are not simultaneously functional in adult crustaceans, but frequently one or the other is retained in a vestigal condition. One of the two sets of glands may be functional during larval stages, while the other type.

Becomes functional for the adult. In some of the 'shelled' crustaceans in which the maxillary glands are functional.one or more loops of the coelomoduct in side the shell fold or carapace.

Such excretory glands are occasionally reffered to as "shell glands".

Both type of head kidneys have a similar basic structure with a mesodermal endsac, and an elorgated tubule opening to the exterior.

The principal nitrogenous waste compound in crustacea is ammonia.

Ammonia is toxic and requires the availability of plenty of water for dilution.this well may be one of the factors hindering wide dissemination of crustaceans terrestrially. There is evidence that the excretory head glands of crustaceans function primarily to removeless soluble wastes and are also for osmoregulation .

Ammonia is excreted principally through as an urea. In some crustaceans, considerable amount of nitrogenous wastes (uric acid urea) appear in the hepatopanereas but the disposal of uric acid is not apparent.

There is some evidence for its retention as solid wastes in the integuments and in some phagocytic cells called nephrocytes in some decapods enzymes are present which convert uric acid to ammonia.

Nephocytes are most common in axial shafts of gills and in the proximal portion of the legs.

The renal organ of crustacea are two pairs of glands, the antennary and the maxillary glands. Both pairs of glands rarely function together, but one succeeds the other. Generally the antennary glands function as kidney's in the larvae while the maxillary glands in the adult. But the reverse is true In the orders decapoda (including prawns) , Amphipoda, schizopoda and dichelopoda.

The excretory system of the adult palaemon consists of :

(1) A pair of antennary or green glands.

(2) A pair of lateral ducts,

(3) An unpaired renal, sac,

(4) The integument.

(1) *Antennary glands :* one antennary gland is enclosed with in the coxal segment of each antenna. It is an opaque white gland as big as a pea-seed or gram-seed, and comprises three parts :

(1) The end sac.

(2) The labyrinth. Or glandular plexus

(3) The bladder.

1. The bean shabed end-sac is the smallest part lying in between the bladder and the labyrirth. Internally, its cavity contains a large blood-lacuna . Its wall is made of two layers and projects in the form of radial septa into the central cavity. The outer thick layers of the wall consists of connective tissue containing numerous small blood-lacunae, while the inner thin layer consists of large excretory epithelial cells.

   The labyrirth or glandular plexus is relatively larger than the end-sac and lies on its outer side. It consists of numerous narrow, branching and greatly coiled excretory tubules embedded in a mans of connective tissue containing blood-lacunae . The tubules are lined by a single layers of large excretory epithelial cells and they open by a single aperture into the end-sac, but by many apertures into the bladder.

   The bladder is the largest of all, lying on the inner side of the end-sac. It is a thin-walled sac made of a single layer of excretory epithelial cells.

   It's inner well is prolonged as a short excretory duct or uretor, which opens to the outside through Small rounded renal aperture, situated on a papilla on the inner of the surface of the coxa of the antenna.

2. *Lateral duct :* A narrow lateral ducts runs posterior from the bladder of each antennary gland.

   The lateral ducts of both the sides are connected by a transeverse connective, in front of the brain. The two ducts run backward along the oesophagus to open into the renal sac.

3. *Renal-sac :* it's a large thin-walled sac lying above the and extending posterior up to the glands. It's well is made of a single layer of flattened excretory epitheial cells.

   The complex nephridia like antennary glands extract nitrogenous waste products and excess water from the blood in much the same manner as the vertebrate kidneys do.

   The end sac excretes compounds of ammonia ,but uric acid and parts. The excretory fluid is collected first into the bladders and then expelled out through the renal apertures.

4. *Integument :*The non living ehitinoul covering or integument is cost off at each mount. When the nitrogenous products is secreted by the body deposited on the integument are also expelled.

   Thus, integument is believed to be an important excretory organ.

The main excretory organs in crabs are the antennary glands and their connections. Besides these, excretion is also performed by the fermant cells of the digestive gland and cells in the gill axis.

Antennary Gland and its Connections

The antennary glands are paired and situated to the right and left side of the body. The first part of the antennary gland is the gland proper, which is situated in the coxal region of the second antenna. It is a small spongy mass of a light green colour having triangular shape. At its posterior end it is connected with the second portion of the gland known as bladder. The bladder is an extensive thin walled sac having many large branches. It can be seen clearly because of its dark brown colour. The third portion of the gland is the ureter which is situated in front of the antennary gland. The ureter is a small narrow tube and opens to the exterior through the operculum which lies on the ventral portion of the coxa of seconds antenna.

Antennary Gland

The antennary gland consists of two portions the end sac and labyrinth. The first portion is the end sac which and situated on the dorsal side. From the floor of the end sac are given off numerous blind prolongations which may be either simple or branched. The epithelial cells lining the end sac are irregularly shaped. Some of these cells are of columnar type while majority of them are of squamous. The nucleus is spherical and is situated at the base of each cell.

Labyrinth

The second portion of the gland is the labyrinth which is large than the end sac. It consists of a means of highly convoluted and branching excretory tubules. The main part of the labyrinth is the renal tube the cavity of which is connected infront with the end sac and behind with the bladder. The roof the renal tube is in close contact with floor of the end sac. From the floor of the end sac numerous branched tubes are given off, the cavity of each tube is being connected with that of the end sac. The wall of each tube is lined by squamous epithelial. The floor of the end sac is closely applied to the roof of the renal tube so that these prolongations push before them the epithelium of the renal tube breaking up its lumen. Invaginations also seen in the ventral and lateral walls of the renal tube giving rise to small partitions known as trabaculae.

In section each excretory tubule appears to be lined by two layers, the inner lining is of squamous epithelium of the renal tube. The epithelial cells of the renal tube show striated appearance and the nuclei of the cells are large and spherical. Each tubule encloses a small lumen, while the inter-tubular spaces between the adjoining tubules are filled with connective tissue containing net work of blood lacunae.

The Bladder

The bladder is a thin walled sac which is more extensive and complicated having many lobes. All these lobes arise from the main vesicle. The main vesicle is situated above the antennary gland and its cavity is continuous with that of the renal tube. From the front side the gland is connected with the ureter. The main lobes found are as follows:

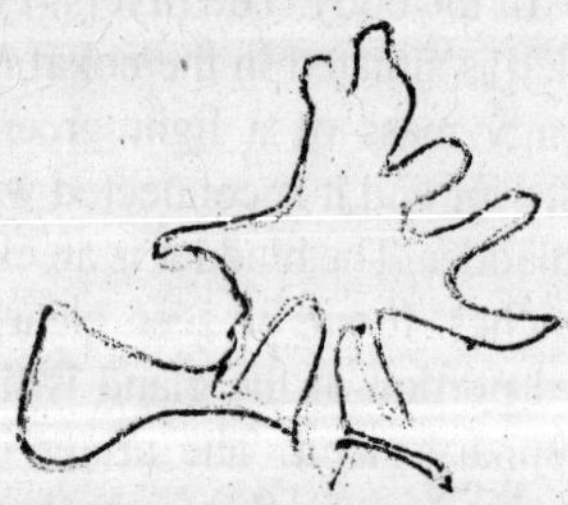

Fig. 5.4.

The main vesicle is situated just behind the outer corner of the stomach and above the antennary gland. On its ventral side it is connected with the ureter.

The epigastric lobe is the most prominent part of the bladder and it arises from the anterior end of the main vesicle. It runs forward and upward along the wall of the stomach. It is broad towards the anterior side and becomes narrower to the posterior end.

The progastric lobe is a small lobe of the epigastric lobe and is situated between the main vesicle and the epigastric lobe. It is closely applied to the front wall of the stomach. The antero-lateral lobe originates from the outer side of the main vesicle and found to be located beneath the digestive gland. The cerebral lobe arises from the anterior side of the main vesicle and it passes above the cerebral ganglia and almost meets the corresponding lobe of the other side. The hepatic lobe arises from the posterior and outer corner of the main vesicle by a very narrow portion which passes beneath the outer adductor muscles of the mandible. This lobe is situated exactly below the digestive gland. The supra hepatic lobe arises from the main vesicle on the inner and posterior side. It is situated on each side of the foregut.

The paragastric lobe takes its origin from the posterior end of the main vesicle very near to the origin of the supra hepatic lobe. The outer wall of this lobe touches the mandibular apophysis.

The esophageal lobe arises from the inner side and posterior corner of the main vesicle. This lobe is divided into two the anterior and posterior lobes which cover the esophageal tube.

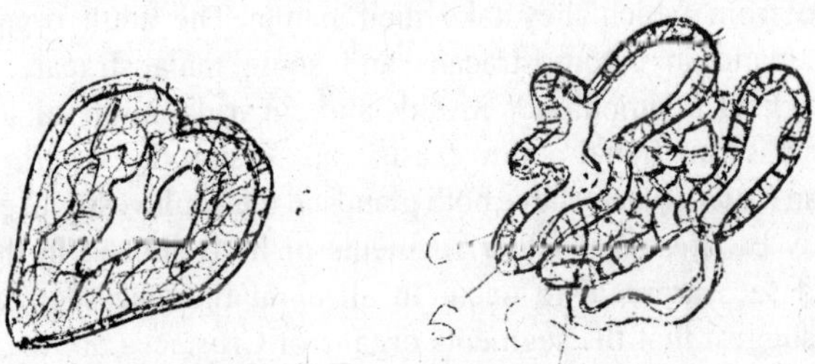

Fig. 5.5.

The wall of the bladder is composed of columnar epithelium. The nucleus is spherical and placed in the centre of each cell. In the outer portion of the cell protoplasm appears to be striated because of its dark stain.

The Ureter

It is situated on the inner side of the antennary gland partly infront and partly below the gland. It is connected with the main vesicle by a narrow opening. It passes downwards and opens to the exterior on the ventral surface of the proximal portion of the second antenna.

Ferment Cells of the Digestive Gland

The digestive gland, in addition to the functions of digestion and absorption, has also the function of excretion. Part of the contents of the ferment cells are of an excretory nature. When the content of these cells passses down the digestive tubule into the midgut the excretory products are separated away and are carried outside along with the faeces.

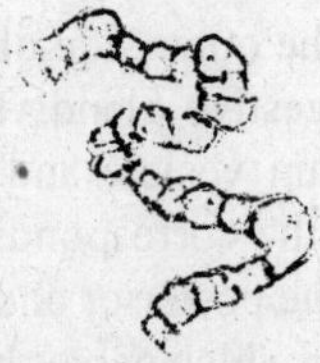

Fig. 5.6.

The Branchial Excretory Cells

The branchial excretory cells are found in the portions of gills in between the afferent and efferent branchial vessels. These cells may play an important role in transport of water and salts in osmoregulation.

Fig. 5.7.

Excretory System of Other Crustacea

The excretory organs of Crustacea are limited to certain segments of the body. Two pairs usually occur during larval and adult life the antennary and maxillary glands, closed tubules opening at the base of appendage from which they take their name. The adult organ is the maxillary gland in entomostracans and some malaostracans and the antennary gland in amphipods mysids and eucarids: in larval stages the other gland is functional a few crustaceans (ostraods *Nebelia* and the mysidacean (*lophogaster*) have both glands in the adult vestigial excretory organs may be present in other segments of body indeed in the mysid paired end sacs are said to occur in all eight thoracic segments such evidence suggest that the segments organs of Crustacea are the remains of a once continuous series such as is found in annelids. Their reduction in number and the tranference of their funtion to other organs and tissues is assoiated with the redution of tissues is associated with the reduction of coelom.

The excretory organs are of the type found generally throughout the artropods. Each consists of an end sac lying in the haemocoel and opening into a narrow excretory canal which discharges to the exterior by a short ectodermal duct. The end sac are the persistent remains of a few of the series of paused, coelomic sacs which make a transient apppearance during development. The canal may be a coelomoduct (*Estheria, Nebalia*), homologus with the genital duct through more frequently it origins from ectoderm.

In branchiopods extensive coils of the excretory canal are lodged in the narrow cavity of the *Carapace* fold, for which reason the maxillary gland is often referred to as the shell gland. In higher malacostracans the canal loses its simple tubular from and at least part of it developes into a compact gland which is most elaborate in decapods; the end sac may remain simple or become subdivided. The initial part of canal forms a spongy mass transversed by a system of fine canals and offering a large surface area to the haemolymph; this is the labyrinth which, with the end sac, constitutes the glandular section of the organ. The disital part from a distensible bladder which may remain as simple as in *Astacus*, or develop lobes and chambers which extened back even into the abdomen, as in crabs Fresh water crustaceans have special adaptations which enable them to excrete a hyposmotic urine. In members of Astracidae this takes the form of an elongated part of the labyrinth, a whitish cord of spongy substance, convolucted upon itself, forming the nephridial canal leading to the bladder: in the freshwater peracaridans *Gammarus pulex* and *Ascellus aquaticus* it is an extra segment which, either by secreting water or absorbing salts, allow their specialised excretory glands (antennary in the former and maxillary in the later) to produce a hyposmotic urine. In other crustaceans, including freshwater species, the modification of thc cxcrctory fluid by the canal of the antennary gland is much less pronounced. Regulation of inorganic salts occurs resulting in a conservation of Ca^{2+} and K^+ and elimination of Mg^{2+} and SO_4^{-2}. The urine is isotonic with the blood, and even in species like *carcinus maenus* in which the blood remains strongly hypertonic when the animal is placed in external salinities as low as 4 per cent seawater (1.2%) the segement organs are not involved in the osrmoregulatory mechanism. No mention has been made of the excretion of nitrogenous waste, since thcsc do not appcar to be a primary concern of these organs: their main function is the maintainace of ionic balances of the body fluids by the selective treatment of ions. The nitrogenous end products are chiefly ammonia with some urea and uric acid and a considerable amount of amino nitrogen, which may be excreted in the ammonotetic animals. Ammonia and urea are lost throny the permeable surface of the gills and a percentage of the former is excreated in volatile forms. Uric acid may be deposited in various pairs of body which act as kidneys of accumulation including the endoskeleton.

Loss of water and salts by way of the anknnary gland is balanced by the uptake of water and ions through gills and Na^+, K^+, Ca^{2+} and $C1^-$ often must be taken up against concentration gradients. It is by this means that crustaceans like other invertebrates, can maintain a concentration of ions in the blood plasma differing from that of a passive equilibrium with the external medium. Decapods and peracaridans living in brackhish or fresh water suffer loss of salts by outward diffusion as well as excretion and make good the loss by achieve absorption from the dilute medium with the expenditure of energy. This uptake of salts takes place over certain areas of gill leaflets in decapods, but apparently over the whole gill in gammarids. Where both the basal and apical surfaces of the cells are highly infolded, with large no of mitochondria in the basal folds to provide energy for the ion pump by which salt is obtained. Cations of sodium and potassium probably being exchanged for hydrogen and anions for hydroxyl.

Artemia the brine shrimp can surive in a range of salinities varying from 10% seawater (8.5%) to crystalline brine (salinity 600%). At and above 25% sea water the blood is hypotonic to external medium and hypertonic below. Hypotonocity is maintained by a process similar to that found in marine teleosts-water with its dissolved salt is swallowed, and the excess salts are excreted by mitochondrial pumps located in certain cells on the epipodities of the ten anterior swimming limbs. There is evidences that salt elimination is also helped by the fact that *Artemia* produce a urine hyperosmotic to the blood.

Products of catabolism are excreted from all animals by diffusion the end product oxidation of proteins, pyrinidines and purines are nitrogen compounds chiefhy ammonia. But ammonia is toxic material (Substance) and need excretion as soon as it formed or it must be converted to some less toxic substances. These substances are either trimethylamine oxide, Urea, or Uric acid. Therefore, the major nitrogenous excreting substances in animal are ammonia, trimethylamine, Urea and Uric acid. Ammonia is very toxic and very much soluble in water.

Urea is less toxic and comparatively less soluble in water. Uric acid is less toxic than either ammonia or Urea. and is almost in soluble in body fluids. Therefore, excretion of a particular type of nitrogen containing excretory substance by the particular animal or particular animal group depends upon its enviroment and mode of life.

Formula

Urea

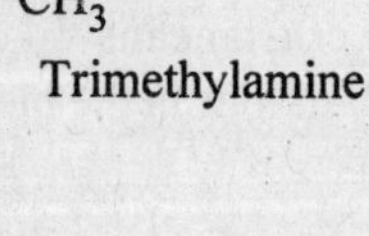

Trimethylamine acide

Ammonia

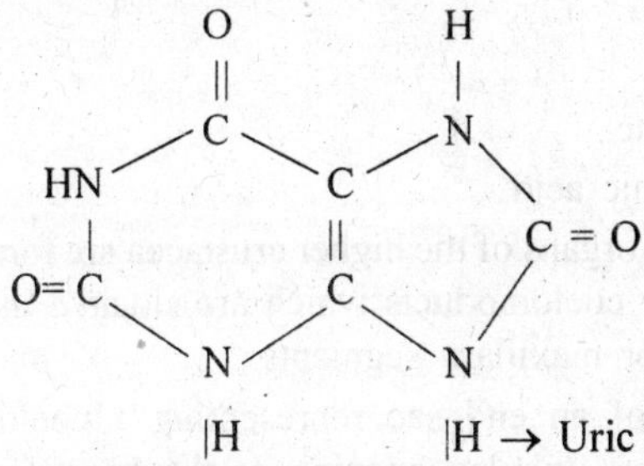

→ Uric acid

1. Marine Invertebrates (Ammoniotelic)

— Most of the marine invertebrates excrete large proportion of their nitrogen as ammonia.

— Ammonia is highly soluble in water and its conceutration in the body fluids of marine invertebrates varies from 0.4 mg per 100 ml.

— Marine animal in unlimited valume of water and their body fluids are mostly isotonic or hypertonic to sea water. Therefore, ammonia which is so soluble in water, readhy diffuses out of their body from higher to lower concentration and never reaches a dangerous concentration.

Various Excretory Products in Animals

The metabolic waste products which are excreted by the animals may be grouped under the following two heads.

1. Respiratory Waste Products

The catabolic waste products of various types of food stuffs are CO_2 and water. In lower animals CO_2 is eliminated directly into the enviroment through the general body surface. In higher animals it is excreted with the expired air through the lungs. Excess of water is eliminated in the from of Urine.

2. Nitrogenous Waste Products

The nitrogenous waste products are derived from the deamination of excess of amino acid tallen in with the dict and also from break down of animals proteins and nucleic acids.

Following are the specific nitrogenous waste products excreted by the animals :

1. Ammonia

2. Urea.
3. Uric acid.
4. Amino acid.
5. Other nitrogenous waste.

— Allantoin and allantonic acid.

— The principal exretory organs of the higher crustacea are formed by a pair of glandular coelomoducts which are situated in the either the antennary or maxillary segments:

— Each gland consists of an end sac representing a coelomic reminant, and excretory tubules opening at the base of the maxilla or second antenna.

— In most forms of tubule is simple, but in decapods it may be expanded with much bolding of the wall to form a labyrinthe at the proximal end and expanded distally to form a bladder.

— The principal problems associated with excretion include:

1. The manner in which the urine is formed.
2. The function of the urine.
3. The mechanism by which urine composition is controlled.

Mode of Urine Formation

— The urine is formed inside the kidneys from nitrogenous waste products, produced during catabolic process in the body.

— No structure comparable with the glomerulus of vertebrates has yet been reported in the crustacean—excrectory organ and there has consequently been considerable controversy as to whether urine is formed by filtration, as in the majority of vertebrates or by some form of secretory process as in the insect malpighian tubules and kidneys of aglomerular fish.

Evidence for Ultra-Filtration

There are two lines of evideuce to support the conclusion that in the decapods urine formation is by process analogous to the ultrafiltration and reabsorption mechanism of vertebrates :

1. Glucose appears in the urine after injection of phlorizin into the blood.
2. The carbohydrate anuline is concentrated in the urine.

— Phlorizin inhibits the reabsorption of glucose by the vertebrates nephron. Hence, as glucose appear in the definitive urine of crayfish and *Homarus* after injection

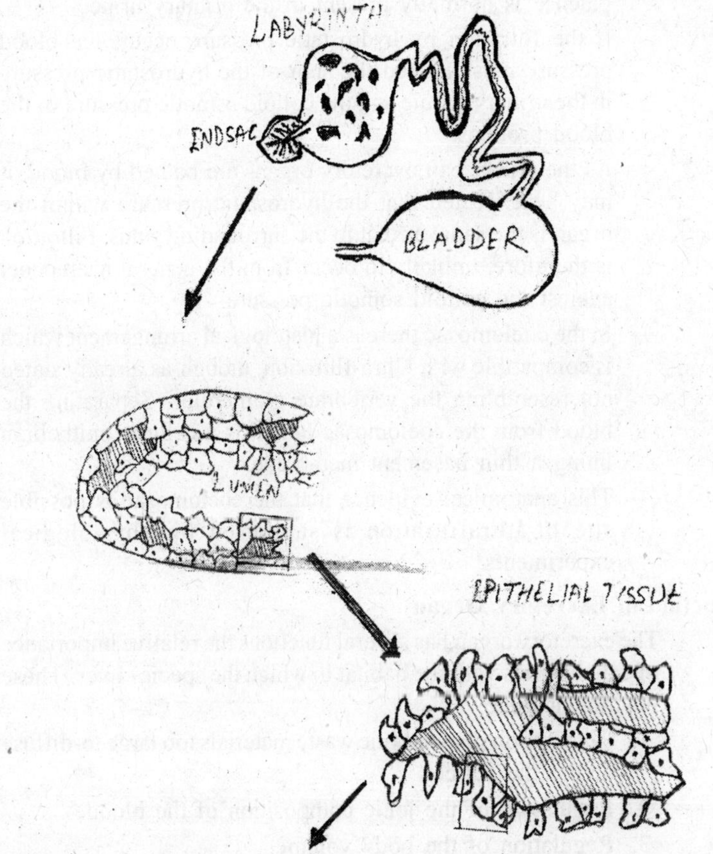

Fig. 5.8 : Diagram of renal organ and the fine structure of the coelomic end sac and potential site of Ultrafiltration (by Redrawh).

with this drug, there must be a strong presumption that glucose is normally present in the primary urine.

— If the filtration by hydrostatic pressure occur, the blood pressure must exceed the sum of the hydrostatic pressure in the urinary tubule and the colloid osmotic pressure of the blood proteins.

— As the crustacean excretory organs are bathed by blood, it may be expected that the hydrostatic pressure within the organ is the same as that in the surrounding sinus. Filtration is therefore, unlikely to occur from the general haemocoel against the colloid somotic pressure.

— In the coelomosac there is a histological arrangement which is compatible with Ultra-filtration, though-as already stated not resembling the vertebrate glomerulus. Separating the blood from the coelomosac lumen is a cellular epithelium lining a thin basement membrane.

— This anatomical evidence that the coelomosac is possible site of ultrafiltration is supported by physiological experiments.

Function of Excretory Organ

– The excretory organ has several functions the relative importance of which varies with the habitat in which the species lives. These function include:

1. Elimination of metabolic waste materials too large to-diffuse across the cuticle.
2. Regulation of the ionic composition of the blood.
3. Regulation of the body volume.

1. Excretion of Waste Materials

— An excretory organ forming Urine by filtration offers a convenient means by which foreign or waste materials. *e.g.,* inuline, for which there may be no available specific transport mechanism, can be readily removed from the blood.

— Some organic substances can be actively secreted into the tubule. This if phenol red is injected into lobster the concentration of dye in the urine, exceeds the concentration of inulin.

— Some 60-87% of the total nitrogenous waste material is in the from of ammonia in marine and fresh-wates forms. Ammonia, being a toxic material, requires a considerable volume of water for its elimination and terrestrial groups, such as the gastropods, reptiles, mammals and birds, tend to release waste nitrogen in the from of less toxic urea or uric acid to conserve water.

2. Ion Regulation

— It has already been noted that the ionic composition of the blood of marine species differs from that of the medium. The very fact that urine is formed at all would contribute to a decrease in the blood concentration of those ions which can only slowly pentrates across the cuticle, even if the urine were similar in composition to the blood.

— Ion such as magnesium and sulphate, which usually have a lower lavel in the blood than in the medium, tend to have a high concentration in the urine.

— Increase in the magnesium concentration of the medium are followed by an increase in the magnesium concentration of the Urine but little change in blood concentration in *Caracinus, Hemigrapsus* and *Cancer*.

Table 5.1 : The ionic composition of the blood and Urine of *Carcinus* in sea water and after 48 hr and 96 hr in water-saturated air.

Na mm/l	In sea water	48 hr	96 hr.
Urine	481 ± 37.8	459 ± 30.8	391 ± 50
Blood	525 ± 10.9	538 ± 18.9	543 ± 19.5
K mm/l			
Urine	12.4 ± 0.57	121.1 ± 0.57	13.4 ± 0.79
Blood	12.7 ± 0.85	12.4 ± 1.39	13.6 ± 0.66
Ca mm/l			
Urine	17.2 ± 1.25	21.3 ± 2.24	22.3 ± 5.03
Blood	14.3 ± 1.41	14.6 ± 1.10	15.2 ± 2.21
Mg mm/l			
Urine	89.4 ± 35	121 ± 47	192 ± 37
Blood	21.2 ± 2.9	20 ± 2	19.4 ± 3.3

3. Regulation of Body Volume

— Some compensation for increased blood volume may be possible by decreasing the volume of the bladder, but the possibility of expansion within a rigid exoskeletion is limited.

— The volume of crustacea must, therefore, be regulated within narrow limits if fluctuation of the internal hydrostatic pressure are to be avoided. An effective mechanism is involved in volume regulation is clearly indicated by the fact that the size increase in crabs. *Callinnectus* is constant at molt, no matter what the gradient of salinity between their blood and medium.

Table 5.2 : The osmotic pressure and chloride coneentration in blood and in the different regions of the crayfish antennal gland-(from Raigel).

	Blood	Coelo mosac	Laby rinth	Capsule Proximal Tubules	Bladder
O.P. Mean ± S.D	200 ± 10	208 ± 22	181 ± 11	169 ± 7, 121 ± 14	18.8 ± 1.4
Cl	184 ± 24	168 ± 20	162 ± 21	132 ± 24, 80.8 ± 10	3.37 ± 1.8

— The excretory organs play an obvious role in regulating the fluid volume.

Osmoregulation in Aquatic Animals

The aquatic animals can be divided into-three categories:

1. Having body-fluid isotonic to the surrounding medium.
2. Having body-fluid hypertonic to the surrounding medium.
3. Having body-fluid hypotonic to the surrounding medium.

In crustacean, there is no problem in regulation of water contents or osmotic pressure in the body. The salt concentration inside and outside of the body is the same, their osmotic pressure will also be identical and there is—neither inflow or outflow of water from the body. Such animal are know as having isotonic body fluids.

— Marine crustacea do not need to expend any energy in order to maintain the composition of their body fluids. The blood, though isotonic with medium and generally similar in ionic—composition, differs in between and there is an even greater difference between the composition of the cell sap and that of sea water.

Crustaceans exhibits all of the possible patterns of osmotic regulation describes by Prosser (1973) and Vernberg and Vernberg (1970). Some animals are stenohaline, tolerating only a narrow range of external osmotic concentrations, while others are Euryhaline, able to function with a wide range of external salinities. The dividing line between 'steno' and 'Eury' is well high to impossible to define.

The relationship between internal and external or molality is usually expressed in terms of osmoconfirmity and osmoregulation.

PATTERNS OF OSMOREGULATION

A. General Considerations

Decrease in concentration of the external medium result in diffusive movement of water and ions into and out of the animal across any permeable surfaces. When in dilute sea-water an animal with a rigid exoskeleton can swell very little before internal hydrostatic pressure reaches intorerable levels. Thus, reduction of the osmotic gradient would minimise the influx of water and concomitant changes in volume and pressure.

The mechanism by an animal achieves a steady balance in the face of diffusive losses fall into two categories.

I. Limiting mechanism that results in minimizing the leaks, either by reducing the gradients that derives diffusive flux, or by reducing the area of permeability of the exchange surface, and

II. Compulsory mechanisms that a countermovement of solute equal in magnitude to the diffusive loss.

B. Limiting Mechanism—Osmoconfirmity

One of the most direct means of minimizing the diffusive movements of ions and water is to reduce the concentration and osmotic gradients between the blood and external movement of medium. This pattern partially or complete osmoconfirmity, is widely exhibited in marine and brackish water crustaceans.

Osmoconformers reduce the osmotic gradient by decreasing osmoregularity of hemolymph when it dilute sea water. However, this decrease represents the same problem to internal tissue that changes in external concentration present to hemolymph - influx of water followed by swelling.

C. Compensatory Mechanisms—Osmoregulation.

By maintaining a hyperosmetic internal concentration when in a dilute medium, osmoregulations place less of a burden on their internal tissues. However, the problem of osmotic influx of water remains and can be overcome by reducing permeability of water, increasing efflux of water via the urine, and increasing uptake of salts from the dilute medium. Water gained osmotically can be lost either lymph which also reduce loss of salt.

In both situation, hyper - and hypoosmatic regulation, specialized boundary epithelia, notably gills, gut and excretory organ, are responsible for active transport and salts to or from the medium.

MECHANISM OF OSMOREGULATION

A. Hyperregulation

An organism living in fresh water or in sea water more dilute than its hemolymph is faced with diffusive gain of water and loss of ions.

These movements may be minimized by reduction in permeability to water or ions or both, but compensatory absorption of Na^+ and Cl^- from the external medium is also necessary to maintain steady state balance.

B. Hyporegulation

For there marine organisms that maintain hemolymph concentrations below that of the external medium, the vectors of diffusive salts and water movement are just the reverse of the fresh water hyperregulation water is passively lost across the body surface in response to the osmotic, and ions are gained. In addition, osmotic water permeability is generally lower than in osmoconformers and weak hyperregulators. Low urine output help to minimise water loss.

In freshwater crustaceans and marine species that regulate osmolarity of blood, water and salt movements may be minimized by a reduction in the area and permeability of the exchange surfaces. No animal can maintain complete impermeability, however, and the epithelia that allow exchange of gases, elimination of waste, and uptake of nutrients also present potential pathways for leaks. Reduction in permeability can decrease but not eliminate the energy expenditure of active compensatory transport.

Early workers suggested that osmoconformers are much more permeable to ions than regulators, a conclusion supported by later experiments. It also appears that freshwater crustaceans, like freshwater teleosts, exhibit a significant reduction in permeability compared to euryhaline or stenohaline marine species, and that permeability may be modified by a species in longterm acclimation to more dilute media. In the congeneric amphipods *Gammarus duebeni, G. pulex, G. lacustris, G. zaddachi, and G. tigrinis,* reduction in Na^+ permeability accompanies the penetration of the genus into more dilute environments.

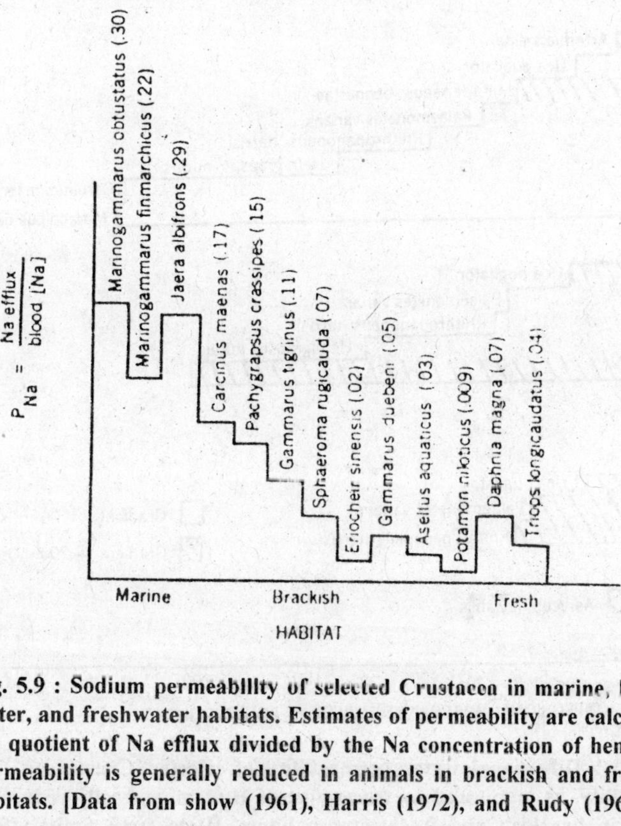

Fig. 5.9 : Sodium permeability of selected Crustacea in marine, brackish-water, and freshwater habitats. Estimates of permeability are calculated as the quotient of Na efflux divided by the Na concentration of hemolymph. Permeability is generally reduced in animals in brackish and freshwater habitats. [Data from show (1961), Harris (1972), and Rudy (1966).]

Short-term permeability changes are more difficult to assess. In osmoconforming crustaceans, there is little alteration in permeability when the external medium is diluted Moderate regulators like *Carcinus maenas* (Shaw, 1961a) likewise show little change in permeability, but in the stronger regulators, such as *Palaemonetes* spp., the rate of salt loss across the body surface appears to be reduced with short-term exposure to more dilute media.

Water permeability, like ionic permeability, is greatest in marine osmoconformers. These animals, which reduce the osmotic gradient between blood and medium, appear to lack means of altering permeability in acclimation to a new salinity. In contrast, hyperregulating crustaceans exhibit a reduced water permeability when exposed to more dilute media.

Outlines the pattern of diffusive water permeability in animals of a similar weight range (2-30 g).

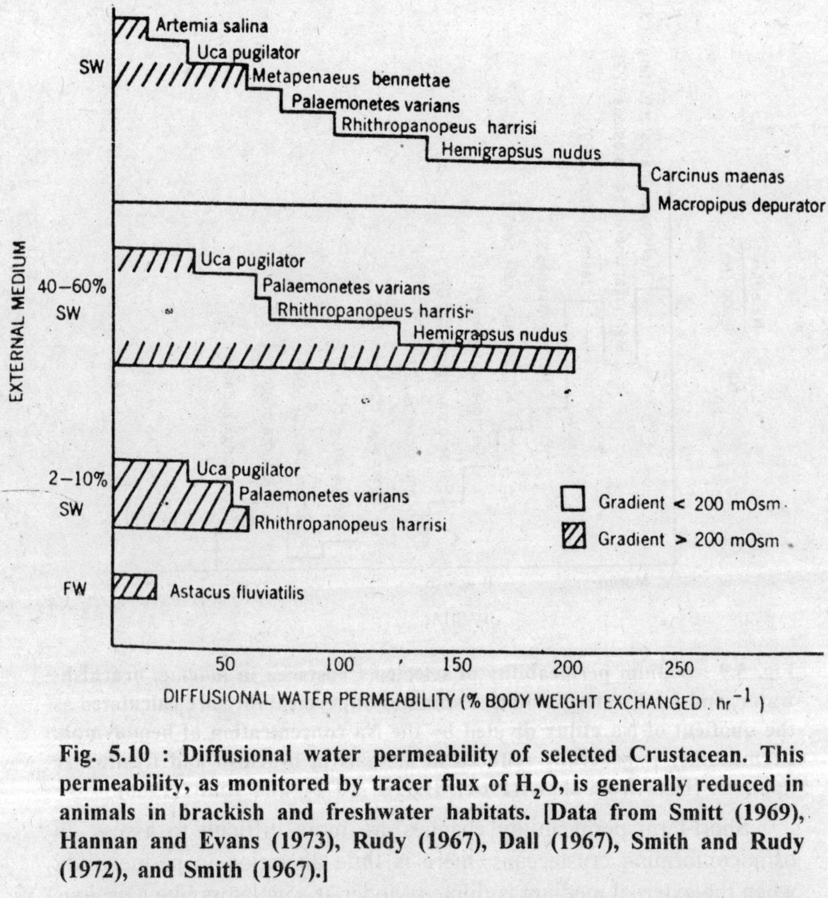

Fig. 5.10 : Diffusional water permeability of selected Crustacean. This permeability, as monitored by tracer flux of H_2O, is generally reduced in animals in brackish and freshwater habitats. [Data from Smitt (1969), Hannan and Evans (1973), Rudy (1967), Dall (1967), Smith and Rudy (1972), and Smith (1967).]

When *Rhithropanopeus harrisi*, an excellent hyperosmotic regulator at low salinities, is in seawater isosmotic to blood, approximately 145% of the body water is exchanged per hour. In 45% seawater, diffusional exchange is limited to 70% hr^{-1} and in 1% seawater the rate drops to 10% hr^{-1}. A similar pattern is seen in *Uca pugildtor* and in *Callinectes sapidus* in dilute seawater. These species are capable of maintaining a strong gradient between hemolymph and medium when in dilute seawater. The moderate hyperregulator, *Carcinus maenas*, exhibits a Higher diffusional water permeability, although there is reduction in exchange from 236% hr^{-1} in seawater to 176% in 40% seawater. In freshwater crayfish *Astacus* spp. diffusional water permeability is even more significantly reduced.

Sphaeroma serratum, an intertidal isopod, shows a decrease in water permeability from 820% when in seawater to 200% hr^{-1} in 50% seawater; this decrease occurs within 30 sec and is apparently brought about by the decrease in external Na and Cl. Possible mechanisms for the decrease in apparent water permeability are discussed by Thuel (1978). By elimination of other possibilities, he concludes that in *S. serratum* there is some unknown, instantaneous change in the permeable membranes themselves when animals are placed in seawater with reduced amounts of Na^+ and Cl^-.

Characteristics of Ionic Fluxes

Salt fluxes and observed flux ratios for some hyperregulating crustaceans are presented in Table 4.3. These measurements support the hypothesis first advanced by Krogh (1939), that hyperregulators actively absorb both Na^+ and Cl^- from dilute solutions. The unidirectional influx of Na^+ in these animals displays saturation kinetics. Among hyperregulators there is a wide range in both l_{max} and K_m, the affinity of the carrier for sodium. Sodium uptake in *Gammarus duebeni*, for instance, has a maximal rate of 57m moles Na^+ liter $blood^{-1}$ hr^{-1}, *G. zaddachi*, of 66 m moles, and the freshwater *G. lacustris* of 7m moles. Part of the variation in l_{max} can be attributed to size differences. Smaller animals with large surface to volume ratios have, in general, higher uptake capacities to balance a larger sail loss per unit weight. In addition, there is a general trend toward reduction of both l_{max} and K_m in freshwater organisms, which reflects the ability of these forms to limit salt loss by reducing permeability and renally reabsorbing ions.

From extensive examinations of salt absorption in hyperregulating decapods and amphipods, Shaw and Sutcliffe noted that affinity of the carrier system for Na^+ is correlated with Na^+ concentration of the animal's normal habitat. Animal inading dilute media have carrier system with a higher sodium felinity.

The correlation between net Cl^- uptake and excretion of HCO_3^{--} in *Astacus leptodactylus*. Dimox an inhibitor of carbonic anhydrase inhibited both Na^+ and Cl^- transport. Injection of Dimox produced a 71% reduction in unidirectional influx of Na^+ and a corresponding in hibition of net H^+ efflux but NH_4^+ efflux reduced 75%. Net chloride movement was inhibited due to a 200% increase in unidirectional efflux. In addition, a reduction in Cl uptake at low salinities, combined with an increased CO_2 content of hemolymph, was found in *Pachygrapsus crassipes* treated with Diamox (Burnett *et al.,* 1981).

Table 5.3 : Fluxes and Potentials in Hyperregulating Crustaceans.

Species	Ion	C_{in} (mmoles liter)	C_{out} (μ moles/ 100 g/hr	I_{in} (mV)	I_{out}	P.D.	Reference
Austroprotamoblus pallipes	Na	302	0.4	38	35.4	+4.1	Bryan (1960a) Shaw (1960)
Palaemonetes antennarius	Na	117	0.5	345	150	–33	Potts and Parry (1965)
Palaemonetes varians	Na	204	9	1800	1400	–32	Potts and Parry (1964b)
	Cl	250	11	2800	2400		
Gammarus duebeni	Na	250	10	875	995	–118	Lockwood and Andrews (1969) Lockwood (1961) Sutcliffe (1967b)
Sphaeroma rugicauda	Na	200	10	92400	13000	–24	Harris (1972)
Callinectes sapidus	Na	302	1.5-3	688	583	–2 to –4	Cameron (1978a)
	Cl	297	1.5-3	961	824		
Uca sp.	Na	399	39	1300	1640	+4.7	Baldwin and Kirschner (1976a)
	Cl	380	49	3280	2170		

Only a few measurements of the TEP have been made in hyperregulating crustaceans adapted to dilute media. Bryan (1960a,b) found that the crayfish *Astacus* sp. had a TEP of –4 to –6 mV (inside negative) in tap water. The potential of the isolated gill in 1% Ringers was –40 mV. These authors also note a smaller negative potential, –5 mV, when gills were bathed on both sides with Ringers solution. Confirmation of electrogenic Na uptake and Na^+/K^+ exchange awaits further examination of the effect of removal of K^+ and ouabain inhibition on TEP and ion flux.

METABOLIC AND ENZYMATIC CORRELATES OF TRANSPORT IN GILLS

Most studies of the relationship between metabolism (measured as oxygen consumption) and osmoregulation have been carried out on intact animals or in isolated gills. In general, crustaceans that osmoregulate increase their metabolic rate when in dilute media, while those that osmoconform reduce their metabolic rate under these conditions. The reasons for these changes in metabolic rate are both behavioral and physiological and are not well understood.

Isolated gills also show a higher metabolic rate when the animal from which they were taken is hyperregulating then when it is isosmotic, particularly when the animal has been acutely stressed by low salinity. Long-term acclimation to low salinity often involves a decrease in metabolic rate of isolated tissues.

It is likely that a part of the increased oxygen consumption is due to activity of the cation-activated ATPase. This enzyme, which is known to be involved in cation transport in tissues of many vertebrates, has also been studied in gill and antennal glands of isopods and decapods, and in larvae and adults of *Anemia salina*. The enzyme is similar to that found in transporting tissues of vertebrates, in that it is localized in the plasma membrane fraction of cells, has a requirement for Mg, Na, and a counter ion such as NH_4 or K, and is inhibited by quabain.

In hyperosmoregulators, such as *Callinectes sapidus*, *Carcinus maenas* (Mantel and Landesman, 1977), *Eriocheir sinensis* (Pequeux and Gilles, 1977), *Panopeus herbstii, Thalamita crenata* (Spencer *et al.*, 1979), crayfish (Horiuchi, 1977), and *Sphaeroma serratum* (Phillipot, 1972), activity of the enzyme in gills increases as the animal is acclimated to a dilute medium. Whether this increase is due to synthesis of more enzyme or to increased catalytic activity of a constant number of sites is uncertain in decapods, although induction of enzyme synthesis is known to occur in *A. salina*, as will be discussed below. In *C. sapidus*, osmolality of hemolymph reaches a new equilibrium within 12 hr after transfer of the animal to dilute salinity. Towle *et al.*, (1976) reported that ATPase activity also increases dramatically within a few hours after transfer. However, Neufeld *et al.* (1980) found that maximal activity was reached only after 1-2 weeks of acclimation, in conjunction with increased activity in the medial portions of the gill. These are the regions in which ultrastructure also shows increased development of salt-transporting cells. It is possible that both short-term activation and long-term synthesis of the enzyme are important in acclimation.

The relationship between enzyme activity and salinity of acclimation is not always direct, particularly in terrestrial and semiterrestrial crabs. Thus, in *Metopograpsus tukuhar* and *Cyclograpsus henshawi,* maximal enzyme activity is found at 75% and 25% sw, respeclivcly, although both animals can be acclimated to lower salinities. In *Uca pugilator*, enzyme activity increases with acclimation to dilute medium, but maximal activity is not attained until 100 hr after transfer to the new medium, whereas Ihe osmotic concentration is stabilized after only 10 hr. Acclimation to hyperosmotic media does not change the activity of the enzyme.

Osmoconforming crabs such as *Libinia emarginata* and *Calappa hepatica* possess ATPase activity at a level of about 10% that in osmoregulators. Activity of the enzyme increases slightly or not at all when Ihe crabs are acclimated to dilute seawater.

Both in osmoregulators and in semiterrestrial crabs, ATPase activity is greater in ihe more posterior than in the anterior gills, and there is a larger increase in activity of posterior gills on acclimation of the animal to dilute seawater. These differences are correlated with differences in fine structure of anterior and posterior gills, as mentioned. In the osmoconformer *L. emarginata,* there is no difference in enzyme activity between anterior and posterior gills. No studies have yet been carried out on fine structure of the gills in osmoconformers.

The Na-K ATPase best studied in Crustacea is that in the nauplius larvae of *Artemia salina*. This enzyme is found both in the larval salt gland or neck organ and in even greater amounts in the larval gut. The appearance of enzyme activity in developing embryos is correlated with the appearance of the Na-regulating system; the increase in activity is due to de novo synthesis of the enzyme on cellular membranes. The enzyme has recently been isolated and partly purified; it consists of large and small subunits similar to those of other ATPase preparations. Both subunits are glycoproteins, and there are isozymes of both subunits. The isozymes could represent developmental stages of the enzymes, or perhaps different enzymes present in the neck organ and gut.

Recently, a HCO_3-stimulated ATPase was found in mitochondrial and membrane fractions of the gills of *Uca minax*. This enzyme is inhibited by SCN; however, its activity remains constant regardless of the salinity to which the animals are acclimated. Although it is likely that this enzyme is a normal constituent of the cell membranes in gills of *U. minax*, its role in acclimation to dilute media is uncertain; it may play a role in $ClHCO_3$ exchange discussed above.

It is likely that the Na-K activated enzymes play a role in uptake of salt from dilute media, since activity does increase dramatically in gills of osmoregulators exposed to dilute media. However, changes in oilier enzymes and metabolic constituents also occur. Content of lipids is greater in posterior than in anterior gills of the hyprrregulator *Eriocheir sinensis*; phospholipids in particular increase in concentration when the animals are adapted to fresh water. Similar results were found in *Callinectes sapidus*, but not in the osmoconformer *Libinia emarginata*. Phospholipids are known to be necessary for the function of ATPase and it seems that increased phospholipid concentration is associated with increased activity of the enzyme. Studies on relationship of lipids to transport in crustaceans are just beginning; they should be continued with consideration of lipid functions as related to energy production, activity of enzymes, and possible changes in fluidity and permeability of the membranes themselves.

Changes in patterns of energy metabolism, other than those related to lipids, may also be important. Conte (1980) has found in *Artemia salina* that active transport of Cl is a major driving mechanism fors salt efflux in both larval and adult salt-transporting tissues. In the nauplius, levels of ATP in cytoplasm of the neck organ decrease is salinity increases; that is, as the transport system is activated. In addition, breakdown of stored glycogen is simulated. The presence of ouabain produces a decrease in ion transport and an inhibition of glycogen breakdown. Conic thus proposes that in the larval salt gland, energy for transport is provided by a facultative anaerobic shunt, in which NCO_3 enters the cell (perhaps in exchange for Cl) and is incorporated into C-4 dicarboxylic acids by means of the enzymes phosphoenolpyruvate carboxykinase (PEPCK) and malate dehydrogenase (MDH). These enzymes are abundant in cytoplasm of the transporting cells and have recently been purified. This model will provide a tool for further study and integration of metabolic changes with transport processes.

Ideally, the information on ultrastructure, kinetics, and enzymatic properties could be integrated into a theoretical model of ion transport by the gill. However, significance of the ultrastructural changes found after acclimation is not clear, since physiological information is still lacking to answer the following questions: What is the osmotic concentration of the fluid in the subcuticular spaces of the apical region? What are the gradients of ions in the basal and lateral channels? What barriers, if any, do the cuticle and basement membrane present to passage of ions and water?

Kinetic data do not yet permit precise location of the various pumps, although the most recent information on ouabain inhibition of transport suggests the involvement of a carrier at the apical surface adjacent to the medium. This contrasts with the situation in fishes.

What could be the function of ATPases in gills with respect to hyper and hypoosmotic regulation? In hyporegulators, the enzyme might function to transport Na out of the animal. Evidence in marine fishes indicates dial Na is extruded into the extracellular spaces and from there into the medium, probably through junctions between the cells. In *Fundulus heteroclitus,* the enzyme is localized in the basolateral regions of the chloride cells, where it can move Na from intracellular space into the hypertrophied basal tubular system, which is directly connected to the extracellular space. This is thought to provide an energy gradient to move Cl actively. In nearly all systems that have been studied the enzyme is localized on the basal and lateral surfaces of cells, regardless of the direction of Na transport across the epithelium as a whole. The enzyme is situated in such a position as to pump Na out of cells into the intercellular space.

In hyperregulators, the enzyme is probably involved is probably involved in uptake of sails. However, location of the ouabain-sensitive pump on the apical surface of the gill does not provide for a parallel model to that in fishes, and subcellular localization of the enzyme in a hyperregulating crustacean has not yet been achieved. Thus the mechanisms of Na and Cl entry and extrusion cannot yet be clearly defined in gills of crustaceans, although essential components of other transporting systems are known to be present.

THE GUT

1. Morphology and Fine Structure

Basically, the gut is a tube of epithelial cells specialized regionally for different functions. The foregut and hindgut are lined with cuticle that is shed at ecdysis. The midgut contains one to many diverticula or ceca that function in absorption, secretion, and storage. The multiple functions of the gut and the ultrastructural features necessarily present for its nutritive role make evaluation of its role in osmotic and ionic regulation difficult. Indeed, the two processes may be linked. As Malley (1977b) points out, secretion of digestive fluid, which is ultimately derived from hemolymph, of a particular osmotic or ionic composition may aid not only in the process of digestion, but also in regulation of the composition of hemolymph.

Fine structure of the foregut of *Cancer magister and Homarus amoricanus* has recently been studied by Mykles (1979). The epithelium consists of a single layer of cells, covered by cuticle, interdigitating on the apical and lateral surfaces. Mitochondria, rough endoplasmic reticulum, and apical junctions are present. In addition to its obvious function related to ingestion and trituration of food, there is some evidence that the anterior portions of the foregut may be involved in movements of ions and water.

The midgut in decapods consists of a tall columnar epithelium with a well-developed apical (luminal) border of microvilli. The cells resemble those of other transporting epithelia discussed above and contain, in addition, many vesicles that appear to fuse with the apical membranes. Anterior and posterior ceca are present; see Smith (1978) for a discussion of the origin and histology of these structures. The posterior cecum, erroneously called the hindgut cecum by many authors, was said to be involved in osmoregulation in grapsid crabs (Heeg and Cannone, 1966). Fine structure of this cecum has been described in *Pachygrapsus crassipes* by Mykles (1977); it shows cells that are typical of absorptive or secretory epithelia.

Recent studies on function of the anterior and posterior midgut ceca of *Cancer magister* by Holliday *et al.*, (1980) have shown that the ceca can produce a small volume of isosmotic fluid, both *in vivo* and *in vitro*, and they also release some enzymatic activity that may function in the midgut. The rate of fluid secretion does not increase when *C. magister* is exposed to a hypoosmotic medium, suggesting that it is not important in regulation of ions or water when the crab is in a dilute medium. However, these authors suggest that the ceca may be important in uptake of water at molt. Clearly, further study of this diverticulum is needed to reveal whether its function is digestive, osmoregulatory, or both. This question can be resolved only on the basis of both physiological and ultrastructural studies, both *in vivo* and *in vitro*. Of particular interest would be the ceca of hyporegulators.

Structure of the hindgut has been described in the isopods *Oniscus asellus* (Witkus et at., 1969), Dynamene bidentata (Holdich and Ratcliffe, 1970), and *Porcellio scaber* (Holdich and Mayes, 1975), as well as in *Homarus americanus* and *C. magister* (Mykles, 1979). Epithelial cells in these tissues resemble other ion and water-transporting cells. Clarification of the function of all of these regions of the gut will come only with further studies.

There have been only a few studies on fine structure of the gut in lower crustaceans. The alimentary epithelium of *Calanus holgolandicus* has been examined by Ong and Lake (1969). They found that the midgut epithelium of this copepod bears salt-transporting cells with a typical morphology. The work of Farmer (1980) described below provides physiological evidence that this tissue may be important in movement of ions and water. The midgut of *Artemia salina* also plays an important role in osmotic and ionic regulation, as discussed above. Fine structure of this tissue, as studied by Hootman and Conic (1974), reveals an extensive brush border on the luminal surface of the cells and a highly infolded basal margin. Enzymatic evidence for functions of these cells in transport of ions will be discussed below.

1. Mechanisms of Action

The gut has been characterized as a site of ion-dependent fluid absorption in both hypo- and hyperregulaling crustaceans. Both oral and anal drinking have been observed in a wide range of species.

Net water absorption in intermolt animals was demonstrated qualitatively by Croghan (1958d) in *Artemia salina,* by the progressive concentration of the impermeant dye, phenol red, in the gut as the brine shrimp absorb the external medium. *Geddes* (1975c) found that both oral and anal drinking contribute to osmoregulatory balance in *Parartemia zietziana.*

Quantitative estimates of the drinking rate may lie obtained from measurement of the concentration of radiolabeled markers in the gut. Data are available for a number of hyporegulators (Table 5.4). Dall and Smith (1977) provide a critical discussion of thc various substances that can be used as markers; they feel that complexes of EDTA (diaminoethane tetra-acetic acid) labeled with ^{15}Cr or ^{58}Co currently offer the optimal results. Since urinary volumes are low in most hyporegulators, the drinking rate has been used as a first approximation of osmotic water permeability in these animals.

Croghan (1958d) suggested that water uptake in *A. salina*, as in the teleosis, is linked to active sail absorption by the gut. His analysis of gut fluid content indicated that fluid absorption is against an osmotic gradient of up to 600 MOsm. These findings were questioned by Geddes (1975), who demonstrated that there is net fluid absorption by the gut of *P. zietziana*, but that gut fluid is isosmotic to hemolymph. More recently, isosmotic fluid transport has been shown in isolated perfused preparations of *Macrobrachium rosenbergiii*, a hyperosmotic regulator (Ahearn *et al.*, 1977; Ahearn and Tornuist, 1977; Ahearn, 1978a,b; 1980), and in *Homarus americanus,* an isosmotic animal (Mykles, 1981). These authors agree that water absorption is solute dependent, but they differ in their findings of the individual ionic requirements. Mykles

(1981) found that net fluid absorption in the perfused lobster gut is Na but not Cl, dependent. He reported a small potential, 0.26 mV, serosal positive, which is ouabain insensitive. Farmer (1980) also noted the presence of Na-dependent fluid absorption in the gut of the copepod *Acartia tonsa.* She found that amiloride inhibits the accumulation of labeled dextran in these zooplankters. Ahearn has examined fluid transport in both intact whole animals and in isolated preparations of midgut of *M. rosenbergii.* In the intact animal, Na and Cl concentrations of the gut are isosmotic to hemolymph, but K is concentrated in the gut. In isolated preparations perfused with Ringer's solution, net absorption of Na, Cl, and water and net secretion of Koccur. The preparation exhibits a ouabain-sensitive potential, 1.63 mV, serosal positive. Ahearn postulated that Na and Cl entry into the epithelium across the mucosal surface is linked by a common carrier that is also Ca dependent, and that electrogenic Na/K exchange across the basolateral surface generates the positive potential and facilitates net Na transport. Net uptake of water from the lumen to blood depends on the presence of Na and Cl in the lumen.

Table 5.4 : TEP of Crustaceans Hyporegulating in Seawater.

Species	TEP(mV)	Reference
Artemia salina	+23	Smith (1969a)
Palaemonetes varians	–1:1 to –40	Potts ond Parry (1964b)
Palaemonetes pugio	+3.3	Lucu (1977b)
Pachygrapsus crassipes	–1.8	Rudy (1966)
Carcinus maenas	–1.5	Mantel (1967)
Gammarus duebeni	–4.5	Lockwood and Andrews (1969)
Sphaeroma rugicauda	–1.5	Harris (1972)
Sphaeroma serratum	–3.9	Harris (1972)
Acartia tonsa	–10	Farmer (1978)
Uca sp.	–0.07	Baldwin and Kirschner (1976a)

In *Panulirus* argus, water movements and salt concentrations in various parts of the gut were measured by Malley (1977b). This animal is thought to have a low drinking rate, 1.5 ml-kg^{-1}.day^{-1}, compared to another isosmotic animal, *Homarus americanus*. However, the substance used by Malley to study drinking rate, polyethylene glycol, is known to be distasteful to lobsters, and thus the rate of drinking under normal circumstances may have been underestimated. Fluid in the proventriculus of *P. argus* is a mixture of imbibed seawater and secreted digestive juice. It is possible that some isosmotic reabsorption of salt and water occurs in the proventriculus: In the hindgut of *P. argus*, Ca and Mg are concentrated, probably by reabsorption of monovalent ions and water, and possibly by additional direct secretion. The net function of the gut in this animal seems to be uptake and absorption of small amounts of Na, K, Cl, SO_4 and water, and a route for net loss of Ca. Similar processes occur in *P. longipes*.

The intestine of some terrestrial and semiterrestrial crabs has a Na-activated ATPase that might be important in uptake or excretion of salt. In the terrestrial coconut card *Birgus latro*, enzyme activity in the gut is higher in crabs given fresh water to drink them in those given access to seawater (Towle, 1981).

The gut of hyporegulators is also thought to be the site of extrusion of salts. In species of *Uca*, fluid in the foregut is more concentrated than blood, urine, or fluid in the gill chamber when crabs are hyporegulating in 100% seawater. Dall (1967) found that Na and Cl levels in the proventriculus of *Metapenaeus bennettae*; are higher than those of the hemolymph. One-tenth of the total Na efflux and one-third of the Cl efflux from the animal are from abdominal region rather than cephalothorax injection of salt load substantially increases abdominal efflux of both ions while efflux from cephalothorax is depressed.

Table 5.5 : Crustacean Drinking Rates in Seawater.

Species	Drinking rate (µl/100 g/hr)	Reference
Artemia salina	2,000	Smith (1969a)
Uca pugilator	600	Hannan and Evans (1973)
Penaeus duorarum	1,730	Hannan and Evans (1973)
Metapenaeus bennettae	700	Dall (1967)
Metopograpsus gracilipes	600-700	Dall (1967)
Macrophthalmus crassipes	600-700	Dall (1967)
Acartia tonsa	13,800	Farmer (1980)

In drinking crustaceans it is hormonal dependent at the time of ecdysis changes in water uptake related to molting water uptake is not related to salinity.

Table 5.6 : Effects of Temperature and Season on Osmotic and Ionic Regulation

A. Osmotic regulation		**(All values in mOsm/kg**			
Ligia oceanica Medium		(Todd, 1963)			
		Hemolymph concentration (mOsm)			
		Summer		Winter	
		5°C	15°C	5°C	15°C
100% sw		1190	1090	1295	1110
75% sw		1140	1065	1235	1095
50%		805	905	1165	960

Parartemia zietziana					(Geddes, 1975b)
Medium (4540 mOsm) Temp. (°C)	5	10	15	20	30
Hemolymph (mOsm)	450	415	395	400	465

Penaeus aztecus			(Williams, 1960)
Medium (mOsm)	Hemolymph (mOsm)		
	8°C	16C	28°C
415	610	590	650
655	750	720	795
790	880	785	840
935	940	845	815

B.Ionic regulation		(All values in mmoles/liter
Sphaeroma serratum		(Charmantier, 1975)
Winter animals at field temp. of 8.5°C		
Acclimation temp.	85°C	20°C
$Na_0 = 300$	Na_1 410	$Na_1 = 365$
Summer animals at field temp. of 25°C		
Acclimation temp.	85°C	20°C
$Na_0 = 300$	Na_1 320	$Na_1 = 365$

Paranephrops zealandicus				(Wong and Freeman, 1976b)
	Na	Cl	K	mOsm
Summer	255	260	4.4	550
Winter	270	280	4.2	580

Callinectes sapidus						(Engel *et al.*, 1974)
		15%			100% sw	
	12°C	20°C	28°C	12°C	20°C	28°C
Na	380	280	280	520	510	610
K	9	8	9	16	17	15

(Contd...)

Mg	7.5	6	9.5	20	14	18
Cl	400	300	320	540	550	580

Hemigrapsus nudus (Dehnel, 1962, 1966; Dehnel and Stone, 1964)

Summer: field salinity 35% sw, temp. 20°C

Winter: field salinity 75% sw, temp 5°C

All animals acclimated and measured at 15°C.

Medium	Summer	Winter	Summer	Winter
	Hemolymph		Urine	
Osmoliility (mOsm)				
110	645	610	650	415
230	715	765	700	770
920	1115	1050	1115	840
1150	1205	1325	1230	1020
Chloride (mmoles)				
60	240	315	280	420
125	290	380	410	500
495	500	505	680	730
620	560	605	780	840

A clear seasonal effect is seen in regulation both of osmolality and Cl in *Hemigrapsus nudus.* All animals were acclimated to 15°C. Summer animals maintained osmolality of hemolymph within a narrower range than did winter animals, but winter animals produced a dilute urine in all salinities. Concentration of Cl in both blood and urine were higher in winter than in summer animals. These differences undoubtedly arise from different acclimation history of the animals in terms of field salinity and temperature.

Interesting as these seasonal and temperature effects are, there is presently no explanation of mechanisms that can account for them. Obviously, in any laboratory experiment, the final state of acclimation will depend on the seasonal and field history of the animal. Many of the conflicts among data from various laboratories can be ascribed to differences in field history of the animals. However, a fuller understanding of the mechanisms involved in osmotic and ionic regulation is necessary before the influences of season and temperature can be explained.

REFERENCES

Bousfield, (1969) at the northern limit of its distribution. II. Temperature-salinity tolerances. *Estuarine Coastal Mar. Sci. 6*, 225-230.

By. Lockwood. A.PM. (1968). Aspect of Physiology of Crustacea. Chapter Excretion of Crustacean.

E.L. Jordan, The Biology of Invertebrates, McGraw Hill Book Company, (1952).

Geddes. M. C. (1973). Salinity tolerance and osmotic and ionic regulation in *Branchinell, australiensis* and *B. compacta* (Crustacea: Anostraca). *Comp. Biochem.* Physiol. 45A 559-569.

Geddes, M. C. (1975a). Studies on an Australian brine shrimp *Parartemia zietziana Sayce* (Crustacea: Anostraca). I. Salinity tolerance. *Comp. Biochem. Physiol.* 51A, 553-559

Geddes, M. C. (1975b). Studies on an Australian brine shrimp *Parartemia zietziana Sayce*

(Crustacea: Anostraca). II. Osmotic and ionic regulation. Comp. Biochem. Physiol. 51A, 561-571.

Geddes, M. C. (1975c). Studies on an Australian brine shrimp *Parartemia zietziana Sayce* (Crustacea: Anostraca). III. The mechanisms of osmotic and ionic regulation. *Comp Biochem. Physiol.* 51A, 573-57H.

Gifford, C. A., (1962). Some aspects of osmotic and ionic regulation in the blue crab *Callinrctcs sapidus*, and the ghost crab *Ocypode albicans. Bull. Inst. Mar. Sci.* 8, 97-125

Gilles, R. (1970). Osmoregulation in the stenohaline crab *Libinia emarginata Leach. Arch. Internal. Physiol. Biochim.* 78, 91-99.

Greenaway, P., and MacMillen, R.E. (1978). Salt and water balance in the terresterial phase of the inland crab *Holthuisana (Austrothelphusa) transversa* Martens (Parathelphusoidea; Sundathelphusidae). *Physiol. Zool.* 51, 217-229.

Gross, W. J. (1955), Aspects of osmotic regulation in crabs showing the terrestrial habit. *Am. Nat.* 89, 205-222.

Gross, W. J. (1957). An analysis of response to osmotic stress in selected decapod Crustacea. *Biol. Bull. (Woods Hole, Mass.)* 112, 43-62.

Gross, W. J. (1958). Potassium ad sodium regulation in an intertidal crab. *Biol. Bull.* (*Woods Hole, Mass.*) 114, 334-347.

Gross, W. J. (1959). The effect of osmotic stress on the ionic exchange of a shore crab. *Biol. Bull.* (*Woods Hole, Mass.*) 116, 248-257).

Gross, W. J. (1961). Osmotic tolerance and regulation in crabs from a hyper-saline lagoon. *Biol. Bull.* (*Woods Hole, Mass.*) 121, 290-301.

Gross, W. J. (1964a). Water balance in anomuran land crabs on a dry atoll. *Biol. Bull.* (*Woods Hole. Mass.*) 126, 54-68.

Gross, W. J. (1964b). Trends in water and salt regulation among aquatic and amphibious crabs. *Biol. Bull,* (*Woods Hole, Mass.*) 127, 447-466.

Gross, W. J. (1967). Glucose absorption from the urinary bladder of a crab. *Comp. Biochem. Physiol.* 20, 313-317.

Gross, W. J. and R. L. Capen (1966). Some functions of the urinary bladder in a crab. *Biol. Bull.* (*Woods Hole, Mass.*) 131, 272-291.

Gross, W. J. and Holland, P. V. (1960). Water and ionic regulation in a terrestrial hermit crab. *Physiol. Zool.* 33, 21-28.

G.S. Carte, A Zoology of the Invertebrates, Reinhold, Book Corportation, (1961).

Graham, Functional Anatomy of Invertebrates.

Harris, R.R. (1972). Aspects of sodium regulation in brackish-water and a marine species of the isopod genus *sphaeroma. Mar. Biol.* 12, 18-27.

Harris, R.R. (1975). Urine production rate and urinary sodium loss in the fresh water crab *Potamon edulis. J. Comp. Physiol.* 96, 143-153.

Harris, R.R. (1977). Urine production rate and water balance in the terrestrial crabs *Gecarcinus lateralis* and *Cardisoma guanhumi. J. exp. Biol.* 68, 57-67.

Heeg, R.R. and Micallef, H (1971). Osmotic and ionic regulation in *Potamon edulis,* a fresh water crab from *Malta. Comp. Biochem. Physiol.* 38A, 769-129.

Heit, M. and Fingerman, M (1975). The role of an eyestalk hormone in the sodium concentration of the blood of the fiddler crab, *Uca pugilator. Comp. Biochem. Physiol.* 50A, 277-280.

Hopper, B. T. (1978). Changes in blood serum prolein levels during the molting cycle of the lobster, *Homarus gammarus* (L) *J. Exp. Mar. Biol. Ecol.* 28, 293-296.

Hill, B. J. (1971). Osmoregulation by an estuarine and A marine species of *Upogebia* (Anomura, Crustacea). *Zoologica Africana* 6, 229-236.

Holdich, D. M., and Mayes, K. R. (1975). A fine-structural re-examination of the so-called midgut of the isopod *Porcellio*. Crustaccana 29, 186-192.

Holdich, U., and Ratclilfe. N. (1970). A light and electron microscopic study of the hindgut of the herbivorous isopod *Dynamene bidentata* (Crustacea: Pericarida). Z. Zelltorsch. 111, 209-227.

Holliday, C. W. (1978a). Aspects of antennal gland function in the crab. Cancer *magister* (Dana). Ph.D. Dissertation, University of Oregon Eugene, Oregon.

Holliday, C. W. (1978b). Glucose absorption by the bladder of the crab, *Cancer rnasisler* (Dana). *Comp. Biochem. Physiol.* 61A, 73-79.

Holliday, C. W. (1980). Magnesium transport by the urinary bladder of the crab, *Cancer Magister. J. Exp. Biol.* 85, 187-201.

Holliday, C. W., Mykles, D. L., Terwilliger, R. C., and Oangott, L.J. (1980). Fluid secretion by the midgul caeca of the crab, *Cancer magister. Comp. Biochem. Physiol* 67A, 259-263.

Holm-lensen, I. (1948). Osmotic regulation in *Daphnia magna* under physiological conditions and in the presence of heavy metals. *Kgl. Danske Vikd Selsk, Bid. Medd. Kbh.* 20, 1-64.

Hootman, S. R., and Conte, F.P. (1974). Fine structure and function of the alimentary epithelium in *Artemia salina Nauplii. Cell Uss. Rex.* 155, 423-236.

Hootman, S. R., Harris, P. J. and Conte, F.P. (1972). Surface specialization of the larval salt gland in *Artemia salina* Nauplii. *J. Comp. Physiol* 79, 97-104.

Horiuchi. S. (1977), Characterization of gill sodium-potassium ATPase in the fresh water crayfish *Procambarus clarkii, Comp. Biochem. Physiol.* 56, 135-138.

Horne, F. R. (1966). Some aspects of ionic regulation in the tadpole shrimp *triops longicaudatus. Comp. Biochem. Physiol*, 19, 313-316.

Hunter, K.E., and Rudy, P.P. Jr. (1975). Osmotic and ionic regulation in the Dungeness crab, *Cancer magister* Dana. *comp. Biochem. Physiol.* 51A, 439-447.

Hyman, (1967), The Arthopoda, McGraw-Hills, New York.

Jones, I.I. (1941). Osmotic regulation in several crabs of the Pacific coast of North America. *J. Cell. Comp. Physiol.* 18, 79-92.

Jones, M.B. (1972). Osmoregulation in the *Jaera albifrons* group of species. (Isopoda, Asellota). J. Mar. Biol. Assoc, U.K., 52, 419-427.

Jones, M.B. (1974). Survival and oxygen consumption in various alinities of three species of idotea (Crustacea, Isopoda) from different habitats. *Comp. Biochem. Physiol.* 48A, 501-506.

Jones, M.B. (1975). Synergistic effects of salinity, temperature and heavy metals on mortality and osmoregulation in marine and estuarine isopods (Crustacea). *Mar. Biol.* 30, 13-20.

Kamemoto, F.I. (1961). The effects of eserine on sodium regulation in crayfish. *comp. Biochem. Physiol.* 3, 297-303.

Kamemoto, F.I. (1976). Neuroendocrinology of osmoregulation in decapod crustacea. *Am. Zool.* 16, 141-150.

Kamemoto, F.I. and Kato, K.N. (1969). The osmotic and chloride regulative capacities of five Hawaiian decapod crustaceans. *Pacif. Sci.* 23, 232-237.

Kamemoto, F.I. Kato K.N. and Tucker, L.E. (1966). Neurosecretion and salt and water balance in the annelida and crustacea. *Am. Zool.* 6, 213-219.

Kamemoto, F.I. Keister, S.M. and Spalding, A.E. (1962). Cholinesterase activities and sodium movement in the crayfish kidney. *comp. Biochem. Physiol* 7, 81-87.

Kamemoto, F. I., and Ono, J. K. (1968). Urine flow determinations by continuous collection in the crayfish *Procamhacus clarkii. Comp. Biochem. Physiol.* 27, 851-857.

Kamemoto, F. I., and Tullis, R. E. (1972). Hydromineral regulation in decapod Crustacea. *Gen. Comp. Endocrinol.* 3, 299-307.

King, E. N. (1966). Oxidative activity of crab gill mitochondria as a function of osmotic concentration. *Conip. Biochem. Physiol.* 17, 245-258.

King, E. N., and Schofleniels (1969). *In vitro* preparation of crab gill for use in ion transport studies. *Arch. Inst. Physiol. Biochim.* 77, 105-111.

Kinne O. (l966). Physiological aspects of animal life in estuaries with special reference to salinity. *Neth. J. Sea Res.* 3, 222-244.

Kirschner, L. B. (1979). Conlrol mechanisms in crustaceans and fishes. *In "Mechanisms* of Osmoregulation in Animals" (R. Gilles, ed.). pp. 157-222. Wiley-Interscience, New York.

Kirschner, L.B. Greenwald, L. and Kerstetter, T.H. (1973). Effect of amiloride on sodium transfer across body surfaces of fresh water animals. *Am. J. Physiol* 224, 832-837.

Kleinholz, L. H. (1976). crustacean neurosecretory hormones and physiological specificity. *Am. Zool.* 16, 151-166.

Koch, H.J. (1934). Essai d'interpretation de la soi-disant reduction vitale des sels d'argent par certaines organes d'Arthropodes. *Ann. Soc. Sci. Med. Nat. Brux. Ser. B.* 54, 346-361.

Koch, H.J. and Evans, J. (1956a). On the influence of lithium on the uptake of sodium and potassium by the crab *Lriocheir sinensis* (M. Edw). *Meded. Vlaamse Acad Kl. Wet.* 18, 3-9.

Koch, H.J. and Evans, J. (1556b). On the absorption of sodium from dilute solutions by the crab *Eriocheir sinensis* (M. Edw). *Meded. Vlaamse Acad. Kl. Wet.* 18(7), 3-15.

Koch, H.J. and Evans, J. (1956c). Influence of a basic dye, thionine, on the absorption of sodium by the crab *Eriocheri sinensis* (M. Edw). *Meded. Vlaamse Acad Dk. Wet.*, 18(8), 3-11.

Koch, H.J. and Schicks, E. (1953). The active absorption of ions by the isolated gills of the crab *Eriocheri sinensis* (M. Edw). *Meded. Vlaamse Acad Dk. Wet.*, 16, 3-16.

Kormanik, G.A. and Cameron, J.N. (1981). Ammonia excretion in the sea water blue crab (*Callmectes sapidus*) occurs by diffusion and Na^+/NH_4 exohange. *J. Comp. Physiol.* 141, 457-462.

Lockwood, A. P. M. (1959). The osmotic and ionic regulation of *Asellus aquaticus* (L). J. Exp. Biol. 36, 546-554.

Lockwood, A. P. M. (1960). Some effects of temperature and concentration of the medium on the ionic regulation of the isopod *Asellus aquaticus* (L.). J. *Exp. Biol.* 37, 614-630.

Lockwood, A. P. M. (1961). The urine of *Gammarus duebeni* and *C. pulex.* J. Exp. *Biol.* 38, 647-658.

Lockwood, A. P. M. (1964a). "Animal Body Fluids and Their Regulation." Harvard University Press, Cambridge, Massachusetts.

Lockwood, A. P. M. (1964b). Activation of the sodium uptake system at high blood concentrations in the amphipod *Gammarus duebeni. J. Exp. Biol.* 41, 447-458.

Lockwood, A. P. M. (1967). "Aspects of the Physiology of Crustacea." Freeman, San Francisco, California.

Lockwood, A. P. M. (1977), Transport and osmoregulation in Crustacea, In "Transport of Ions and Water in Animals" (B. L. Gupta, J. L, Oschman, R. B. Morcton, and B. I. Wall, eds), pp. 673-707. Academic Press, London.

Lockwood, A.P. M. and Andrews, W. R. H. (1969), Active transport and sodium fluxes at molt in the amphipod *Gammarus duebeni. J. Exp. Biol.* 51, 591-605.

Lockwood, A. P. M., and Croghan, P. C. (1957). The chloride regulation of the brackish and fresh water races of *Mesidotea entomon* (L.) *J. Exp. Biol.* 34, 253-258.

Lockwood, A. P. M., Croghan, P.C. and Sutclifle, D. W. (1976). Sodium regulation and adaptation to dilute media in crustacea as exemplified by the isopod *Mesidotea entomon* and the amphipod *Gammarus duebeni: In* "Perspectives in Experimental Biology," J. Zoology (P. Daview, 'ed.), 93-106. Pergamon, Oxford.

Lockwood, A. P. M., and Inman, C. B. E. (1973a), Changes in the apparent permeability to water at moult in the amphipod *Gammarus duebeni* and the isopod *idotea linearis. Comp. Biochem. Physiol.* 44A, 493-952.

Lockwood, A. P. M., and Inman, C. B. E. (1973b)1. Water uptake and loss in relation to the salinity of the medium in the amphipod crustacean *Gammarus duebeni. J. Exp. Biol.* 58, 149-163.

Lucu, C. (1969), Magnesium regulation in decapod crab *Carcinus mediterraneus. that Jugoslav.* 5, 219-221.

Lucu, C. (1977a). Sodium balance and salinity tolerance of the nysid *Leptomysis mediterranea. In* "Physiology and Behaviour of Marrine Organisms" (D.S. Mclusky and A.J. Berry, eds), pp. 95-103. Proc. 12th Europ. Sympos. Mar. Biol. Stirling, Scotland.

Lucu, C. (1977b). Sodium kinetics in the shrimp, *Palaemonetes pugio.* II. Sodium fluxes and electro-chemical potentials. *J. comp. Physiol.* 115, 207-214.

Lucu, C. Roesijadi, G. and Anderson, J.W. (1977). Sodium kinetics in the shrimp, *Palaemonetes pugio.* I. Steady-state and non-steady state experminents. *J. Comp. Physiol,* 115, 195-206.

Lucu, C. Siebers, D. and Sperling, K.R. (1973). Comparison of osmoregulation between Adriatic and North Sea *Carcinus. Mar. Biol.* 22, 85-95.

Lutz, O.P. (1969). Salt and water balance in the West African fresh water/land crab *Sudanonautes africanus africanus* and the effects of desiccation. *Comp. Biochem. Physiol.* 30, 469-480.

Lynch, M.P. and Webb, K.L. (1973). Variations in serum constituents of the blue crab, *Callinectes sapidus:* total serum protein. *Comp. Biochem. Physiol.* 44A, 1237-1249.

Lynch, M.P. Webb, K.L. and Van Engel, W. A. (1973). Variations in serum constituents of the blue crab, *Callinectes sapidus:* chloride and osmotic concentration. *Comp. Biochem. Physiol* 44A, 719-734.

McFarland, W. N., and Lee, B. D. (1963). Osmotic and ionic concentrations of shrimps of the Texas coast. *Bull. Mar. Sci.* 13, 391-416.

McLusky, D. S. (1967). Some effects of salinity on the survival, molting and growth of *Corophium volutator* (Amphipoda). *J. Mar. Biol. Assoc. U.K.* 47, 607-617.

McLusky, D. S., and Heard, V. E. J. (1971). Some effects of salinity on the mysid Praunus *flexuosus. J. Mar. Biol. Assoc.* U.K. 51, 709-715.

Mackay, W. C., and Prosser, C. L. (1970). Ionic and osmotic regulation in the king crab and two other north Pacific crustaceans. *Comp. Biochem. Physiol.* 34, 273-280.

MacMillen, R. F. and Greenaway, P. (1978). Adjustments of energy and water metabolism to drought in an Australian arid-zone crab. *Physiol. Zool.* 51, 230-240.

Maetz, J. (1974). Aspects of adaptation to hypo-osmotic and hyperosmotic environments. *In* "Biochemical and Biophysical Perspectives in Marine Biology" (D. C. Malins and J. R. Sargent, eds.), pp. 1-167. Academic Press, New York.

Maissiat, I. (1972). Contribution A I'ctude du role endocrine de la glande antennairc de I'Oniscoide *Porcedio dilatalus* Brandt. Efffet de son ablation sur l'equilibrc hydro-mineral. C.R. Soc. Biol. 166, 916-919.

Malley, D. F . (1977a). Salt and water balance of the spiny lobster *Panulirus argus:* the role of the antennal glands. *J. Exp. Biol.* 70, 221-230.

Malley, D. F. (1977b). Salt and water balance of the spiny lobster *Panulirus argus*: the role of the gut. *J. Exp. Biol.* 70, 231-245.

Mantel, L. H. and Landesman, J. (1977). Osmotic regulation and Na^+ K-activated ATPase in the green crab, *Carcinus meanas,* and the spider crab *Libinia emarginata. Biol. Bull. (Woods Hole, Mass.)* 153, 437-438.

Marshall, A.T. and Cheung, W.W.T. (1973). Studies on water and ion transport in hompoteran insects: ultrastructure and cytochemistry of the cicadoid and cercopoid hindugt. *Tissue Cell* 5, 671-678.

Martin, J.I.M. (1978). Le metabolisme hydrique chez *Cancer irroratus* (Crustace decapode). *Lethys* 8, 193-196.

Mason. C.A. (1970). Function of the perit ardial sacs during the molt cycle in the land crab *Gecarcinus lateralis. J. Exp. Zool.* 174, 381-390.

Milne, D.J. and Ellis, R.A. (1973). The effect of salinity acclimation on the ultrastructure of the gills of *Gammarus oceanicus* (Segerstrale, 1947). (Crustacea: Amphipoda). *Z. Zellforsch.* 139, 311-318.

Morse, H.C., Haris, P.J. and Dornfeld, E.J. (1970). *Pacifastacus leniusculus:* Fine structure of arthrobranch with reference to active ion uptake. *trans. Am. cicrosc. Soc.* 89, 12-27.

Muramoto, A. (1981). Effects of eyestalk extracts and ecdysterone on water intake through the anus of the crayfish. *Comp. Biochem. Physiol.* 69A, 197-203.

Sutcliffe, D. W. (1968). Sodium regulation and adaptation to fresh water in gammarid crustaceans. *J. exp. Biol.* 359-380.

Sutcliffe, D. W. (197la). Regulation of water and some ions in gammarids (Amphipoda). I. *Gammarus duebeni* Lilljeborg from brackish water and fresh water. J. Exp. Biol 55, 325-344.

Sutcliffe, D. W. (1971b). Regulation of water and some ions in gammarids (Amphipoda). II. *Gammarus pulex (L.) J. Exp. Biol.* 55, 34.5-3.5.5.

Sutcliffe, D. W. (1971c). Regulation of water and some ions in gammarids (Amphipods). III. Three euryhaline species. *J. Exp. Biol.* 55, 357-369.

Sutcliffe, D. W. (1974). Sodium regulation and adaptation to fresh water in the isopod genus Asellus. *J. Exp. Biol.* 61, 719-736.

Sutcliffe, D. W. (1975). Sodium uptake and loss in *Crangonyx pseudogracilis* (Amphipoda) and some other crustaceans. *Comp. Biochem. Physiol.* 52A, 255-257.

Sutcliffe D. W. (1978). Water chemistry and osmoregulation in some arlhropods, especially

Malacostraca. In "Fresh-water Biological Association—Fortysixth Annual Report." pp. 57-69. Freshwater Biological Association of the U.K.

Sutcliffe, D. W. and Shaw, J. (1967). The sodium balance mechanism in the fresh-water amphipod *Gammarus lacustris* Sars. *J. Exp. Biol.* 519-5211.

Sutcliffe, D. W., and Shaw, J. (1968). Sodium regulation in the amphipod *Gammarus duebeni* Lilljeborg from freshwater localities in Ireland *J. Exp. Biol.* 48, 339-358.

Talbot, P. Clark W. H. and Lawrence, AlL. (1972). Light and electron microscopic studies on osmoregulalory tissue in the developing brown shrimp *Penaeus aztecus. Tissue Cell* 4, 271-286.

Taylor, E. W. Butler, P. J. and Al-Wassia, A. (1977). The effect of a decrease in salinity on respiration, osmoregulation and activity in the shore crab, *Carcinus maenas* (L.) at different acclimation temperatures. *J. Comp. Physiol.* 119, 155-170.

Telford, M. (1968a). The identificaction and measurement of sugars in the blood of three species of Atlantic crabs. *Biol. Bull.* (*Woods Hole, Mass.*) 135, 574-584.

Telford, M. (1968b). Changes in blood sugar composition during the molt cycle of the lobster *Homarus americanus. Comp. Biochem. Physiol* 26, 917-926.

Telford, M. (1968c). The effects of stress on blood sugar composition of the lobster *Homarus americanus. Can J. Zool.* 46, 819-826.

Theede H. (1969). Einige neue Aspekte beider osmoregulation von *Carcinus maenas. Mar. Biol.* 2, 114-120.

Thompson L.C. and Pritchard, A. W. (1969). Osmoregulatory capacities of *Callianassa and Upogebia* (Crustacea: Thalassinidea). *Biol. Bull.* (*Woods Hole, Mass.*) 136, 114-129.

Thuet, P. (1978). Etude des flux de diffusion de I'eauen fonction de la concentration du milieu exterieur chez I'isopode *Sphaeroma serratum* (Fabricius). *Arch. Int Physiol. Biochim.* 86, 289-316.

Todd. M.E. (1963). Osmoregulation in *Ligia oceanica* and *Idotea granulosa. J. exp. Biol* 40, 381-392.

Towle, D.W. (1981) Transport-related ATPases as probes of tissue function in three terrestrial crabs of Palau. *J. Exp. Zool.* 218, 89-95.

Towle, D.W. Palmer, C.E. and Harris, D.L. (1976). Role of gill sodium ion and potassium ion dependent ATPase in acclimation of blue crabs *Callinectes sapidus* to low salinity. *J. Exp. Zool.* 196, 315-321.

Tullis, R.E. and Kamemoto, f.I. (1974). separation and biological effects of CNS factors affecting water balance in the decapod crustacean *Thalamita crenata. Gen. Comp. Endocrinol.* 23, 19-28.

Tyson, G. E. (1968). The fine structure of the maxillary gland of the brine shrimp, *Artemia salina*: the end sac. *Z. Zeilforsch.* 86, 129-138.

Tyson, G. E. (1969). The fine structure of the maxillary gland of the brine shrimp, *Artemia salina*; The efferent duct. *Z. Zeiltmxh.* 93, 151-163.

Van den Bosch, P. (1976). Neurosecretion et regulation hydroelectrolytique chez *Artemia salina.* Experientia 32, 228-229,

Venkatamariah, A., Lakshmi, G. and Gunter, G. (1974). Studies on the effects of salinity and temperature on the commercial shrimp, *Penaeus aztecus lves*, with special regard to survival limits, growth, o\ygen consumption and ionic regulation. *U.S. Army Engineer waterways experiment station-contract report H-74-2.*

Vernbcrg, F. J. and Vernberg, W. B. (1970). "The Animal in the Environment." Holl, Rinehart and Winston, New York.

Wall, B. J. (1977). Fluid transport in the cockroach rectum. In "Transport of Ions and Water in Animals" (B. L. Gupta *et al.*, eds), pp. 599-612. Academic Press, London.

Warner, G. P. (1977). "The Biology of Crab," pp. 149-157. Elek Science, London.

Weber, R. E. and Spaargaren, D, H. (1970). On the influence of temperature on the osmregulation of *Crangon crangon* and its significance under estuarine conditins. *Neth. J. Sea Res.* 5, 108-120.

Werntz H.O. (1963). Osmotic regulation in marine and freshwater gammarids (Amphipoda). *Biol. Bull.* (*Woods Hole, Mass.*) 124, 225-239.

White, K.N. and Walker, G. (1981). The barnacle excretory organ. *J. Mar. Biol. Ass.* U.K. 61, 529-547.

Whitney, J.O. (1974). The effect of external salinity upon lipid synthesis in the blue crab *Callinectes sapidus* Rathbun and in the spider crab *Libinia emarginata* Leech. *Comp. Biochem. Physio.* 49A, 433-440.

Wong I.M. and Freeman R.F.H. (1976a). Haemolymph concentrallons of two species of New Zealand freshwater crayfish in relation to the concentration of their external media. *Comp. Biochem. Physiol.* 55A, 13-16.

Wong. T.M. and Freeman R.F.H. (1976b). Seasonal and thermal effects on the concentration of the haemolymph in the New Zealand freshwater crayfish *Paranephrops zealandicus* White. *Comp. Biochem. Physiol.* 55A, 17-22.

Young A.M. (1978). Desiccation tolerances for three hermit crab species *Clibanarius vittatus* (Bosc), *Pagarus pollicaris* Say and *P. longicarpus* Say (Decapoda, Anomura) in the North inlet estuary, South Carolina, U.S.A. *Est. Coast. Mar. Sci.* 6, 117-122.

Young A.M. (1979). Osmoregulation in three hermit crab species, *Clibanarius vittatus* (Bosc), *Pagurus longic arpus* Say and *P. pollicaris* Say. *Comp. Biochem. Physiol.* 63A 377-382.

Zanders, I.P. (1978). Ionic regulation in the mangrove crab *Coniopsis cruentata. Comp Biochem. Physiol* 60, 293-302.

6

MUSCULATURE

The muscle fibres of custaceans are multinucleated and striated. In general they run from the origin of the cuticle to the insertion on a apodeme (in tucking of the cuticle) of the next segment.

TYPES OF MUSCLE FIBRES

1. *Slow fibres* : they have long sarcomeres and produce slow powerful contractions that can be maintaineed over a considerable length of time *e.g.*, closer muscle of crusher claw of *Homarus*.
2. *Fast fibres* : They have short sarcomeres and produce rapid, easily fatigued contractions *e.g.*, abdominal muscle fibres during tail flip escape response.

Muscle work by sliding filament mechanismwhich causes shortening of muscles and tension development. Tension development may due to development of linkage between actin, myosin and paramyosin.the stage of tension declines following cessation of stimulus at the rate which depends on the presence of chemical agent called as relaxing or plasticizing substance. The muscles are innervated by 1-8 axons. six of which are excitatory and 1 or 2 inhibitory. The excitatory axons have small diameter and show little fatigue during sustained nerve stimulus the axons innervating fast fibres are large in diameter, fatigue quickly upon stimulation and have a targe component synaptic vessel. the function of the muscles depends upon the properties of fibres, it could be made mostly of fast fibres or mostly of slow fibres or a mixture of both. repetitive stimulus causes tetanic action tropomyosin along with actin,myosin and free A.T.P. may act as plasticoizer inducing changes in muscle viscosity. In crustacea however thesis no evidence for an increased viscosity following excitation many anticholinesterases prevent stimulation action of subsequent application of acetylcholine but once contraction has developed these agent fail to cause relaxation the transmitter substance at the synapse of excitatory axons appear to be :

(i) glutamate and that of inhibiting synapse

(ii) *GABA amino butryric acid.*

EXPLAINED EXAMPLE OF MUSCLES OF CRAYFISH

Most noticeable are the muscles of the abdomen. Of these, the relatively slender extensors originate dorsolaterally in the thorax, a shot distance behind the cervical groove, and run back through the pericardial sinus, to insert on the six terga of the abdominal somites. The much more powerful flexor muscles originate near the floor of the thorax, along most of the length of the upper surface of an exoskeletal invagination. the endophragmal. shelf; in the abdomen the fibers become arranged in groups and follow a complexly twisted course, inserting on the six sterna and in the telson. These muscles are able to act as flexors because their points of insertion lie ventral to the exoskeletal hingesd between the successive abdominal somites. The hinges are placed on the sides of the abdomen: at the level where the terga and pleura meet. If the exoskeleton is carefully clipped away around one of these hinges, a small ball-and-socket type of joint will be disclosed. The great length of the abdominal flexors probably favors the strength of their contraction, while their subdivision into separately twisted bundles may serve to reduce the friction between fibers.

On each side of the stomach is the large adductor muscle of the mandible. This originates dorsally by a large head attached to the exoskeleton. and quickly tapers to a strong tendon running anterioventrally, to insert about midway along the posterior border of the mandible. Contraction of the two adductors rotates the mandibles in such a way that they bite together.

Connecting the stomach with the exoskeleton are a number of paired muscles. Largest of these are three pairs chiefly responsible for the action of the teeth in the gastric mill (q.v.), namely the anterior, posterior, and posterlateral gastric muscles. These are all dorsal in position. A number of pairs of much smaller, dilator muscles also run between the stomach and various parts of the exoskeleton. In the wall of the stomach itself numbers of small, intrinsic muscles run between various ossicles. These serve chiefly to bring about movements whereby the teeth and other internal parts of the gastric mill are kept clean.

Under the endophragmal shelf will be found the proximal muscles of the pereiopods. Most of them have their origin on parts of the endophragmal skeleton (q.v.). Two small muscles insert on the proximal rim of each coxa, four insert on the proximal rim of the basis of each large chelipod, and two insert on the proximal rim of the basis of each of the other pereiopods. It is of interest to note that the muscle used in autotomy is the levator anterior basis in the large chelipod, and simply the levator basis in each of the other pereiopods. This muscle can best be found by cutting a window in the dorsoanterior part of the coxa, enlarging the window if necessary after the muscle has been seen. The autotomizer muscle in the chelipod inserts by two separate tendons.

The more distal muscles of the pereiopods both originate and insert within the appendages. Some of them have provided material for important studies of neuromuscular physiology.

MUSCULAR SYSTEM IN CRAB

The muscu;ar system Crab can be studied into two groups :

(1) the muscles of the cephalothoracic region and its appendages, and

(2) muscles of the abdomen.

Muscles of the Cephalothorax

The chief muscles which play an important role in bringing out the movement of the optic penducle and first antenna are the flexor and Extensor muscles. The movement of the optic peduncle is enectad by ventral flexor and a dorsal extensor muscle. Similarly, in the first antenna the basal segment of the protopodite has dorsal extensor and ventral flexor muslce. In case of second antena the operculum, which is supposed to be the coxopodite of this appendages, has the extenson and flexor muscle. But here these muscles help in raising and closing the operculum. The mucles of the flagellum are very small, and delicate.

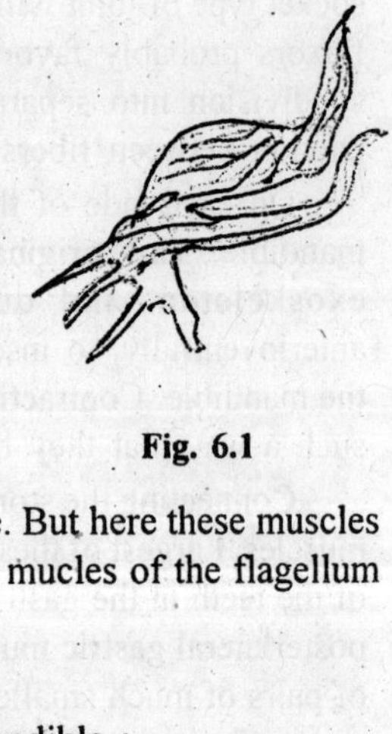

Fig. 6.1

Muscles of the Mandibles

It helps in closing and opening of the mandible :

1. External and internal adductor muscles, and
2. External and internal abductor muscles.

The external adductor muscle arises from the anterior region of the carapace under the lateral spines. This muscle is inserted on a iong tendon attached to the outer part of the mandibular apophysis whereas the internal adductor muscle is a short broad muscle and is inserted on the extremely long narrow tendon

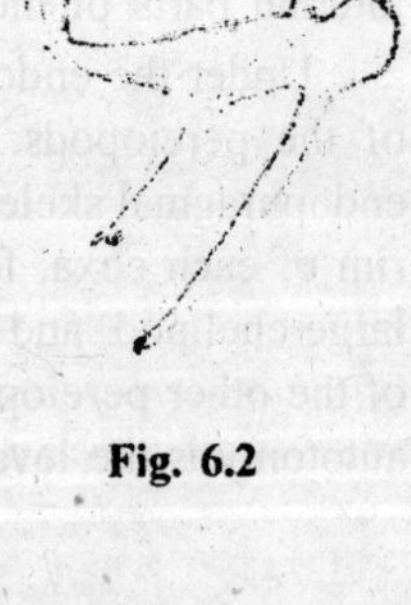

Fig. 6.2

attached to the posterior margin of the mandible. The external abductor muscle also arises from the posterior hepatic region of the carapace. It passes downward and is inserted on small narrow tendon attached to the posterior side of the mandibular apophysis. The internal abductor muscle arises from the rod-like portion of the first post-oral endopleurite. This mucles is attached to the outer part of the apophysis.

Muscles of the First and Second Maxilla

In the first maxilla there are two extensors and two flexor muscles which help in the the protogastric region of the carapace and are inserted on the outer and inner part of the coxopodite of the first maxilla. Similarly, the outer and inner extensor muscles arise from the first post-oral endopleurite, They are aiso inserted on the same point of the coxopodite as the corresponding flexor muscle. In the second maxilla also there are two extensors and two flexor muscles. There are again two sets of the scaphognathite. The two extensor muscles pull the scaphcgnathite downward whereas the two flexors raise it up to its original position, in hs natural from the scaphognaihite does not remain but but always found in curved position. This is done by a set of accessory muscles which are situated on the anterior wall of the scalphognathite. The flexor muscles with scaphognathite take their origin from the anterior or surface of the last cephalic endopleurite and are inserted on the anterior wall of the base of the scaphognaihite whereas the extensor muscles lie beneath the flexors and are inserted on the posterior wall of the scaphognathite.

Muslces of the Maxillipeds

First Maxilliped

The extensor and flexor muscles help in the movement of this appendage. The extensor muscle arises from the epimeron of this sornite and is attached to the outer and posterior margin of the coxopodite. The flexor muscle arises from the posterior surface of the last cephalic endopleurite and is inserted at the base of coxopodite. The muscles of the flagellum are large and powerful. The extensor muscle of the flagellum lies on the inner sider whereas the flexor lies at the base of the flagellum.

Second Maxilliped

The extensor muscle of the second maxilliped arises from the ineer side of the second thoracic epimeron whereas the flexor rnuscle arises from the second thoracic eoirrieron wi;ereas the flexor muscle arises from the anterior side of the second throracic endosternite. The muscles of the flagellum are similar as described in the first maxilliped.

Third Maxilliped

Coxopodite : The extensor muscle of the coxopodite of the third maxilliped arises from the inner side of the third thoracic epirneron and is attached to the outer side of the coxopodite by means of a narrow tendon. The flexor muscle arises from the anterior side of the third endostemite. It is inserted on a broad tendon at the inner side of the coxopodite.

Basi-ischium : The extensor muscle is very small and arise from the third thoracic endosternite. It is inserted on the ventral side of the bashischium. The flexor muscle takes its origin very near to the extensor miuscle and is attached to the dorsal side of the basi-ischium. There is one flexor and one extensor muscle for each of the remaining segments but the muscies are very small. There are two small muscles in the exopodite, the dorsal extensor and ventral flexor. There is a large extensor muscle occupying the whole region of exopodite and is attached to the outer edge of the flagellum. No definie flexor muscle is noted here.

Muscles of the Chela

Coxopodite

The extensor muscle which arises from the fourth thoracic pleural muscle chamber passes forward and downward and inserted on long narrow tendon from the posterior side of the coxopodite. The flexor muscle is somewhat a larger muscle and arises from the fourth thoracic sternal muscle chamber. It is inserted on a broad tendon on the anterior side of the coxopodite.

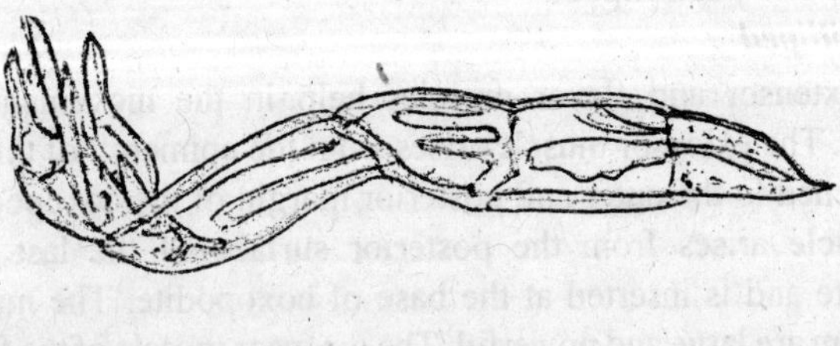

Fig. 6.3

1. extensor muscles, and
2. flexor muscles,
3. inserted on the basi-ischiopodite.

The anterior extensor muscle arises from the inner wall of the fourth thoracic pleural muscle chamber and is inserted on a long narrow tendon situated on the anterior side of it. The posterior extensor is small muscle which arises again from the fourth thoracis pleural muscle chamber and is insertcd on a small tendon which lies above to that of the anterior extensor muscle.

The anterior flexor muscle arises from the fourth thoracic sternal chamber and is inserted on a broad tendon situated on the ventral side of the basi-ischium. The greater posterior flexor is inserted on an extremely long tendon which is situated behind the tendon of the anterior flexor muscle. The lesser posterior flexor is inserted on a small tendon and lies just near to the greater posterior flexor.

Meropodite

The extensor muscle of the meropodite is small one and arises from the posterior wall of the ischium. It is inserted on a small tendon on the ventral side of the meropodite. There is no definite flexor muscle.

Carpopodite

These are two muscles, anterior one is the extensor muscle arises from the posterior side of the meropodite and is inserted on a long tendon situated on the dorsal side of the carpopodite. This tendon lies in the dorsal part of the merus. The flexor muscle arises from the anterior side of the meropodite and is inserted on tendon which is situated on the ventral region of the rneropodite and take its origin from the anteroventral side of the carpopodite.

Propodite

There are two muscles present, posterior extensor and anterior flexor. The posterior extensor arises from the posterior wall of the carpopodite and inserted on a broad tendon at thc posterior portion of the propodite whereas the flexor muscle arises from the anterior wall of carpus and is inserted on the tendon situated at the anterior side of the propodite.

Dactylopodite

There is a dorsal and ventral flexor muscie. The dorsal extensor is a very small muscle and arises from the dorsal of the posterior wall of the propodite. It is inserted on a narrow tendon which is flexor muscle is a large muscle which arises from the ventral side of the propodite and is inserted on a very broad tendon which is attached to the ventral side of the dactylopodite. The flexor muscle occupies greater portion of the propodite.

Muscles of the First Walking Leg

There are two extensor muscles and one flexor. The extensor muscles of the coxopodite arise from the fifth pleural muscle chamber. They are inserted on two long narrow tendons on the dorsal side of the coxopodite. The flexor muscle arises from the upper portion of the fifth sternal muscle chamber and it passes forward and downward. It is inserted on the anterior part of the coxopodite.

The extensor muscle of the basi-ischiopodite arises from the fifth thoracic endosternite and is inserted on a long tendon on the dorsal of the basi-ischium. The flexor muscle arises from very near to the extensor muscle of the fifth thoracic endosternite and its insertion is on a broad tendon on the ventral side of the basi-ischium. The muscles of the other segments of all limbs are very similar to those described in the chela.

Fig. 6.4

Muscles of the Second Walking Leg

The muscles are very similar to that of the first walking leg. The extensor of the coxopodite arises from the sixth pleural muscle chamber whereas the flexor muscle arises from sixth sternal muscle chamber. The muscles of the basi-ischiopodite arise from the sixth sternal muscle chamber.

Muscles of the Third Walking Leg

The extensor muscle takes its origin from trie seventh pleural muscle chamber and the flexor muscle from the seventh sternal muscis chamber. The muscles of the basi-ischiopodite arise from the seventh sternal muscle chamber.

Muscles of the Fourth Walking Leg

Here the muscle chamber is not divided into pleural and sternal region. So, it is called as pleurosternal rnuscle chamber. All the muscles of the coxopodite and basi-ischium arise from the same muscle chamber. There are two extensors and two flexors muscles attached to the coxopodite.

Fig. 6.5

Muscles of the Abdomen

In the crab, *Barytelphusa cuicularis,* though there is a partial fusion between the third, fourth and fifth somites, it has not affected much on the structure of the male abdominal musculature and there is no significant difference between the male and female abdominal muscles. The only difference being in size and shape of the abdomen. However, each abdomnal segment possesses two sets of muscles, dorsal extensor and ventral flexor.

Musculus Ventralis Superficialis Thoraco Abdominalls

This muscle takes its origin from the last thoracic somite and passes backward and is inserted infront of the line separating the first and second strerna of the abdomen. Although the muscle is paired it appears as a single band of muscle fibre. This helps to pull the abdomen towards the thorax.

Musculi Ventralis Superficialis Abdminis

All these muscles are paired unlike the extensors. These muscles are arranged regularly on the ventral side in accordance with the segmentation of the abdomen. The first pair of muscles arises from the posterior end of the musculi ventralis superficialis thoraco abdominis and is inserted on the sclerotized ridge marking the junction between second and third segments. The muscles of the second pair arise from the line separating the second and third abdominal sterna and are inserted on a line separating the third and fourth abdominal sterna. The third paired originates from the line separating the third and fourth sterna and is inserted on a line separating the third and fourth, the fourth and fifth sterna. The fourth pair of muscles starts from the middle of the line separating the fourth and fifth abdominal sterna and is inserted on two cartilagenous projections on the sides in the middle region of the fifth abdominal segments. These muscles do not extend the whole length of the segment. The fifth and the last pair of muscles arises from the middle line separating the fifth ans sixth abdomnal sterna and is inserted at about the middle region of the sixth sterna. There is ventral muscle connecting the sixth segment with the telson in either sex. All the sternal muscles of male and female are more or loss similar in their origin and insertion. Partial fusion between the third, fourth and fifth somites has not affected much on the development of the musculature. The muscles connected with the opening and widening of the anus are musculus dilator ani. The main pair of muscles connected with this is inserted on the ventral side of the telson and a small pair is inserted on the dorsal wall of the telson. These two pairs help to open the anus by their contraction.

Musculus Dorsalis Superficialis Thoraco Abdominalis

This is a paired muscle which arises from the posterio-lateral edge of the anthrophragm of the eighth thoracic somite and is inserted on the anterior protion of the tergum of the first abdominal somite. This muscle functions in reverse direction to that of the flexor muscles. When this muscle contracts the rectanglar portion of the tergum is pulled forwards so that the tip of the abdomen will be pulled out wards, thus widening the distance between abdomen arid thorax.

Musculus Dorsalis Superficialis Abdominis

Female

There are five bands of dorsal superficialis muscles which arise at the base of the eighth thoracic somite and are inserted on sclerotized ridges in the middle region of the line separating the first and second abdominal segment. The strands of the muscles are short and broad and all the five bands of muscle lie very close to each other. The next set of five muscle bands arises from a calcareous region just behind the line on which the previous muscles are inserted. These muscles are inserted on the sclerotized ridges in the middle line separating the second and third abdominal terga. The third sat of five muscle bands arises just behind the posterior end of the previous muscle bands and are inserted in between the third and fourth terga. In the fourth abdominal taroa there are four sets of four sets of muscle bands which arise approximately from the middle region and are inserted in between the fourth and fifth terga. In the fifth abdominal terga there are again four sets of muscle bands which arise from the middle region and are inserted in the middle of the line separaiing the fifth and sixth terga.

Male

Although, there is a partial fusion between the third, fourth and fifth abdominal somites, the number of muscles remains the same. All the sets of muscles of the male abdominal terga are present at the respective places as found in the female. However, the muscles in male are narrow and are very transparent.

Musculus Dorsalis Superficialis Telsonis

This is the last and paired muscle. The muscles are long and slender with the anterior ends slightly broader. This muscle arises from the

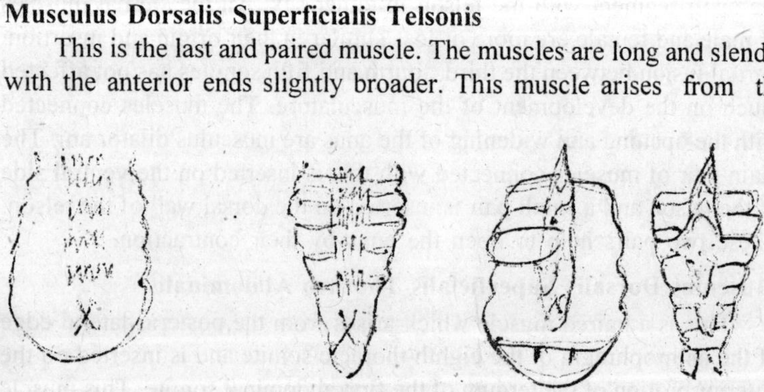

Fig. 6.6

anterior half of the sixth terga on both the sides of the middle line. These muscles become narrow as they pass posteriorly and converge at one place and are inserted on the anterior border of the teloon which receives all its power of motion from the muscle, as no flexor exists in either sex for the teison.

MUSCULATURE

Muscles of the Foregut

The musculature of the foregut in *Barytelphusa* falls into two groups:

(1) The extrinsic muscles, and (2) The intrinsic muscles. The extrinsic muscles are those which originates on some part of the skeletal system outside the foregut and are inserted to the ossicles lying in the wall of the foregut. The intrinsic muscies run between the ossicles of the gastric armature within the foregut itself.

Extrinsic Muscles

The anterior gastric muscles extend between the pterocardiac ossicle and the anterior region of the carapace. It is in the form of two bands which diverge considerably at the anterior end, the inner posteriogastric muscles arise from the underside of the posterior region of the carapace. Each muscle passes downward forwards inserting on the pyloric ossicle. The outer posterior gastric muscles also arise from the under side of the posterior region of the zygocardiac ossicle. The above three sets of muscles help in the working of the gastric mill.

The following muscles help to dilate the foregut. The upper anterior dilator muscle arises from the cephalic sternum immediately behind the orbit and inseried ontne anterior side of the cardiac chamber. A pair of lower anterior dialator muscles arises from the upper side of the epistoma and are inserted on the lower part of the foregut. These are smali pairs, of muscles. A pair of anteriolateral dilator muscles arises from the outer edge of the roof of prebranchial chamber and are inserted on the lateral wall of the cardiac region above the oesophagus. A pair of posteriolateral dilaor muscles originates from the roof of the prebranchial chamber and are inserted on the anterior side of the posterolaterai cardiac plate. There are two pairs of dorsal pyloric dilator muscles which arise from the under side of the carapace very close to the outer posterior gastric muscle and are inserted on the ossicles of the dorsal wall of the pyloric foregut. There are two pairs of ventral pyloric dilator muscles. The outer pair of muscles. The outer pair of muscles is much shorter and originates from the endopleurite of the first maxillary segment and is inserted on the ventral side of the pyloric region. The inner pair of muscles is a long one and arises at the base of the mandibular apophysis. It is inserted on the anteroinferior pyloric ossicle in the ventral wall of the pyloric chamber.

Intrinsic Muscles

The cardiopyloric muscles arise from the posterior border of the mesocardiae ossicle and are inserted on the posterior side of the propyloric ossicle. This is composed of three bands, one median and two lateral. There are three pairs of lateral cardiac muscles, the dorsal, middle and ventral. The dorsal pair arises from the dorsal border of the zygocardiac ossicle. The middle one arises between the upper region of the infra cardiac and the prepectinal ossicles. The ventral one is short and arises from the side of the infralateral cardiac ossicle and passes across the upper region of the posteriolateral cardiac plate. The posterioinferior cardiac muscle is a broad sheet of muscle lying tre ihe posterior wall of cardiac region. On both the sides of it are attached the infralateral cardiac ossicles. The anterior cardiac muscia is a median muscle and arises from the front poriion of the mesocardiac ossicle. At the base, it is divide into two main branches which are atiached separately to the front wall of the forgut. The above two sets of muscles aci as constrictor for cardiac region of the foregut. The pyloric constrictors are formed by several thin bands of muscles between the postpectinal and infralateral cardiac ossicles.

In Crustacea there is no evidence for increased viscosity following excitation but a background of motor activity at a low frequency reaches certainly after present in the same action. There may be simple rules regarding the extent to which exitro and plasticizens neurons sure together or certain reflexes may involve them to different extent in regular manner.

Many anticholinesterases present the stimulation action of subsequent application of acetyl choline but once the contraction has developed those agents fail to cause relaxation.

REFERENCES

1. Bishop, S.H., (1976), The Physiology of the Nitrogen Metabolism and Excretion, Regulation of Intracellular, Amino Acid Concentration. Vol. 1, pp. 414-431.
2. Diwan, A.D., (2000), Indian Freshwater Crab Central Institute of Fisheries Education Versova, Mumbai.
3. Linda H. Mantel, (1983), Internal Anatomy and Physiological Regulation, Vol. 5, Academic Press, New York, London, pp. 90-118.

7

RESPIRATORY SYSTEM

Although, Crustacean body is enclosed in a stong skeleton it enought with stand the pull of muscles, no special respiratory surface is required by animals below a certain size; thus in *Calanus* gaseous exchange occurs over the general surface of the body. When gills become necessary they are derivatives of the appendages. Respiratory pigments develop in small species which live in an environment deficient in oxygen and in larger forms wliicli have elaborate gills. Thus, *Daphnia* and *Apus* have haemoglobin in the blood, the amount varying inversely with the amount of oxygen in the water, and haemocyanin characterizes the decapod, but not the nondecapod malacostracans, which are almost always smaller. In addition to appendages the carapace fold figures largely in the respiratory mechanisms of such groups as possess one. Arising from the maxillary segment, it stretches over the thorax and whilst the outer wall is thick the inner one is thin; the cavity which it encloses is ventilated by the movements of appendages. The inner wall is well vascularized and forms the chlef respiratory surface in mysids, land crabs and barnacles. It is usually simple, but in barnacles is produced to form a single gill on each side, washed by a flow of water, set up by beating of the cirri, which enters the mantle cavity right and left of the head and passes out posteriorly.

Gills are typically lateral outgrowths of the basal region of thoracic legs, known as epipodites. The simplest is a lobe or lamella as in branchiopods and *Anaspides*, bathed by currents produced by these limbs as they perform locomotory or feeding movements. In amphipods, the habit of crawling on one side and the absence of a carapace to protect respiratory organs make lateral gills impractical, and the epipodites are median outgrowths in females they arise ventral to the oostegites (oost), which are modified from a second series of epipodial plates. They lie in a channel between the limbs deepened by coxal plates projecting from the margins of the thoracic segments, and ventilated by the beating of the pleopods, the first three pairs of abdominal appendages. A current of water is passed over the gills even when the slirimp is not moving,

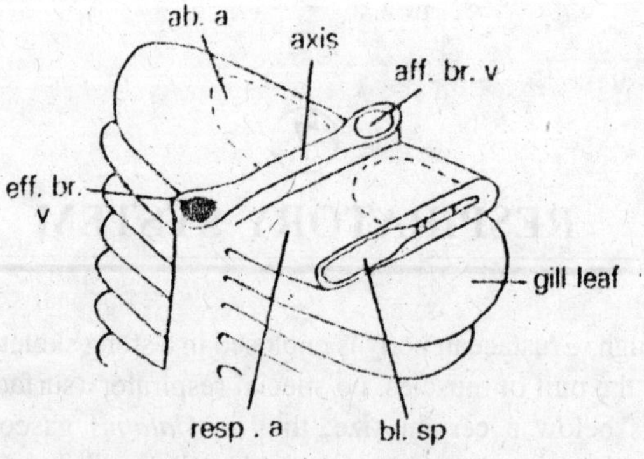

(A) Respiration in Crustaceans.

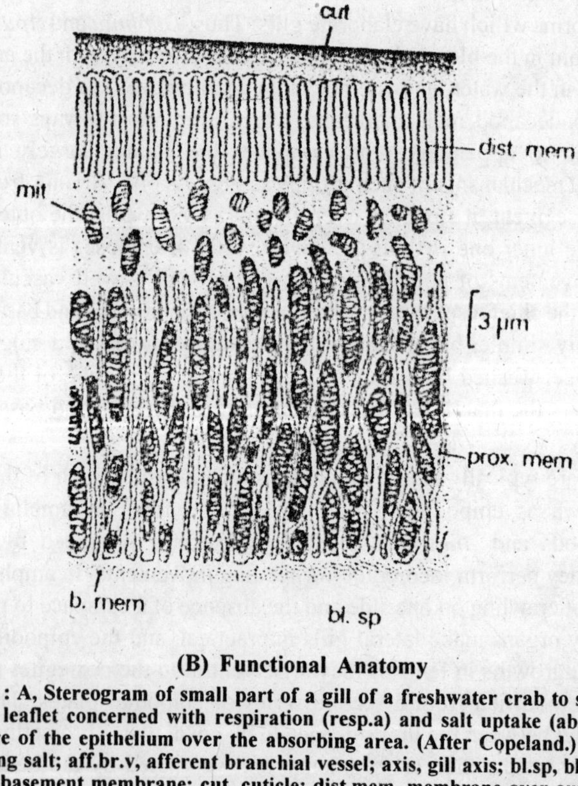

(B) Functional Anatomy

Fig. 7.1 : A, Stereogram of small part of a gill of a freshwater crab to show areas of each leaflet concerned with respiration (resp.a) and salt uptake (ab.a). B, fine structure of the epithelium over the absorbing area. (After Copeland.) ab.a, area absorbing salt; aff.br.v, afferent branchial vessel; axis, gill axis; bl.sp, blood space; b.mcm, basement membrane; cut, cuticle; dist.mem, membrane over outer surface of cell; eff.br.v, efferent branchial vessel; gill leaf, gill leaflet; mit, mitochondrion; prox.mem, membrane at inner surface of cell.

facilitated by the longer posterior thoracic legs raising the hinder end of the body. In the littoral sandhopper, *Talitrus saltator*, the gills function only in moist air kept in motion by pleopods, and the animals drown in water.

The eucarid Malacostraca have. more complex gills with a larger area for respiratory exchange. Each consists of an axis which bears numerous branches. In euphausiids they undergo further division to give feathery structures not covered by the carapace but hanging freely at the sides of the body (one attached to each thoracic leg) and washed by the swirl of water set up by the exopods during swimming. In dccapods, a group which includes the largest crustaceans, typically benthic, the gills are enclosed in a branchial chamber ventilated by the scaphognathite propelling water forward with a rapid sculling movement. The scaphognathitc of the prawn *Palaemon serratus* maintains a rate of 294 beats per minute at 17°C, though this is slow compared with the thoracic appendages of calanoid copepods which range from 60-2640 beats per minute. Perhaps because the gills are less efficient when enclosed they then increase in number. The maximal number on each side of a segment is four: two arthrobranchs attached to the articular membrane between the coxopodite of the appendage and the body, one pleurobranch dorsal to this on the body wall, and, on the leg itself, one podobranch, which may be joined to an cpipodial plate carried by the coxopodite. Probably all arise as outgrowths of the limb, for a dorsal migration from this position to the more protected one beneath the branchiostegite occurs in the development of *Penaeus* and *Crangon*. The axis of the gill bears numerous lateral branchcs which may be filiform or, in crabs, flattened plates. In both euphausiids and decapods the axis contains the main afferent and efferent vessels linked by channels running through the leaflets. The cuticle covering the respiratory surface is extremely thin and in *Carcinus maenas* is penetrated by invaginations which extend into the underlying epidermal cells and so come close to the blood. It is probable that they increase the efficiency of the gill.

The degree of isolation of the gill chamber varies considerably within the decapods and may be correlated with the mode of life. It is least, as might be expected, in swimming macrurans in which the entrance is a narrow slit along the margin of the branchiostegite, protected against the entry of silt by setae along the edge of this fold and on the coxopoditcs of the limbs. It is greatest in crabs where the cephalothorax is flattened from above downwards and the lateral portions of the branchiostegite bend abruptly inwards and are closely applied to the ventrolateral region of the thorax, except near the bases of the legs. This gives freedom for their movement as well as a scries of inhalant openings guarded by setose tracts. The branchiostegite now forms a lateral margin to the cephalothorax, thickened, often armed with teeth, and projecting in the direction of habitual movement. The adaptations, which primarily protect the gills from the dangers of silting associated with benthic life on a wave-washed shore, not only enable the decapod to exploit more silty or sandy situations but provide preadaptive features for colonization of the supralittoral zone and terrestrial environment. The portable shelter of hermit crabs, typically a gastropod shell, enhances the isolation of the gill chambers, and the carapace becomes membranous behind the cervical groove where it is covered by the shell.

In *Carcinus* and *Cancer* there are nine gills in each branchial chamber. Except for the podobranch of the second maxilliped which is directed horizontally backwards, they curve upwards and inwards from their points of attachment so that their tapering tips converge along the dorsomedian wall of the chamber. The gills are closely applied side by side and separate a shallow hypobranchial space towards the body wall from an cpibranchial space under the *branchiostegite*. The latter leads to the shallow cxlialant passage or prcbrancliial chamber in the cephalic region where the movement of the scaphognathite (scaph) to and from the roof directs water towards an anterior exhalant opening, bordered behind by the cndopoditc of the first maxilliped. Water enters the hypobranchial space, mainly at the base of the clieliped, passes between the gill lamellae to the epibranchial space and then anteriorly to the exhalant opening. In *Carcinus maenas* the entrance channels comprise a relatively large anterior opening (inhalent) in front of the chelipod and a narrow slit above each leg (inhalent I), including the chelipod. The size of the channels is affected by the position of the appendages: when a leg is moved backwards and upwards the opening above it is naturally almost closed. The size of the anterior inhalant opening is regulated by the basal segment and epipoditc of the third maxilliped (epip 3) in such a way as to allow maximal intake of water and use of all the gills at times when these appendages are in use, as in feeding, or at other times, when oxygen is short. If the third maxillipeds are at rest the broad bases of their epipodites and coxopoditcs block the anterior part of the opening and bar the passage of water forwards towards the more anterior gills. If they are moved apart water gains access to these gills. The respiratory flow is also influenced by changes in the beat of the scaphognathite and the activity of the exopods (flagella) of the maxillipcds flick to and fro extremely rapid and so reinforce the exhalant current from the gill chambers, turning it outwards away from the mouth. This occurs when the crab feeds. After this the mouth parts clean themselves, are folded away and the oral field almost completely enclosed by the third maxillipeds.

The two scaphognathites may change both the rate and the direction of the water current, which is typically postero-anterior. They work independently and either may cease beating for a while or change the rate and force of its beat. In a more or less regular fashion the effective beat is reversed and water is drawn through the exhalant openings into the epibranchial space for periods lasting a few seconds; the current then reverts to normal. Reversal always occurs in *Corystes* when it burrows into sand and is maintained until it emerges. In deoxygenated, shallow water *Carcinus* can raise the normally exhalant openings to the surface and draw through them better oxygenated surface water or air which bubbles out in front of the coxae of the chelipods.

Despite the fringe of setae at the openings to the branchial chambers a certain amount of particulate matter passes through the filter and might close the passages between the gill leaflets were it not for the activity of the epipodites (sometimes called mastigobranchs) on each maxilliped. These are long, flexible and stoutly setose, and can reach almost every part of the gills, moving constantly over their surface and brushing it clean. The epipodite of the first maxilliped lies above the gills and sweeps their outer surface, those of the second and third lie in the hypobranchial space and sweep the inner surface.

In macruran prawn *Caridina, laevis* feathery bristles of exopodite of first peleopod act as sieve from preventing ingress of debris into branchial cavity with respiratory current.

In other decapods epipodites have a similar function though they are not so well adapted. In these the branchiostegite fits less tightly. As a consequence a pereiopod can be inserted into the gill chamber to clean the gills, making the development of epipodites less essential, and the removal of silt by the reversal of the respiratory current more effective. In *Galathea* and *Porcellana*, both inhabiting silty places on the shore, the last pereiopod is specialized for this function and is not used for walking; it may be found folded closely against the body, cleaning the grooves of the carapace or brushing the gill lamellae. To this effect the two terminal segments of the limb are beset with stiff setae, some serrated, giving it the appearance of a bottle brush. In *Porcellana* there are no epipodites and in the larger species, *P. platycheles*, a very thick curtain of bipectinate setae guards the inhalant openings to the gill chamber; in *Galathea* epipodites are present and the setal filter less developed. In both, podobranchs, the most ventral gills of the malacostracan series, are missing, perhaps because they would be too easily silted.

The respiratory organs are situated in the gill chambers between the branchiostegite and the body, one on each side. Each chamber is large but narrow and contains eight gills, three epipodites and lining of the branchiostegite. All of them function as respiratory organs. The gill chamber is in communication with the exterior along its. anterior, ventral and posterior borders, but is closed along its dorsal side.

The thin lining of branchiostegite contains blood spaces and thus forms a respiratory surface. Three pairs of epipodites, the leaf-like outgrowths of the coxae of three pairs of maxillipeds, are covered with thin cuticle and act as primitive gills. The gills have been named according to their places of attachement, there is one pair of podobrachs as each is attached to the coxae (prodomere) of the second maxilliped. Two pairs are arthrobrachs, each pair is attached to the arthrodial membranes of the second and third maxillipeds. The remaining five pairs are pleurobranchs, since they are attached to the pleura of the thorax. Out of the eight gills the seven become exposed after has removal of the brachiostegite, the gill number eight remains hidden below the second. The part of the gill by which it is attached to the thorax wall is called a gill root. Each gill has a long base or axis bearing two rows of gill plates the gill plates of each row are arranged like leaves of a book such a gill is called a phyllobrach. In each gill root run three blood vessels, two are lateral channels which carry non-aerated blood to the gill, and one is the median channel carrying aerated blood from the gill.

Mechanism of circulation of water in the gill chamber, the scaphognathite vibrates constantly and thus, produces a constant current of water in the gill chamber, the water enters the gill chamber along the posterior and ventral margins of the branchiostegite, after passing over the gills, epipodites and over the lining of the brachiostegite. Thus, a constant current of water enters the gill chamber and the blood in these structures becomes aerated. Most of the time the direction of the water current remains posterior to anterior, but occasinally the direction is reserved to flush out debris that may have collected on the gills. Since the blood is oxygenated in the gill plates, it is believed that the blood in circulating through the two sets of plates would be exposed to aeration twice.

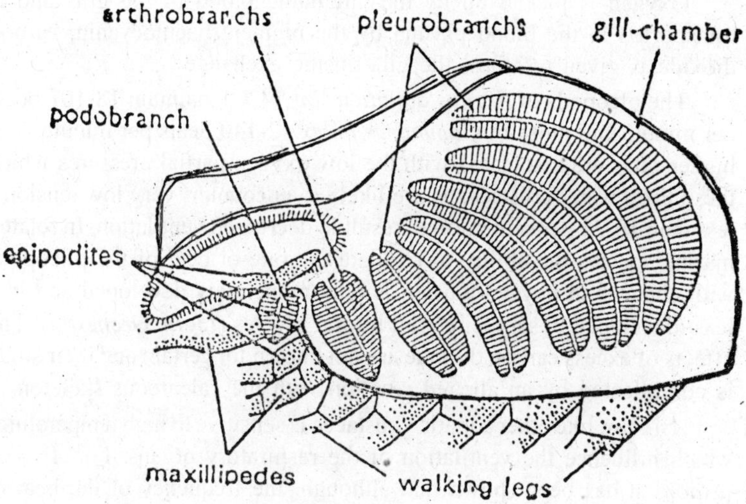

(A)

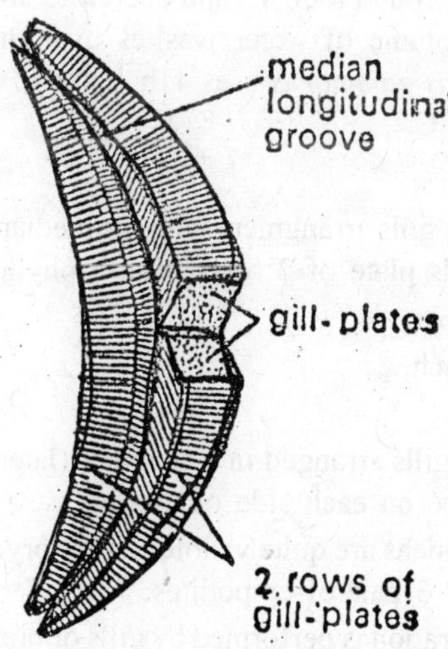

(B)

Fig. 7.2 : *Palaemon* : A. Left gill chamber exposed to show the gills. B. A phyllobranch.

Oxygen is picked up by the circulatng blood of the gills and is transported in the blood plasma by the pigment haemocyanin, carbon dioxide is given off form the gills in the exchange.

The pleopods of *Asellus aquaticus* (at 14°C) maintain 18-167 beats per minute and those of *Gammarus pulex* 12-140 beats per minute. The higher rates are associated with the low oxygen partial pressures which these Freshwater crustaceans are likely to encounter; very low tensions, less than 10-25 per cent normal, result in decreased ventilation. In related marine species the ability to regulate the rate ol beat of the pileopods with changes in oxygen tension is only moderately developed and it is least developed in semiteirestrial littoral species (*Ligia, Orchestia*). The effects of excess carbon dioxide are not known for certain and their study is complicated by an alleged reaction with the calcareous skeleton.

Astacus, like other aquatic Crustacea is sensitive to high temperatures which influence the ventilation of the respiratory organs. For *Astacus astaous* it has been shown that although, the frequency of the beat of the scaphognathite increases from 33-72 per minute over the temperature range 6-30°C, and results in an increased volume of water flowing through the gill chambers, the volume of water affected by a single stroke is maximal around 8-10°C, and decreases above this temperature. A considerable volume of water waslies over the gills of the larger decapods. *Homarus gammarus* uses $41h^{-1}$ at 5°C with a linear increase to $11.81h^{-1}$ at 21°C.

Phyllobranch

It is prawn the gills arrangment with the median longitudinal groove. Gill plate and gills plate of 2 rows from a phyllobranch.

1. Gill chamber,
2. Pleurobranch,
3. Epipodite.

Eight pair of gills arranged in a chamber (lateral) in prawn is called gills chamber cone on each side of thorax).

— Pleurobranchs are quite visible respiratory system consist of 2nd and other 3 pair of epipodites.

In crabs respiration is performed by gills or branchiae as a outgrowth from walls of thorax. Gills are present in gill chamber on each side of cephalothorax. Branchial cavity has epimera (Its innerside is calcified) on innerside and lateral flaps of *Carapace* on dorsal and outerside. In diogenes branchiostegite in thin and membranous. Branchial chamber has inhalent and exhalent openings for continuous flow of water. There is too long setae to check water flow in branchial chamber. The posterior region of branchial chamber a slit opening from last walking leg know as posterior inhalent aperture—second opening is infront of coxopodite of chela leads to branchial chamber from anterior side. The opening in front is guarded by coxa and epipodite of third maxilliped. The epipodite has numerous setae. The epipodites never allow any send or foreign particle in branchial chamber or gill surface. The prebranchial cavity chamber at the side of mouth has, a very narrow thin sloping down of roof of *Carapace* space, which diverges anteriroly outside by exhalent chamber. The respiration current of water is because of vigrous action of scaphognathile, current flows in through inhalent aperture and goes out by exhalent aperture. If current is reversed is because to relax the fatigued muscle of scaphognathite. Epipodite of first maxilliped is the largest one and occupies the whole length of branchial chamber lying upon the gills. Epipodite of second maxilloped is situated below the gills on dorsal surface of thoracic epimera and epipodite of third maxilliped lies below the gills on the ventral and outer side of second maxillped. The epipodite of second and third maxillipeds extends up to the fourth and fifth thoracic somites.

Gills are of three categories :

(i) podobranch—directly arises from an appending;

(ii) arthrobranch—arises between an appendages and epineron.

(iii) pleurobranch—directly arises from epimeron.

Table 7.1

Thoracic appendages	Podo-branch	Pleuro-branch	Arthro-branch
1st maxilliped	0	0	0
2nd maxilliped	0	0	0
3rd maxilliped	0	0	0
1st pereiopod	0	1	0
2nd pereiopod	0	0	2
3rd pereiopod	0	0	2
4th pereiopod	0	1	0
5th pereiopod	0	1	0

The first gill is pleurobranch and arises from fourth thoracic somite. The second and third are arthrooranchs from fifth thoracic somite. The third gill is always longest smong the seven gills. The fourth and fifth gills are arthrobranchs and arise at one place from sixth thoracic somite. The sixth and seventh gills are again pleurobranchs but have different arising. The sixth arising from seventh thoracic somite and seventh from the eighth.

Histology of the Gill

The branchial axis is roughly triangular in cross section and bears two rows of thin lamellae. Each gills is pyramidal in shape and its apex pointing upwards. Along the outer side of each gill runs the different branchial vessel and on the inner side is situated the efferent branchial vessel. Each gil! lamella has an outermost layer of thin cuticle. In transverse section each gill is also triangular in shape snd efferent vessel lies at the apex and afferent vesse; lies at the base. The space between these two vessels is the branchial septum. In the regions of the branchial septum with excretory cells. Gill lamella has outer lining of epimeral cells, a narrow space of lamellar sinus. Each lamellar sinus has a contact with affeent branchial sinus on the outerside and different branchial sinus on innerside in his blood is purified in these spaces blood flow.

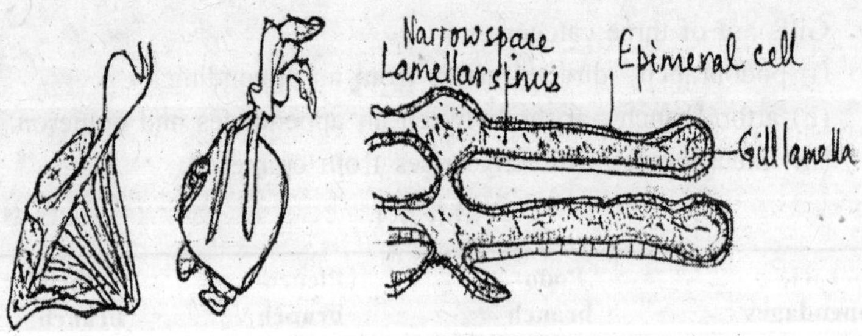

Fig. 7.3 : Crayfish gill.

CRAYFISH RESPIRATORY SYSTEM

On each side of the thorax a gill chamber is present, bounded externally by the branchiostegite, internally by the body wall, and to a large exetent closed ventrally by the appendages. Small chuks between the appendages open into chamber ventrally, but the only considerable openings are one at the ventroposterior angle of the branchiostegite and a channel which leads out near the anterior end of the scaphognathite. In a living specimen the rapid beat of the scaphognathite can be observed after cutting out a narrow strip of the branchiostegite just posterior and parallel to the ventral end of the cervical groove. A strong current of water is driven forward in the channel leading out form the gill chamber, by the alternate beating of the posterior and anterior portions of the scaphognathite. Water enters the gill chamber mainly through the posterior opening. as can be demonstrated by introducing from a pipette a few drops of carmine suspension near this opening. Once every few minutes the scaphognathite reverses its beat for a few strokes so as to wash loose silt or other debris, which may have settled on the gills. Hair-like setae on the inner wall of the branchiostegite perhaps aid in cleaning the gills that brush against them.

The gills are of the type known as trichobranchs, each consisting of numerous filaments arranged around a central axis, somewhat like a bottle brush. Depending on their position of attachment, three series of gills are distinguished, podobranchs, arthrobranchs, and pleurobranchs. On a single thoracic somite the following gills may be present: one podobranch, arising from the coxa of the appendage; two arthrobranchs, arising one just ventroanterior to the other, both from the membranous joint which attaches the coxa to the body; and one pleurobranch, arising higher on the side of the body. In addition there may be a leaf-like lamella, or mastigobranch, acconmpnying the podobranch, and a tuft of coxal setae (sometimes termed a setobranch). Not all these structures are present on all thoracic somites. The actual distribution is shown in the table 7.2.

Table 7.2

	Podo-branch	Mastigo-branch	Anterior Arthro-branch	Posterior Arthro-branch	Pleuro-branch	Coxal Setae
2nd maxilliped	1	1[1]	1[3]			
3rd maxilliped	1	1[1]	1	1[4]		1
1st pereiopod	1	1[1]	1	1[4]		1
2nd pereiopod	1	1[1]	1	1[4]	1[2.5]	1
3rd pereiopod	1	1[1]	1	1[4]	1[2.5]	1
4th pereiopod	1	1[1.2]	1	1[4]	1[2.6]	1
5th pereiopod					1[2]	

[1]Not in crayfishes of the southern hemisphere.

[1]Not in Cambarinae.

[3]Not in *Ilomarus*; rudimentary in some crayfishes of the southern hemisphere.

[4]Rudimentary or absent in some crayfishes of the southern hemisphere.

[5]Rudimentary or absent in most crayfishes.

The axial stems of the podobranchs perhaps represent modified epipodites, especially since in most crayfishes of the southern hemisphere a few gill filaments occur in the epipodite of the first maxilliped. It has even been suggested that the arthrobranchs and pleurobranchs are derived from epipodites of precoxac which have fused with the body. However, at least in their appearance, it is the mastigobranchs rather than the gills which most closely resemble epipodites. In crayfishes the lamella of each mastigobranch is longitudinally pleated and is fused along most of its length with the axis of the accompanying podobranch. In lobsters the mastigobranchs and podobranchs are separate. No function for the mastigobranchs is known. The coxal setae probably help to close on the gaps between appendages which lead up into the gill chamber.

Air contains 20 to 40 times more oxygen than can be present in air-equilibrated natural waters, and has one-thousandth of the density of water,

Although, both factors should considerably reduce both the amount and energy cost of ventilation, aerial gas exchange in crustaceans does pose its own particular problems and has produced some unique solutions.

(A) Structural Modification

Due to the higher and more stable oxygen content of air, diffusion across simple respiratory surfaces may actually increase on air exposure. Hemolymph oxygen tensions rise during air exposure in the gooseneck barnacle *Pollicipes* presumably resulting from enhanced oxygon uptake across nonsclerotized portions of the external cuticle. The more specialized aquatic crustacean respiratory organs, however, tend to collapse in air, suffering drastic reduction in both ventilation and perfusion of the lamellae as well as reduction in surface area. Consequently, gas exchange capability is severely limited, being reduced by a factor of threefold by air exposure in *Callinetes sapidus* in *Cancer productus.* These effects, however, are not universal. Many liltoral crabs show an increased sclerotization of the gills and wider spacing of ihcir lamellae lhan do aquatic species. These factors presumably prevent collapse of the gills in air in *Carcinus maenas*, and allow oxygen consumption to be maintained on transition from water to air exposure *Carcinus* may actually tolerale air exposure for several days but is probably rarely exposed for more lhan a few hours between tides in a normal habitat. A tendency toward thickening of the gill surface and toward reduction of both the number and surface area of gill lamellae continues with increasing terrestriality of habitat in both anomuran and brachyuran crab species for review). These changes are often interpreted as providing protection from excessive water loss across the gill surface. Chitin alone, however, does not constitute a particularly effective water barrier; both insects and arachnids have, for instance, added oil or wax monolayers to achieve their considerable impermeability. The observed changes in gill structure may thus be more associated with strenghthening the gill lamellae, thereby ensuring their separation in air, rather than with water conservation. Whatever the function, chitin is relalively impermeable to gases and the reduced gills of the more terrestrial decapods have become very diffusion limited gas exchangers. Certainly in many of the more highly terrestrial species, accessory gas exchange structures have developed to supplement exchange across the gills have recently attempted to quantify the role of these structures in aerial respiration. Both the size of the branchial chambers, and vascularization of the chamber walls are increased with increasing "terrestriality," and the surface area of the branchial walls may be further increased by partial "perforation," or complex infolding of the walls. These modifications finally culminate in exceedingly complex "respiratory trees" seen in the branchial chambers of such animals as the highly terrestrial Coconut or Robber Crab *Birgus latro*. The gills of *Birgus* were much reduced in area and could be surgically removed without apparent impairment. It could observe no reduction in oxygen uptake, he did note apparent CO, retention especially during exercise. In other crustacean groups, where the gills are not incorporated in a "branchial cavity," gas exchange may be supplemented by the development of internal structures such as the air-filled sacs penetrating the pleopod gills of many terrestrial isopods. These sacs may ramify complexity within the appendage of such forms as *Porcillio* so as to merit comparison with some insect Tracheal systems. Active ventilation of these structures has not been observed, but passive ventilation of the type described for insect tracheal systems by Levy and Schneiderman (1966) may possibly occur.

(B) Vcntilatory Modification

In both anomuran and branchyuran air breathers, the scaphognathite pump seems to function as well in aerial as in aquatic ventilation. This is probably assisted by the sealing effects of water held by capillary effects between the fringing hairs and between the hairs on the "walls" of Ihe scaphognafhile channel (R. E. Young, personal communication). Movements of the sea-phognathite of *Coenobita clypeatus* in air generate pressure waveforms similar to those generated by the scaphognathite of water breathers Flow of air in both the forward and reversed modes has been demonstrated in the crab *Carcinus maenas*. In most terrestrial crabs, movements of the scaphognathite in air are sporadic, but conditions of severe hypoxia or hypercapnia cause virtually continuous scaphognathite beating. In the case of *Birgus latro*, ventilation volumes of up to $1000 cm^3 min^{-1}$ have been observed (wt. range 1-2 kg: Cameron and Mecklenburg, 1973). In these continuously ventilating animals, stroke volume apparently changes little over the frequency range of 50-300 beats.min^{-1}, indicating that the scaphognathites seem to function equally efficiently pumping either water or air. Air flow rates for several species of decapods are included. Although, the scaphognathite can apparently function well as an air pump, movements of the branchiostegites have been observed in several species and may function as an accessory ventilatory device (branchiostegal pump). McMahon and Burggren (1979a) noticed a marked increase in the incidence of large pressure fluctuations in the branchial cavity of *Coenobita* during hypoxic exposure and attributed these to branchial pumping. These authors noted that pumping movements always occur at a particular point in the scaphognathite cycle, a situation which they likened to the movements of the branchiostegites during reversed pumping in the marine macruran Homarus americanus. In both cases the movements are apparently caused by the action of the epimeral attractor muscles. Similar movements have been noted for the crab *Birgus latro* and a possible ventilalory function. The movements of the branchiostegites in this species but were unable to detect associated changes in branchial air flow.

Air flow rates and percent oxygen extracted ($Extr_A$) are lower in air breathers than their equivalents in aquatic forms. Whether the oxygen is taken up via gills or via accessory branchial structures the structures must remain moist, since oxygen uptake decreases 30-fold if the chitin is allowed to dry. Several mechanisms for keeping the gills moist are observed. The evolution of our breathing may have involved bubbling of air through the branchial chambers with the scaphognathite beating in the reversed mode and tin- anterior apertures in air, as is seen in response to hypoxic exposure in the crab *Carcinus* and in the crayfish *Orconectes*. Although, studies of the ventilatory flow patterns are not available for most species, this mode of ventilation may persist in many temporary air breathers, as the majority of these maintain considerable water stores in the gill region at the base of the branchial cavity. The ventilalory air stream could either be bubbled through this water, thus oxygenating the water in contact with the gills, as occurs in hypoxic

Circinus maenas, or could be directed against the branchial wall or through other accessory air breathing structures. In *Birgus* and *Gecarcinus*, and several other more fully terrestrial crabs, the branchial chambers usually do not contain much water. Nonetheless these respiratory sites must still be kept moist, perhaps by glandular secretion or in the case of the shell-dwelling land hermit crab *Coenobita* by periodic wetting using the "shell" water stores. The necessity to maintain a moist surface for respiration may contribute substantially to the animal's water loss, although C.F. Herreid (quoted in Bliss, 1968) has reported that water loss across the gills may be no greater than that occurring across the rest of the body surface.

(C) Oxygen Consumption

Individual Carcinus maenas showed an immediate increase in oxygen consumption from settled, submerged levels of air exposure and levels of oxygen consumption of air breathing species are generally higher than those of aquatic insects tested at equivalent temperatures. The aerobic metabolic scope of the air breathing crab *Cardisoma guanhumi* (0.12 $mmol.O_2.kg^{-1}.min^{-1}$) is greater than that of a cold water crab *Cancer magister* (0.05 $mmol\text{-}kg^{-1}$), but not appreciably greater than that of the aquatic crayfish Pacifaslaci.is (0.11 $mmol.kg^{-1}.min^{-1}$) at nearly equivalent temperature. This latter comparison suggests that the greater availability of oxygen in air may be offset by the reduction and coarsening of the gills, which is observed in many terrestrial species. An increase in the ability of gill and/oi accessory systems to function in air is seen in the more terreslrial cruslaceans as presented. While *Cardiosoma guanhumi* has clearly retained bimodal breathing, the morc aquatic *Callinectes sapidus* clearly has liltle ability to respire in air, whereas the more terrestrial *Gecarcinus* has apparently lost the abilily to maintain oxygen uptake in water.

Factors modifying oxygen uptake in air are similar to those of aquatic forms. Increases in ventilation art" noted during activity in *Cardisoma guanhumi* and in *Coenobita clypeatus* (McMahon and Burggren, 1979a), and in response to hypoxic or hypercapnic exposure in several species (*Birgus*: *Cecarcinus* and *Cardiosoma*; *Coenobita clypeatus*; *Ocypode quadrata*). Elevated venlilalion occurs by an increase in the duration and frequency of scaphognathite pumping as in aquatic forms, bill in some, it may be supplemented by branchiostegal pumping as discussed above. Measured values for oxygen extraction from air are generally less than 5%, considerably lower than values from water breathers (generally above 20%,, but extraction may be increased to levels approaching 20% in response to hypoxia in *Birgus, Gecarcinus* During emergence into air the normally aquatic crayfish *Austropotamobius pallipes* ventilates the branchial chambers with air at very low rates (5 mmol.kg^{-1}.min^{-1}). Under these conditions extraction from air may rise as high as 25%.

REFERENCES

Alspach. G.S., Jr. (1972). Osmotic and ionic regulation in the Dungeness Crab *Cancer magister,* Dana. Ph.D. Thesis, Oregon State University, Corvallis.

Angersbach, D., and Decker, H. (1978). Oxygen transport in crayfish blood: Effect of thermal acclimation, and short-term fluctuations related to ventilation and carcdiac performance. J. Camp. Physiol. B 123B, 105-112.

Ansell, A, D.(1973). Changes in oxygen consumption, heart rate and ventilation accompanying starvation in the decapod crustacean *Cancer pagurus. Neth. J. Sea Res.* 7, 455-475.

Arudpragasam, K.D., and Naylor, E. (1964a). Gill ventilation and the role of reversed respiratory currents in *Carcinus maenas* (L.). J. Exp. Biol. 41, 299-307.

Arudpragasam, K. D., and Naylor, L (1964b). Gill ventilation volumes, oxygen consumption and respiratory rhythms in Citrcinus maenas (L.). J. Exp. Biol. 41, 309-321.

Arudpragasam, K.D., and Naylor, E. (1966). Patterns of gill ventilation in some decapod Crustacea. J. Zool. Soc. London 150, 401-411.

Ashby, E. A., and Larimer, J. L. (1965). Modification of cardiac and respiratory rhythms in crayfish following carbohydrate chemoreception. *J. Cell. Comp. Physiol.* 65, 373-380.

Alwood, H. L. (1973). An attempt to account for the diversity of crustacean muscles. Am. Zool. 13, 357-378.

Babula, A., Bielawski, J., and Sobieski, J. (1978). Oxygen consumption and gilt surface area in Mesidothea entomon (L) (Isopoda, Crustacea). Comp. Biochem. Physiol. A 61A, 595-597,

Battelle, B. A., and Kravitz, E. A. (1978). Targets of cotopamine action in Ihe lobster: Cyclic nucleotide changes and physiological effects in hemolymph, heart and exoskeletal muscle. *J. Pharmacol. Exp. Ther.* 205, 438-448.

Batterton, C.V., and Cameron, J.N. (1978). Characteristics of resting ventilation and response to hypoxia, hypercapnia and emersion in the blue crab *Callinectes sapidus* (Rathbun). *J. Exp. Zool.* 203, 403-418.

Belamarich, F.A. (1963). Biologically active peplides from the (H'ricardial organs of the crab, *Cancer borealis, Biol. Bull. (Woods Hole, Mass)* 124, 9-16.

Belamarich, F. A., and Ferwilliger, R.C. (1966). Isolation and identification of cardio-excitor hormone from the pericardial organs of *Cancer borealis Am. Zool.* 6, 101-106.

Belman, B. W. (1976). New observations on blood pressure in marine *Crustacea J. Exp. Zool.* 196, 15-37.

Belman, B.W., and Childress, J.J. (1976). Circulatory adaptations to the oxygen minimum layer in the bathypelagic mysid *Gnathophausia ingens. Biol. Bull. (Woods Hole, Mass),* 150, 15-37.

Bennett, A. F. (1978). Activity metabolism in the lower vertebrates, *Anna. Rev. Physiol.* 40, 447-469.

Bergmiller. E., and Bielawski. J. (1970). Role of the gills in osmotic regulalion in the crayfish *Astacus, leptodactylus, Comp. Biochem. Physiol.* 37, 83-91.

Berlind, A., (1976), Neurohumoral organ extracts effect the ventiation oscillator in crustaceans. *J. Exp. Zool.* 195. 165-170.

Berlind, A., (1977). Neurohumoral and reflex Control of scaphognathite beating in the crab *Carcinus maenas, J. Comp. Physiol. A.* 116A, 77-90.

Berlind, A. and Cooke, J. M. (1976). Release of a neurosecretory hormone as peptide by eletrical stimulation of crab pericardial organs. *J. Exp.* 53, 679-686.

Blatchford, J.G., (1971). Hemodynamics of *Carinus maenas (L.). Comp. Biochem, Physiol. A.* 39A, 193-202.

Childress,). j. (1971). Respiratory adaptations to the oxygen minimum layer in the bathypelagic mysid *Gndthophausia ingens. Biol. Bull.* 141,109-121.

Cochrane, D. M. (1935). The skeletal musculature of the blue crab *Callinectes sapidus Rathbun, Smithson. M.W. Collect.* 92, No. 9, 1-76.

Cooke, I. M., and Goldstone, M. W. (1970). Fluorescence localization ofmonoamines in crab neurosecretory structures. *J. Exp. Biol.* 53, 651-668.

Copeland, D.E., and Fitzjarrel, A.T. (1968). The salt absorbing cells in the gills of the blue crab, (*Carllinectes sapidus,* Rathbun) with notes on modified milochondria. *Z. Zelliorsch. Mikresk Anat.* 92, 1-22.

Coyer, P.E. (1979). Heart and one gill-bailer rhythm phase couple in *Cancer borealis* and *Cancer irroratus. Comp. Biochem. Physiol. A* 62A, 677-683.

Cumberlidge, N., and Uglow, R.F. (1977a). Heart and scaphognathite activity in the shore crab *Carcinus maenas (L.) J. Exp. Mar. Biol. Ecol.* 28, 87-107.

Cumberlidge, N., and Uglow, R.F. (1977b). Size, temperature, and scaphognathite the frequency dependent variations of ventilation volumes in *Carcinus maenas (L.) J. Exp. Mar. Biol. Ecol,* 30, 85-93.

Cumberlidge, N., and Uglow, R.F. (1978). Heart and schaphognathite activity during the digging behaviour of the shore crab, *Carcinus maenas (L.) J. Physiol. Behav. Mar. Org., Proc. Fur. Symp. Mar. Biol. 12th* 1997, pp.23-30.

deFur, P.L., and Mangum, C.P. (1978). The effects of environmental variables on the heart rate of invertebrates, *Comp. Biochem. Physiol.* A 62A. 283-294.

Dejours, P. (1975). "Principles of Comparative Respiratory Physiology." Am. Llsevier, New York.

Dejours, P. and Beekenkamp, H. (1977). Crayfish respiration as a function of water oxygenation, *Respir. Physiol.* 30, 241-251.

Diaz, H., and Rodriguez, G. (1977). The branchial chamber in terrestrial crabs: A conparative study, *Biol. Bull.* 153, 485-504.

Dyer, M.F., and Uglow R.F. (1978a). Gill chamber ventilation and scaphognathite movements in *Crangon, crangon (L.) J. Exp. Mar Biol. Ecol.* 31, 195-207.

Dyer, M.F., and Uglow R.F. (1978b). Heart and scaphognathite beat behaviours in Laboratory held *Crangon, crangon (L.) J. Exp. Mar Biol. Ecol.* 32, 209-218.

Eddy, F.B., (1970). Acid-base balance in rainbow trout *(Salmo gairdneri)* Subjected to acid stresses. *J. Exp. Biol.* 65, 159-171.

8

CIRCULATORY SYSTEM

Parts of this system can be seen by dissection to be encountered in an internal dissection. Thus, it seems desirable to describe the circulatory system now, although, some parts will not be seen until later. The internal dissection is best done with the animal under water, particularly when dissecting very soft structures, such as the hepatopancreas and gonads.

If a living specimen is being used, the circulatory system may be color-injected. This is best done by boring a hole mid-dorsaliy in the carapace, halfway between the posterior border and the cervical groove, and using a syringe needle just large enough to fill the hole. This method usually fills the pericardial sinus and branchiocardiac sinuses, as well as the heart and arteries. If it is desired fo avoid having the heart surrounded by the color mass, it is possible to inject the heart directly, using a very fine needle, after first having cut a small window through the carapace and dorsal wall of the pericardial sinus.

To see the heart and neighboring arteries, it is necessary to remove the *Carapace* and the dorsal exoskeleton of the head. For this, suitable lateral scissors cuts should be made and the exoskeleton then carefully loosened from underlying tissues with a sharp scalpel, starting at the posterior margin of the carapace. Considerable care is necessary not to damage certain arteries and muscles located close under tho hypodermis. The dorsal exoskeleton of the abdomen sliould be removed in the same way, starting at the posterior end.

The heart lies in a space called the pericardial sinus. It is suspended by three elastic ligaments on each side and is stretched laterally by these ligaments during diastole. Blood enters the heart from the pericardial sinus through valved slits, termed *ostia*, of which thery are a dorsal, a lateral, and a ventral pair. Thin-walled arteries, whose connections with the heart are very easily broken, leave thy heart anteriorly and posteriorly. An internal valve is located at the start of each. From the anterior dorsal margin of the heart a slender ophthalmic artery runs forward in the midline, close under the; hypodermis. On the anteriodorsal surface of the stomach, not far behind the eyes, this artery expands slightly to form a structure termed the cor frontale, then forks into brandies which supply the brain, eyestalks, and antennules. Two tiny muscles, the posterior basal ocular muscles, arise by tendons from-the dorsal integument, close on either side of the midline, a short distance posterior to the cor frontale. These two muscles run ventroanteriorly, converging towards each other, pass through the wall of the cor frontale, and emerge to join a median tendon which runs venterally to its insertion on the integument between the antennal coxac. The belly of each muscle lies within the cor frontale, and the swelling of these two muscles when they contract-must exert pressure on the surrounding blood. It is thought flint the whole arrangement acts as an accessory heart, increasing the flow of blood to the brain, eyes, and antennules. A further remarkable feature of the cor frontale is that it also encloses a part of the stomato-gastric nervous system. A tiny pair of anterior basal ocular muscles arises between the eycstalks and inserts on the same median tendon as does the posterior pair, but without traversing the cor frontale.

A large antennary artery leaves the heart on each side close to the origin of the ophthalmic artery and runs lateroanteriorly to supply the gonad, branchiostegite, body wall, thoracoabdominal muscles, hepatopancreas, stomach and its muscles, adductor muscle of the mandible, antennal gland, antenna, antennule, eyestalk, rostrum, and brain. From the ventroanterior margin of the heart a pair of hepatic arteries originate, each running a course parallel to the more dorsal antennary artery, and supplying the hepatopancreas. Ventroposteriorly a median dorsal abdominal artery leaves the heart. The beginning of this artery is swollen as a thin-walled bulb, from which a pair of arteries arises supplying the gonads and genital ducts. The root of one of these arteries is much enlarged, and from this root a large branch, the sternal artery, which will be considered presently, passes vertically downwards. From the thin-walled bulb the dorsal abdominal artery runs posteriorly, dorsal to the intestine, giving off a pair of lateral segmental arteries in each abdominal somite to supply the intestine, posterior parts of the hepatopancreas, abdominal muscles, pleopods, uropods, and telson. The sternal artery passes either to left or right of the intestine and pierces the ventral nerve cord, below which it divides into a small, ventral abdominal artery and a larger, ventral thoracic artery. The ventral abdominal artery supplies the nerve cord, the last two pairs of pereiopods, and a small portion of the ventral part of the abdomen. The ventral thoracic artery supplies the nerve cord, first three pairs of pereiopods, the mouth parts, and portions of the antennal glands.

All the arteries branch repeatedly, the fine, ultimate branches sometimes being termed capillaries, because they often have a diameter as small or even smaller than of a mammalian capillary. However, such a use of the term seems inadvisable, since there is no proof of physiological exchanges through the walls of these vessels, nor do they connect with attenules as do true capillaries in a closed blood system. Instead, they open into tissue spaces, in which the blood can bathe the tissue cells directly. Some of these spaces are no more than the chinks between cell. These, however, are continuous with wider spaces, termed sinuses as well as with the main cavity of the body. All these spaces develop firm the primary body cavity, and since they contain blood they are collectively spoken of as the hemocoel. In the hemocoel the blood follows a more or less definite course of circulation, eventually reaching, in the thorax, the sternal sinus, which surrounds the ventral nerve cord and ventral thoracic artery. From here the blood enters the gills, passing up the outer side of each gill in an afferent branchial sinus, then descending on the inner side of the gill in an efferent branchial sinus. A small branch from the afferent sinus leads out to a superficial plexus in each gill filament, from where a channel starting in the center of the filament returns the blood to the efferent sinus. From the gills the blood on each side of the body passes to the pericardial sinus through six branchiocardiac sinuses (five in the lobster). These run dorsally within the body wall, which in this region is very thin and membranous, being protected by the overlying gills and branchiostegite. From the pericardial sinus the blood enters the heart, as already described.

The nervous supply to the heart is described under the nervous system, but if only a single specimen is available and it is desired to see these nerves, they should be looked for before the dissection has progressed too far.

The blood may have a slightly bluish tint imparted to it by a respiratory pigment, hemocyanin, dissolved in the plasma. Three types of blood cells are present:

(1) explosive corpuscles, which initiate blood-clotting by an explosive cytolysis;

(2) thigmocytes, which phagocytize foreign particles and which further contribute to bloodclotting by throwing out long, Fine processes;

(3) α-mococytes, which are much less phagocytic than the thigmocytes, but which display considerable amoeboid activity, as can be observed in a gill filament of a living animal. Blood allowed to flow out under water usually coagulates rapidly, the clot having the consistency of a soft gel. The blood cells are derived, at least in part, from a pair of blood-cell-forming organs on the anteriodorsal surface of the cardiac stomach, as well as from small nodules in the connective tissue in various parts of the body. The organs on the stomach are not easily seen, other than histologically. They are thin, semitransparent, and formed of discontinuous lobules in the crayfish, but are more solid and continuous in the lobster.

In crab (freshwater) blood is transparent blue coloured with colourness lymph. It contains types of cells amoeboid semitransparent with distinct nucleus and granular protoplasm and globular with big granules cells.

The cmstacea, like other arthropods, have an open vascular system in which the circulatory fluid is not confind to vessels and so separated by membranes from the interstitial or lymphatic fluids; for this season it is often called haemolymph. The open system is furthur differentiated from a closed one by a lower arterial pressure and higher blood volume. The haemolymph unlike that of trachaete arthropods but like that of mollusc. Transport oxygen as well as nutrients and waste. In most crusctaceans the circulation is mainatained by a systemic heart, dorsally placed and assisted by movement of appendages and other organs, and sometimes by one or more accessory hearts, local dilalation of a blood vessel with muscle which decrease their volume on contraction. Barnacles have only this type of heart, situated between the adductor muscler of shell and oesophagus, where the acceleration of venous flow due to the beating of cirri is least.

The heart is single chambered situated in a pericardial space by suspensory ligaments and by arteries space is itself a blood sinus separating the heart (with which it communicates through ostia) from the rest of gahenocoelomic cavity primitively, the heart is tubular, as in the Anostraca (*Chirocephalus, Artemia*) where it extends through the thorax and abdomen and has a large number of paired segmentally arranged ostia. The heart may also be elongated in the lower malacostraacans but the ostia are not segmental and may be rednced to a single pair (Anaspides). Since blood enters the pericardial cavity manily from the respiratory organs the heart tends to become restricted to the part of body where these organs occur; thus in mysides amphipods and eucarids is large though short, with the walls thickened by several layers of crossing muscle strands; in this it contrasts with tubular heart of lower forms with a single muscle layer.

System is affected by the cardiac muscle which, on contraction, expel blood into the arteries where there are valves preventing back flow. Pressure in the pericardial cavity is lower and as a consequences blood is drawn into it from visceral sinuses or branchial vessels. The heart then tends to return to a resting shape but a diastolic mechanism is provided by the suspensory ligaments which distend the heart and open the ostia by their elastic pull. This mechanism is adequate to maintain circulation in a low- pressure system and even during the increase in blood volume and body size at time of ecdysis. In the majority of crustaceans the heart in a dorsal cardiac ganglion or scattered through the cardiac nerves. In branchiopods evidence suggests that the pace maker is muscular.

Charges in the volume of pericardial cavity influence the role of cir culation and movements of the body and of the pericardial septum bring these about. In ostracoda, copepods and Eumalacostraca alary muscles origimate on the lateral body wall and extend medially into the septum ending a short distance from the mid line which it crosses as a nonmuscular elastic memberane. By their contraction they flatten the normal convex pericardial floor and so enlarge the cavity and draw in blood from surrounding space in higher crustaceans are the branchiopericardial sinuses or efferent branchial veins from the respiratory organs and the action of *alary* muscles speed up the circulation through these and the channels in the gills. In ostracods, copepods, cladoceran the pericardial cavity is less completely closed and may have large openings to the visceral sinuses.

In lower crustaceans only an arterior arota leaves the heart. It leads over the gut towards the brain and may open directly into the haemocoel without branching, as in branchiopods and copepods. The blood flow as in other arthropods and annelids is anterior in the dorsal and posterior in the ventral channel and oxygenation occurs in gills, appendages or carpace just before the haemolymph returns to the heart by way of the sinus system. Malacostraca are larger and have appendages often with specialised gills which can support the body; associated with these changes is the development of an arterial system more like that of *Astacus*. Their arteries have elastic nonmuscular walls and in certain place branch to give capillary beds. The arterial blood is discharged into lacunae and then collected into venous sinuses which are spaces typically bounded by a membrane. It ultimately flows to the sternal sinus and so to the gill and heart. Anterior with a variable number of lateral versds segmentally arranged; or the lateral vessels may arise from the aortal. branches from the segmental vessels supply the join a longitudinal artery associated with the ventral nerve cord and musculature. The ventral vessel not present in decapods where special respiratory organs occur with segmental veins return blood from them to the pericardium and the series of segmental arteries tends to be interrupted.

The arrargement of vessels is modified in the decapods to give a more direct arterial supply to the ventral region of the body and in particular, to the thoracic appendages which are of increasing importance in these benthic forms. Thus much of the blood which passes posteriorly from the heart enters the stemal artery. This artery is perhaps derived from fusion of a segmental pair. It passes ventrally to the sub neural vessel, now known as ventral vessel from which segmental branches pass to all the appendages of the cephalthorax except antennules and antenna.

Then adaptation of isopod species to aquatic amphibious and terrestrial habitats is associated with an increasing efficiency of the circulatory system. This culminates in land isopoda, which have a relatively larger heart, stronger alary muscles and strictly defined often vessel like lacunae, which with the true vessel from an almost complete circulatory system ensuring a better regulated and swifter transport of blood.

Despite the possession of an open vascular system some of the higher malacostraca. are able to cast off injured appendage without loss of blood or damage of soft tissues. In crab, chelipods and pereiopods have each a preformed breakage planc high on the limb near the line of separation between basopodite and is cheliopodite. Here a ventricle invagination of the epidermis produces a two-layered septum, which leaves a small opening through which run blood vessel and nerve but no muscle. The Autotomy of a limb at this plane is reflexly initiated and is brought about by extrereme contraction of a muscle (the automizer muscle) which is at other times involved in the normal activity of the limb. The outer layer of the septum is cast off with the limb the inner persists and its opening sealed by a stump. A sheet of new cuticle is secreted over the whole surface. Regeneration of cuticle begins at the next molt.

The Blood System

The main body cavity of arthropods is a haemocoel, true coelomic spaces being restricted to the cavity of the excretory organ and genital ducts. There is no development of an interstitial fluid and lymphatic system separate from the blood, as the blood itself comes into direct contact with the tissues.

Since the blood thus combines the function of interstitial and circulatory fluids, it is usually described as haemolymph.

- Blood serve primarily as an equilibrating agent.
- Blood serve as a vehicle for the transport the about body of food materials, respiratory gases, metabolic excretory products and hormones.
- Blood pressure affects the shape of the body. (*e.g., Apus, Artemia*)
- Limb movements also are governed by blood pressure.
- In addition to its circulating role, it acts as a hydrostatic skeleton, giving shape and mechanical rigidity to the body. (*e.g., Artemia*).
- In larger crustacean, the skeleton function is taken over by cuticle and the haemolymph is restricted to a complex system of vessels and lacunas.

The Circulating Pathways

- A dorsally situated heart is present in most crustacean though a true heart is lacking in the cirripedes in many copepods and Ostracods.
- So, the function of heart is served by somatic muscles.
- Alternatively, circulation of the body fluid may be brought about by movements of the body (upward and downward).
- The direction of blood flow is controlled by values. So that blood tends to pass forward in a medium sinus and posteriorly in lateral sinuses.
- Thus, even in absence of the heart, an essentially one-way circulation of blood can be achieved.
- Lacking of heart, than true circulation seems unlikely though peristaltic movements of the gut may provide a measure of mixing of blood, (*e.g., Cyclops, Trebius*)
- A heart, where present, may take various forms, ovoid in Calanoid copepods.
- Tubular in amphipods, isopods and stomatopods. Rhomboidal in Brachiura.

Heart-Neurogenic (Except-Branchiopoda)—(and Arterial Heart)

- The walls of the single heart chamber are pierced at intervals

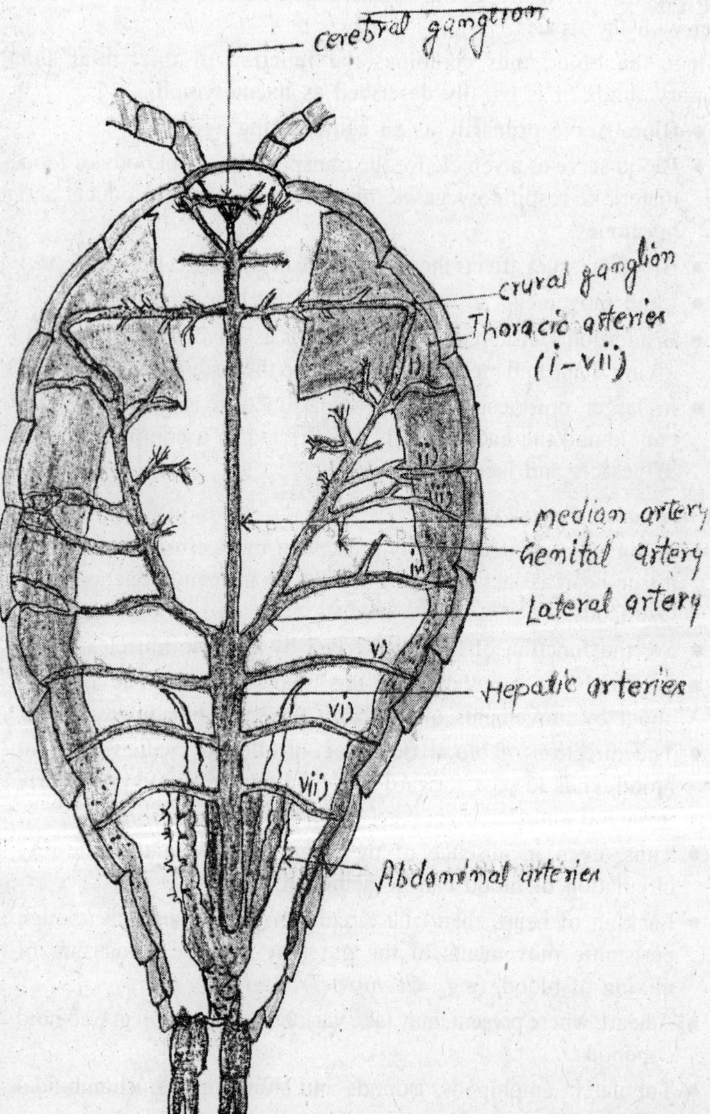

Fig. 8.1 : A heart and arterial system of the isopod *Ligia*.

by ostia. The pericardial cavity to refill the heart after a contraction.

- Ligaments, taking their origin in the pericardium and ventral septum, serve tossting the heart within the pericardial cavity.

Working of heart

- When the heart contracts, most of the contained blood is driven anteriorly into the peripheral circulatory systems.
- In most decapods, amphipods and isopods the circulatory system is largely composed of well defined vessels.
- In arteries lead from the heat, though these differ from their vertebrate equivalents in being composed only of endothelial lining and connective tissue.
- In some small forms, the main artery may open into the haemocoel without branching, (*e.g.,* Branchiopoda).

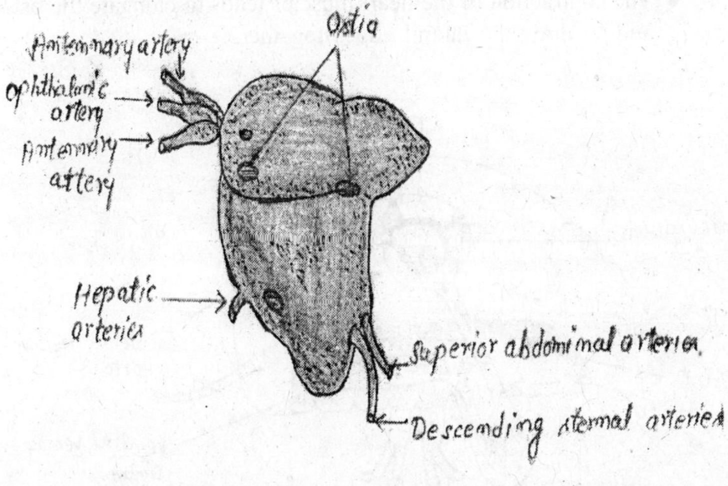

Fig. 8.2 : The Heart of *Eupagurus* (showing the compact arrangement typical of decapod hearts and also showring the astia).

- The arteries subdivide; and the thickness of the connective tissue progressively decreases, *i.e.,* like capillaries of invertebrates, (*e.g.,* Decapoda)
- The capillaries ultimately open into lacunae, spaces ramifying between the tissues where there is little or no barriers separating the haemolymph and general body cells.

- Blood leaving lacunae is passed into sinuses which, although often large and ill-defined, differ from the lacunae in having an endothelial wall separating than from the general body tissues.
- In cases where the diameter of sinus in such as to localize the blood flow they are sometimes termed veins.
- In the lobster, study of the fate of dyes injected into various parts of the body indicates that all blood returning to the heart must first have passes through the gills.

Major circulating routes in decapoda

In the smaller crustacean there are usually fewer well defined vessels and the lacunae are proportionately more extensive than in the larger forms.

The heart beat

- When the heart muscles begin to contract the ostia must be closed if blood in not to escape in the wrong direction.
- The contraction of the heart muscles tends to elongate the ostia and so draw the lateral edges together.

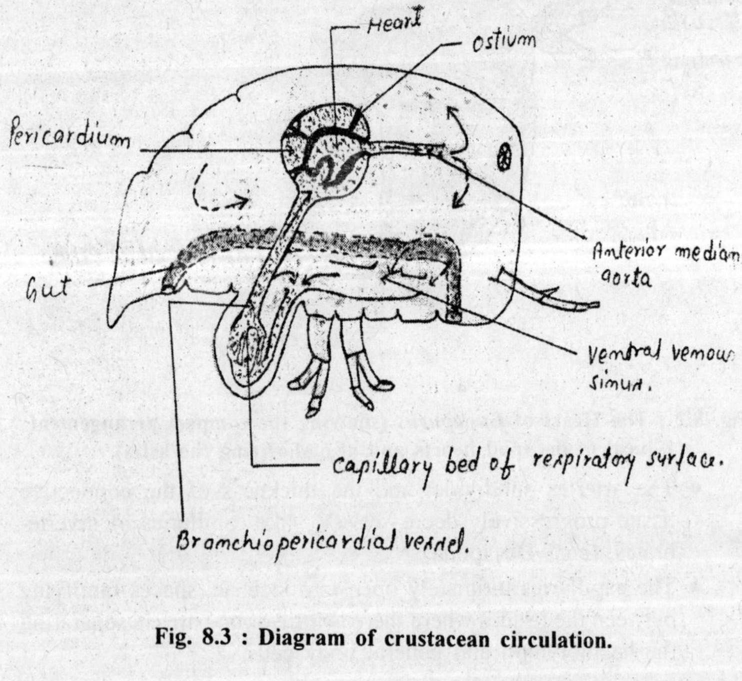

Fig. 8.3 : Diagram of crustacean circulation.

- Complete closure is effected by the shortening of the ostial muscles which run round the rim of the parts.
- When muscular ostial values are also present, as in the decapods, these are shut by increasing internal pressure in the heart as the systolic contraction begins. Values, preventing backflow of blood into the hearts are present at the cardio-aortic junction.

Systolic pressure in heart	→	Pressure in Aorta	→	Valve open	—Blood→	Aorta

- The pericardial walls from an essentially rigid structure so that as the heart volume decreased during the expulsion of blood the pressure in the precardium naturally falls.
- Blood in, therefore, drawn in from the branchio-pericardial vein.
- When the delivery stroke is completed the heart muscle relaxes the cardio-aortic values shut and the ostia open.
- The elastic ligaments supporting the heart, which are stretched during contraction, then pensively re-expand the heart while simultaneously haemolymph flows in through the ostia. If the ligaments are cut, re-filling of the heart fails to occur.

Blood composition

- Crustacean blood or hemolymph is composed of plasma and blood cells.
- The blood cells are frequently amoeboid, and two structural types predominate.
- One which lacks granules and the average size is smaller.
- Two somewhat larger cells contains granules thought to be stored nutrients.
- Besides the storage function crustacean blood also serve to transport nutrients, hormones and respiratory gases.
- Two respiratory pigments occur in the plasma of crustaceans.

Decapod and stomatopod, malacostracans commonly have hemocyanin, where as hemoglobin occurs in many other crustacean.

Transport nutrient and hormones throughout the body. Coagulation in which both cells and plasma participate, is the important function of blood.

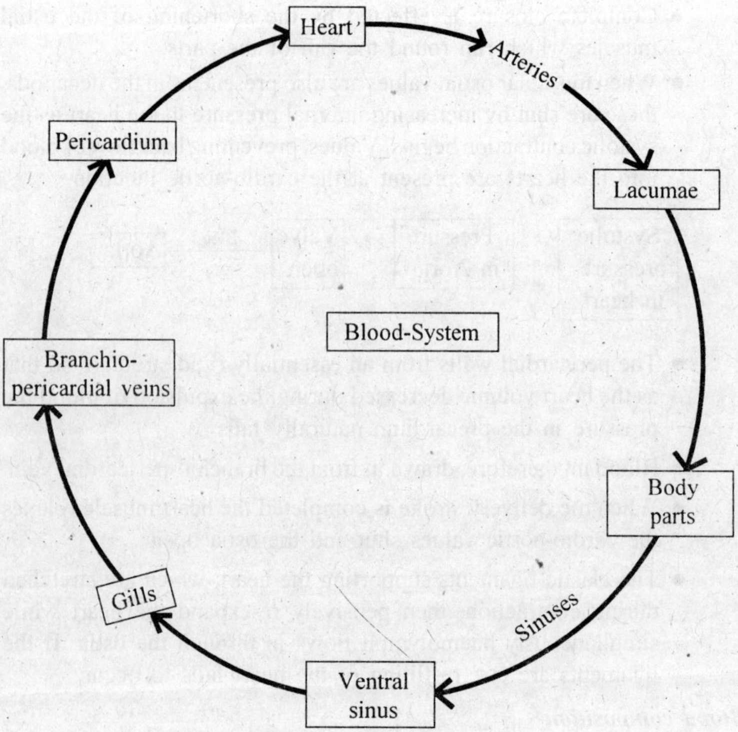

Fig. 8.4 : Below the gut.

Heart of Crustaceans

Sub-class	Heart position
Branchiopods	Extends almost to the end of the uniform trunk, and a pair of ostia is present in each segment except the first and last.
Malacostraca	Heart tends to reduced when the abdominal region is suppressed as in the cladocera.
	Heart tends to become reduced in length and limited to the thorax. Long heart *e.g., Nebalia*
Cladocerans	Heart may be reduced when the abdominal rigion is suppressed.
Small	No heart
Baranchiopoda	Amphipoda elongated heart which is restricted to the thorax and which never ahs more than 3 pairs of ostia.
	Isopoda extends into both thorax and abdomen, especially in those forms with an elongate heart. More posterior position than malacostraca (Fairyshrimps).

Heart Beat

- An open circulating system is less depended on heartbeat and shows greater fluctuation in pressure then a closed system. Pressure rises and falls dramatically with each heart beat, for there are few elastic blood vessels walls to absorb part of the pressure.
- Blood pressure in a resting lobster is 13 cm, H_2O at systole and only 1 cm, at diastole. Blood pressure rise markedly with body activity. Systolic (Contraction) pressure is doubled in an active lobster and diastolic pressure is brought to about the level of systolic pressure in resting animals.
- The stimulus for the beat originating in the heart muscle, itself. The small neurons of heart ganglions are pacemaker and fire spontaneously.
- The large neurons are motor neurons, which follows the pacemakers and do not fire spontaneously unless they are separated from the small neurons.
- With the appearance of inhibitory and accelerator fibers, the machinery needed to efficiently regulate heart beat by feedback controls is achieved.

The crustacean circulatory system, with few exceptions is best described as an open system in which the blood or haemolymph flows freely throughout the hemocoelic cavity. Circulation of blood may result from body movements, muscular and gut contraction , or through the pumping action of a heart. Not all crustaceans are provided with heart, but when one is present it is a simple, single chambered organ lying dorsally above the gut and enclosed in a pericardial sinus. The musculature of the heart may vary from a simple pattern of fibres to a thick layer formed by crossing muscle strands. Blood enters the heart from pericardial sinus through one to many pairs of valved, slit-like openings termed ostia. Blood leaving the heart may directly flow into sinuses and lacunae of the hemocoel, or may be channeled to various body regions by artries and capillaries. When the pumping action of the heart is insufficient for adequate blood flow, heart may be developed in one or more blood vessels. This structure, referred to in the Malacostraca as a cor frontale, is formed from an enlargement of a blood vessel. Pumping action is causd by contraction of outer tangential muscles or internal muscles of blood vessel derieved from muscles having other functions.

Among crustacean groups the development of an arterial system is quite variable. The principal component of the system is the anterior arota, which leads from the heart anteriorly towards the cephalic region. To this posterior, lateral and occasionally ventral artries may be added, particularly in the Malacostraca. The arterial system of a number of representative crustaceans are summarized well by Von Baumann. Veins are lacking in all crustaceans; the venous pathways are through a series of sinuses that return the blood from tissues and respiratory surfaces to the pericardial sinus. Retrograde blood flow is prevented in the artries by cardioarterial valve and arterial valve and in the venous sinuses by venous valve.

In the Branchiopoda, the circulatory system of the Anostraca and Cladocera, differ appreciably in heart structure. In the former the heart is elongate tubular and provided with 14 to 18 pairs of segemental ostia; no artries are present. In contrast the cladoceran heart is globular and with only one pair of ostia. Among the ostracoda, a heart with one pair of ostia is present only in myodecopids. Blood flows from heart directly into the hemocoel in same taxa, but an anterior arota is present in other. Only in the calanoid and myophriod orders of free -living copepoda a short heart is present. It is provided with a lateral pair of ostia, one ventral ostium, and a very short (heart) is present.

As barnacles aortic artery differ significantly from other crustaceans in the morphological structure so does their circulatory system. A definitive heart is lacking in all cirripeds.In barnacles blood leaving the rostral vessel is transported to the peduncle and ovaries by a large peduncular vessel possibly to represent the anterior arota of crustaceans. From the ovarian plexus and longitudinal muscle plexus of the peduncle, blood flows into a mantle plexus which is continuous with the plexuses of the opercular valve. Blood is collected from these plexuses by a pair of actual vessels and transported to the paired scutal sinuses located to other side of rostral sinus. The actual sinuses contribute significantly to cephalic circulation as well as directly to the maxillary gland plexuses and to the gut circulations by means of paired lateral vessels. Blood from the gastric vessel eventually is channelled into a median ventral epineural sinus from which it flows to the cirri, penis and oral (mouth parts). Blood from the oral corners is returned to the rostral sinus. Blood from cirri and penis enters the thoracic peripheral circulation and is finally collected in a pair of dorsolateral vessels which converge in the cephalic region to from the proxomal sinus.A pair of large channels,provided with valves connects the proxomal sinus with the rostral vessel. Blood flows from the rostral sinus into large basal vessel, thought to be equivalent to the penduncular vessel of the *lepadomorphs*, but which doesnot extend to the basis of the barnacle. Instead it branches into two pairs of circumferential vessels from which thoracic and peripheral circulations primarily are derieved.

Among the Malacrostraca the arterial system generally is well developed. In leptostracans the tubular heart extends from the region of the cephalon to the fourth abdominal somite. Three pairs of ostia are present anteriorly and four pairs, three dorsal and one lateral in the thoracic region of the heart. The arterial system consists of a short branching artries, eight pairs of thoracic segemental artries, four pairs of abdominal segemental artries, and a posterior arota

The circulatory system of stomatopods has been considered to represent a generally primitive condition. The tubular heart, which extends from the region of the second thoracic somite to the fifth abdominal somite, is provided with paired segemental ostia and artries. As the large anterior arota leaves the heart it gives off a pair of anterior laterial artries which provide blood to the gut and mouth parts. More anteriorly the arota branches to supply blood to the antennae, antennules, eyes, and supraoesophegeal ganglion. In *Squilla* the first pair of segemental lateral arterial supplies both the first and second pairs of thoracopods. Other variations form the truly segemental pattern are found in the eighth pair which provides blood to the tissues and musculature of the last thoracic and first abdominal somites, and in the thirteenth pair, which supplies both the fifth pair of pleopods and the uropods. Branches of these lateral artries connnect with large ventral sub neural artery. A short posterior arota provides blood to the telson. Blood returning to heart passes first through a large ventral lacunae into the pleopodal gills and then through segementally arranged, dorsally directed venous sinuses to the pericardial sinus. Whereas, in *Squilla* the posterior thoracic appendages are supplied by the preceding segmental artery, the their teeth pair of lateral arteries is the only supplier of blood to the uropods: the forteenth pair provides blood to the lateral margins of the telson.

Among the pericaridans the heart is confined to the thorax in mysids, crutaceans, tanaids and amphipods. It is generally restricted to the abdomen in isopods;however in the terrestrial species, *Metoponorthus pruinosus* an oval heart is located at the level of the fifth and sixth thoracic somites. Among the Mysidacea the heart is provided with two or three pairs of ostia. A cor frontale has been reported both in mysids and in isopods.

The aorta extends the length of the abdomen in amphipods, my sids and some terrestrial isopods, but is lacking in most of the later taxon. In lophogastrial mysids nine pairs of artries leave the heart laterally: however only the first pair and three unpaired artries are present in representatives of the Mysids that have been studied. Five pairs of lateral arteries are typical of isopods. Three pairs are characteristic of amphipods. Although a sub neural artery is present in isopods only in mysids do both a sternal and ventral thoracic artery occur. All three are absent in amphipods. While variations do occur particularly between species of lophogastrida and mysids, other seemingly contradictory statements may result from a lack of uniformity in terminology.

The heart in euphausids is relatively small, sub rectangularly shaped organ with two pair of ostia. Three vessels the single anterior arota and a pair of anterior lateral arteries leave the heart anteriorly to supply blood to eggs and supraoesophageal ganglion. A short distance from the heart each anterior lateral artery gives off a recurrent branch that is directed posterior into the thoracic musculature. In addition to the cephalic appendages, the anterior lateral arteries supply blood to the musculature glands. A pair of hepatic arteries leaves the heart anterioventrally to provide blood to the mid gut gland. Leaving the heart posteriorly are a pair of posterior or lateral arteries the posterior aorta, which may be single or paired, and a ventrally directed sternal artery. The latter also may be paired but only one of the pair is ever well developed. At the level of the ventral nerve cord the stemal artery branches into two or three vessels. The anterior branch the ventral thoracic artery supplies blood to maxillae and anterior six pairs of thoracopods through lateral branches. In those taxas where a small middle branch of sternal artery is retained it provides blood to the sixth pair of thoracopods. The posterior branch the ventral abdominal artery supplies blood to the last two pairs of thoracopods and to the lumniscent organs of the abdomen. Branches of posterior aorta transport blood to the appendages musculature and body tissues in each abdominal somite.

Blood from the pericardial sinus enters the heart through three pair of ostia in most decapods: however through five pairs in the heart of caridian shrimp *Caridina laevis* Heller. The position of these ostia are highly variable. In *C. laevis*, two pair are dorsal two are lateral. In cray fish of the genus *Astacus* and in lobster *Homarus americanus* there is one dorsal one lateral and one ventral pair whereas in pagurids and brachrurans one pair is dorsal or anterodorsal and two pairs are lateral, m the spiny lobster *Panulirus* and in galatheids, two pairs are lateral and one is ventral. As in other crustacean heart cach ostium is provided with a set of valves that prevents the blood from flowing back into pericardial sinus on systole; however, mechanism in galatheids differs from that found in *Astacus* and *Homarus*. The arterial system among the decapods basically all are quite similar. Only in the panaeids is the anterior arota vestigeal although, it usually is more strongly developed in brachrurans and many macrurans than in anomurans. A corfromtale has been reported in the anterior arota in *Astacus* and *Homarus*; in *Panulirus interrupts* and two species of *Cancer*. The anterior arota typically privides blood to eyes and supraoesophageal ganglion, whereas the anterior lateral artries are primarily suppliers of blood to cephalic appendages, the walls of foregut, the musculature, antennal glands and carapace. The hepatic artries are similar in all decapods except the pagurids and possibly some of the thalassinoids where the midgut gland have been displaced to the abdomen. In these taxa the hepatic arteries terminate on the foregut or midgut. The origin of sternal artery is variable. In *Astacus, Panulirus* and *Homarus* it arises from a bulb like swelling immediaty posterior to the heart.

Because many of the oxygen carrying proteins in animal blood do not occur in cells they have physiological functions other then gas transport .They enhance the buffer capacity of blood, they dominate the temperature sensitivity of its acid base status and are fully responsible for its colloid osmotic pressure.

It is avadanced and well adapted for mud and borrowing habitat.

The dorsal crab chambered, quadrangular in shape and is situated in the posterior median aspect or the cephalothoiax in between the two branchial chambers. It is surrounded by thin, pericardial membrane on its dorsal side which forms the pericardial cavity or the blood sinus. The pericardium is a thin transparent membrane. Ths roof and lateral walls of the pericadium are formed from subcutaneous connective tissue of outer body wall, extention of the pericardial septum and rarely longitudinal musculature of the thorax.

The blood enters the heart from the pericardium by three pairs of ostia which are provided with naps opening inward. One pair is situated at the anterior end of the dorsal wall of the heart, ths second pair lies at the posterior end of the dorsal wall just below the first pair and the third lies on the lateral side of the heart. The circulation is of open type and the blood is the hemolyph. It usually contains corpuscles.

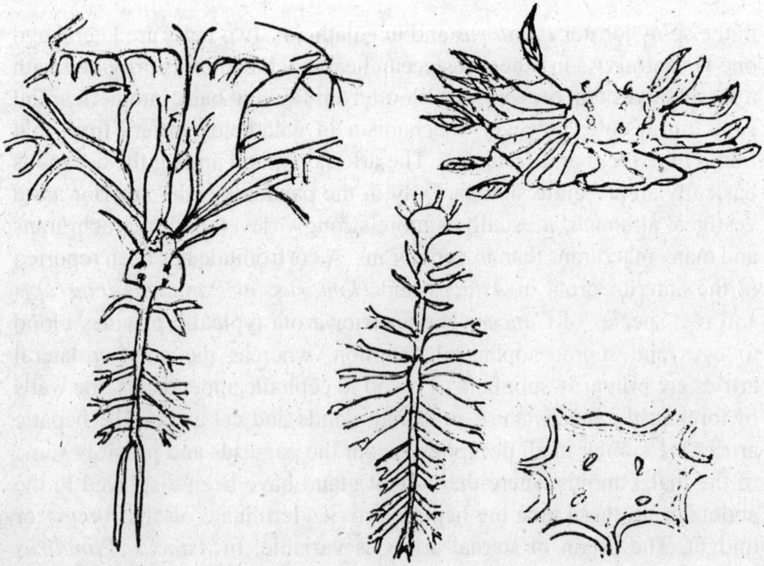

Fig. 8.5.

Arterial System

All the arteries of the typical crustaceans studied in three parts:

1. Anterior artery,
2. Dorsal abdominal artery, and
3. Sternal artery.

1. *Anterior Artery*

Three main trunks run forward from the anterior end of the heart the median is the cephalic artery and the paired lateral arteries.

Median Cephalic Artery : This lies in the median dorsal part fo the cephalothorax. It originates from the middle of heart and passes directly forward at the anterior end.

It goes to a straight course without any division over the pyloric and cardiac portion of the stomach and also between the anterior gastric muscles. At the anterior end of the cardiac forgut it plunges downward and then divides into two branches supplying the eyes and various part of the front region of the head.

Lateral Artery : The paired lateral arteries arise one on either side of the median caphalic artery. Immediately it then curves outward and gives out two branches which supply to the cardiac and pyloric part of the stomach. Further the lateral artery runs for a little distance and then it divides again into two branches, an outer onc is the ovarian artery (spermatic artery) which supplies to the gonad. The other one plunges downward at the anterior end, passes around the front part of the foregut and supplies to the various organs in the head.

Branches of the Lateral Artery

1. *Branch to the Hind-gut :* This branch arises on the inner side of the artery and supplies to that portion of the hind-gut which lies beneath the front part of the heart.
2. *Branch to the Digestive Gland :* This is branch which arises immediately in front of the branch to the midgut. It gives out many branches which supply to the digestive gland and hypodermis.
3. *Branch to the Cardiac Foregut :* This is a large branch given out on the inner side or the lateral artery. It passes through the gonads and breaks up into many branches on the lateral and dorsal walls of the cardiac foregut. It also supplies to the muscles of the foregut.
4. *The Ovarian Artery (Spermatic Branch) :* The lateral artery after giving out a branch to the cardiac foregut further divides into two branches. One is the antennary branch and the other is the ovarian branch This divides into numerous small branches which supply to the gonads.
5. *The Antennary Branch :* It gives out a small branch which supplies to the 'Y' organ. The outer branch dips downward and gives branches to we external adductor muscles of the mandible. It further supplies to the hypodermis and hepatic lobe of the bladder. The inner branch passes inward and supplies to the antennae and the front portion of the head. It also sends branches to the gastric muscles, paragastric lobes end and esophageal lobes of the bladder.

Hepatic Artery : The hepatic arteries originate close together from the ventral portion or the heart slightly external to that of the origin of the antennary artery. Immediately after leaving the heart it dips downward and passes into the deeper parts of the digestive gland. Near its origin it gives off a branch to the hindgut. It divides into two, out of which one supplies to the digestive gland and the other passes beneath the gonads, external adductor muscles and supplies the anterior portion of the gland.

2. *Dorsal Abdominal Artery*

This starts as a median vessel from the posterior end of the heart, it gives off many branches to reproductive organs lying below the periardium and muscles of the basi-ischium of the chela. Just above the hindgut caecum paired branches are given out called as posteroleteral arteries. On vessel supplies to the coiled portion of hindgut caecum and another to the external muscle of coxopodite of the four walking legs. The main artery then divides into two and runs up to the anus giving off few branches to the muscles of the abdomen and hindgut.

Descending Artery : This arises from the posterior end of the heart very near to the origin of the posterior aorta. An extremely wide vessel which directly downwards on the right side of the hindgut. It passes downwards and forwards between the muscles of the fourth and fifth thoracic appendages and finally through the foramen of the thoracic ganglion. Immediateiy below the nerve chain this artery divides into two branches, one short broad posterior branch and another long narrow anterior branch. These are sternal arteries.

Sternal Artory

This is a large and most prominent vessel lying just below the nervous system and between the muscles of the thoracic appendages. At the anterior end the sternal artery becomes narrower and many branches are given off from the main appendages and digestive gland. The arteries supply tn the chela and walking legs arise separately from the sternal artery except in the last two pairs of walking legs. A single vessel goes to appendages into two branches each passes into appendage to the extreme end of it giving off may branches on the way supplying muscles or the limbs. The arteries supplying to the mexilliped arise from the main branch of the sternal artery and further divides into two supplies to the endopodite and other to the exopodite. The sternal artery bifurcates just in front of the first maxilliped, each branch passing forwards and unitng again behind the month, forming a ring. From this ring arises many arteries which supply to the mandibles, first and second maxillae.

Posterior Sternal Artery : This artery continues as a narrow vessel and passes backwards and down the abdomen beneath the nerve cord and the hindgut. It gives few small branches to the hindgut and flexor muscle of tha abdomen.

Blood Sinuses and Veins

The arteries after repeated branching open into small inter-communicating blood spaces which are called as blood sinuses. These lead into larger spaces through which the blood is brought back to the pereicardial sinus via branchiae. There are no definite vessels as such for returning the blood from the body to the heart. All the main organs of the body have sinuses in close contact with them.

Dorsal Sinus

Above and infront of the cardiac foregut there is a distinct blood space called as dorsal sinus. This passes down the front portion of the foregut and is connected with a sinus which lies beneath the stomach. The dorsal sinus divides into two parts near the esophagus.

Sternal Sinus

The main sinuses are sternal which run ventrall along the entire length of the thorax between the leg muscles and beneath the pericardium. These two sternal sinuses are not separated from one another but are connected by smaller sinuses hear and there. The sternal sinus does not communicate directly with the gills but it is connected to the infra-branchial sinuses which opens by five distinct clefts on either side. These are the branchial sinuses.

Branchial Sinuses

The branchial sinuses are five in number. The first branchial sinus starts from the anterior of pericardium. This goes upto the pleural muscle chamber of the second thoracic segment and opens into the infra-branchial sinus. The second branchial sinus passes down the third thoracic pleural muscle chamber, the third into the fourth thoracic pleural muscle chamber, the fourth into the fifth thoracic pleural muscle chamber and fifih opens at the posterior end of the pericardium. All the pleural sinuses and pleural muscle chambers are situated just below the branchiocardiac veins.

Infra Branchial Sinuses

They run along each side of the thorax at the base of the thoracic legs and the gills. From this are given off the afferent branchial sinuses to each of the gills. The different branchial veins collect the pure blood from the gill lamellae.

Branchio-Cardiac Veins

The branchio-cardiac veins collect the pure blood from the efferent branchial veins and convey it to the pericardial cavity. There are five cardiac veins having definite walls. Valves are absent between the branchio-cardiac vein and the pericardium.

The blood after purification in the gills passes into the pericardium via branchio-cardiac veins. From there the blood enters into the hearts through ostia. The blood is pumped out from the heart into the arteries and from there the blood is carried to the various organs of the body. The impure blood is collected by the irregular spaces or sinuses beteeen the various organs. The blood from these sinuses passes into the efferent branchial sinuses. As these sinuses lead into the gill lamellae their the blood is oxygenated, and oxygenated blood leaves the gills by the efferent branchial veins. From the efferent branchial veins it passes into branchio-cardiac veins and finally into the pericardium.

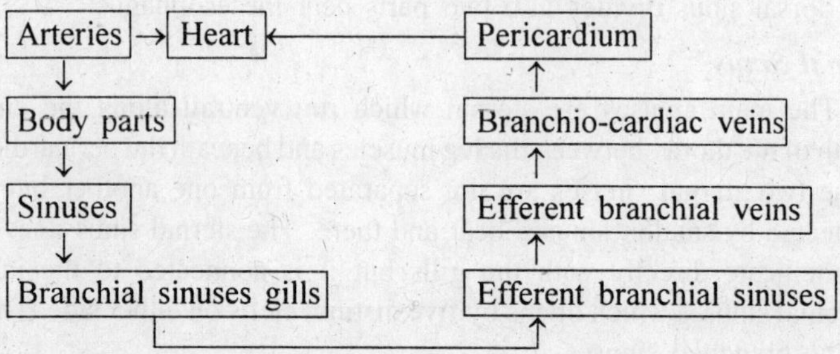

Composition of Blood

Hemoglobin

1. *Structure* : Crustacean hemoglobin exist in the blood as extracellular molecules which are either 630-800 × 10^3 dalton polymers (*e.g.*, in the notostracans and most of the cladocerans) or 220-280 × 10^3 dalton polymers (*e.g.*, in anostracans and conchostracans). In an early study, the intermediate value of 420 × 10^3 daltons to *Daphnia* haemoglobin was assigned, a result which has been recently confirmed.

The relationship between the native polymer and its subunits has only recently become clear using SDS (Sodium dodecyi sulfate)gel electrophoresis, several investigators reported dissociation of the polymer to subunits of about 18-20 × 10^3 daltons, implying that the native molecules are *dodecamers*. This relationship is supported by hypothesis of Felontent, or number of oxygen binding sites to molecular weight However, the native molecules have been described from negatively stained electron micographs as pentagons, which may exist in stacks of two.

Using inhibitors of proteolytic digestion while extracting homogenates of whole animals, Moens and Kondo (1977) were unable to prepare the 19 × 10^3 subunits of *Artemia salina* haemoglobin that has been reported earlier and they suggest that the native 230-250 × 10^3 alton molecule is made up of polypeptide chains each with a molecular weight of 122×10^3. This hypothesis is strongly supported by sedimentation equilibrium data and, somewhat less clearly,by the observation in electron microscope of two stacked discs. Digestion with subtilisin, which clears the covalent bonds, produced $14.8 - 16.5 \times 10^3$ dalton polypeptides that retained the ability to bind oxygen. The basic subunit is a 34×10^3 dalton polypeptide chain of two domains, each with a heme. Essentially same structure appears to occur in *Daphnia pulex* and in several conchostracans.

2. *Bio Synthesis and Lability :* However, the extreme lability of the process was demonstrated many years ago by H.Munro Fox and his coworkers. The hypoxic induction of haemoglobin synthesis in the fresh water cladoceran *Daphnia magna* results in a change five times greater than in man at high altitude. In *Monia macrocopa*, however the oxygen carrying capacity of blood reaches a level of about 2.9ml/100ml, considerably lower than in many annelids and molluscs vertebrates,the hypoxic induction of hemoglobin synthesisis controlled at an early step in formation of the pyrole ring, a basic unit of the iron metalloporphyrin known as heme. The activity of the enzyme α-amino-lavulinic acid synthetase which catalyzes a reaction that produces a pyrole precursor, varies inversely with PO_2. It is unlikely that the complex porplyrin ring is built differently in the crustaceans, and the investigation of this enzyme in cladocerans would be of considerable intrest in understanding the phylogeny of metabolic regulation. Ambient salinity also influences hemoglobin synthesis with an increase occurring in the branchiopod crustacean *Artemia salina* at higher salinities. The genetic basis for hemoglobins produced has been examined recently and the control mechanism is totally unknown and the induction may be trivial or even absent in related species.

Unlike hemocyanin, the level of hemoglobin in the blood doesnot depend on the nutritional state of the animal.

HAEMOCYNIN

Progress in understanding the relationship between structure and oxygen binding of the hemocyanins found in various animal phyla has been especially rapid in recent years. The crustacen hemocynins are high molecular weight macromolecules composed of copper (0.17%), small amounts of carbohydrate (less then 9%) and lipid and polypeptides arranged in a complex hierarchy. They differ from the molluscan hemocyanins in their lower copper content in subunit size and in molecular architecture. Although their copper content and subunit size are similar, the crustacean hemocyanins differ from the chelicerate hemocyanins in the state of polypeptide aggregation. The degree of polymerization of crustacean hemocyanins is of physiological as well as phylogenetic interest. Not only may it influence oxygen binding, but the size and shape of the native polymers may also influence the physical properties of blood which may intum limit oxygen carrying capacity.

Many crab and lobster hemocyanins exist in the blood as dodecamers of about 900×10^3 daltons, which may dissociate into hexamers of about 450×10^3 daltons at high pH. In some species however the hexamer preodominates in the blood and no trace of dodecamer can be found, m others, the predominant polymer is accompanied by very large fractions of the alternative. Therc is more diversity within the decapods or the isopods than between the most of malacostracans and haplocarids.

The association of hexamer to dodecamers requires divalent cations about three mg^{21} per hexamer of callianassa hemocyanin or about the same number of ca^{2t} per hexamer of *Homarus* hemocyanin. No additional mg^{2t} is bound when callianassa hemocyanin forms larger multiples. It is becoming increasingly clear that the association of hexamers to dodecamers also requires a particular type of monomer the function of which may be to crosslink the two hexamers but whether the assembly to larger polymers has also this requirement is not known. The heamocyanin of dendrobranchiate shrimp *Penaeus setiferus* exits in native blood entirely as a hexamer, but isopod hemocyanin may exist predominantly as either the dodecamer or the hexamer. In other respects such as pH and the divalent cation requirement for stability these hemocyanins resemble crab and lobster hemocyanins. Even in ultra structure the molecule from the terrestrial isopod *Ligia* resembles its decapod counterpart. In electromicrography the dodecamers appear as multiples of either one hexagon plus one square or two squares joined side by side. The ultra structure of haplocarid hemocyanin, however, appears to be distinctive in that alignment of the hexagon and the square is slightly skewed

The stability of the eikositetramer in thalassinids also depends on high temperature and an alkaline blood pH with acidity inducing dissociation within the potentially physiological range.

The acid-base distrubances evoked by changes o external salinity result from metabolic adjustments in the cells.

In various euryhaline crustanceans where extracellular osmo-regulation is absent or incomplete, hypoosmotic exposure entails a decrease of the intracellular free amino acid pool by increased deamination. Ammonia is thus rclcascd to the extracellular fluid. According to Mangum *et al.* (1976), experimental evidence suggest, that in *Callinectes*, the increase of blood pH with decreasing salinity results largely from binding of H^+ ions to NH_3 produced by catabolism of cellular free amino acids. The resulting NH_4^+ ions might be excreted in exchange for Na^+. Thus, extracellular acid-base changes elicited by changes of ambient salinity appear linked to extracellular anisosmotic regulation as well as to metabolic adjustments involved in intracellular isosmotic regulation.

Another point of interest is that salinity changes may be accompanied by variations of tiration alkalinity (TA) of water, that is, the sum of concentration of weak acid anions, aminly bicarbonate and borate, in seawater. This variable determines acid-base balance and buffering power of water toward carbonic acid. The higher the TA of water, the smaller would be the increae of PCO_2 in expired water at a given inspired PCO_2 (Dejours *et al.*, 1968). This might explain why a decrease of waler TA without a change in inspired PCO_2 leads to an extracellular hypercapnic acidosis in the crayhsh *Astacus leptodactylus* (Dcjours and Armand, 1980) and in the crab *Carcinus maenas* (Truchot, 1981). These findings pont out the importance of controlling precisely the acid-base state of water in experimental work on crustaceans.

3. *Concentrations of O_2 and CO_2 in the respiratiory medium :*

Variations of the concentrations of O_2 and CO_2 in the respiratory medium (air or watcr) cntail vcntilatory responses which may cause acid-base distrubances. These primary disturbances may be or not be further compensated.

(a) *Oxygen :* Partial pressure of oxygen is a factor of paramount importance in regulation of external respiration in those animals that breathe in water. In crustaceans, a general response to hypoxic ambient conditions is an increase of the ventilatory activity.

There is no apparent common pattern of response to prolonged hypoxia or hyperoxia. Hyperoxia-induced repiratory acidosis is partially compensated after 10 hr in *Carcinus maenas*.

These responses are accompanied by immediate acid-base disturbances of respiratory origin (Table 8.1). In hypoxic lobsters and crabs, pco_2 of hemolymph decreases and pH increases, whereas the inverse changes are observed in hyoroxia but not compensated at all in *Astacus*. A metabolic alkalosis appears superimposed on the hypoxia-induced respiratory alkalosis in *Carcinus* as well as in *Homarus* McMahon *et. al.*, 1978a) In this last species, the respiratory.

Table 8.1 : pH and Values in Hemolymph of Carcinus maenas after 2 Hourse Exposure to Hypoxic Water at 17°C.

Ambient Po_2 (Torr)	Hemolymph pH	Hemolymph pco_2 (Torr)
158	7.79	1.4
83	7.89	1.1
52	7.91	1.0
34	7.97	0.8
13	8.05	0.7

[a]From Truchot, 1975b. Reprinted by permission of Masson, editeur, paris.

A. Acid-Base Balance and the Respiratory Medium

Crustaceans are primarily aquatic animals. However, a number of brachyuran and anomuran species are adapted to terrestrial life, and certain intertidal decapods are periodically exposed lo aerial conditions, or may withstand emersion for long periods.

Water and air are very different respiratory media. In air, the change of partial pressure for a given concentration increment is the same for O_2 and CO_2, In water, by contrast, the solubility coefficient of CO_2 is considerably higher than that of O_2; thus, adding or removing the same quanitity of O_2 and CO_2 changes the partial pressure of CO_2 much less than the partial pressure of O_2. This is just what occurs during steady state respiratory exchange and as a result, the partial pressure of expired CO_2 should be much lower in a water-breather than in an air-breather. As shown by Rahn (1966), the partial pressure of CO_2 in expired water cannot exceed a certain value: this value, 1-3 Torr depending on acid-base properties of the particular water. Consequently, it may be inferred that internal PCO_2 should be lower in water breathers than in air breathers.

However, values for pH in hemolymph are similar in water-breathing and air-breathing crustaceans at the same body temperature. This is related to the finding that the more terrestrial the animal, the higher is the HCO_3^- ion concentration in hemolymph.

In *Carcinus maenas*, an intertidal crab which consumes about as much oxygen in air as in water. The onset of air-breathing is accompanied by a decrease of hemolymph pH due essentially to a rise of PCO_2 from 12 to 5.0 Torr at 1 5°C (respiratory acidosis). Blood PCO_2 then remains constant where as pH and $|HCO_3^-|$ values increase (metabolic compensation). As temperature changes both physico chemcal properties of buffers and physioloical regulatory mechanisms act to assuse not a constnat pH of haemolymph but ratter constant acid hase state as defined by relative alkalinity.

B. Intracellular Acid-Base Regulation

The cell membrane is permeable to both hydrogen ions and CO_2. However, experimental evidence suggests that the intracellular acid-base balance does not confirm passively to the extracellular acid-base state, but is actively regulated.

In crustaceans, intracellular pH has usually been measured in the large isolated muscle fibers of crabs and Barnacles, either by pH microelectrodes or by studying the transmembrane distribution of the weak acid DM0 (5,5'-dimethyl-2,4-oxazolidine dione). The pH measured is probably that of cytosol only and not of other cell compartmcnts. Thc values obtained with these methods agree quite well. Intracellular pH is lower than extracellular pH, ranging from 6.91 to 7.32. Taking into account the measured values of the membrane potential (–40 to –70mV), the hydrogen ion distribution does not correspond to a passive Gibbs-Donnan equilibrium).

Intracellular buffers are mainly proteins, various polypeptides, and organic phosphate compounds. The intracellular buffer value is generally high, 47.3 slykcs for muscle fiber of *Carcinus maenas* exposure of *Carcinus muscle* fibers to CO_2 entails a rapid fall of coil pH followed by a slow recovery will) only minor changes of the membrane potential. The cell pH recovery is more rapid in the same conditions for crayfish (*Procambarus clarkii*) neurons. Such effects can only be explained by postulating the existence of an active process, the so-called "proton extrusion pump," which may be either an outward transport of H^2 ions, or an inward transport of OH^- or HCO_3^- ions. Thus, cellular membrane transport mechanisms appear to act in controlling intracellular acid-base balance. In what manner this regulation interacts, with extracellular acid-base; control remains to be investigated in crustaceans.

In *Callinectes sapidus* there is no Ventilatory responses to 1-2% CO_2 exposure in water a moderate increse of PCO_2 is external water cause a typical hyper capnic acidoses which further, particular compensated metabolically in *Carcinus.* The mangnitude of changes may be dependent on titration alkalinity of buffering power of external water.

C. Acid Base Balance and Physiological State

Exchanges of oxygen and carbon dioxide through the resperatory surpaces or at tissue level may affect the acid base state of haemolymph.

1. Respiratory Mechanism

Gill ventilation of aquatic crustaceans is relatively insensitive to increased ambient PCO_2 Dejours Ventilatory activity decreases notably in hyperoxygenated water in crayfish and crab. This response appears completely independent of internal PCO_2 or pH, since a hypercapnic acidosis occurs concomitantly. Since the difference in PCO_2 between inspired and expired water is always very small due to the high solubility of CO_2 in water, ventilatory changes cannot be very efficient in controlling acid-base balance.

To summarize, as is true in other water-breathers such as fishes, gill ventilation in aquatic crustaceans seems to be regulated primarily to ensure oxygenation of the blood and thus is probably not concerned, with acid-base balance.

2. Carbonate Stores

Mobilizalion of calcium carbonate from the exoskeleton should be accompanied by a concomitant increase of Ca^{2+} concentration in hemolymph Conflicting result have been reported in some brachyuran species. In *Carcinus maenas,* on changes of Ca^{2+} in hemolymph were found during compensation of respiratory acidosis following emersion into air or hyperoxic conditions water.

A rise of hemolymph Ca^{2+} was found during recovery from severe excercise in *Cardisoma carnifex*, suggesting a mobilization of calcium carbonate from the carapace. Possibly, the role of carbonate stores in the exoskeleton and elsewhere may be more important in terrestrial than in aquatic.

Circulating-levels of oxygen in hemolymph recorded for various crustacean types in air vary considerably. Very low levels are recorded for both *Coenobita clypeatus* and *Ocypode quadrata*. In both cases these levels are sufficient to virtually saturate the blood an passant through the gills due to a relatively high oxygen affinity of the hemocyanin. Much higher oxygen tensions may occur in *Birgus* and are associated with a lower hemocyanin oxygen affinity. The marked difference between these two types may be related to the state of development of the accessory respiratory structures. These are particularly well developed in *Birgus*. The associated increase in respiratory area available in *Birgus* may allow much better oxygenation of the blood than is possible with either *Ocypode* or *Coenobita*. A second possible rationale for the presence of a high O_2 affinity pigment in *Coenobita*. This animal, in which ventilation is very intermittent, minimum respiratory water loss would occur if the oxygen trapped within the branchial chamber could be virtually exhausted between ventilatory burster This situation would be clearly enhanced by the presence of a high affinity pigment, which could pick up oxygen at very low oxygen tensions. The relatively high oxygen carrying capacity of the hemolymph of most terrestrial species would also facilitate oxygen, uptake under these conditions. High oxygen capacities also reduce the hemolymph convection requirement and thus perhaps the energy cost of oxygen transfer.

Depletion of prebranchial oxygen tensions during conditions of increased oxygen demand, such as activity or increased temperature, have not been measured in air-breathing forms, but a decrease in prebranchial oxygen tensions is associated with hypoxic exposure in *Ocypode quadrata*.

With the exception of a few isopod species, land crustaceans have not enjoyed the spectacular success seen in other terrestrial arthropod groups. In general they are still dependent on periodic visits to water and must return to water to reproduce; To some degree, limitations of the respiratory system may contribute to this lack of success. Air breathing allows considerable reduction of the air convection requirement, with concomittant reduction in energy expenditure for ventilation. Advantages gained at ventilation, however, are apparently offset by the disadvantages of an open branchial chamher and gill system. The gills, retained as organs in ions and probably, osmoregulation, have of necessity been decreased in area and thickened to reduce water loss or to ensure rigidity. This has reduced their gas exchange efficiency. Gas exchange is supplemented by other structures in several forms, but here too, water conservation is a problem. The lack of a completely internalized respiratory system such as that of insects or arachnids from which respiratory water loss can be minimized, thus, probably restricts terrestrial crustaceans, with few exceptions, to littoral, burrowing, or other moist habitats. They are thus amphibious rather than terrestrial animals.

NEUROMUSCULAR CONTROL OF VENTILATION

For the purposes of this section, the .ventilatory apparatus is taken to include the scaphognathites and the epimeral attractor muscles of decapod crustaceans.

A. Functional Anatomy of the Scaphognathites

The skeleton, musculature, and innervation patterns of the scaphognathites appear to be homologous between macruran and brachyuran decapod crustaceans. The muscles have been variously named, but the simplified scheme of Young (1975) based on the functional organization and bursting patterns is adequate and preferable as the easiest to conceptualize. The rhythmic healing of a scaphognathite is effected by ten muscles subdivided into two groups or four subgroups (five + levators, subgroups L1 and L2; five depressors, subgroups D1 and D2. The movements produced are D1, depression of posterior tip of scaphognathite; D2, depression of anierior lip; 1.1, levation of posterior tip; and 1.2, levation of anterior tip.

Each scaphognathite is innervated by two nerves arising from the sub-esophageal (Mactura) or thoraele (Brachuara) ganglion. The levator nerve (LEN) divides with LENa supplying motorneurons to levator muscle's. LENb supplies two motonearons to ync depressor muscle D2a and also carries the major complement of sensory afferent fibers, including those from the oval organ The depressor nerve also branches with DEN being primarily a motor nerve sending axons to all depressor muscles, except D2a, while SEN contains motor neurons to and sensory fibers from the endites and endopodites. The latter two segments of the second maxilla do not participate in ventilation. Each scaphognathite muscle is innervated by two, three, or four excitatory neurons (Table 8.2). Each scaphognathite muscle is activated by a sharply defined burst of excitatory postsynaptic potentials (EPSP's), and the muscles are of the gradedly responding phasic type. Intracellular recorded EPSP's from three of the muscles (D1a, D2a,b) were of low output type (1-2 mV) and showed a high degree of facilitation. In contrast to most other crustacean skeletal muscular systems, no peripheral inhibitory system hsa been found the distribution of inhibitory motor neurons and noted that they are absent from most purely phasic muscles or the phasic protions of mixed fiber type muscles. The scaphognathite muscles or the phasic portions of mixed fiber type muscles. The scaphognathite muscles thus be classified as phasic.

Table 8.2 : The Muscles in the Macrurans *Homarus* and Greeneries[a] and in the Brachrurans *Callinectcs*[b] *Cancer*, and Nectocarcinus,[a] and Their Homologues with Those in Carcinus[d]

Macrura *Homarus* *Orconectes*	*Callinectes, Cancer*	**Muscle**	**Brachyura** *Carcinus, Cancer* **Number of motorneurons innervating muscle**	**Group**	**Nerve**
CFM	mr4llm (quartus)	D1a	$2^{(3)+}$		
BFM	mr3llm (terius)	D1b	2	Depressor 1	DEN (a)
BTM	mr5llm (quintus)	D1c	2		
ASDM	mr2llm (flexor scaphognathitis)	D2a*	2(2)		
SFM	mr6llm (sextus) mr7llm (septinus)	D2b	2(4)	Depressor 2	
CEM	mr2llm (secundus)	L1a	3	Leavator 1	LEN
ACEM	mrllm (primus)	L1b	3		(b)
PSDM	mrllm (remotor)	L2a	3		
BEM	mpllm (promotor)	L2b	3	Levator 2	
Not figured	not figured	L2b,	2		

[a] Pasztor, 1968.

[b] Cochrane, 1935.

[c] Pilkington and simmers, 1973; Pilkington and Marcfarlane, 1978.

[d] Young, 1975. See also Fig. 9. Muscles specifie to endites and endoposites of the second maxillae do not move the SG proper and were not listed here here (see Young, 1975, for details of these muscles).

* Nerves to muscle D2a travel in LEN.

* *Procambarus darki* and *Orconectes virilus* (moody-Corbett and Pasztor, 1980).

B. Neural Control of Scaphognathitc Beating

Models of neural circuits, real or hypothetical, are often constructed to help visualize the architecture of a system and to direct thinking toward points where further research efforts are required.

From this point of view the organization of the scaphognathite motor patten generator can perhaps most clearly be understood by considering the hypothetical model proposed by Young (1975) and the observations on which it is based Fig. 8.6. This model, when supplemented by the appropriate additional interneuron circuitry, adequately describes the generation of both forward and reversed motor patterns. The following

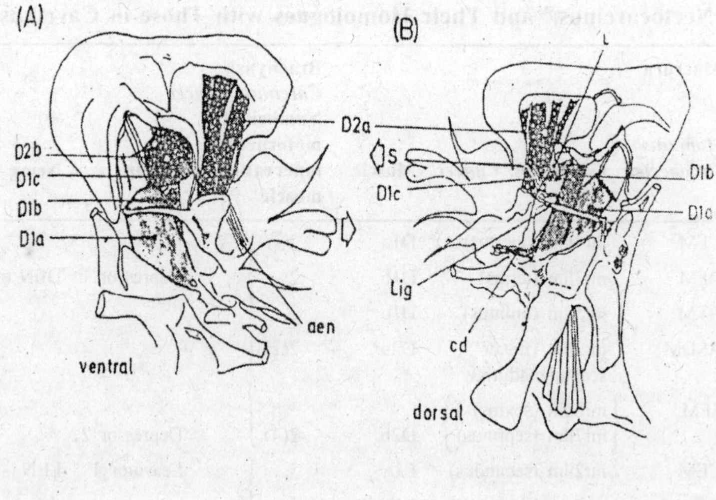

Fig. 8.6 : (A-D) The musculature of the right sc ıphognathite of *Carcinus maenas*. Muscles blocked in similarly belong to the same temporal group. (A and B) Ventral and dorsal views, respectively, showing the five ventral depressor muscles. In (B) all the the dorsal muscles (levators) have been removed. Hatched, group D1; White stipple on black, grop D2. (C and D) The five dorsal levator muscles; dorsal view. These muscles lie dorsal to the ligament (Lig) running between t he articulation points for the promotion remotion axis. Black stipple, Group L1; Black, Group L2. The depressor muscles are blocked in as in (A) and (B). Anterior direction indicated by arrows. (From Young. 1975 with permission).

discussion will consider each level of organization as set forth in the model in sequence.

1. Oscillatory Systems

Mendelson (1971) identified bilateral nonspiking oscillatory neurons which appeared to intiate the scaphognathite motor pattern in

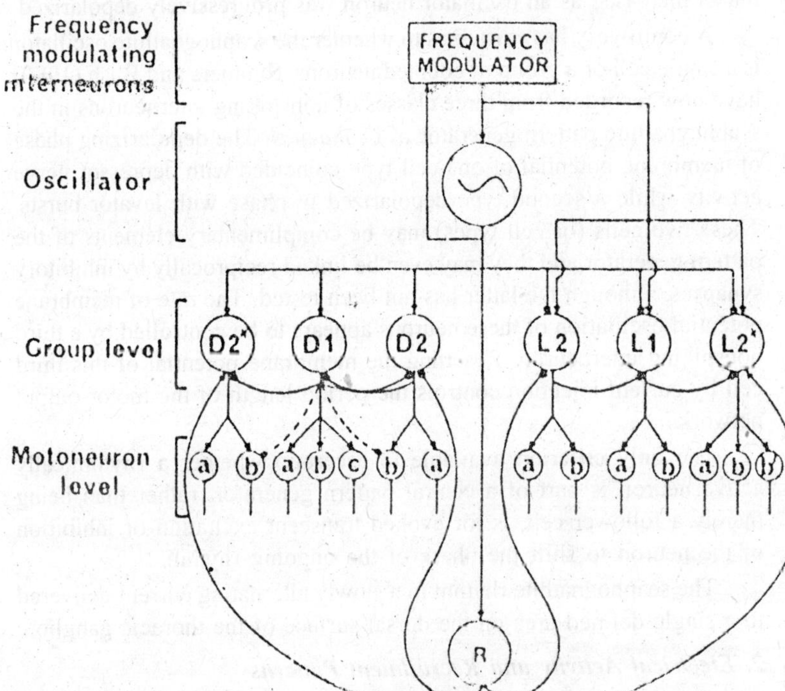

Fig. 8.7 : A tentative model of the scaphognathite motor circuit for *Carcinus maenas*. The model includes a single, driving oscillator in keeping with mendelson's (1971) conclusions the oscillator drives six group level units-three depressor and three levator. Two of these (D2 and L2) are stand by units which are activated only during reversed beating, when D2 and L2 are inhibited. The group level units activate the motor units to the muscles in the four respective groups, inhibition from D2 onto D1 and from L2 onto L1 during forward beating prevents D1 and L1 from firing for the whole depressor and levator sessions, respectively. During reversed beating a similar relationship would hold between D1 and D2 and between L1 and L2. There appears in addition to be inhibition from D1 onto D2 in the forward mode and between D1 and both D2b and D2*b. The main reversal command input by-passes the oscillator, acting at the group level. Activity in the reversal command unit (R) results in a change in the relative thresholds of the groups level units from D1<D2<<D2* to D2*< D1<<D2, and from L1<L2<<L2* to L2*< L1<<L2. (from Young, 1975 with permission).

the subesophageal ganglion of both lobsters and hermit crabs. Intrinsic depolarization and repolarization of an oscillator neuron coincided with bursts of spikes in the depressor and levator nerves, respectively. Induced depolarization or hyperpolarization of an oscillator neuron by intracellular current injection caused the appropriate nerve to fire tonically while its antagonist was silent. Further, the rate of generation of motoneuron bursts increased as an oscillator neuron was progressively depolarized.

A controvesy has existed as to whether the scaphognathite oscillator is a single cell or a group of coupled neurons. Simmers and Bush (1980) have now recorded from three classes of nonspiking interneurons in the scaphognathite pattern generator of *C. maenas*. The depolarizing phase of membrane potential of one cell type coincided with depressor nerve activity while a second type depolarized in phase with levator bursts. These two cells (or cell types) may be complimentary elements of the pattern generator and they may even be linked reciprocally by inhibitory synapses, although the latter has not been tested. The rate of membrane potential oscillation of these neurons appears to be controlled by a third nonspiking interneuron. Resetting the membrane potential of this third cell by current injection controls the period length of the motor-output network.

The only criterion available to ascertain whether a rhythmically active neuron is part of a central pattern generator, rather than being merely a follower cell, is for evoked transcent excitation or inhibition of the neuron to shift the phase of the ongoing rhythm.

The scaphognathite rhythm to a slowly alternating current delivered to a single defined area on the dorsal surface of the thoracic ganglion.

2. Electrical Activity and Recruitment Patterns

As shown earlier, the scaphognathite muscles are activated by a burst of EPSP's during normal movements of the appendage. Muscle response to nervous input shows facilitation at normal spike frequencies within a burst, but the muscles fatigue rapidly when stimulated at frequencies of >100 Hz. Each of the muscles may be activated by a single motorneuron at times, but at other times they receive input from two or more units. The force of scaphognathite beating, measured either in terms of muscle tension or of the hydrostatic pumping pressure generated during a scaphognathite beat cycle, has not as yet been correlated with either the frequency of impulses within a burst or with the number and/or identity of the active motorneurons.

The scaphognathites can pump water or air in either a forward or reversed direction. The bursting and recruitment patterns of the approximately 35 to 39 motorneurons innervating the 10 scaphognathite muscles have been reconstructed by recording the total motor output pattern in the nerves DEN and LEN along with the simultaneous emg activity in two or more muscles Young 1975; Wilkens and young, 1975). The recruitment patterns and burst durations during both forward and reversed pumping are shown for *Carcinus maenas* (Fig. 8.7). Essentially similar patterns are seen in the other decapods for which data are available. During forward pumping the muscle groups are recruited in the sequence:

D1, D2, L1, L2, D1, ...

while during reversed pumping (ho muscles remain grouped (*e.g.*, D1a, b, and c) as in forward pumping, but the recruitment sequence of groups is altered. In the marine macrurans *Homarus americanus*; *Nephrops norvegicus*: Young, 1978) and scaphognathitcs only reverse during the levation half of a full beat cycle, and the altered effective stroke is made possible by the deletion of the D2 group while the bursts from the L2 group may be delayed or inhibited:

D1, D2, LI, L2, D1, - LI, (L2), D1, D2, ...

reversal beat forward pumping

Immediately following the half cycle reversal beat the forward sequence is resumed.

Brachrurans and the crangonid and pandalid shrimps on the other hand, are capable of long sessions of sustained reversed, pumping, and the transition from forward to reversed as well as the reversed recruitment sequence as seen in *Carcinus maenas* follows the sequence:

D1, D2, L1, L2 D1, - - L2*, L1, D1*, D1, L2*, ...

Reversed pumping

Note that reversing also begins with the deletion of the D2 group of motorneurons and, as indicated by asterisk, the neurons supplying the D2 and perhaps also the L2 muscles during forward pumping are silent and replaced by two new sets of motorneurons. These additional motorneurons burst in advance of the D1 and L1 units. In addition, one muscle, L2b, is active only during reversed pumping. For additional details refer to the study by Young (1975).

Both Young (1975) and Wilkens (1976) propose the existence of interneurons (R, in Fig. 8.6) that induce reversals of beat. These neurons would exert their effects on the order of recruitment at the level of

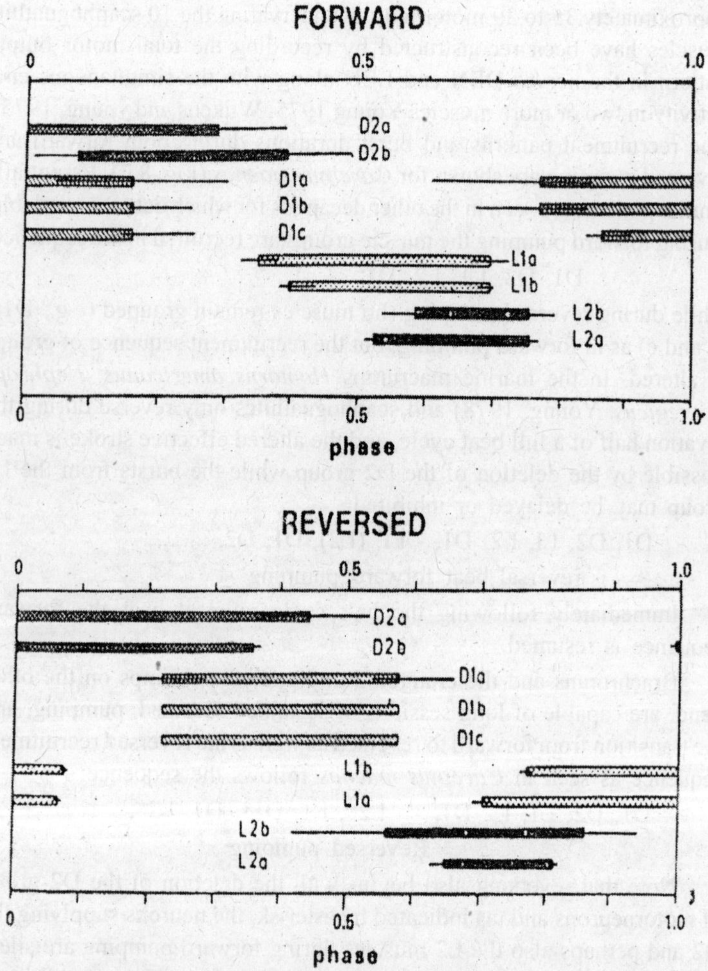

Fig. 8.8 : Average normalized bursting patterns in the main ventilatory muscles during forward and reversed beating in *Carcinus maenas.* The start of the D2a burst marks the start of each cycle. Bare blocked in similarly indicate the mean postitions in the cycle, of bursts in muscles from the same temporal group. Horizontal lines indicate the maximum and minimum normalized starting and stopping times for the various muscles (From Young, 1975).

interneurons controlling groups of motorneurons. Whatever the circuitry, the addition of particulate matter to the seawater bathing a crab or the systemic injection of 5-hydroxytryptamine (5-HT) decreases the interval between, and occasionally increases the duration of, scaphognathite reversal sessions. This reversal response can be blocked by the prior injection of a postsynaptic antagonist of 5-HT. Berlind found that dopamine, noradrenaline, octopamine, and y-aminobutyric acid had no effect on scaphognathite reversed beating behavior, and in a different study, picrotoxin, which exerts powerful effects on scaphognathite pattern generation, did not prevent reversals in lobsters.

Two more observations may be taken to support the model presented in Fig. 8.6. First, antidromic stimulation of LEN or DEN does not alter the bursting patterns or the period length of the antagonist group of moioneurons in Cancer products. For example, stimulation of LEN' which contains the D2a motorneurons does not affect the timing in the other D1 or D2 motorneurons. Although, not a proof, this indicates that the interactions between motorneurons are not necessary for pattern generation, since the neurons do not appear to be synaptically interconnected. Rather it suggests that the hypothesized group level interneurons are synaptically interconnected to effect pattern generation. It must be cautioned that antidromic spike propagation up the scaphognathite motorneurons fails at a point distal to the integrating segments; however, small amplitude electrotonic depolarizations presumably capable of effecting synaptic communication between motorneurons are still a possibility.

Second, injection of picrotoxin solutions into *Homarus americanus* (to produce a systemic dose between 8×10^{-8} and 4.5×10^{-6} M) led to disruption of the normal motor output to the scaphognathite. The phase separation of the starts of the bursts in the D1 and D2 muscles was reduced by 49% from that shown in Fig. 8.7; that for the L1 and L2 muscles by 16%. The durations of the D1 and L2 bursts increased by 64 and 48%, respectively, such that there was complete overlap in activity of D1 and D2 and partial overlap of L1 and L2 motorneurons. On the other hand, the phase separation between the starts of the levator and depressor portions of the cycle was not altered by picrotoxin. Such data suggest that the oscillators excite or inhibit depressors and levators equally and that picrotoxin sensitive "inhibitory" synapses are then necessary for the temporal ordering among the group level interneurons.

3. Bilateral Coordination Between Scaphognathites

In several crustaceans the two scaphognathites show bilateral phasic coordination, in which phase constancy alternates with periods of drift during which the scaphognathites demonstrate the independence of their respective pattern generators During drifting coordination, the two appendages may beat at markedly different rates. It is not known at what level transganglionic communication is effected, *e.g.*, communication between oscillators, between components of the pattern generators, or by sensory neurons; however, all motoneuron somata are located on the side of the ganglion from which the scaphognathite nerves arise, and both Wilkens and Young (1975), in a macruran, and Pilkington and MacFarlane (1978), in a brachyuran, have identified axonal collaterals presumably arising from motorneurons that cross the ganglion. The fact that bilateral phase constancy exists may reflect the evolutionary derivation of the scaphognathites from synchronously beating filter feeding appendages.

A number of physical perturbations are known to alter bilateral coordination. Restraining one scaphognathite, which reduces the rate of burst generation to the appendage, causes wild drift in bilateral phasing. Artery of the oval organ (a scaphognathite sensory receptor) may actually increase slightly the strength of bilateral coupling, whereas imposed sinusoidal movements of one scaphognathite reduce coupling.

C. Sensory Control of Scaphognathite, Function

The sensory system of the scaphognathites consists of innervated hair sensillae, epidermal receptors, and the oval organ, a specialized receptor located on the dorsal surface of the scaphognathite at the axis of dorsoventral flexion. The oval organ is innervated by two large sensory fibers which discharge in bursts of large amplitude spikes in phase with scaphognathite movements.

Afferent neural feedback from a scaphognathite is not necessary for the generation of the motor pattern, but the interburst interval of motoneuron bursting to a scaphognathite is reduced following amputation or immobilization of that appendage. This suggests that this supportive role can be demonstrated in dying preparations, *e.g., when an animal is dissected to gain access to the CNS, the resulting blood loss and/or asphyxia leads to a disruption of scaphognathite function.* The loss of activity is characterized by short sessions of scaphognathite beating interrupted by quiet periods, and as the preparation ages, the sessions of activity become shorter and less frequent. It during this period of decline, nerve LEN, which contains most of the scaphognathite sensory fibers, is stimulated with trains of pulses delivered at the same rate as the intrinsic output, the system will continue to generate motor bursts for prolonged periods of time (Table 8.3). This sustaining influence docs not occur during stimulation of DEN.

Table 8.3 : The Sustaining Effect of Stimulating Nerve LEN on the Length of Sessions of Continuous Burst Generation of the Scaphognathite Motorneurons in a Dying Preparation of *Cancer productus*[a]

	Average length of bursting sessionsperiods (sec)	**Intersession silent (sec)**
Unstimulated	19.1 ± 1.3 (N = 11)[b]	24.3 ± 0.9 (n = 12)
Stimulation by trains of pulses	53.0 ± 12.6 (N = 8)	18.1 ± 0.6 (n = 7)

[a]Trains of pulses wore delivered to LEN at a repetition rate similar to the intrinsic rate of burst generation. Pulse frequency within trains was 100 Hz, train duration 50 msec, pulse duration. 0.1 msec.

[b]Mean ± SLM. (Number of occurrences.)

Further, the scaphognathite motor output of *H. americanus* can be entrained to imposed dorsoventral movements applied along the radial axis of the appendage. This ipsilateral coupling tendency between an imposed movement and motor output indicates that sensory reafference exerts a phasic influence on the pattern generator as well as the sustaining role described above. Taken together, these data support the notion that sensory feedback from the scaphognathite exerts some of its effects at the level of the oscillator neurons.

In addition to reafference, *a wide variety of optical, mechanical, chemical, and osmotic stimuli are known to alter scapnognathite rhythms. Stimuli applied within a gill chamber often cause ipsilateral scaphognathite responses only, while stimuli applied at almost any other site on the animal generally cause bilateral responses. The most commonly seen responses are decreases in pumping rate, reversed pumping and pause, and more rarely, increases in pumping rate. Novel stimuli are generally the most effective in causing a response, and in all cases reported, the scaphognathite responses adapt rapidly.*

D. Control of Motor Output to the Scaphognathite

From the variable patterns of ventilation and the wide variety of sensory stimuli which affect scaphognathite rhythms, it is obvious that the *autogenous pattern generators are themselves controlled. Among the control mechanisms known are direct metabolic matching of* f_{sc} *to oxygen availability, higher level neuronal modulation of the pattern generators, and hormonal modulation.*

1. *Metabolic Matching*

Metabolic matching is used here to describe the linear relationship between f_{sc} and PO_2 seen in *Homarus americanus* during severe hypoxic stress. At the onset of hypoxia, when the bath PO_2 has fallen to about 50 torr, lobsters show a compensatory increase in f_{sc} but below this critical level the f_{sc} decreases linearly with falling oxygen tension. When severe hypoxia (PO_2 <12 torr) has been attained, the scaphognathites beat intermittently, alternating with apnea, and the heart beats slowly. If at this time aliquots of aerated seawater are injected into the branchial cavities, the heart rate begins to increase in 20 sec followed, in an additional 10 sec, by increases in f_{sc} These delays are equivalent to the transport time of blood from gills to heart and then to subesophageal ganglion. The suggestion made here that O_2 has a direct effect on neuronal oscillators is consistent with known direct effects of O_2 on oscillatory neurons in Aplysia.

A compensatory incresae in scaphognathite pumping rate at the onset of hypoxia has been seen in several crustaceans. Hypoxia is one of the few stimuli which induces increases in f_{sc} almost all other stimulus modalities decrease f_{sc}, or cause reversed pumping or pausing. These observations imply some ability of these decapods to sense the onset of hypoxia. In addition to the proposed intraganglionic sensors and receptors are identified in the podobranch gills of *Procambarus clarkii* that fire at a low and irregular rate (1-5 spikes/min) when superfuscd with aerated water, but whose firing rates increase markedly as the superfusate is made hypoxic (Fig. 8.8). Maximum firing frequency of 30 spikes/min occurred at a PO_2 of 13-15 torr. This is the only report of "oxygen receptors" in crustaceans, although, they are known from the merostomatid *Limulus polyphemus*.

2. *Neuronal Modulation*

Higher level command interneurons modulate scaphognathite function, and these interneurons are in turn activated by a wide variety of sensory input of external and presumably internal origin. Mendelson

(1971) identified command interenurons in lobsters and hermit crabs which caused the oscillators lo cycle more rapidly, whereas Wilkens *et al.*, (1974) identified interneurons in the crab *C. magister* which induced predominantly inhibitory responses (decreased amplitude and frequency of forward pumping, reversed pumping and apnea, in order of increasing threshold). It thus appears that neuronal modulation occurs at the oscillators ns well as other levels within the pattern generator.

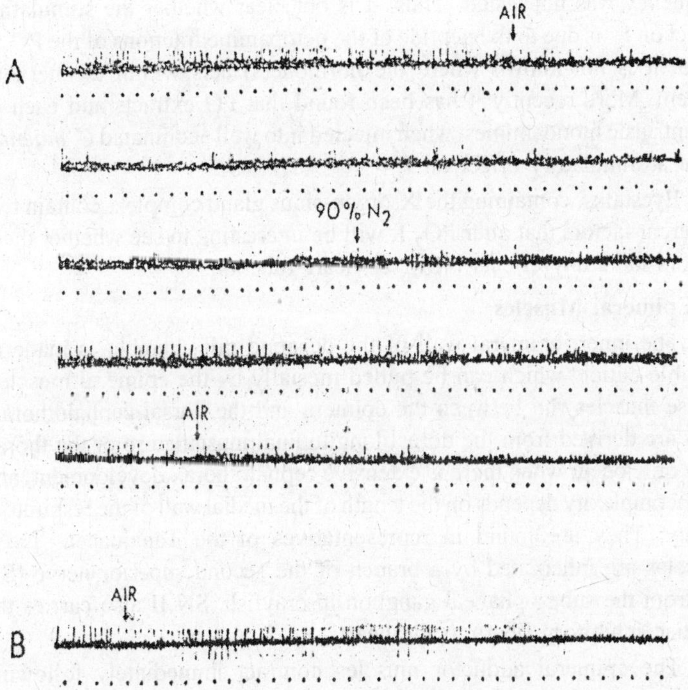

Fig. 8.9 : Afferent podobranch nervous activity recorded from an excised gill of *Procambarus clarkii* during normoxic and hypoxic exposure. (A) Trace starts during hypoxic exposure. At the indicated points, air or a 90% N_2–10% air mixture was continuously bubbled through the water in the experimental chamber. (B) Afferent activity as severe hypoxia at start of trace is progressively alleviated. A single unit fires briefly during the transition from severe hypoxia to normoxia. The other units respond to hypoxia as described in A and become almost inactive as normoxia is approached. Time marked at 5-sec intervals in all traces.

3. *Hormonal Modulation*

A system of neurosecretory neurons rnnke up the two pericardial organs (PO) located in the pericardial sinus lateral to the heart and their anterior ramifications (AR), which, branch among the scaphognathite muscles. The PO's are known to contain two peptides: 5-HT or dopamine extract's of the PO of crabs (*C. maenas* and *Libinia emarginata*) and crayfish (*Orconectes sp.*), when injected into intact animals, cause an increase in f_{sc} which lasts for 2-10 min. 5-HT or dopamine, when injected alone, did not increase f_{sc} while the effect of octopamine on beat frequency was not tested. Thus, it is not clear whether the stimulating effect on f_{sc} is due to the peptide or the octopamine fractions of the PO_2's. Also, it is not known where the hormone(s) acts within the nervous system. More recently it has been found that PO extracts and each of the intrinsic monoamimes, when injected into well acclimated *C. maenas*, have a stimulatory effect on f_{sc}.

Eyestalks, containing the X-organ-sinus gland complex, contain two different factors that alter PO_2 It will be interesting to see whether these factors have any effects on f_{sc} or heart function.

E. Epimeral Muscles

The inner, epimeral wall of the decapod gill chamber is made of flexible cuticle which can be pulled mesially by the epimeral muscles. These muscles run between the epimera and the dorsal cephalothorax, they are derived from the dorasl longitudinal musculature of the thorax they only occur when there is extensive cephalothorax development, and their complexity depends on the length of the medial wall of the respiratory cavity. They are found in representatives of the Tanaidacea. These muscles are innervated by a branch of the second superior nerve (SN II) from the subesophageal ganglion in crayfish. SN II also carries the cardiac inhibitory fibers to the heart.

The epimeral adductor muscles contract immediately following reversed pumping beats in *H. americanus.* This contraction pulls the epimeral wall inward and by enlarging the branchial cavity helps reestablish the branchial negativity characteristic of forward water pumping in this-species. This ventilatory pattern of the terrestrial hermit crab *Coenobita* sp. occasionally contains large amplitude negative pressure excursions not caused by scaphognathite movements. It is tempting to speculate that these arise from epimeral muscle contractions. The epimeral attractor system may play a more important role in the ventilatory process than has been previously recognized.

CONTROL OF HEART FUNCTION

The crustacean hearts are neurogenic with a cardiac ganglion intrinsic to the heart serving as the autogenic pacemaker. Heart rate (f_H) is determined by the rate of bursting of this ganglion and each heart beat is viewed as a brief tetanus. The amplitude or force of contraction is dependent on spike frequency within each burst and the total number of spikes.

Heart rate and cardiac output vary in response to sensory stimulation, locomotion, and possibly to internal factors. These patterns of activity may represent modulations of cardiac ganglion burst generation. Included as modulating agents which exert effects at the level of this ganglion are stretch activation direct metabolic effects of O_2 and intermediary metabolism, the cardioregulator nerves, and hormonal factors.

A. Metabolic Effects

Nerve tissue in general is known to be of the highly oxidative-type whose normal function shows a strong dependence on the continued availability of oxygen and metabolic substrates. The cardiac ganglion is no exception and responds to hypoxia in a manner analogous to dial shown by the scaphognathite oscillators. It is found that heart rate in two species of crayfish was strongly correlated to PO_2 over the range 0-150 torr however, raising PO_2 above 150 torr resulted in no further increase in f_H. During short duration exposures to hypoxia, this direct relationship between PO_2 and f_H has been seen in a variety of species including *Procambarus simulans*, *Carcinus maenas*, *Homarus americanus* and in the isolated hearted of *Cancer magister* and *Procambarus clarkii*. In the latter study, cardiac arrest occurred within 10 min when hearts were perfused with saline at PO_2, = 5 torr and within 1 hr if perfused with glucose-free aerated saline. In contrast, hearts perfused with aerated saline containing glucose conlinued to beat over 6 hr. Further, the metabolic inhibitors 2,4-dinitrophenol (DNP) and potassium cyanide (at concentrations of 10^{-6}) to 10^{-4}M), when added to aerated perfusate, cause decreases in f_H over a similar time course as does hypoxia.

The frequency and amplitude of heart beat declined markedly during hypoxia both in situ and in isolated crab hearts. The recovery upon readmission of oxygen was approximately 5.8 × faster for *in situ* (22 sec) than for isolated hearts (127 sec). The quicker *in situ* recovery indicated roles for receptors sensitive to oxygen and for the CNS; however, it is inclined to believe that the lag in recovery in situ is too long to represent direct nervous mediations, but would be similar to the transport time for blood from gills to heart. The slower uptake of oxygen by saline perfusion media over that of hemocyanin containing blood could explain the in vitro delays.

Since these early reports we have learned that animals may require as long as 72 hr to become acclimated to experimental conditions and that upon acclimation f_H is much lower than in recently handled animals. Upon initial exposure to hypoxic water, the f_H of well-acclimated animals may remain unchanged or show slight tachycardia. Since the f_H observed at any movment is the result of a complex interplay between metabolic, nervous, and hormonal influences, it is not easy to resolve these disparate results; however, it is safe to assume that all of the earlier studies were made on unacclimated animals whose heart rates were near maximal. At high f_H the scope of heart response is reduced to near zero in which case the only response to hypoxia would be the observed parallel reduction in f_H.

The electrogenic Na^+ pump, by affecting the membrane potential of the pacemaker cells of the isolated lobster cardiac ganglion, is capable of modulating the discharge rate of this ganglion. Anoxia and 10^{-4} M DNP, by decreasing the activity of the electrogenic pump, caused an increase' in tonic discharge rate from this ganglion. They predicted this would result in increased heart rate in intact systems; however, their isolated ganglion preparations did not generate discrete bursts of spikes as occurs in situ. Where hypoxia, which would also reduce the electrogenic pump activity of the cardiac ganglion, causes bradycardia.

B. Neuronal Modulation

The cardiac ganglion is directly influenced by the cardioregulatory nerves, inhibitors (SN II), and accelerators (SN III), which arise from the anterior ventral nerve cord. In decapod crustaceans these nerves are collected into a single pair of dorsal nerves within the pericardial sinus, which enters the heart and arborizes among the neurons of the cardiac ganglion. Of the axons in Sn III, one enters the cardiac ganglion end in the PO_2, the dorsal muscles of myocardium, or the suspensory ligaments, which support the heart.

The changes in heart function evoked by direct stimulation of these nerves. The cardioaccelerators increase heart rate when stimulated at 5-10 Hz or greater, and the heart does not adapt to prolonged stimulation. The thershold stimulating frequency for the inhibitor stimulation. The inhibitor action is dominant over that of the accelerators when both are stimulated simultaneously. He also showed that the accelerator transmitter decays slowly while the inhibitor transmitter is inactivated relatively rapidly. PO hormones could not have gained access to the perfused hearts used. The prolonged accelerator effects observed in intact animals could reflect cardioaccelerator hormone effects as well as cardioaccelerator input.

In vital preparations, the cardioregulator neurons fire tonically with periodic bursts of higher frequency spikes superimposed. The spike frequency within bursts are similar to those found effective in altering f_H in response to direct stimulation. In *Procambarus clarkii,* the cardioaccelerators fire tonically at 2-3 spikes/sec but occasionally reach 12 spikes/sec; peak cardioinhibitor burst frequency may reach 40-50 spikes/sec and produced cardiac arrest (presumably in diastole).

An inhibitory stretch reflex in which stretch of the dorsal pericardial wall induces long duration cardioinhibitory discharges and concomitant cardiac arrest. A role for this reflex in the life of a crayfish is not known.

Command interneurons located in the circumesophageal connectives have been identified which cause either tachycardia or bradycardia and has a lower threshold than cardiac arrest with the latter response often requiring the simultaneous stimulation of several command interneurons. The cardioinhibitor firing rate varied linearly with stimulation frequency of certain command interneurons. The command interneurons themselves show tonic activity which is greatly augmented by stimulation of mechanoreceptors on the anterior carapace in *Cancer magister*. Finally, a reciprocity has been demonstrated between the cardioinhibitor and accelerator nerve firing rates during command fiber and sensory stimulation wherin stimulation wherein stimulation of an accelerator command interneuron causes an increase in Sn III and decrease in Sn II firing. The opposite effect, an increase in SN II and decrease in SN III, occurred during the inhibitory reflex which follows application of glucose solutions to specific regions within the gill chambers..

No physiological functions have been demonstrated for the anterior median aortic nerve, which travels from the stomatogastric ganglion toward the heart ostial valves, or for the axons in Sn III, which/supply the dorsal muscles and suspensory ligaments of the heart.

C. Hormonal Modulation

The modulation of heart beat by hormones has been previously reviewed by Maynard (1960) and Frontali and Gainer (1977). Extracts from a variety of tissues (brain, eyestalks, tritocerebral commissures, and PO) contain cardioactive substances in both decapod and stomatopod crustaceans. The PO's in particular are considered to be directly involved in cardioregulation also scaphognathite regulations. The physiology of the PO including excitation-secretion coupling and hormone biochemistry information is available on the cardiac effects of hormones from other sources.

Octopaminc and in lesser quantilites, 5-HT are contained within neurons whose cell bodies are located at the branch points of the second roots of the thoracic ganglia in lobsters. These nerves possess two neurosecretory release sites, one close to the cell bodies in the second roots and the other distal where axons from these cells contribute to the PO's. Both release sites contribute to hemolymph, which is upstream from the heart, suggesting that these amines are involved in regulation of heart function. These second root neurons appear to receive synaplic input synaptic input from the heart, suggesing that these amines are involved in regulation of heart function. These second root neurons appear to receive synaptic input from sensory nerves, and the cell within one root are electronically coupled to one another and become spontaneously active above 14°C. The cellular origin of the PO peptides and dopamine is not known.

When assayed on intact perfused hearts, PO extracts increase the rate and amplitude of contractions. 5-HT is the most active amine, being five to twenty times more potent than octopamine in *H. americanus*, Astacus leptodactylus, and Eriphia spinitrons; octopamine even at 5×10^{-9} M caused a persistent increase in amplitude and frequency in *Astacus and Eriphia* and has a threshold in the lobster at about 10^{-7} M. The lobster hearts responded poorly to octopamine over the range of 10^{-7} to 10^{-4} M, whereas the above results were dramatic. Dopamine is as potent as 5-HT on *Astacus*, but five to ten times less potent on *Eriphia* hearts. In this latter study, the acceleratory effect of octopmaine was preceded by brief inhibition, whereas 5-HT and dopamine caused acceleration only. Until they are available in pure form it will not be possible to establish doseresponse relationships for the peptide components of the PO's; however, one of the peptides has been tentatively identified as proctolin (a putative peptide-neurotransmitter in insects. Elicits an inotropic response from crab hearts whereas the other as yet uncharacterized peptide elicited predominantly chronotropic responses.

Cooke and Hartline (1975) assayed the effects of 5-HT and PO extracts on the isolated cardiac ganglion of *H. americanus*. 5-HT increased burst rate (hence f_H in intact animals) at any point of application; however, when applied locally, the small pacemaker cells were found to be more responsive than the large follower cells. In contrast, PO extracts (presumably the peptide fractions at the dilutions assayed) were more specific and increased burst rate and burst length only if applied to the small cells.

Possibly related to the above hormone-induced responses, it has often been difficult to evoke tachycardia by stimulation of cardioacceleratory nerves or to identify command interneurons which induce cardiac acceleration, an observation consistent with the observations that f_H remains elevated for several hours to days following handling and surgical procedures. It is tempting to speculate that many such "stressful" manipulations stimulate the release of PO cardioaccelerator factors which then cause the heart to beat at or near its upper limit. This would be expected to mask other accelerator responses.

Although, more difficult to quantify, the role of hormones may prove to be as important as direct neuronal input in the control of heart function. Much work remains to be done in this area.

HEART-SCAPHOCNATHITE COORDINATION

Under a variety of natural and experimental settings we observe what appear to be coordinated behaviors between the heart and scaphognathites (Table 8.4). The behaviors most often recorded are parallel reductions in frequency, or apnea and cardiac arrest. At the onset of unilateral scaphognathite pumping in quiescent *C. magister*, the heart rate in a majority of animals begins to vary with the scaphognathite rate.

In *H. americanus*, heart and scaphognathite rhythms may at times become relatively or absolutely coordinated showing the characteristic lock and drift patterns of gliding coordination, but the strength and incidence of heart-scaphognathite coordination is weaker than that served between the two scaphognathites. In *Nephrops norvegicus*, highly correlated covariation of heart and scaphognathite rates ($f_{sc} = 2 \times f_H$) occur during both normoxia and hypoxia. The cardioinhibitory nerves provide at least part of the route through which coordination is effected. Prolonged bursts of impulses (ca. 20 spikes/sec) in the cardioinhibitor nerve produce cardiac arrest simultaneously with apnea. The heart adapts

Table 8.4 : Evidence for Heart-Scaphognathite Co-ordination as Derived from Observations Where Both System Were Monitored Simultaneously.

Species	Condition or animal at time of obsenation and stimuli applied if any	Heart	Scaphognathites	References
Homarus americanus	Animal tethered, no stimuli	Simultaneous onset and termination of cardiac arrest and scaphognathite apnea. During pauses, periodic and simultaneous escape of both systems.		McMahon and Wilkens (1972)
H. americanus	Hypoxia ($PO_2 < 30$ mrnHg), animal restrained	Bradycardia	Unilateral beating at low frequency	McMahon and Wilkens (1995)
H. americanus	Restrained, unstimulated	Occasional periods of relative and absolute coordination with phase constancy between heart and scaphognathite beats.		Young and Coyer (1979)
H. gammarus	Experimental reduction of hemocyanin in blood	Tachycardia	Increased frequency	Spoek (1974)
Nephrops norvegicus	Restrained, pericardial chamber exposed to reveal cardioregulator nerves	Bursts of spikes in cardioinhibitor nerves, bradycardia	Sudden decreas in f_{sc} simultaneous cardioinhibitor bursts	Young (1978)
Procambarus simulans	Chemical stimuli applied to one gill chamber	Bradycardia or arrest	Homolateral apnea	Larimer (1965)
	Tactile stimuli to animal	Bradycardia	Bilaterial acceleration	Ashby and Larimer (1965)

Orconectes rusticus	Free	Tachycardia	Increased frequency	Wilkes and McMahon (1979)
Crangon crangon	Locomotor activity	Moderate tachycardia	80% acceleration	Dyer and Uglow (1978b)
Cancer magister	Stimulation of command interneurons in circumesophayeal	Bradycardial < arrest	Decreased f_{sc} < reversed pumping < apnea	Wilkens *et al.* (1974)
		(Responses ranked in order of increasing thresholds)		
C. magister	Unrestrained and quiescent	Bradycardia or arrest occurring simultaneously with reversed scaphognathite pumping or apnea Covariation of f_H and f_{sc} during unilateral scaphognathile beating.		McDonald *et al.* (1977)
Cancer pagurus	Animal quiescent	Rapid heart beat associated with renewed scaphognathite activity following pauses		Ansell (1973)
Cancer borealis and *C. irroratus*	Restrained and quiescent	Phase coupling between heart and one scaphognathite. The coupling phase may remain constant during altered beating rates.		Coyer (1979)
C. productus	Restrained with mask over anterior	Spontaneous and simultaneous onset and termination of heart and scaphognathite beating. Cardiac arrest often occurred during scaphognathite reversed pumping		McMahon and Wilkens (1977)

(Contd...)

Carcinus maenas	At rest	Arrest coincident with reversed pumping or apnea.		Cumberlidge and Uglow (1977)
C. maenas	Blockage of inspiraton openings to gill chambers	Arrest	Reversed pumping	Cumberlidge and Uglow (1978)
C. maenas	Animal at rest	Increased diastolic ventricular pressure, blood pressure in leg sinus' infrabranchial sinus, and branchial vein during reversed scaphognathite pumping.		Blatchford (1971))
C. maenas	Emersion response to hypoxia	Tachycardia	High incidence of reversed scaphognathite pumping, alternating with apnea	Taylor *et al.* (1973)
(Merostomata, included for reference)				
Limulus polyphemus	Hypoxia	Bradycardia	Decrease in gill plate ventilation rate	Watson and Wyse (1978)
	Normoxia	Covariation of heart and gill activity, occasionally with phase coordination.		

to continued inhibitor discharge whereas apnea is maintained until the inhibitor spike frequency falls below ca. 12 spikes/sec.

These data suggest that a common system might govern heart scaphognathite coordination. Command interneurons in the circum-esophageal connectives that influence f_H in crayfish, the scaphognathite pattern generator in lobsters and hermit crabs and both systems in *C. magister* have been identified. In the crab, increased activity in, these interneurons following, stimulation of various anterior sensory receptors was found by en passant recording techniques. Different patterns of heart-scaphognathite activity could be recruited when single interneurons were stimulated at different frequencies. Taken together these studies demonstrate at least some of the neuronal pathways that coordinate heart and scaphognathite activity.

The neuronal oscillatory systems controlling heart and scaphognathites have been shown to be dependent on metabolic energy This direct relationship between organ pumping rates and PO_2 may passively serve to match the two systems so as to maximize the efficiency of gas exchange across the gills.

Finally, hormonal coordination of heart and scaphognathite rhythms seems a distinct possibility. All of the known biogenic amines and peptides of the PO's increase f_H and often amplitude of heart beat as well 5-HT decreases f_{sc} during forward pumping while increasing the occurrence of reversal sessions dopamine and octopamine does not alter reversal incidence, and no comment on the effects of these two amines on frequency of forward beating is available.

Tenton and Cameron (I978) found no ventilatory responses to l-2% CO_2 Exposure in water. A moderate increase of PO_2 in external water causes a typical hypercapnic acidosis which is further partially compensated metabolically in *Carcinus*. Except in very addic waters, natural changes in PCO_2 are probably slight; however, very small modifications of external PCO_2 or P^H or both can lead to large changes of internal acid-base balance in crayfish. In this case again, the magnitude of these changes may be dependent on titration alkalinity and buffering power of external water.

Acid-Base Balance and Physiological State

1. Influence of Respiratory Gas Exchanges

Exchanges of oxygen and carbon dioxide through the respiratory surfaces or at the tissue level may be expected to affect acid-base state of hemoglobin.

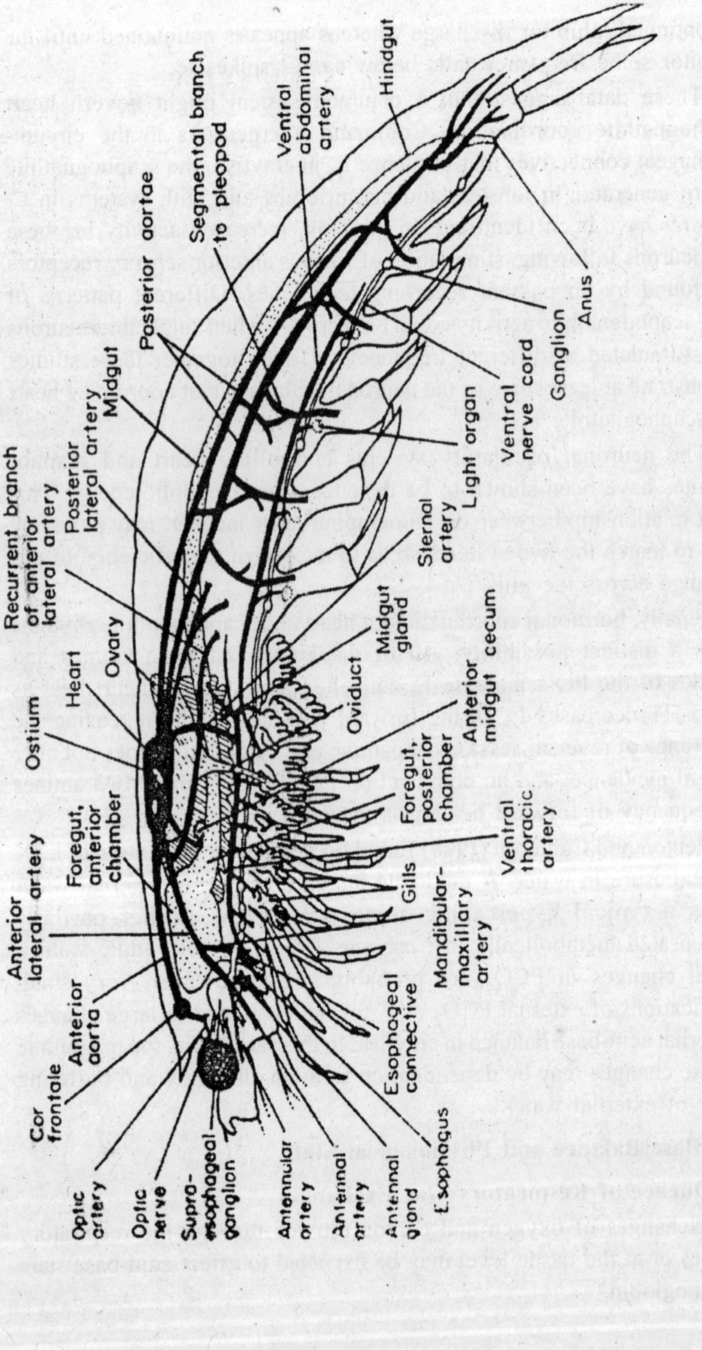

Fig, 8.10 : Diagrammatic euphausiacean with musculature removed to show major organ

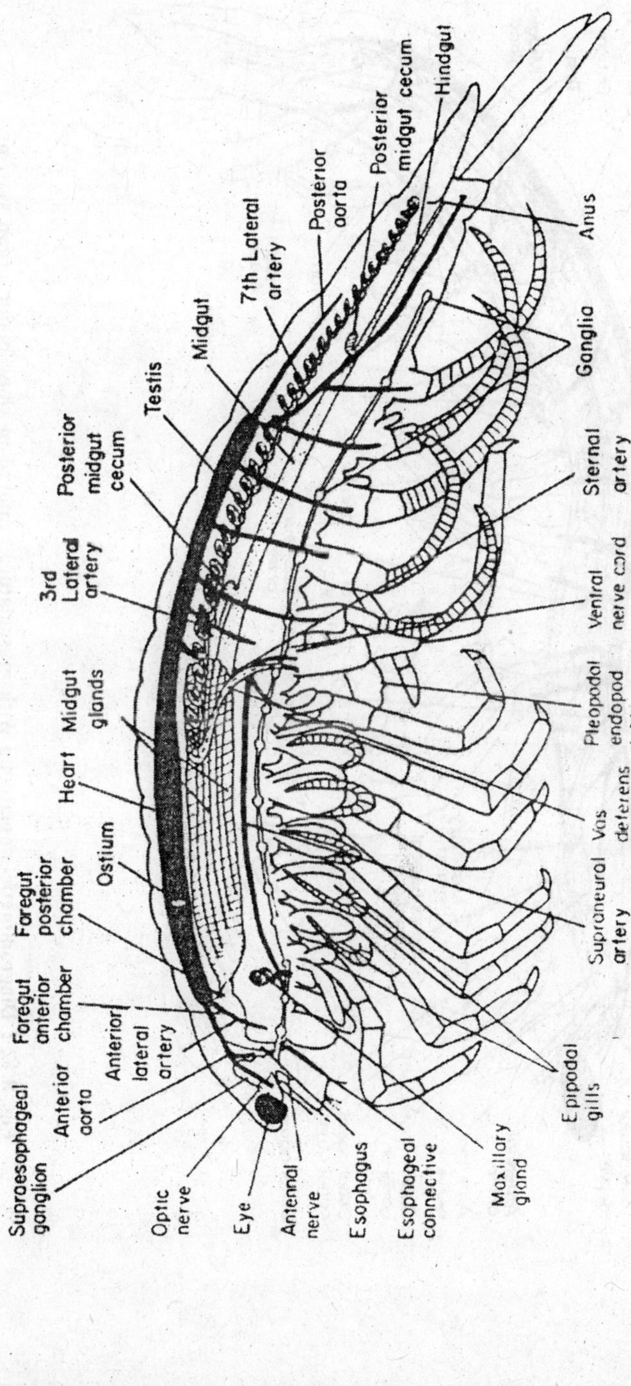

Fig. 8.11 : Diagrammatic *Anaspides* with musculature removed to show major organ systems.

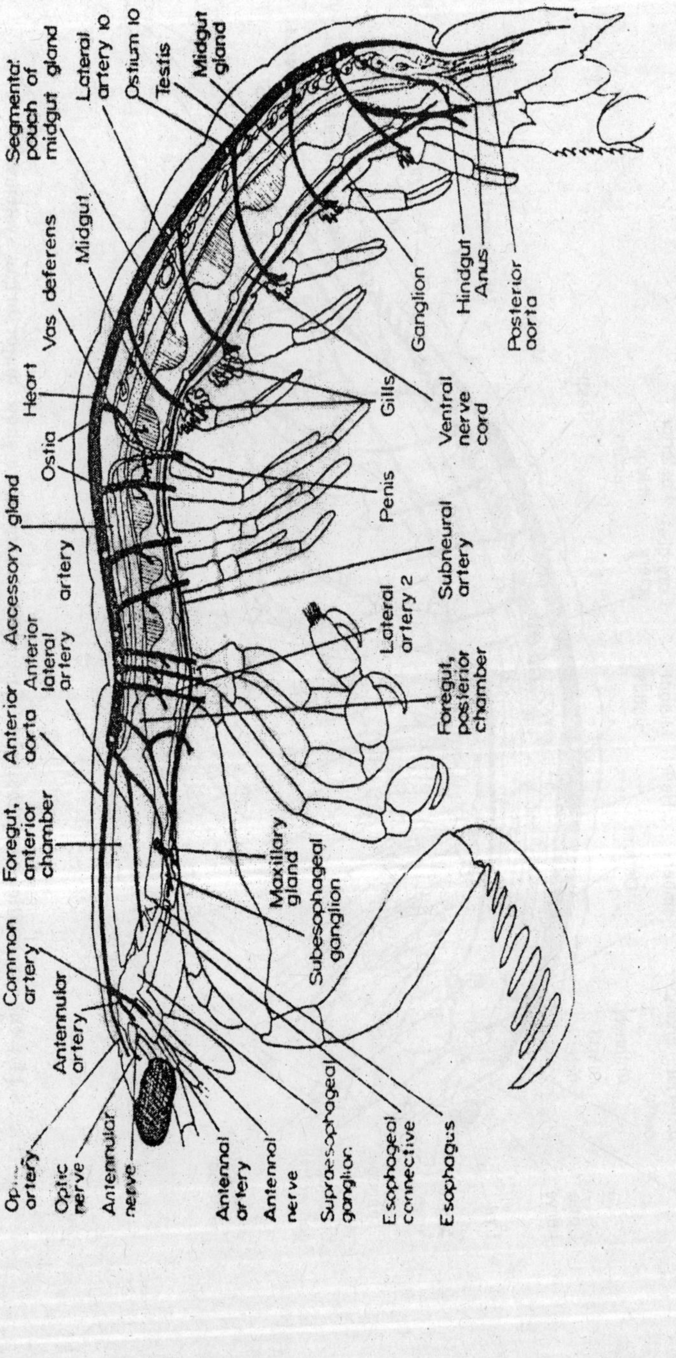

Fig. 8.12 : Diagrammatic stomatopod with musculature removed to show major organ systems.

Patsy A. McLaughlin

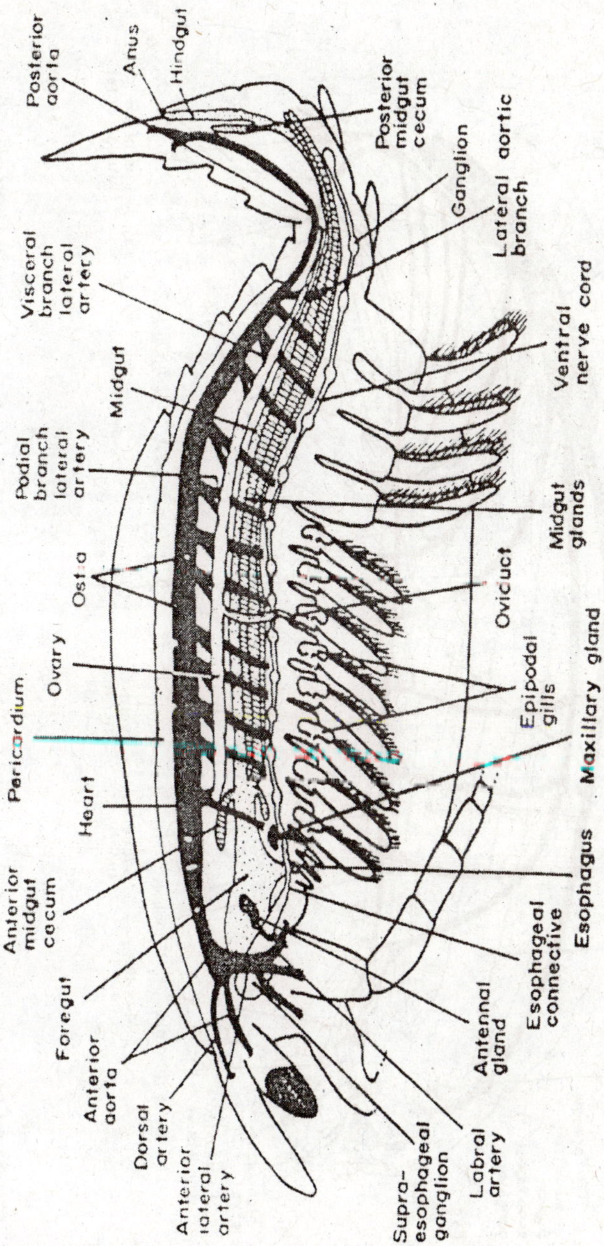

Fig. 8.13 : Diagrammatic leptostracan with musculature removed to show major organ system. (After McLaughlin, 1980.)

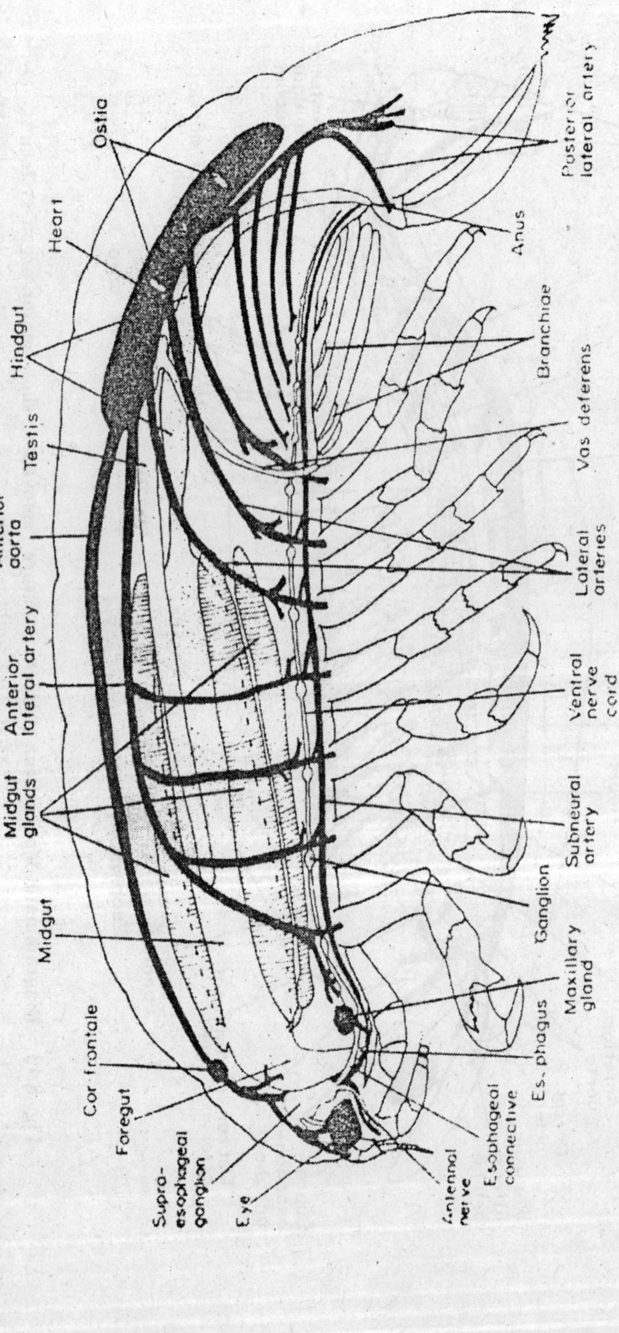

Fig. 8.14 : Diagrammatic flabelliteran isopod with musculature removed to show major organ systems. (After McLaughlin, 1980.)

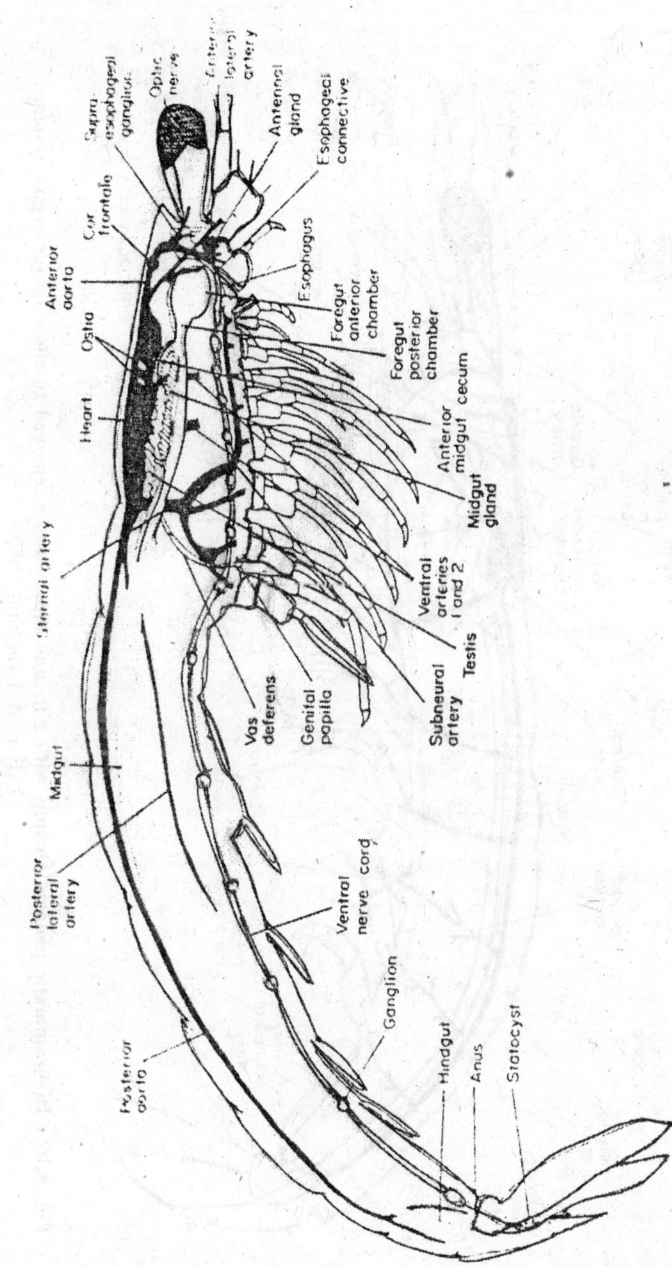

Fig. 8.15 : Digrammatic mysid with musculature removed to show major organ systems. (After McLaughlin 1980.)

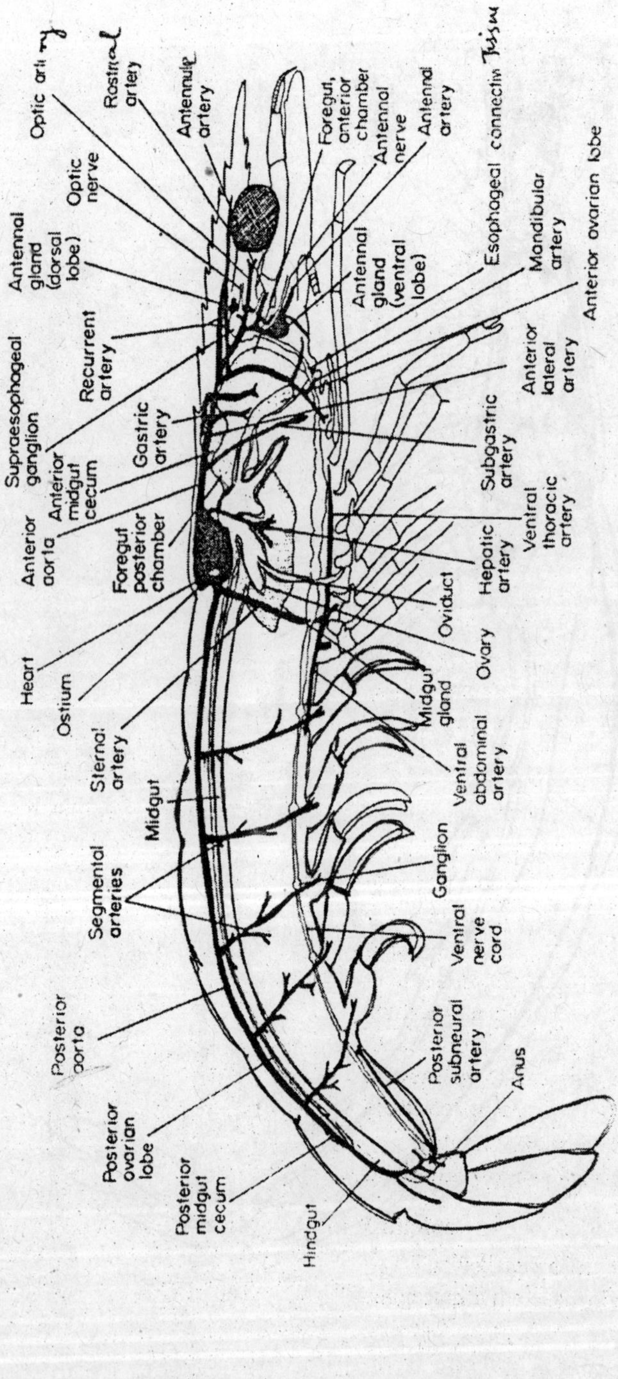

Fig. 8.16 : Diagrammatic penaeid shrimp with gills and musculature removed to show major organ systems. (After McLaughlin, 1980.)

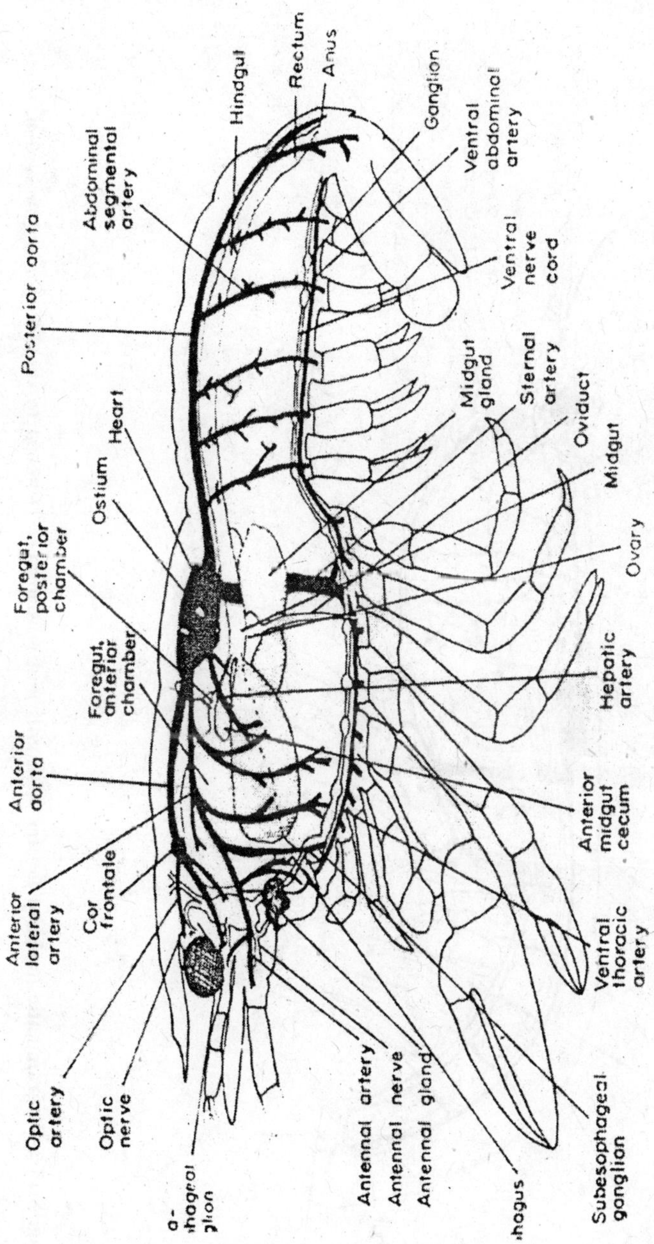

Fig. 8.17 : Diagrammatic astacidean crayfish with gills and musculature removed to show major organ systems. (After McLaughilin, 1980.)

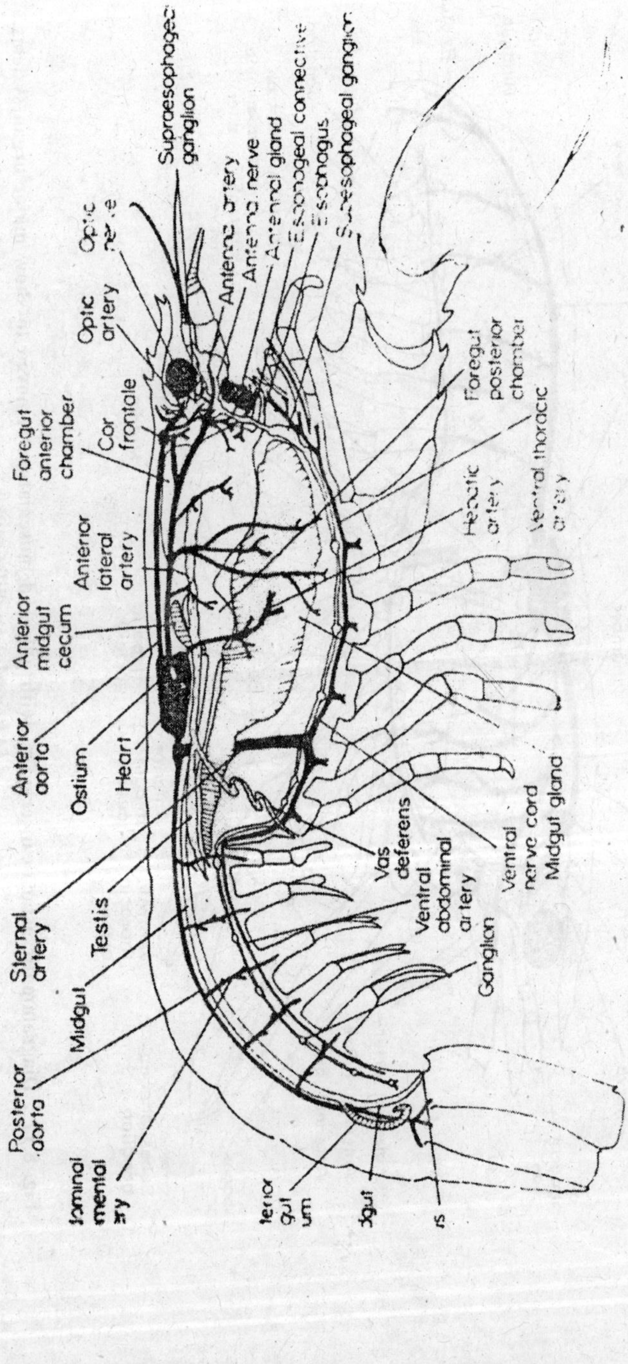

Fig. 8.18 : Diagrammatic homarid lobster with gills and musculature removed to show major organ system.

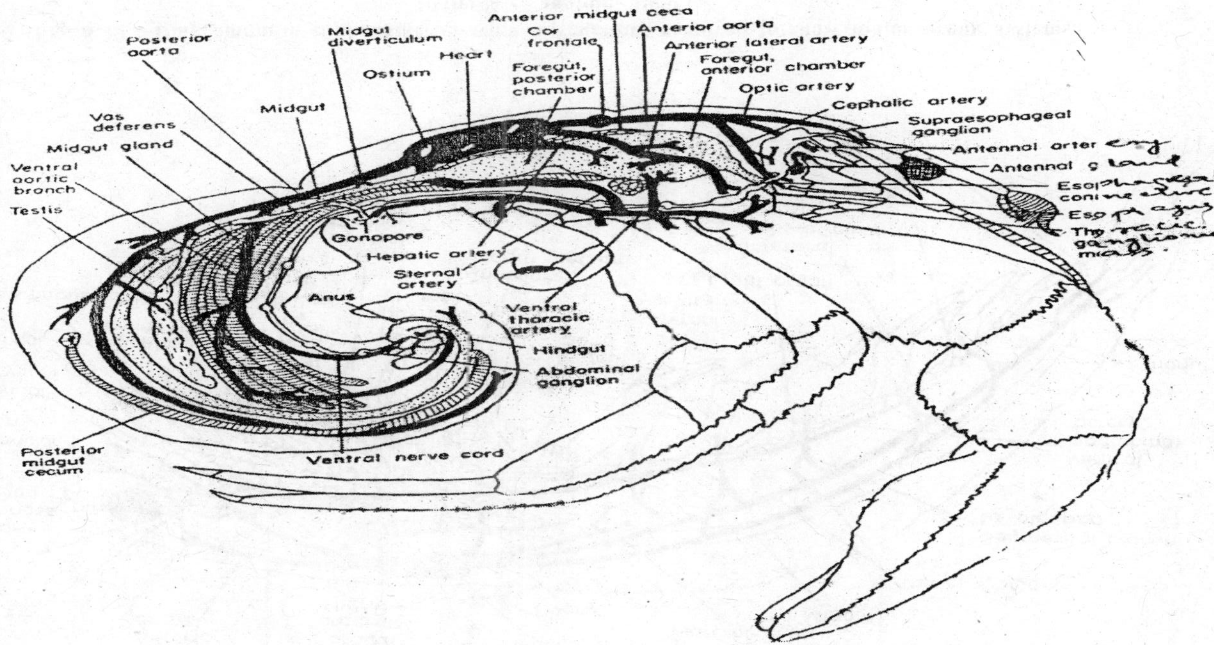

Fig. 8.19 : Diagrammatic pagurid hermit crab with musculature removed to show major organ systems.

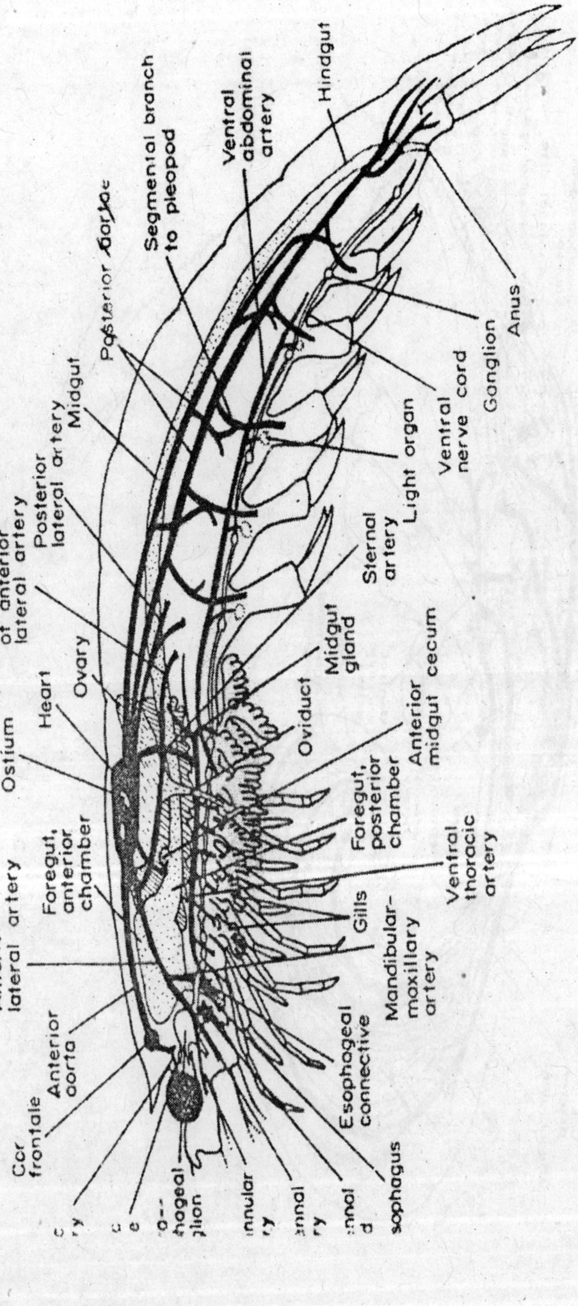

Fig. 8.20 : Diagrammatic euphausiacean with musculature removed to show major organ systems. (After McLaughlin, 1980.)

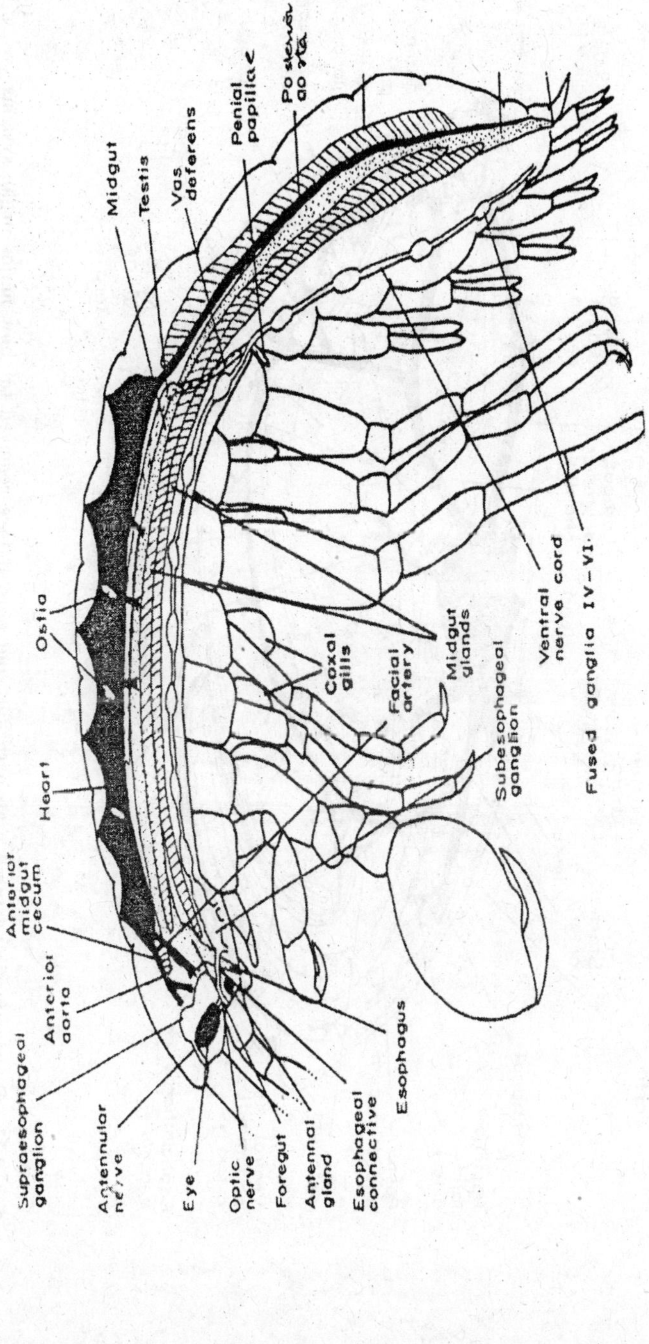

Fig. 8.21 : Diagrammatic with musculature removed to show major organ systems.

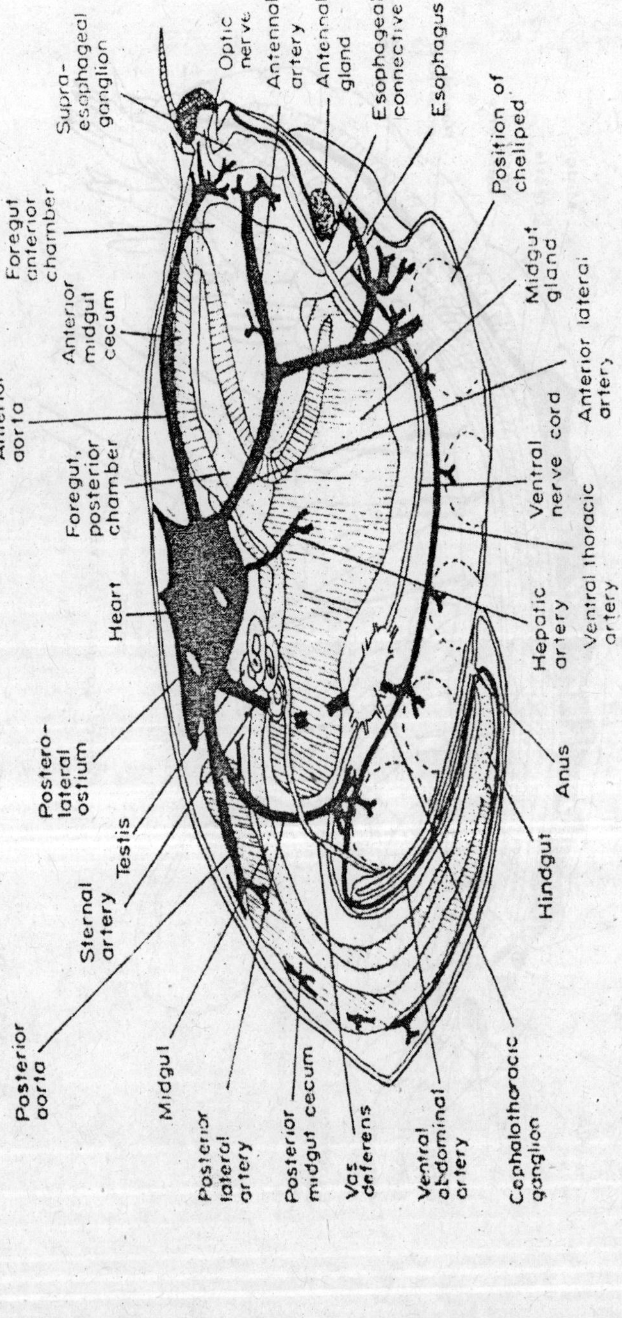

Fig. 8.22 : Diagrammatic portunid crab with gills and musculature removed to show major organ systems. (After McLaughlin, 1980.)

REFERENCES

1. Alexandrowicz, J.S., and Carlisle, D.B. (1953). Some experiments on the function of the pericardial organs in Crustacea. J. *Mar. Biol. Assoc. U.K. 32, 175-192.*
2. A.P.M. Lockwood, Oiver Body. Aspects of the Physiology of Crustaces. 595.3 Edimburgh and London.
3. Borradáila, L.A., Notes upon Crustacean Limbs Amer. Mag. Nat. Hist. 17:193.
4. David Mc Farland (1981), Animal Behaviour, Oxford University Press, Oxford.
5. Evas P.D., Insect Physiology, Academic Press, New York, 1994.
6. H.H Kombuber (1979), Handbook of Crustacea, Academic Press, New York.
7. Jame and H. Wilmoth, c. 1967. Biology of Invertebrata. Prentice-Hall, Engliwood Chiffs, New Jersday.
8. Kaentmer Alfred:, Invertebrate Zoology, Pub.-John Wileny and nomn, New york. London. Sydney, Joronto, Edited by Herbertus, Levi and Lormar Levi.
9. Linda H. Mantel, (1983), Internal Anatomy and physiological regulation, Academic Press, New York.
10. Paul A. Meglitsch, 1963 Invertibrate Zoology Oxford Univercity Press. Oxford.
11. Sinha, N. P., and Dejours, P. (1980). Ventilation and blood acid-base balance of the crayfish as function of water oxygenation (40 to 1500 Torr). Comp. Biochem. Physiol. A 65A, 427-432.
12. Stewart, P. A. (1977). Independent and dependent variables of acid-base control. *Respir. Physiol.* 33,9-26
13. Taylor, A. C. (1977). The respiratory responses of *Carcinus maenas* (L.) to changes in environmental salinity. J. Exp. Mar. Biol. Ecol. 29, 197-210.
14. Taylor, E.W., and Buther, P.J. (1978). Aquatic and aerial respiration in the shore crab, *Carcinus maenas* (L), acclimated to 15°C. *J. Comp. Physiol.* 127, 315-323.
15. Taylor, E.W., and Wheatly, M.G. (1979). The behaviour and respiratory physiology of the shore crab, *Carcinus maenas* (L). at moderately high temperatures. J. Comp. Physiol. 130, 309-316.
16. Truchot, J.P. (1973b). Tixation et transport de oxygene par le sang de *Carcinus maenas*; Variations en rapport avec diverses condittions de temperature et de salinite. Neth. J. Sea. Res. 7, 482-495.
17. Truchot, J.P. (1973c) Regulations de l equilibre acide-base du sang de

Carcinus maenas (L). Crustacea Decapode Brachyura) maintenu dans l air a diverses temperatures. C. R. *Hebd. Seanses Agriculctuer Sci.* 276, 2865-2898.

18. Truchot, J.P. (1975a). Blood acid-base changes during experimental emersion and remmersion of the intertidal crab, Carcinus maenas (L). Respir. Physiol 23, 351-360.
19. Truchot, J. P. (1975b). Action de phypercapnie sur petal acide-base du sang chez le crabe Carcinus maenas (L). (Crustace Decapode). C. R. Hebd. Seances Acad. Sci. 280, 311-314.
20. Truchot, J. P. (1976a). Carbon dioxide combining properties of the blood of the shore crab *Carcinus maenas* (L). Carbon dioxide solubility coeticient and carbonic acid dissociation, constants. J. Exp. Biol. 64, 45-57.
21. Truchot, J. P. (1979). Mechanisms of the Compensatioin of Blood Respiratory Acid-base Disturbances in the Shore Crab, *Carcinus maenas* (L). J. Exp. Zool. 210, 407-416.
22. Truchot, J. P. (1981). The effect of water salinity and acid-base state on the blood acid-base balance in the euryhaline brab, *Carcinus maenas* (L.). Comp. Biochem. Physiol. A 68A, 555-561.
23. Weber, R. E., and Hagerman, L. (1981). Oxygen and carbon dioxide transporting qualities of hemocyanin in the hemolymph of a natant decapod *palaemon adspersus*. J. Comp. Physiol. 145, 21-27.
24. Weiland, A. L., and Mangum, C. P. (1981).The influence of environmental salinity on hemocyanin function in the blue crab, *Callinectes sapidus. J. Comp. Physiol.* 132, 305-341.
25. Wilkes, P. R. H., DeFur, P. L., and McMahon, B. R. (1980). A new operational approach to PCO_2 determination in crustacean hemolymph. Respir. Physical. 42, 17-28.
26. Winterstein, H. (1954). Der Einfluss der korpertemperature aufdas saure-basen-Gleichgewicht im Blut. *Arch. Exp. Pathol. Pharmakol.* 223, 1-18.
27. Wolvekamp, H.P., and Kruyt, W. (1948). Experiments on the carbon dioxide transport by the blood of the edible snail (*Helix pomatia* L), The common crab (*Cancer pagurus L*) and the common lobster (*Homarus valuaris M. E.) Arch. Neerl. Physiol.* 28, 620-629.
28. Woodbury, J.W. (1974). Regulation of P^H *In* "Physiology and Biophysiec" (T.C. Ruch and H. D. Patton, eds.), Vol.-II, pp. 480-524. Saunders, Philadelphia, Pennsylvania.

9

NERVOUS SYSTEM

Dissected animal (Crayfish) submerged in alcohal to make the nerves white.

The cerebral or supraesophageal ganglion lies in the midline, just ventral to a point between the eyestalks. From this a number of nerves radiate out, of which four pairs are larger than the others, namely:

(1) the nerves to the antcnnulcs;

(2) the optic nerves from the eyes, each expanded within the eyestalk into a series of four optic ganglia; these latter will be seen when the dissection for the sinus gland is carried out;

(3) tegumentary nerves from epidermal sense organs on the dorsal surface of the head;

(4) the nerves to the antennae, with small branches to the excretory bladders.

Larger than any of these nerves are two circumesophageal connectives, which run from the cerebral ganglion to the subesophageal ganglion, located behind the esophagus underneath the anterior end of the endophragmal shelf. A slight swelling on each connective as it passes to one side of the esophagus marks the position of the connective ganglion. Two.minute nerves belonging to the so-called stomatogastric system leave this ganglion. One of them passes forwards and medially, to meet its fellow from the other side under the stomach. Here a minute median branch is received from the cerebral ganglion, and another branch passes over the anterior surface of the stomach towards its anterodorsal surface. Here, this branch pierees the ophthalmic artery, expands as a tiny ganglion inside the cor frontale, and then leaves the artery to proceed posteriorly on the dorsal surface of the stomach. Other branches of the stomatogastric system, too minute to be easily seen, ramify over the surface of the stomach and esophagus, pass outwards to certain areas of the integument, and extend back dorsally as far as the pericardial sinus.

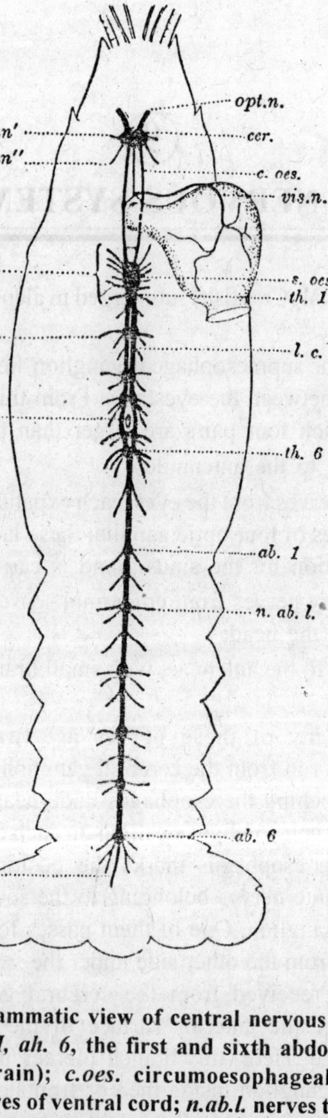

Fig. 9.1 : A semidiagrammatic view of central nervous system of *Astacus*. From Borradaile. *ab. I, ah. 6*, the first and sixth abdominal ganglia; *cer.* cerebral ganglion (brain); *c.oes.* circumoesophageal commissure; *l.c.* longitudinal commissures of ventral cord; *n.ab.l.* nerves to abdominal limbs; *n.an.'* nerve to antenoule; *n.an."* nerve to antenna; *n.ch.* nerve to chelipod; *n.m.* nerves to limbs adjoining the mouth; *opt.n.* optic nerve; *s.oes.* sub-esophageal ganglion; st.ar. sternal artery; *th. I, th. 6*, first and sixth thoracic ganglia; *vis.n.* nerve to proventriculus; *vis.n.'* nerve to hind gut.

If the esophagus is gently pulled forwards, a slender tritoccrebral commissure may be seen, running just behind the esophagus from one connective to the other.

To see the regulatory nerves of the heart, which issue from the subesophageal ganglion, a special method of preparation has to be used. First, the esophagus should be cut, just above the connectives. Then the stomach and the hepatopancrcas should be carefully pulled forwards and removed. The adductor muscles of the mandibles also should be removed. This procedure should reveal the cardiac inhibitors, a pair of nerves which emerge from a squarish median hole in the endophragmal shelf, just behind the esophagus and just in front of the most anterior point of origin of the flexor abdominis muscles. Each of these nerves passes outwards and backwards, crossing the extensor abdominis muscle, then ascends dorsoposteriorly inside the body wall to the pericardial sinus, where it breaks up into an internal plexus. The pair of cardiac accelerators emerges through a second median hole in the endophragmal shelf, which is much larger and more rounded than the first one. This second hole, however, is covered by the flexor abdominis muscle, the anterior origin of winch should be carefully cut, at least on one side, and the muscle cautiously bent back. Each accelerator nerve follows a course parallel to the more anterior inhibitor and eventually joins the same pericardial plexus.

Details of the pcricardial plexus can be made out only with great difficulty. It is known, however, that it connects dorsolatcrally with the heart. Impulses arriving via either accelerator or inhibitor fibers serve only to modify the beat of the heart. "Normally, this beat is maintained at a steady rate by a spontaneous, rhythmic discharge of impulses from a ganglionic group of scattered nerve cells on the inner dorsal surface of the heart. There are about sixteen such cells in the crayfish, nine in the lobster. They can be seen by slitting open a fresh heart midventrally, pinning it out flat, and staining for about 10 hours in a refrigerator with 0.005 per cent methylene blue in physiological saline, acidulated to pH 4.0-5.0 with hydrochloric acid. Since the subesophageal ganglion and succeeding thoracic ganglia lie under the endophragmal shelf, the beginner may find it easiest to identify the ventral nerve cord first of all in the abdomen, and then work forwards. To do this, the left and right flexor muscles in the abdomen should be separated with the migers, aided slightly with a scalpel. On looking down between these muscles in the midline, the central nerve cord will then. be seen, lying just above the slender ventral abdominal artery. In each of the first five abdominal somites the, cord is slightly expanded as a ganglion, each of which gives rise to three pairs of lateral nerves, the third pair appearing to come off slightly behind the ganglion. In the sixth abdominal somite a terminal ganglion is present which supplies thc telson as well as the last somite.

On following the nerve cord forwards into the thorax it will be found to dip down into the sternal sinus. Here it can be traced farther forwards to the subesophagcal ganglion by gradually cutting or chipping away the endophragmal shelf. The two most posterior thoracic ganglia are very close together, just behind the point where the sternal artery passes between the two separated halves of the nerve cord. In front of this separation, three distinct ganglia are present, one in each of the somites bearing the first three pairs of pereiopods. In the somite bearing the third maxillipeds a ganglion is present indistinctly divided from the subesophageal ganglion. This latter represents the fused ganglia of the somites bearing the mandibles, maxillules, maxillae, and first two pairs of maxillipeds.

The ventral nerve cord may now be lifted out from the animal and placed in water in a petri dish for study under the binocular microscope. Roots of the lateral nerves will show up clearly, and it will be seen that the cord and its ganglia really are paired, though the left and right divisions are for the most part rather completely fused. In the connectives between ganglia only nerve fibers are present, whereas in the ganglia nerve cells also are present. Among the nerve fibers it should be possible to distinguish four especially large ones, located near the midline on the dorsal surface. These are the so-called giant fibers, which conduct impulses more rapidly than do the other fibers. Their stimulation brings about the "escape reflex," in which the antennae, perciopods, and pleopods are thrust forwards, while the flexor muscles of the abdomen contract strongly to produce a powerful flip of the tail. If not seen in a whole mount, the giant fibers can be located in a histological cross section.

The innervation of the appendages will not be described here, but certain unique features about the nerves in the pereiopods deserve special mention. Like most nerves, each of these nerves contains many (upwards of 10,000) sensory fibers, but unlike most nerve's each contains extremely few motor fibers and is further remarkably in containing a few special inhibitory fibers. Determinations of the number of efferent fibers in the nerve as it passes through the ischium have indicated about twelve motor and three inhibitory fibres. These have to supply the great nausele fibre in seven different muscles which they do by repeated branching. Morever, each of fibres in a single muscle may be supplied by as many as four motor nerve fibres and some may by two inhibitory nerve fibres. The efferent nerve fibers a fact which makes possible their isolation.

Nervous System of *Squilla*

Take nicely preserved specimen and pinned on the dissection dish. Remove the middle portion of tergum remove muscles and for bodies and expose a portion of the nerve cord is the abdomen and trace backward and in anteries region. Expose the brain carefully after tracing complete nerve cord trace various branches. Expose and study the following ganglia and nerves.

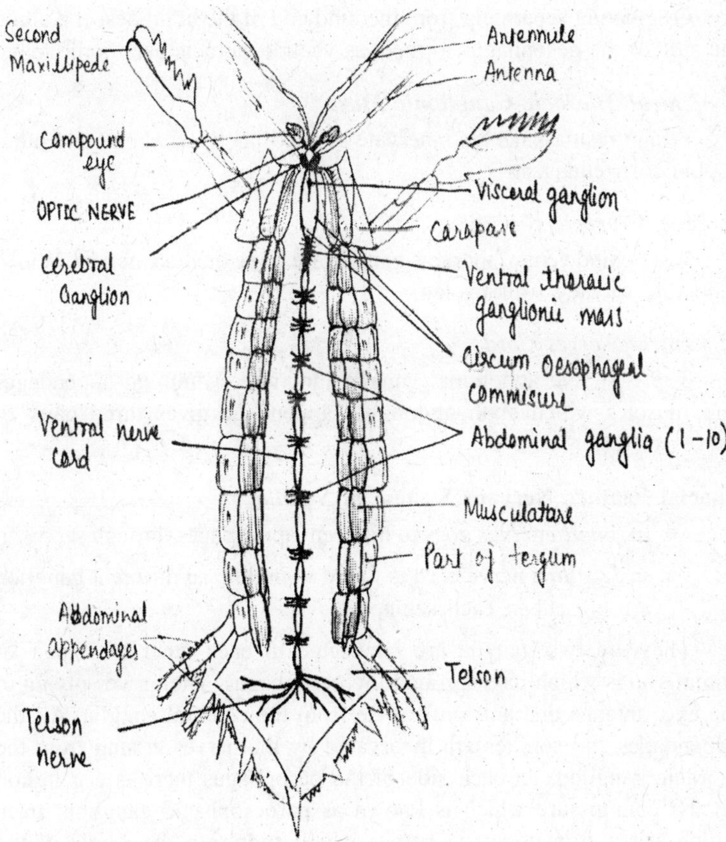

Fig. 9.2 : *Squilla* : Nervous System.

1. Brain or Supra-Osesophageal Ganglion

From brain following nerve originate:

(a) *A Pair of Antennutary Nerves :* They originate from brain and innervate antennules.

(b) *A Pair of Optic Nerves :* They originate from brain innervating eyes.

(c) *A Pair of Antennary Nerves :* Originating from the ventral surface of brain and innervating antennae.

2. Cercum-Esophageal Commissures

Originating separately from the hind end of the brain deseend along the side of the oesophagus to meet as ventral thoracic ganglionic mass.

3. Ventral Thoracic-Ganglionic Mass

From this mass nerves innervate to maxillipede, chelipod and other cephalic appendages.

4. Free Thoracic Ganglia

Just behind ventral thoracic ganglia are three thoracic nerves. Which innervate to each walking leg.

5. Ventral Nerves Cord

The first five abdominal ganglia innerrate swimming appendages and muscles which sixth abdominal appendage gives three pairs of nerves to telson.

Special Feature Nervous System of *Squilla*

- In *Squilla* nerves goes to different appendages through segment.
- In *Squilla* a nerve cord is found in middle an different ganglion are present in each segment.

They are two ganglia are connected to each other by a pair of commissures which passes round the oesophagus. The nerves plying to the eye, antenna and antennule arise from the cerebral ganglia. All the appendages and somites are innervated by the nerves arising from the thoracic ganglion. At each side of the oesophagus there is a ganglion on the commissure which is known as paroesophagal ganglion, from which arises stomatogastric nerves which supply to the whole of the foregut region.

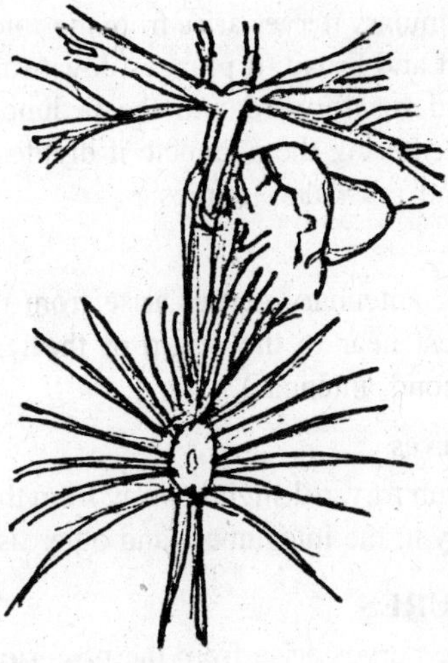

Fig. 9.3.

CEREBRAL GANGLIA

The cerebral ganglia are situated at the anterior end in the middle line under the carapace just above the epistoma. Each ganglion is roughly rectangular in shape forms a complex nerve cells and nerve fibres. The nerves supplying to the principle sense organs and the commissures with their offshoots arise from the cerebral ganglia. The following nerves arising from the cerebral ganglia are given below.

Optic Nerve

A pari of optic nerves arise from the dorsal side of the brain at its two anterior corners. They run straight way into the eyestalks. Immediately after passing into the optic peduncle the nerve becomes enlarged to form the optic ganglion at the distal cylindrical portion of the eyestalk. From this ganglion many nerve branches are given off which supply the various pans of the compound eye.

Oculomotor Nerve

A pair of nerves arise from the brain immediately behind the optic nerve. These follow the course of optic nerve arid finally supply to the eye muscies and adjacent parts.

Antennulary Nerve

A pair of antennulary nerves arise from the anterior side of the brain. Each nerve is short and broad. It plunges downward as soon as it arises from the brain and rus forward into the peduncle of the antennuale. Immediately after entering the peduncle it divides tfito three branches: tactile, auditory and muscular nerves.

Antennary Nerve

A pair of long antennary nerves arise from the lateral side of the cerebral ganglia just near to the origin of the tegumentary nerve and enters into the second antenna.

Tegumentary Nerves

These nerves run forward slightly away from the brain. The branches of this nerve supply to the integument and other tissues of the cephalon.

THE COMMISSURES

A pair of large nerves arise from the posterior side of the cerebral ganglia. These nerves pass posteriorly on each sids of the esophagus ana are connected with the ventral nerve mass in the thoracic region. On either side of the esophagus each commissure dilates into a ganglion known par-esophageal ganglion or commissural ganglion. These two commissures are connected together by transverse nerve behind the esophagus which is called as post-esophaceal connectives. From the par-esophageal ganglion stomatogastric nerves arise and supply to complete foregut.

STOMATOGASTRIC SYSTEM

The stomatogastric system consists of nerves mainly arising from the par-esophageal ganglion. They fuse together to form the stematogastric nerve which supply to the various regions of the foregut. There are also posterio-lateral nerves arising from the posterior region of the par-esophageal ganglion. The two main nerve arising from each par-esophageal ganglion fuse together with the similar nerves of the other side to form the mam stomatogastric nerve. This nerve runs upward towards the anterior part of the stomach. It enlarges to form the first stomato foregut and then bifurcates into two lateral gastric nerves which supply to the muscles of the cardiac stomach. These lateral gastric nerves give rise to internal branch which joins again with the stomatogastric to form a large ganglion. The stomatogastric nerve again bifuracates into two. Each branch is known a posterior gastric nerve which supplies to the posterior region of the pyloric foregut. These two nerves again unite to form a ring From this ring three nerves are given off one to the hindgut, second to the integument and the third to the digestive gland. The. posteriolateral nerves arise from the inner side of the par-esophagsal ganglion. These nerves innervates into the posterior drigfof muscles of the oesophagus.

THORACIC GANLIONIC MASS

The thoracic ganglionic mass lies over the sternal artery on the endosternal plates of ths third and fourth thoracic somites. It contains nerve elements showing the post-oral cephaothoracic ganglion and abdominal ganglia. In the centre of this fused ganglionic mass there is a foramen through which passes the descending artery. The thoracic ganglionic mass is formed by combination of many other smaller ganglia. Five pairs of ganglia supplying nerves to the appendages are readily goes in the posterior region of the thoracie ganglion. The abdominal ganglia are all distinguishable at the posterior end. The anterior portion of the thoracic ganglion is narrow whereas posterior portion is circular. From the thoracic ganglionic mass various nerve arise as given below.

1st Nerve

A paired nerve arises just to the lateral side of the commissures on the ventral nerve mass. It runs forward parallel to the commissures and divides into three branches, each ens inner/ating the mancible, first maxilla and second maxilla respectivsly.

2nd Nerve

This is also a paired nerve arising just behind the first nerve, It passes forward slightily outward and supplied to the various parts of the first maxilliped.

3rd and 4th Nerves

These are also paired nerves arising separately from the ventral nerve mass. Each nerve arises one behind the other. They supply to their appendages of the respective somites, the second and third maxilliped.

Nerves 5 to 9

Those are also parted nerves arising separately from the lateral side of the ventral nerve mass. These nerves are large compared to the previous nerves. The fifth nerve supplies to the cheliped and the rest four nerves supp'y to the four walking legs.

Abdominal Nerves

Abdominal nerves two which, arise from the posterior end of the ventral nerve mass. Each nerve passes backward along the madian line of the thoracic region and along the median line of the thoracic region and the ventral region of the abdomen. Many small branches supply to the segments of the abdomen arises from each nerve.

NERVOUS SYSTEM OF PRAWN

It consists of (1) the central nervous system including the brain connected with a ventral ganglionated nerve cord through a pair of circumoesophageal commissures, (2) the peripheral nervous system including the nerves, and (3) the sympathetic nervous system.

1. Brain or Supra-esophageal Ganglia

The brain lies at the base of the rostrum, anterior to the oesophagus and surrounded by a thick mass of fat. It is a bilobed structure derived from the fusion of several ganglia. On each side, the brain gives off :

(1) an antennulary nerve to the antennule of its side, into which it sends a statocystic branch to the statocyst,

(2) a stout optic nerve to the compound eye,

(3) an ophthalmic nerve to the muscles of the eye-stalk,

(4) an antennary nerve to the antenna, and

(5) a slender tegumental nerve to the labrum.

2. Circum-Esophageal Commissures

Posteriorly, the brain gives off a pair of stout nerves, the circum-esophageal commissures. These run backwards and downwards, and after encircling the oesophagus, unite ventrally with the sub-esophageal ganglia. The latter form the indistinguishable anterior part of the ventral thoracic ganglionic mass. The two commissures are crossed over, just behind the oesophagus, by a double bridge of tough connective tissue, called the endosternite. Each commissure bears a small commissural ganglion near its anterior end, sends a small mandibular nerve to the mandible of its side and a slender transverse connective to the fellow of the opposite side near its posterior end.

3. Ventral Thoracic Ganglionic Mass

Like the segments, the segmental nerve ganglia of the cephalothorax also become fused to form an elongated ventral thoracic ganglionic mass, lying mid-ventrally on the floor of the cephalothorax. This mass represents the fusion of eleven pairs of ganglia and gives off laterally

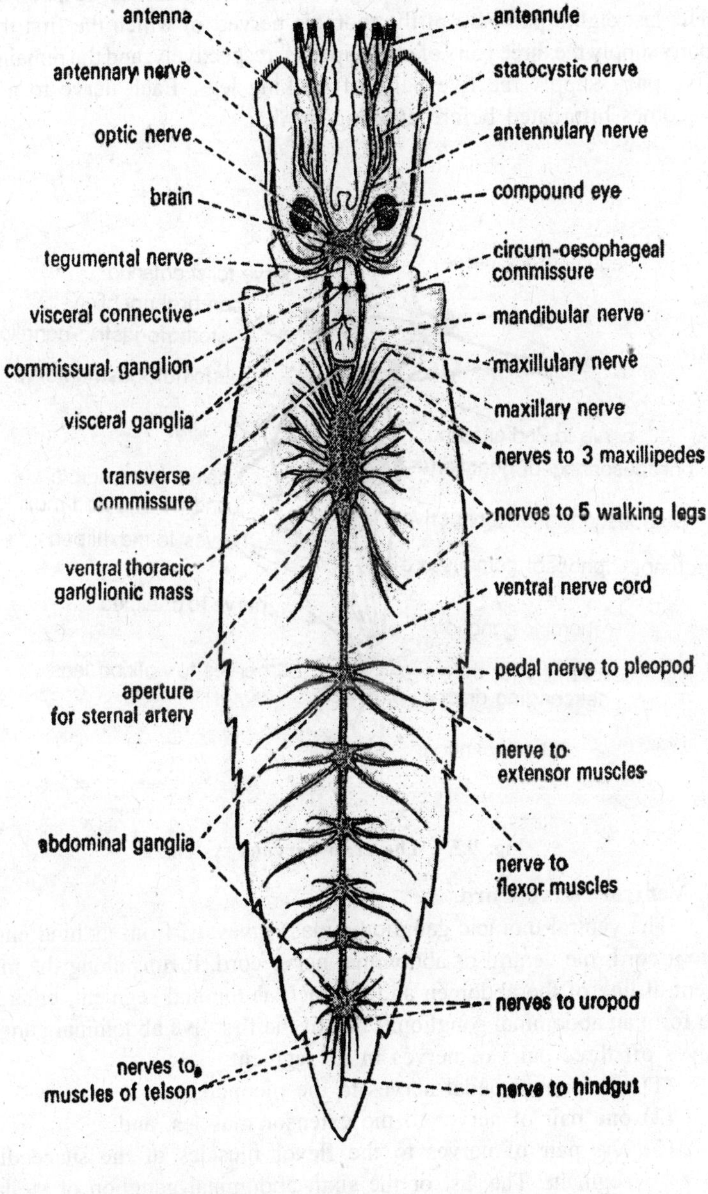

Fig. 9.4 : *Palaemon*. Nervous system in dorsal view.

eleven pairs of nerves. The first three pairs are of the cephalic nerves, supplying the mandibles, the maxillulae and the maxillae, respectively. The last eighth pairs are of the thoracic nerves, of which the first three pairs supply the three pairs of maxillipedes, respectively; and the remaining five pairs supply the five pairs of walking legs. Each nerve to a leg becomes bifurcated before entering the leg.

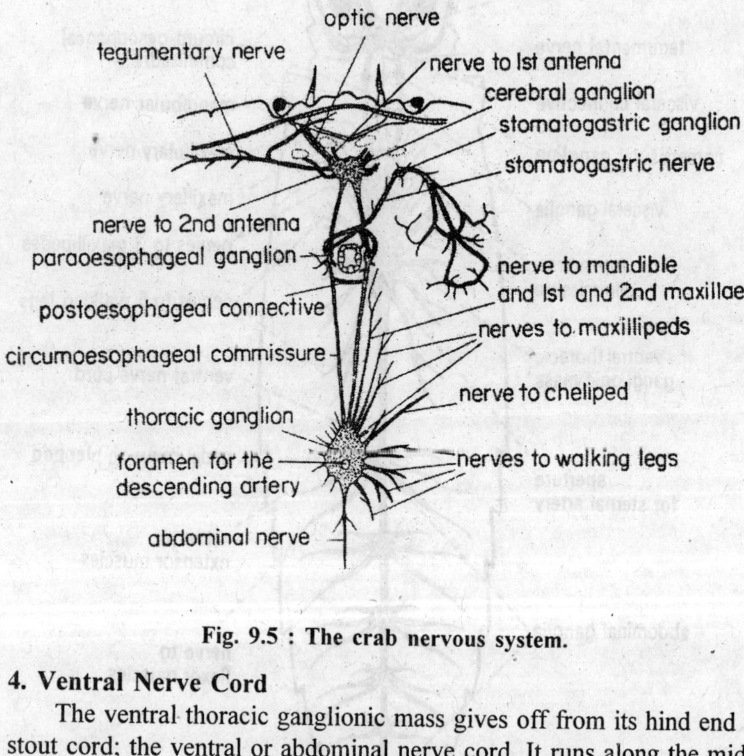

Fig. 9.5 : The crab nervous system.

4. Ventral Nerve Cord

The ventral thoracic ganglionic mass gives off from its hind end a stout cord; the ventral or abdominal nerve cord. It runs along the mid-ventral line of the abdomen and, in each abdominal segment, enlarges to form an abdominal ganglion. Each of the first five abdominal ganglia gives off three pairs of nerves in its segment:

(1) one pair of pedal nerves to the pleopods,

(2) one pair of nerves to the extensor muscles, and

(3) one pair of nerves to the flexor muscles of the succeeding segment. The last or the sixth abdominal ganglion or stellate ganglion is comparatively large, composed of several fused

ganglia, and supplies two pairs of nerves to the flexor muscles, two pairs to thc uropods, two pairs to the tel son and a single median nerve to the hind-gut.

5. Sympathetic Nervous System

The sympathetic or visceral or autonomic nervous system comprises a few ganglia and nerves. A small nerve, arising mid-posteriorly from the brain, bears two visceral esophageal ganglia lying one behind the other. The first ganglion is joined with the two commissural ganglia through a pair of connectives. The second ganglion gives off two pairs of nerves to the walis of esophagus and the cardiac stomach.

Neurons of the cord : Three methylene blue studies of the decapod nerve cord appeared at the end of last century by Alien, Bethe and Retzius. Neurons of *Astacus* are figured by Retzius (1890) in one of the classical works of invertebrate neurology. The eariy studies were designed to show the type of neurons and the fact that they do not an astomose. Less information is available as to peripheral connections, central regions of synaptic contacts, and three-dimensional relationships. Retzius found that most all bodies are unipolar except that characteristic multipolar ones occur at the anterior end of each ganglion. Many of Ratizus motor cells 2 interneurons are probably equivalent to those described by Alien (1896) and again many of the long interneurons are equivalent to those picked out physiologically by Wiersma and Hughes. Short interneurons confined to one ganglion in cercumference.

Physiological observation suggest that in certain cases at least the cell body either does not give electrical responses or else does so only become they are conducted electronically to it from the axon and there fore some probably has only a tropic function. The most important topographical feature of the neurons from the physiological point of view is the position of the dendrite and axon arborizations.

Neurons of the embryonic lobster serve as examples of types which presumably recur throughout the decapodes if not all Crustacea (Alien 1894,1896) on the basis of Alien's work, four classes of intemeurons LAN be distinguished.

(i) Cell body in the brain or in the ventral ganglion, axon descends to the posterior end of the cord having arborizations in all or most of the ganglion on its path.

(ii) Cell body in a ventral ganglion axon ascends to the brain.

(iii) Celi bodies in one ganglion, axon ascends or descends to the adjacent ganglion.

(iv) Cell body in one ventral ganglion axon ascends giving off collateral fibres in the adjacent ganglion and ends in the subjacent one.

Thoracic motor neuron : Each thoracic ganglion of *Homarus* gives rise to main nerve bundles, the anterior mainly to the limbs, the other mainly to the body wall. The anterior nerve is really double and soon divides.

Cell body anteriolateral axon to anterior nerve. Similar neurons are found in thoracic segment 1 to 11 and evens to the muscles of the 2nd antenna from the cerebrum.

- Cell body ventrolateral; axon takes a peculiar twist in the neuropile and passes out by anterior nerve; only one such neuron in each segment.
- Cell body small median; axon leave by an anterior root with numerous ramification in the neuropile, 2 or 3 such per ganglion
- Cell body small median, axon makes a complete loop and passes out by posterior root and has a branch which probably joins with similar brand of the other side.

Abdominal motor neurons : The axon may or may not decussate and may sun out by a nerve of the anteriorly adjacent segment. Those runs in the anterior nerve of the ganglion mainly supply the appendages. Part of the axon of each neuron runs 11 with its partner. There are the giant motor fibres to the abdominal muscle with the familiar arrangement of synapses.

Sensory fibres : sensory neurons generally have peripheral cell bodies. Alien found a no of different central fibres with no ascribable cell bodies, and presumed them to be sensory.

Neuron types in a crab (Carcinus) were described by Bethe (1897) from methylene blue preparations of the ventral condensed ganglion.

- Interneural cell body at the sub-esophageal end of the ganglion or in the brain with one axon descending and having arborization in each segmental region.
- Interneural, descending axons from the brain and ending in a restricted area in one segment
- Interneural, median cells with an H-shaped axon, which arborizes on both sides in many segments.
- Interneural limited to one segmental area.
- Motorneurons with one efferent axons in different segmental nerves, usually with widespread dendrites.

Neuron arrangement : Cell bodies lie mainly anteriorly. Neuropile is diffliscly spread, mixed with through—going axons and lies mainly posteriorly.

The fibres running through the ganglion are as follows :

- Fibres without connections in the ganglion.
  (1) From a superior esophageal nerve to the stomatogastric nerve, there are probably motor axons having cell bodies in the commissural isolateral ganglion.
  (2) From one superior esophageal nerve to the other with a thin branch to the gastric ganglion.
  (3) Similar ones without the latter branch.
  (4) From the inferior esophageal nerve to the stomatogastric nerve; probably motor. Axons to the esophagus from the commissural ganglion.
  (5) To the anterior esophageal from the inferior esophageal nerve.
  (6) From the superior esophageal nerve via a small median nerve.
  (7) Thick fibres which come from the inferior esophageal nerve with a branch down each superior esophageal nerve,
- Fibers without cell bodies with branch in the neuropile.
  (1) Fibers which run in via one superior esophageal nerve and out by other.
  (2) Fibers from the gastric ganglion, which run down the stomatogastric nerve and send a branch, down each of the superior esophageal nerve.
- Neurons with cell bodies not having arborizations in the neuropile of the esophageal ganglion.
  (1) Having a single fiber in the inferior esophageal nerve.
  (2) Unipolar cell having a T-shaped fiber to each of the inferior esophageal nerve.
  (3) Unipolar cell having a fiber each of inferior esophageal nerve and and ones in the inferior median to the brain.
  (4) Unipolar cells have 5 branches 2 in each of the inferior esophageal nerves and one in the inferior median nerve to brain.
- Neurons having arborizations in the neuropile of the esophageal ganglion.

(1) Cells having arborizations in the posterior region of the ganglion, an axon in the small median nerve and in of the inferior esophageal nerves, probably motor in function.

(2) Unipolar cells having a branch each of the superior esophageal nerves and in the stomatogastric nerve with arborization of collaterals from these where they pass through the ganglion

(D) Commissural ganglion : In contrast to the esophageal and gastric ganglia the commissural ganglion has the characteristic appearance of a crustacean central ganglion with a dense distinct neuropile surrounded by unipolar cell bodies.

(A) Through-going fibers having no collaterals in the ganglion.

(B) Fibers with arborization in the commissural ganglion.

(C) Neurons [with cell] of the commissural ganglion.

(D) The gastric ganglion : A fiber which run through Ihe ganglion.

(1) Large fiber from the stomatogastric nerve to muscle on both side of the stomach.

(2) Motor axon probably with cell in the esophageal ganglion and arborization the gastric ganglion.

(3) Afferent fibers of a few bipolar or tripolar sensory cell which may be in the ganglion or on the wall of the alimentary wall.

Orlov formed 2 or 3 large cell of this type.

Neural Organnization in the Optic Lobes

Ouring development of optic lobes. In *Penaeus duorarum,* the internal medulla and terminal medulla have been shown to develop from brain and not from paired disks of the common proliferatioin zone for retina lamina and external medulla. The previously accepted view that external and internal medulla of malacostracans develop from the single medulla in *malacostracans aneestors* is therefore impossible. *Artemia salina* and *Triops comcriformis* have less to the emaniciation of a satisfactory theory cencering compound eyes and optic neurophils and also to an explanation of the formation optic chiasmata between optic neurophil of *malacostracans*. The common proliferation zone is located anteriomedially in the eye, cells grow forward and around the outer surface of the eye to from ommatidia across the eye to from the lamina and at 90° to these resterorly to give rise to the external medulla.

The chiasmata from as a consequence of the way in which axons grow along the prepositions neurophils and subsequent movement of the neuraphil later in development.

The central nervous system of crustacean is ladder like in its more primitive form a condition retained by some anostracans and cladocerans. Ganglia are formed from embryonic pairs joint across the mid line by commissure and longitudinally by connective. Anteriorly several (probably three) ganglia fused to form the brain are cerebral ganglion. The two anterior division of the brain (the protocerebral and the deutrocerebral) are thought to stem from primitively preoral unsegmented structure. The third division of the brain (the tritocerebrum) is derived from the first post oral ganglion.

The Brain Or Cerebral Ganglion

(1) *Anterior median Nerves :* Motor neurons to the muscles controlling the fused median joint of the eye stalk assembly and proprioceptors around the proximal joints of the eyestalk (deutro cerebral origin).

(2) Connectives intemeuron between brain and esophageal ganglion.

(3) Inter neuron between the brain and optic ganglion in the distal joint (eyecup) of the eyestalk assembly (protocerebral origin).In some forms the optic nerve also contains some motor neuron (originating in the deutrocerebum) to the eye muscles.

(4) *Oculomotor nerves :* Motor neuron to the eye muscles and sensory neuron from mechano receptors on and around the eye stalk assembly (deurocerebral origin) Given by Sandeman 1964 and Mellon 1977.

(5) *Antennulary nerves :* Motor neuron to the antennules, sensory neuron from mechano receptors and chemoreceptors on the antannule and from the statocyst in its basal joints (deutro cerebral origin).

(6) *Antennary nerves :* Motor neuron to the antaenna and sensory neuron from proprio receptors and mechano receptors hair of the antennae (trito cerebral origin).

(7) *Tegumentry nerves :* Almost Holly sensory nerves from homolateral receptor on the anterior portion of the carapace (tritocerebral origin).

Neural Organisation of the brain descriptive anatomical studies of the brain exist in the three broad areas

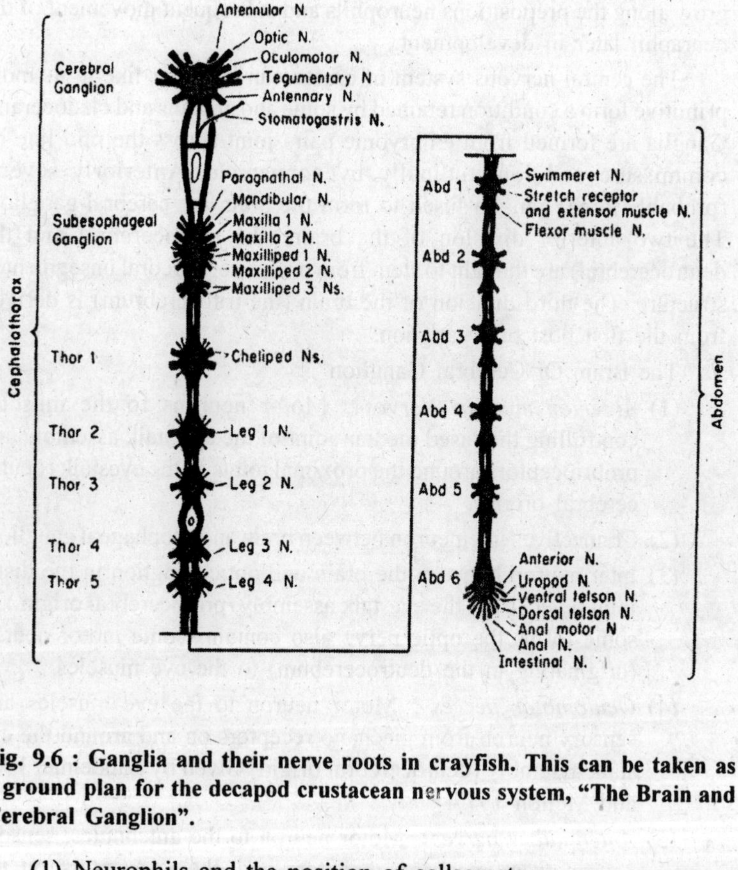

Fig. 9.6 : Ganglia and their nerve roots in crayfish. This can be taken as a ground plan for the decapod crustacean nervous system, "The Brain and Cerebral Ganglion".

(1) Neurophils and the position of cellsomata

(2) Neuronal tracts linking neurophils

(3) Form and projection of individual neuron.

(1) Neurophil of the Brain

Main neurophil can be found in most Decapodes. These were first tabulated by Helm (1928) who based his account on his own studies and those of a number of other anatomist till date. His neuro nomenclature has been adopted here, but anglicized. The neurophils are

(a) Protocerebrum : anterior optic neurophil (Al) Posterior optic neurophil (A2) Protocerebral bridge (A3) Central body (A4)

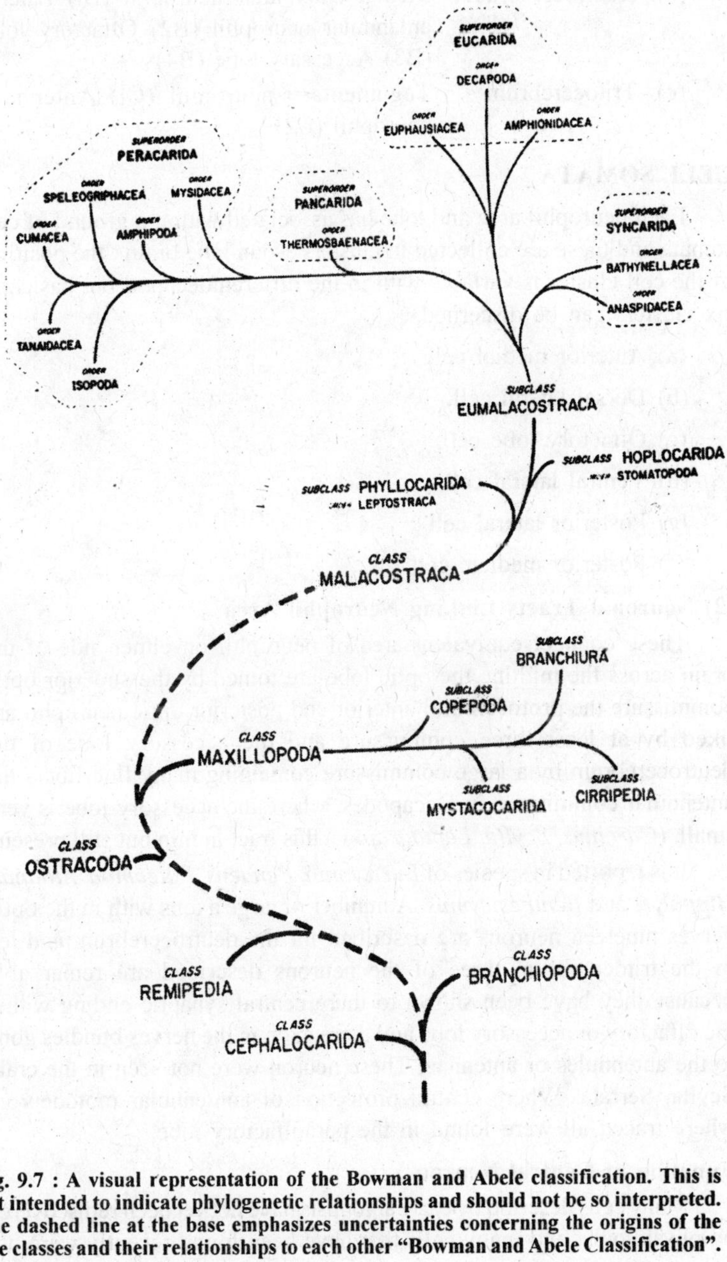

Fig. 9.7 : A visual representation of the Bowman and Abele classification. This is not intended to indicate phylogenetic relationships and should not be so interpreted. The dashed line at the base emphasizes uncertainties concerning the origins of the five classes and their relationships to each other "Bowman and Abele Classification".

(b) Deutrocerebram : Medial antannular neurophil (Bl) Lateral antannular neurophil (B2) Olfactory lobe (B3) Accessary lobe (B4)

(c) Tritocerebrum : Tegumentary neurophil (Cl) Antennary neurophil (C2)

CELL SOMATA

Each neurophil area and lobe has associated with it a ground of cell somata and these are collected in clusters around the brain . the position of the cell cluster is variable with in the differentdecapod but basically six clusters can be discerned.

(a) Anterior medial cell.

(b) Dorsal lateral cell.

(c) Olfactory lobe cell.

(d) Ventral lateral cell

(e) Posterior lateral cell

(f) Posterior mediate cell.

(2) Neuronal Tracts Linking Neurophil Area

These connect equivalent area of neurophil on either side of the brain across the midline the optic lobe are joined by the superior optic commissure the protocerebral anterior and posterior optic neurophil are liked by at least three commissure and the accessory lobe of the deutrocerebrum by a large commissure containing many fine fibers the antennular commissure in Decapodes, where the accessory lobe is very small. (*Carcinus, Scylla, Leptograpus*) this tract in thin but still present.

It is reported in species of *Euricyonia, Penaeus, Paleomon, Athanas*, *Hippolyte* and *Benthesicymus*. A number of large axons with in the optic nerves nineteen neurons are described for the deutrocerebrum and ten in the tritocerebrum three of the neurons described are remarkable because they have been shown to there central synaptic ending within the olfactory or accessory lobe and there axon in the nerves bundles gone to the antennules or antennae. These neuron were not seen in the crab, Scylla, Serrata. Where central projection of anntennular motoneuron, where traced all were found in the paraolfactory lobe.

Function of Central Neuron

Many cells describd by early antomists have not yet been rediscovered by physiologist and many cells that have been found recently were not

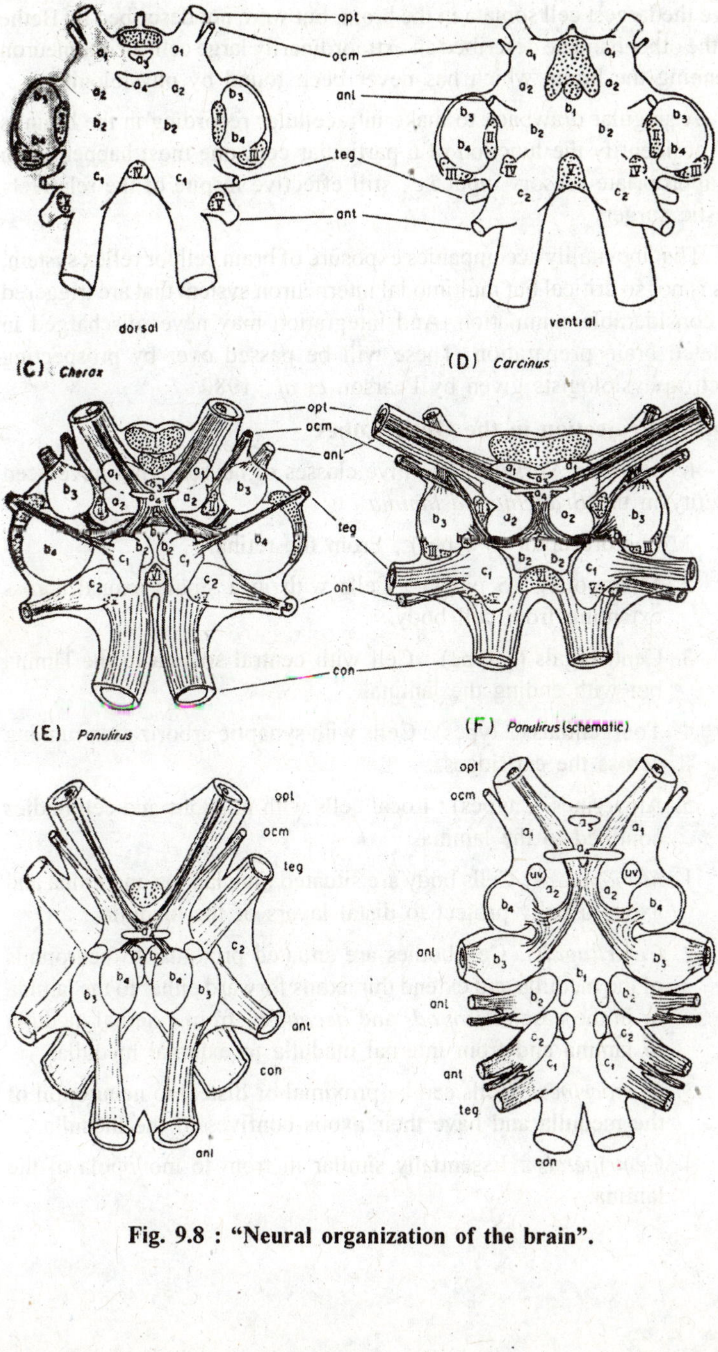

Fig. 9.8 : "Neural organization of the brain".

seen previously. The eye with drawal motorneuron in *Carcinus e.g.*, have the largest cell somata in the brain, but were not described by Bethe on the other hand he described an extraordinarily large cell, an interneuron scanning the brain, which has never been found by physiologist

A singular drawback to make intracellular recording in the brain is that to identify the function of a particular cell, one most happen upon an appropriate sensory input *i.e.*, still effective inspite of the relatively drastic surgery.

That inevitallly accompanies exposure of brain cell for reflex system, this is not so critical but multimodal interneuron system that are triggered by considerable summation. And integration may never discharged in isolated brain preparation. These will be passed over by prospecting electro-physiologists given by Pearson *et al.*, 1980.

Neuron Projection in the Optic Lobes

Accorting to Nassel 1973 : Five classes of neuron types have been itevitity in the *Branchiopod lamina.*

1. Photoreceptor (3 types) : From the retina.
2. Monopolaris (5 types) : Cells with in a single axonal across extenting from cell body.
3. Centrifugals (1 type) : Cell with central somata to the lamina but with ending the lamina.
4. Tourgentials (2 types) : Cells with synaptic arborization running across the cartridges.
5. Amacrines (2 types) : Local cells with in axons and cell bodies contined to the lamina.

1. *Monopolars :* Cells body are situated blue lamina and retina and axons usually project to distal layers of the medulla.
2. *Centritugals :* Cell bodies are situated proximal to neurophils of the medulla and extend thir axons forward either to the lamina or in the case of *isopods* and *decapods.* from external medulla to lamina and from internal medulla to external medulla.
3. *Demacrines :* Cells can be proximal or disteal to neuurophil of the medulla and have their axons confives to the medulla.
4. *Centripetals :* Essentially similar in from to monopola of the lamina.

5. T–Shaped Neurons : In branchiopods, cells bodies lies between medulla and protocerebrum, and axons extend distally to and in the medulla and proximally to and in the protocerebrum.

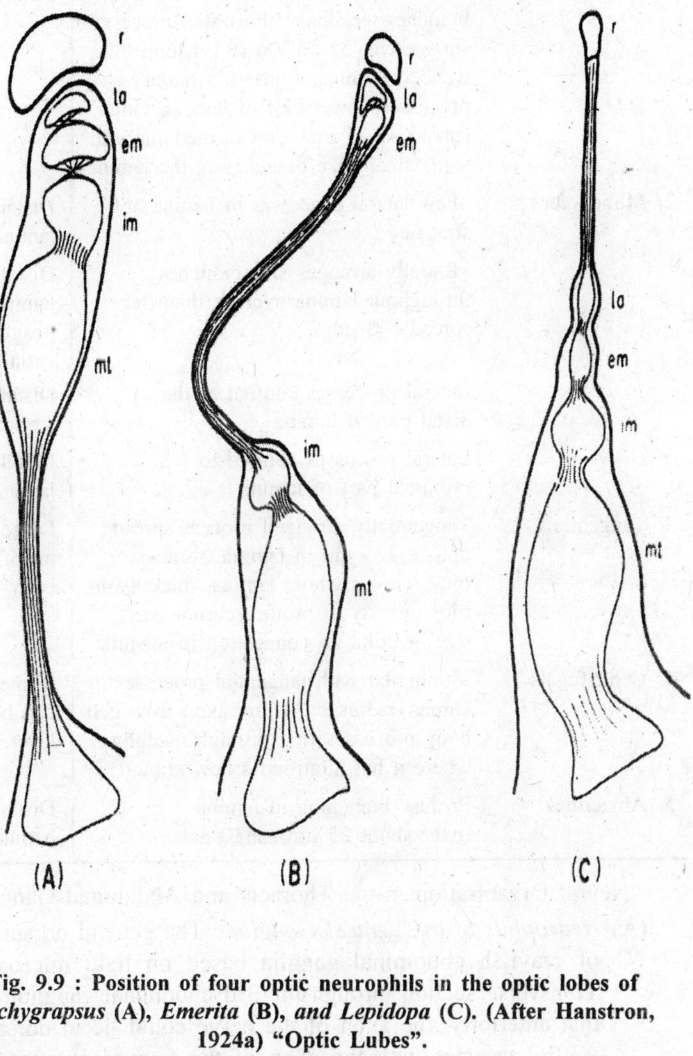

Fig. 9.9 : Position of four optic neurophils in the optic lobes of *Pachygrapsus* (A), *Emerita* (B), *and Lepidopa* (C). (After Hanstron, 1924a) "Optic Lubes".

Table 9.1 : Branchiopod Lamina

Cell Type	Variation and characteristic	Position of cell body
Artemia		
1. Photoreceptors	Terminates in distal lamina, enlarged branches terminals; thin ratial process spreas over 35-55 μm (RT_1) long thin webbed terminal trunk-thin radial processes, lower half of lamina, radial spread 37-75 μm. End in medulla with some small size branches in the lamina.	Retina
2. Monopolars	–Few lateral processes in lamina not stratified	Distal to lamina
	–Ratially arranges size branches throughout lamina average diameter spread = 8 μm.	Distal to lamina position variable
	Lateral processes confied to the distal part of lamina	Distal to lamina.
	Lateral processes confied to proximal part of lamina	Distal to lamina
3. Tangentials	Tangentially arranged process spread upto ± 125 μm in lamina, sometimes cavess whole lamina, thick axon runs directly to protocerebrum out size medulla no connection in medulla.	Large in medulla cell body layer.
4. Centrifugals	Monopolar with tangential processes in lamina–radius ± 70 μm, axon from cell body processes first through medulla where it has stratified arborization.	In medulla cell body layer.
5. Amacrines	Profuse branching in lamina (over about 25 μm)–anaxonal	Distal to lamina

"Neural Organisation in the Thoracic and Abdominal Ganglia"

(A) *Neurophils of the Ventral Ganglia :* The general organization of crayfish abdominal ganglia based on light microscopy. Transverse section through the 3rd abdominal ganglion show that anteriorly, the axon of the nerve could lie in dorsal and ventral vasicles with the axon of the four giant interneuron (lateral and medial giants) that lie dorsally, further into the ganglion two ventral cell body clusters and large commissure (anterior ventral commissure) appear and above the commissure is a fine fibres neurophils. The dorsally situated card axons and giant axon are still present.

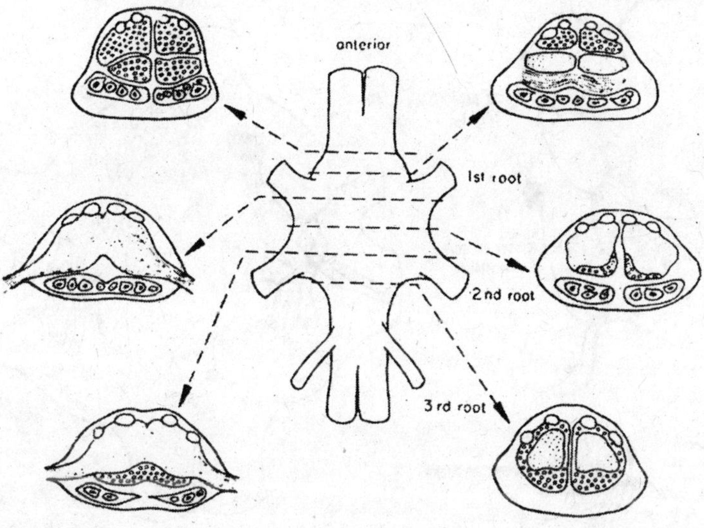

Fig. 9.10 : Neural organization of the third abdominal ganglion of the crayfish (*Procambarus clarkii*). shown by transverse sections at different levels. In the diagrams, the large dorsal circles are median and lateral giant fibers; the small circles are longitudinally running axons; the slippled area is the neurophil; the lined areas are commissures; and the ventral circles with solid centers are cell somata. (After Kendig, 1967.)

(B) *Neuron Projection in Ventral Ganglia :* The position of a number of neuron somata in thoracic ganglia of the American lobster has been determines by allowing cobalt chloride to perfuse back into to ganglion along various nerve goods (Wilson and Sherman, 1975). While not as selective as the implacement of single physiologically identified cells. This method demonstrates the presance of some 47 somata related to musculature of the appendages and to the thoracic body musculature. Distingtion between the ebberent to these two areas be made becouse the first lateral root contains all efferents to the appendages, and the second root contain to the body musculature, cell somata are gathered into anteriolateral, posteriolateral, and posteriomedial

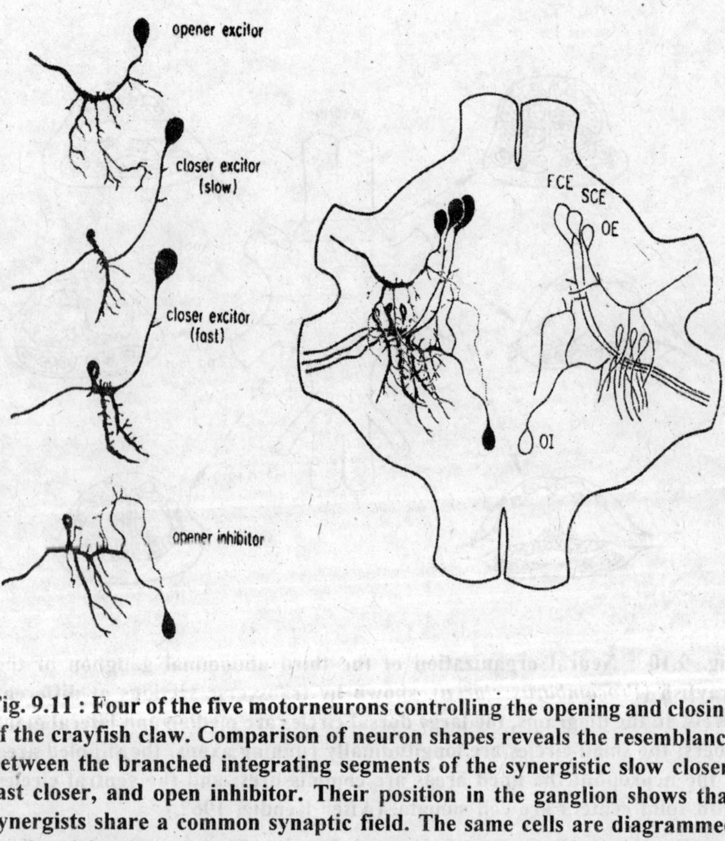

Fig. 9.11 : Four of the five motorneurons controlling the opening and closing of the crayfish claw. Comparison of neuron shapes reveals the resemblance between the branched integrating segments of the synergistic slow closer, fast closer, and open inhibitor. Their position in the ganglion shows that synergists share a common synaptic field. The same cells are diagrammed on the right side of the figure to bring this point out more clearly. FCE, fast closer, excitor; SCE, slow closer, excitor; OE, opener excitor; and OI, opener inhibitor. (After Wicns, 1976.)

clusters between anteriolateral and posteriolateral clusters. Four motoneuron somata of the appendages are located in the posteromedial cluster. As in crayfish and spiny lobsters, many neurons show central branching patterns that parallel one another. In the lobster it has not been established that the paralleled neurons are synergists in particular on the nervous system control. The fact that a substance causes a reaction which is similar to some normally occurring nervous phenomenon does not establish this substance as a transmitter substance. A number of criteria should be fulfilled before this can be established, among others the following:

(1) The substance or. a precursor should be present in the neuron from which the suspected transmitter is released.

(2) The substance should be present in the extra-cellular fluid in the region of the activated synapse.

(3) When applied to the post-synaptic structure, the substance,. . should mimic the action of the transmitter.

(4) A mechanism for removal or inactivation of the substance should be present, either through enzymatic inactivation or through specific uptake or reabsorption.

In addition, pharmacological agents have helped to clarify mechanisms by interfering with synthesis, release, removal, inactivation, or target, but they do not provide direct evidence for transmitter function.

The crucial demonstration of a transmitter substance consists in proving that, at the arrival of an action potential at the pre-synaptic ending, the substance is released in sufficient quantity to produce the observed physiological effect on the post-synaptic structure.

It is indeed difficult to meet all these criteria, and very often the technical obstacles have been insurmountable. As a result much work on suspected transmitter substances, especially in invertebrate animals, has been unsatisfactory because it has not progressed much beyond the demonstration of an effect of some suspected agent. Therefore, present information about transmitter substances is not very adequate and is insufficient for generalizations. Nevertheless, two substances should be mentioned here because they are invariably recognized as being highly active in many invertebrates.

5-*Hydroxytryptamine* often written 5HT, is found in many inver tebrate neurons, notably of molluscs and arthropods. 5HT is not restricted to nervous structures, however, but the concentration in other tissues is usually lower. Large amounts of it are found in certain animal toxins, such as those from *Octopus*, the snail *Murex*, and the venom of wasps. The heart'of bivalve molluscs is quite sensitive to 5HT, and there is good evidence that this substance is a normal transmitter for acceleration of the mollusc heart. The effect of 5HT on vertebrate blood pressure (which it increases) was known before the substance was chemically identified. It was then given the name *serotonin*, a name which is still in use. 5HT is also suspected of being a transmitter within the vertebrate central nervous system, but whether it has a normal role invertebrates still remains uncertain.

Gamma-aminobutyric acid, often written GABA, is another substance suspected of being a transmitter. It occurs in certain regions of the mammalian central nervous system, both in the brain and spinal cord. When applied to crustacean synapses, it mimics the action of inhibitor neurons, and it may well be a normal transmitter. Although, it is found in the vertebrate nervous system, it is again impossible to say with certainty whether it has a normal function as a transmitter substance.

Some amino acids, notably *glutamic acid* and *aspartic acid*, have also been suggested as transmitter substances. They occur both in vertebrate and in invertebrate nervous tissue, but substantial evidence for their normal role is still lacking. One major difficulty in establishing amino acids as transmitter substances is that they also occur normally in the animals, and their mere presence, therefore, carries no particular significance. Another problem is that it is extremely difficult to establish an inactivation mechanism. specially for amino acids as related to their suspected role as transmitter substances, for all the amino acids are regular participants in normal metabolic pathways.

Control and Integration in Invertebrates

In the more highly; organized invertebrates the control of physiological function is under both nervous and endocrine control. As in vertebrates, the nervous system Serves for rapid communication which is essential for actions that are related to escape, feeding, mating, and so on. As in vertebrates, an endocrine system produces hormones which control slower processes, such as growth, maturation, and many other metabolic functions. The nervous system has a direct and primary role in hormone production and the connection between the nervous and endocrine systems seems to be even closer than in vertebrates.

The Function and Role of Nervous Systems

We have seen that in vertebrates the central nervous system coordinates a large number of separately controlled functions, both nervous and endocrine. The apparent contrast between nervous control (usually rapid) and hormonal control (usually slow to very slow) breaks down under closer scrutiny. Not only is the transmission from neuron to neuron, and from neuron to effector, usually of a chemical nature, but the central nervous system itself is a major producer of hormones and exerts much of its over-all regulatory function in this way.

When we examine the role of nervous systems in the less highly organized invertebrates, we will need a more precise definition of the term *nervous system* than we used for vertebrates, where the meaning ot the term is self evident. A nervous system is an assembly of neurons which is specialized for repeated and organized transmission of information from sensory receptor sites to neurons, or between neurons, or from neurons to effectors (*e.g.*, muscles, glands, etc.). This definition prealudes the existence of a nervous system in unicellular animals (protozoans), although, functions within such cells may be coordinated with the aid of conducting organelles.

It is, on the whole, reasonably easy to recognize a neuron or nerve cell, mostly based on its membrane potential and the ability to produce repeated action potentials. However, it may be difficult to recognize whether a 'neurosecretory' cell is truly a modified nerve cell or not. Its location in association with nerve cells is helpful, and further information can be gained from recognition of action potentials, structural characteristics, and the function of similar structures in related species.

Invertebrate neurons have the same characteristics as vertebrate neurons. They are able to produce *action potentials* which consist of changes in membrane permeability and potential. Such action potentials are of an *all-or-none* nature and can be propagated in a nerve fiber, or axon. Typical neurons are able to produce repeated brief action potentials which last from a fraction of a millisecond to several milliseconds. and such neurons occur in all major phyla except protozoa and sponges. The transfer of information from one neuron to another takes place at *synapses* which usually show evidence of *chemical transmission*. As in vertebrates, the nerve membranes of the two neurons most often remain spatially separated and transmission is achieved through the release of an intermediary *chemical transmitter substance*. It appears, however, that there may be direct electrical transmission at certain synapses where the 'nerve membranes of the two cells become contiguous, a situation that also may occur, although rarely, at some vertebrate synapses. Typical *post-synaptic potentials* seem to be universally present in at least four major phyla, molluscs, annelids, arthropods, and chordates. These may be *excitatory post-synaptic potentials* (EPSP) or *inhibitory post-synaptic potentials* (IPSP). In general, the excitatory post-synaptic potentials consist in a decrease in membrane polarisation (in the direction of complete depolarization), and inhibitory potentials in an increase in membrane potential. These general characteristics, then, are in principle similar in invertebrate and vertebrate neurons.

The function of the central nervous system is to integrate the behaviour of the whole animal. This occurs by simple reflexes and by more complex chains of integration in which information from a variety of sources may combine with centrally programmed activity patterns to determine behaviour Peripheral ganglia concerned with local integration occur on the heart (cardiac ganglion) and stomach (stomatogastric ganglion) but both are subject to central control. The optic ganglia are accumulations of neurones which analyse information from the eyes and pass it on to the central nervous system.

Longer term control is achieved by means of neurosecretory cells which are scattered throughout the nervous system. Particularly important locations are in the optic ganglia, the brain, the post-esophageal connective and the thoracic ganglion. The hormones secreted by these cells control many of the rhythms in the lives of crabs, such as colour change and activity which often operate by tidal and diurnal cycles; the longer term rhythm of the molt cycle is also controlled by neurosecre.

The functioning of these extremely efficient sense organs is coordinated by the nervous system; in the Higher Crustacea this is reminiscent of the primitive type of nervous system found in the Segmentaria, to which the annelids also belong. For example, the nervous system of the crayfish indicates clearly that it has arisen from a segmented body form, for lateral branches are given off by the longitudinal nerve cords at absolutely regular intervals. The true crabs are, in this instance, the most greatly modified forms. Here, several such segmental sections join to form a uniform nerve knot (ganglion), especially in the case of the subesophageal ganglion which provides innervation to the oral appendages. A smaller ganglion above the mouth sends nerves to the compound eyes and antennae; in the higher Crustacea it has reached the same level of development as in the insects. The cephalic sense organs are thus made extremely efficient, since they are innervated from three "brain" centers (lamina ganglionaris, medulla externa, and medulla interna). An "olfactory center" is also well developed; the efferent nerves of the chemical sensory receptors from the first antennae terminate in this center. In addition to these centers, the crabs also have a nerve network beneath the epidermis, which receives stimuli from the mechanoreceptor hairs and elsewhere.

The high level of efficiency of this nervous system explains certain extremely highly developed behavior patterns, such as orientation, prey capture, and courtship. This efficiency is illustrated by certain measurements made on the speed of transmission of stimuli in the American crayfish *Orconectes*. In the two pairs of giant fibers which this animal has, excitation can progress at a rate of 20 m per second from anterior to posterior, thus making possible the rapid backward beat of the abdomen which is usually necessary for successful escape. Such an efficient nervous system also makes possible the performance of certain behaviour patterns which are reminiscent not only of the insects, but of birds as well; these are illustrated in the courtship behavior of certain crabs.

The large axon also allowed penetration by microelectrodes and the measurement of the resting and action potentials under experimental conditions. Such nerve fibers showed resting potentials of 50-80 mv, action potentials of 90-130 mv, and could produce up to 3×10^5 action potentials. This organic materials in the axoplasm were shown to have only minor importance in the resting and action potentials when it became possible to replace the axoplasm with preparation has thus proved of great value and has been studied by many workers. The results were summarized at a symposium held at Woods Hole in 1965.

It was once thought that the unequal distribution of ions across the nerve membrane could be largely explained in terms of Gibbs-Donnan equilibria, perhaps due to the physiologists' desire to keep "vital" phenomena out of any explanation. With the advent of labeled tracers it became clear that Na^+ was being pumped out of the axon and K^+ was being pumped in, both pumps working against the concentration gradient. The Na^+ efflux was inhibited by 2 HIM cyanide or 0.2 mM dinitrophenol (DNP) measured the phosphate in squid crab axoplasm and showed that cyanide or DNP decreased the concentration of ATP and arginine phosphate found that the Na^+ efflux from cyanide-poisoned axons was increased following microinjection of ATP, phosphoethopyruvate. or arginine phosphate. There were three Na^+ extruded for each high-energy bond split

There were certain differences between the effects of injecting ATP and arginine phosphate on the recovery of the Na^+ efflux from squid nerve. Injecting ATP restored the Na^+ efflux but that this efflux was not coupled to an influx of K^+. On the other hand. injecting arginine phosphate would bring about a coupled Na^+/K^+ movement of ions.

In crab nerve showed the presence of an ATPase in microsomes. The ATPase in intact nerve is dependent for its activity on the presence of Na^+ inside the nerve and K^+ outside the nerve. The ATPase was partially inhibited by quabain and completely inhibited by cyanide or DNP. Between 2.7 and 4 Na^+ were extruded for each energy-rich phosphate bond split.

So far there has been no evidence of the functioning of the Na^+ pump in the extruded axons. This may be due to the fact that removal of the axoplasm also removes some high-energy compounds in addition to ATP and arginine phosphate.

In crab axons the mitochondria are often arranged at the periphery of the axon near the membrane perhaps indicative of an active role. The role of the Schwann cells in the maintenance of ionic concentrations has not yet been fully investigated and these too may play a metabolic role.

The main disadvantage of invertebrate CNS preparations is that the roots are not separated into sensory and motor branches. Thus it is necessary to trace the nerve fiber out to its ending in order to be sure of the sensory or motor nature. This makes the study of invertebrate reflexes more difficult than that of vertebrate reflexes.

The general state of our present knowledge of invertebrate neurochemistry will be shown by the following examples.

In *Homarus amercanus,* inhibitory and motor axons in meq can be dissected 120 mm long → 40 μ.

Table 9.2 : Analysis of Axoplasm, Valuses in meq/kg

K	260
Na	152
Ca	13
Mg	23
Cl	145
Aspartate .	138
Glutamate .	35
Taurine .	65
Glycine .	5
Alanine .	33
Isethionate .	–
Total P .	45

Table 9.3 : Evidence for the role of glutamate and GABA as transmitter agents at nerve-nervc and nerve-muscle junctions.

1. Occurrent	Peripheral nervous system *Homarus* (Kraviz *et al.*, 1963)	Inhibitory axon of *Homarus* (Kraviz, Kuffler and Potter, 1963; Kravitz and Potter, 1965)
2. Active on Neurones or Muscle Crustacea	Crustacean muscle (Van Harreveld and Men delson, 1959) Crayfish muscle (Takeuchi and Takeuchi, 1964) Opener Muscle of *Cambarus* (Robbins, 1959)	Astacus muslce (Boistel and Fatt 1958. Opeener muscle of *Cambarus* (Robbins, 1959). Neuromuscular junction of *Homarus* (Grundfest, Reuben & Rickles, 1959). Crayfish muscle (Takeuchi and Takeuchi, 1965 a,b,) Crayfish central nervous system (Furshpan and Potter, 1959) Crayfish stretch receptor (Kuffler and Edwards, 1958). Stretch recetor of *Cambarus* (Hagiwara, Kusano and Saito, 1960).
3. Enzymes for Synthesis Crustacea	From glucose/oxoglutarate	Glutamic acid decarboxylase: peripheral and central nervous tissue of *Homarus* (Kravitz. 1962).
4. Enzymes for Breakdown: Crustacea	Including glutamic acid decarboxylase: peripheral and central nervous tissue of *Homarus* (Kravitz, 1962).	Including GABA transaminase
5. Release from Nerve Endings: Crustacea	Leg of *Carcinus*(Kerkut, Shapira and Walker, 1965)	Including GABA transaminase
6. Antagonists: Crustacea	(GABA) Crayfish opener muscle (Robbins, 1959)	(Picrotoxin) Crayfish opener muscle (Robbins, 1959) Lobster neuromuscular synapse (Grundfest, Reuben and Rickles, 1959).

REFERENCES

1. Abbott N.J. (1971); The organization of the cerebral ganglion in the shore crab, *Carcinus manenas*.
2. Bowerman, R.F. and Laminer; (1974) structure and function in the nervous system of invertebrates.
3. Chapple W. and Hearney (1974); The morphology of the fourth abdominal ganglion of the hermit crab.
4. Elobsson R. (1969); The Development of Compound eyes of *Penaeus fuorarum* with remark on the nervous system.
5. Haustrom B (1947); The brain, the sense organ and the incretory neurosecretory cells of the head in the crustacea malaconstraca.
6. Krieger K.R. 1980. Zoology Ueber das, central nervous system, des flusskrebses, Z. Wiss.
7. Theodore Holmes bullock, G. Adrian Horridge, Structure and function in the nervous system of invertebrates.

10

SENSE ORGANS

The sense organs of Palaemon are : (1) The compound eyes, (2) The statocysts, (3) The Tactile organs, and (4) The olfactory setae.

COMPOUND EYES

The prawn has a pair of black and hemispherical eyes, inserted at the base of the rostrum, one on each side in an orbital notch. Each eye is borne upon a short movable and two-jointed stalk. The eyes are peculiar in structure, being composed of a large number of visual elements, the ocelli or ommatidia. Such eyes, made up of hundreds or thousands of individual elements, so are termed the compound eyes and are met with in many Arthropods.

The outermost layer of the eye is a portion of cuticle known as cornea, which is chitinous and transparent. The cornea is divided into a large number of squares or facets, giving the appearance of a graph paper. Below each facet lie's one ommatidium inside the eye. The ommatidia are arranged radically and are alike in structure, and each consists of many cells arranged along the central axis.

Each corneal facet thickens in the centre to form a biconvex corneal lens. Under the lens lie two corneagen cells, which are modified epidermal cells and secrete a new cornea as soon as the old one is cast off in molting. Beneath the corneagen cells lie elongated cone-cells or vitrellae which constitute a transparant, homogeneous crystalline cone. The inner ends of the cone-cells are long and tapering.

The part of the eye, from cornea up to the extreme ends of cone-cells, is known as the dioptrical region, which focusses the light upon the inner sensitive part or receptor region of the eye.

The inner ends of cone cells lie upon an elongated, spindle shaped rod, the rhabdome, which has a transversely striated appearance. The rhabdome is secreted and surrounded by a group of seven elongated retinal cells. The rhabdomes and the retinal cells together form the receptor region of the eye. Their inner ends rest upon a basal membrane beyond which they are continuous with the nerve fibres of the optic ganglion which is connected with the brain by the optic nerve.

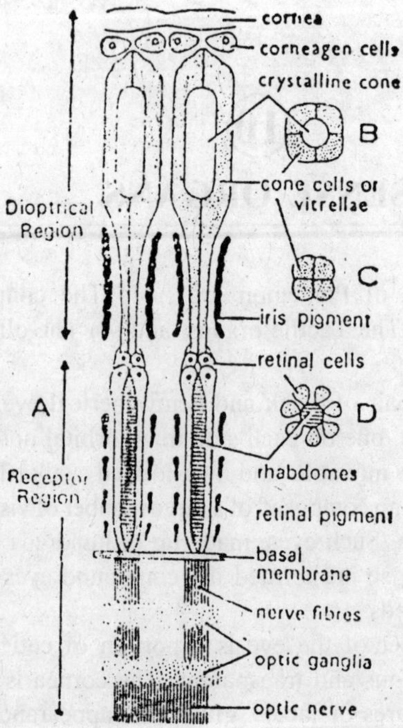

Fig. 10.1 : *Palaemon*. Structure of eye A-Two ommatida in L. S. (Semi- diagrammatic). B-T.S. less of an ommtidium through cone cells. C-T.S. through basal ends of cone cells. D-T.S. through retinal cells.

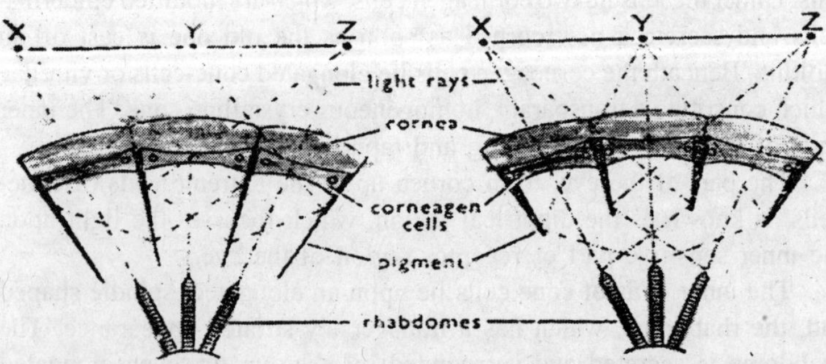

Fig 10.2 : *Palaemon*. Diagrammatic representation of image-formation by a compound eye. A-In bright light (dayvision). B-In dim light (night vision).

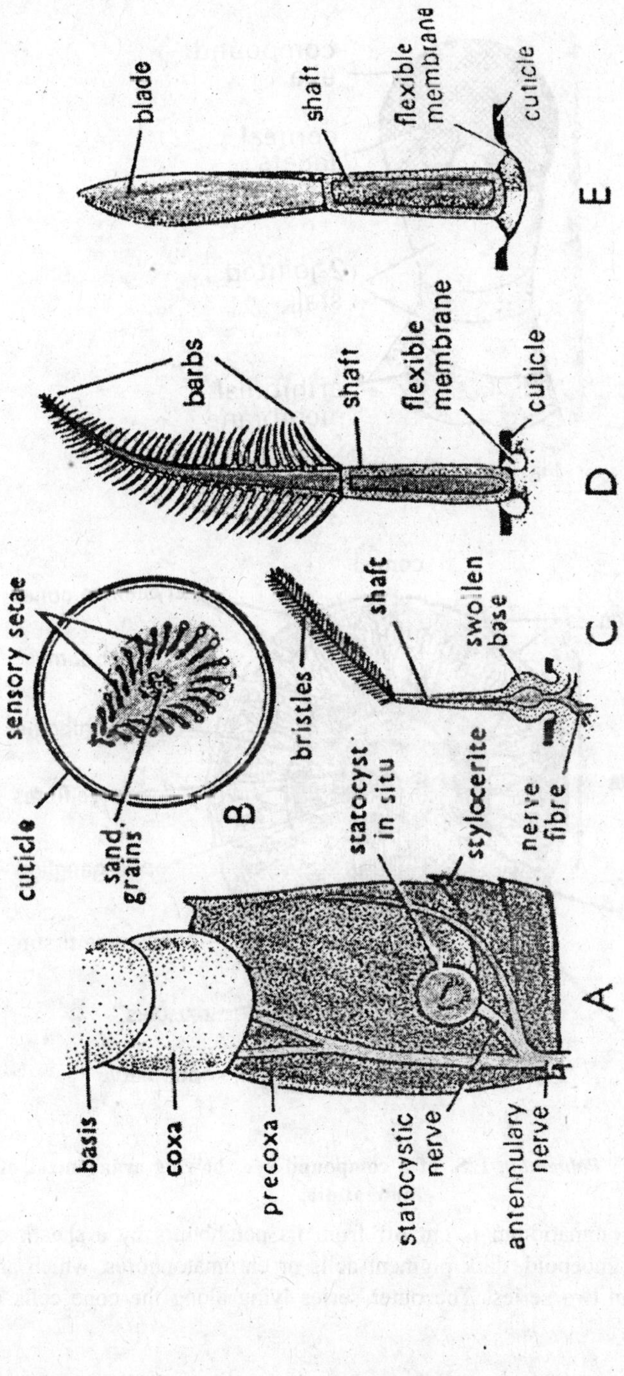

Fig.10.3 : ***Palaemon.*** **A-Statocyst in situ. B-T.S. of statocyst. C-A Single receptor seta, (highly magnified). D-A tactile seta. E-An olfactory seta.**

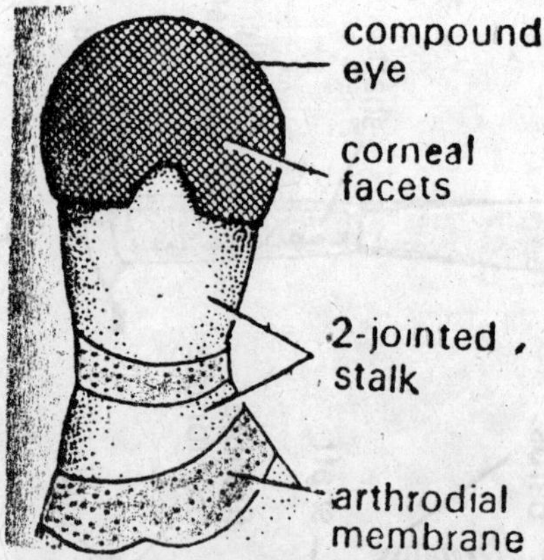

Fig. 10.4 : *Palaemon*. A compound eye.

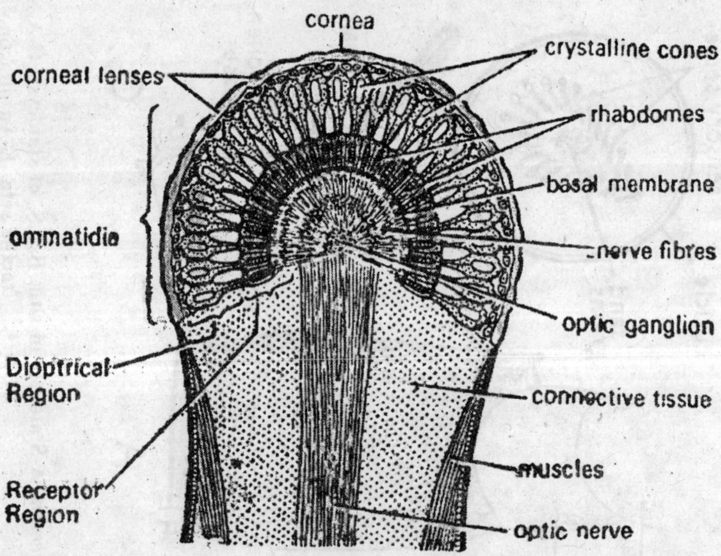

Fig. 10.5 : *Palaemon*. L.S. of a compound eye showing arangement of ommatidia.

Each ommatidium is cut off from its neighbours by a sheath of movable, amoeboid, dark pigment cells or chromatophores, which are arranged in two series. The outer series lying along the cone cells is called the iris pigment and the inner series separating the rhabdomes is called the retinal pigment. The amoeboid pigment cells take up different variations in the intensity of light.

Working of Eye

Working of compound eye is very complex although, it is defirenet in focusing ability and elarity of image, such an eye is unsurpassed the pitching up motion and for peripheral vision. As they are stalked the photoreception mounted on movable stalk it can be move on head in much and some manner as antenna of radar and gives animal almost 360° vision. Each ommatidium is capable of producing a separate image of small part of object seen. Therefore, in prawn and other arthropods posseessing compound eye, the image of object viewed consist of several small pieces. The vision effected is said to be mosaic vision because of similarity to mosaic net work the nature of compound image formed vary according to different intensity of light

In bright light during day image pigment cells spread in such a way that they completely isolate optically adjecent ommatidia. No light can pass through the chamber falls from one visual unit to other. In this condition the rays of light which strike the cornea optically are absorbed by pigmented cells without producing a visual effect only those rays of light which fall perpendicular upon cornea can travel through ommatidium and reach rhabdomeres to form a point of image. As a result a complete image formed as a mosaic of several components placed the slightest movement is to readily be detected.

Each ommatidum response to a fragment of total field and that these fragmentary images are fitted together into a single general image the sharpness of images depend upon number of ommatidia involved degree of isolation from one another. In dimlight or at might the pigment migrate in such a way that neibouring ommatidia in longer remained optically isolated but work union. In this condition even oblique after passing through the number of ommatidia in their way. As a result over lapping of adjacent points of image occurs so the continuous image called super imposition image is obtained. It is not sharp but this animal gets some sort of idea of object moving about in surroundings.

Nerve fibres leading form inner of retinal cells joint to from optic nerves which carry impulses electrochemical waves of energy to brain where they interpreted and registered as a mental image. The object viewed by prawn is not inverted as it is invertebrates and conseqvently does not have right in brain.

Sense Organs : A pair of stalked compound eyes is present, each composed of upwards of 10,000 photorcceptor unitss called ommatidia. Under a binocular microscope the exoskeleton of the eye can be seen to be modified into many corncal lenses or facets, mostly square in shape, each being the outermost portion of a single ommatidium. The inner parts of these ommatidia and their connections with the optic nerve can be made out in a general way in a histological preparation, or in an eye bisected longitudinally with a razor blade (if a fresh specimen is being used the eye should first be immersed briefly in boiling water). To ascertain the whole structure, however, would require a more prolonged microscopical study of histological preparations.

Each ommatidium is surrounded along its whole length by a sleeve of cells containing black pigment and near its inner end by some cells which contain a reflecting pigment. In a light-adapted eye the black pigment is distributed along most of the length of the cylindrical sleeve, so that light entering the ommatidium is not able to pass into neighbouring ommatidia. In a dark-adapted eye the black pigment is concentrated at the two ends of the sleeve, leaving the reflecting pigment exposed, and permitting liglit to pass from one ommatidium to another. These differences in the distribution of black pigment are easily seen in eyes slit open longitudinally after the animals have been killed by immersion in water at 80°C. for 10-15 seconds. The two different states also can be detected in intact animals by briefly illuminating the eye, in the dark, with a bright beam of light. If the eye is liglit-adapted, it will appear black, especially towards the periphery. If the eye is dark-adapted, it will still appear black around the periphery, but in a large circular area in the center it will be a brilliant orange-red. This color results from the fact that the exposed reflecting pigment near the inner end of each ommntidium is reflecting the light back through the central structure tenned the rhabdome, containing the pigment, visual red.

Distal to the rhabdome, between it and the conical facet, lies the crystalline cone, which plays the principal role in the optics ot the ommatidium.

Even after loss of the eyes, crayfishes are able to avoid light. The sensitive tissue concerned is in the sixth abdominal ganglion, although the actual receptor cells have not been identified. Possession of such a caudal photorcceptor is perhaps of value to the crayfish when backing into a supposedly dark crevice; if a second opening into the crevice should be present, light entering through this opening might act as a stimulus and warn the crayfish of the unsafe nature of the retreat.

Identified as probable chemoreceptors arc certain small, blunt-tipped setae, with thin walls, and without lateral sctules, known usually as aesthetascs. These receptors occur on some of the mouth parts, and more particularly on the ventral surface of the more distal segments of the outer flagella of the antennules. In this latter position the setae are in small tufts of three or four, two tuft and to each segment, and each seta has a characteristic node about halfway along its length. Chemosensitivity has been demonstrated on various other parts of the body, but the nature of the receptors in such locations remains uncertain.

Setae which respond to touch, and perhaps sometimes to water currents, are abundant on various parts of the body. These setae are small, pointed, usually feathered with sctules, and at the base each has a thin-walled, bulb-like swelling of the exoskeleton, permitting free movement of the seta in any direction.

A specialized use of such setae is met with in the statocysts, one of which is located in the proximal segment of each antennule. Each of these statocysts consists of a sac, opening to tlip outside, within which are numerous sand grains and the sensory setae. The external opening is a minute slit, concealed by a tuft of setae, located on the dorsal surface of the proximal segment, about one-third of the distance from the anterior end. In large specimens a fine pin can be pushed through the opening into the membranous sac, which occupies most of the proximal two-thirds of the segment. To see this sac it is necessary to cut away the exoskeleton on the two sides of the segment, using a razor blade. The sac should then be cut open dorsolaterally, to disclose the sand grains and the rows of minute sensory setae, arranged in a horseshoe-shaped fashion on the floor of the sac, and, in the lobster, in a small tuft projecting down from tine dorsal surface. A binocular microscope is required to sec the details. Most or all of the sand grains are cemented to the setae, though many of the delicate strands of cement are likely to become broken during the dissection. Like all other sensory setae, the ones here are supplied by nerve fibers, which in this case come from a small branch of the antennulary nerve, entering the sac posteriorly. If a crayfish commences to molt in captivity, it may be possible to repeat Kreidl's classical experiment, by placing the animal in-filtered water with a supply of minute iron filings on the bottom. The lining of the membranous sac, with its contained sand grains, is cast at the molt, and after the molt the animal rubs its antennules on the bottom to refill the sac. In this case iron filings, instead of sand grains, will enter the sac. If the animal is now blinded by black lacquer painted over the eyes, and is suspended by threes back upwards in the water, it can be induced by a magnet to carry out righting movements with its legs. Instead of responding to the direct pull of gravity, it now responds to the resultant of this force and the force exerted by the magnet.

Proprioceptors presumably must be present to make possible the maintenance of posture and of muscle tonus, particularly in the pereio-pods. The only sense receptors of tins type for which we have any clear evidence are some so-called movement receptors, located in the membranous joints between successive segments of the legs. These receptors have been demonstrated physiologically in a number of decapods, but are little known morphologically, and have not yet been looked for in the crayfish or lobster. They respond to movement, but not to tension or pressure.

THE COMPOUND EYES

The eye can be divided into two portions an apical hemispherical black portion is the ommatidial portion, end a proximal cylindrical optic stalk lying in the medulia terminalis. The former black portion is the cornea which is the outer most surface of the eye. Of the two segments of the eyestalk or peduncle the basal part is short and thick whereas distal portion is covered by a soft chitin. This allows movement between the two parts and thus the eyestalks are moveable. Each ccmpound eye is made up of terge number of ommatidia. Each ommadtidum has a definite arrangement and various kinds of cells are arranged in definite order. Basement membrane on the inner side of the ommatidium is continuous with the hypodermis is also. The cornea is composed of thin layer of chitin, which is continuous with the thick layer present of integument covering the distal part of the peduncle. The cornea consists of closing packed numerous small hexagonal prisms.

Structure of the Ommatidum

Each ommatidum can be divided into two portions proximal and distal. The outer part of both these regions are covered by a pigment. Just below the cornea there is a layer of corneal hypodermis. Below this are the cone cells the distal region of which encloses the crystallie cone. The cone cells are surrounded by pigment on all the sides. The pigmented retinulae are situated at the proximal end of the cone cells. These retinulae are surrounded by rhabdome.

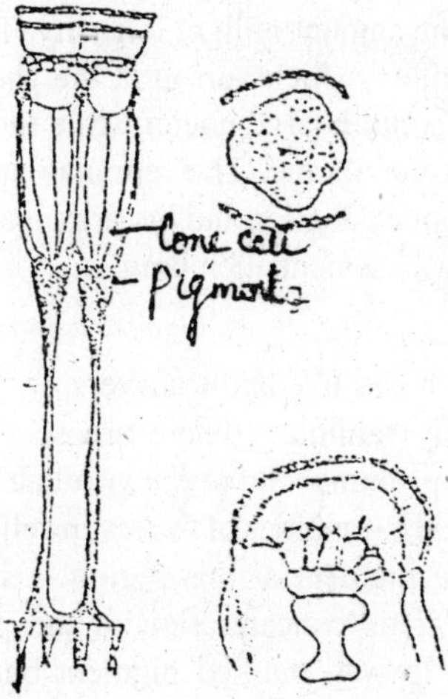

Fig. 10.6.

Corneal Hypodermis

It lies immediately below the cornea. It consists of two cells which secrete new cornea as soon as the old one is discarded at the time of molting.

Cone Cells and Crystalline Cone

Beneath the corneal hypodermis there are four tall cells called as the cone cells. These cells become narrow as they pass inward and distally they enclose the crystalline cone. Which is secreted and nourished by the cone cells. Distaliy prolongations of the cone cells appear to pass between the comeal hypodermal cell and proximal between the retinular cells and envelop the rhabdome. The cone cells are surrounded by two pigment cells which are continued inward as fine processes and ultimately they come into contact with the retinulae. The part of the ommatidium starting from the comeal facet to the inner ends of the cone cells is known as diptrical region as it is concerned with the focussing of light received each ommatidium and inner is the receptive region.

Retinular Cells

The crystalline cone and rhabdome ere surrounded by heavily pigmented retinular cells. They are of two series the distal and the proximal. The distal retinal cells are two and always placed on either side of the cone cells. They are deeply pigmented and contractile and their size varies with the amount of light intensity. The proximal retinular cells are seven in number which surround the rhabdomer Their dista! ends are swollen and contain large nuclei while the proximal end tapers towards the basement membrane. The retinulae are in contact with the nerve fibres from the optic ganglion and the proximal end of the rhabdome is in contact with the basement membrane.

Rhabdome

The rhabdome is a rod-like and transversely striated structure. It is completely covered by retinulae. There are seven retinular cells and these extend from the proximal end of the vitrellae to the inner basement membrane. All the distal extremities of the seven cells completely surround the rhabdome and the pigment at this region is so much concentrated as to hide the latter. This concentration of the pigment of the entire ommatidia gives rise to well marked pigment band. The rhabdome is composed of four parallel rods and these four rods are the ends of the four cone celts.

OTOCYST

The otocyst is situated at basal segment of the antennule. It is a simple saclined by a layer of chitin which is continous with the outer chitinous integument. The stalk of the otocyst is at the extreme posterior end. The otocyst opens to the outside by narrow longitudinal slit en the dorsal surface and this opening is guarded by different setae. It is stated that in the larger crab the otoliths are absent from the sac and the sac is completely closed. However, in the younger crab the otocyst cyst opens to the exterior and otoliths are present in the sac. The otolith may consists of sand drains usually cemented loosely toaether by glandular secretion of the cyst wall thread setae. It has similar function as of statocyst of prawn.

Hooked Setae

Those are found on the posterior wall of the otocyst. The distal portion of the shaft of this setae bears fine barbs. The base of the shaft sunk into the socket-like depression.

Thread Setae

These are found on the antero-lateral wall of the sac. They are largest and longest setae present in the otocyst. The barbs are completely absent from the shaft. The base of the seta lies in cup-shaped depression on the wall of the sac. All the setae are innervated by the otocyst nerve.

Olfactory and Tactile Setae

The sensory setae of *Barytelphusa cunicularis*, are of two types: tactile setae and olfactory setae. The tactile setae are present on the various parts of the body On the antennule, antenna and on most of the other appendages. A tactile seta is characterized a long tapering shaft which is hollow from insids and without any barbs. Each seta is innervated by a single nerve fibre and single nerve cell.

The olfactory setae are very short and blunt each seta is divided into proximal and distal portion. At distal end there is ambranous covering which is extremely thin. The proximal portion is somewhat swollen. The olfactory setae differ from the tactile setae in having large number of nerve fibres. Such type of setae are usually found on thc exopodite and endopodite of the antennule. They are extremely small and are present on the.rings of the exo and endopodite.

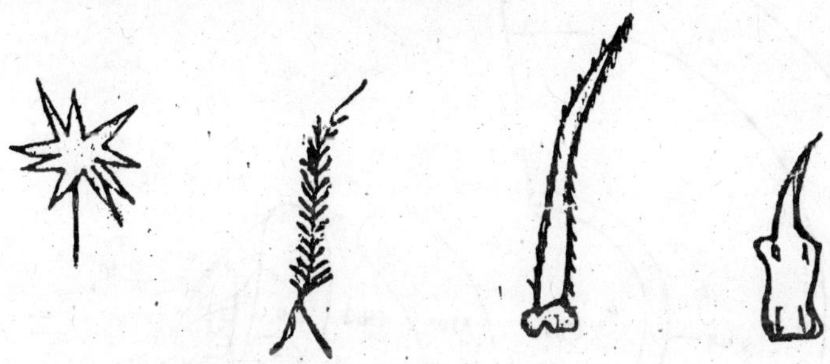

Fig. 10.7

THE CRAYFISH VISUAL SYSTEM

1. Plan of the Crayfish Visual System and Terminology

The eyestalks of the crayfish contain three successive columnar neurophil regions underlying the compound retina and which are connected by crossed projections of retinotopic fibres, *i.e.*, the First and second optic chiasmata. Central to these columnar neurophils lie lateral portions of the protocerebrum which are drawn out into the eyestalks. These are connected to cerebral centers by the so-called optic nerve.

The neurophils in the crayfish and insects is so clear that it was fell a radical change in the terminology for the crayfish and other Crustacea was justified (NASSL, in preparation). Thus, the insect terminology is simply applied here to crustacea.

Furthermore, the term medulla terminalis will be avoided and replaced by lateral protocerebrum or by appropriate terms for neurophil regions or glomeruli in this lateral region of the brain. This is called for because the lateral protocerebrum comprises a conglomerate of delineated neurophil regions that serve not only vision but also several other modalities. Amongst the visual regions there are at least four fourth-order visual neurophils which by analogy with insects are termed optic loci. The term optic nerve will still be used even though it is not fully appropriate considering dial it also refers to chemosensory, mechanosensory, motor and multimodal neurons. Its structure is a compound one, consisting of several fibre bundles that project from the lateral brain into its most central regions.

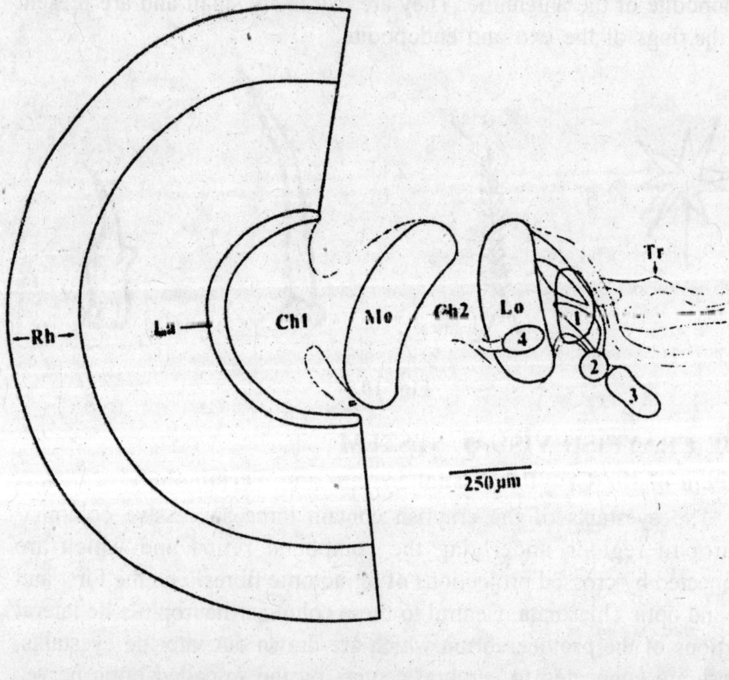

Fig. 10.8 : Schematic drawing of the crayfish retina and optic lobes in the eyestalk. as seen in horizontal cross-section. The receptor layer of rhaboms (Rh) is shown to the left. The dioptric elements of the ommatidia reside external to this layer. The eyestalk three columnar neurophils; the lamina (La). the medulla (Me). and the lobula (Lo). Between these the axons from chiasmata which reverse and then re-reverse the anterioposterior order of the retinal mosaic (Ch and Ch). At least four neurophil regions are recognized (L4) as receiving axons from the ipsilateral columnar neurophils. These are termed the optic foci. Tracts of axons desived from the lobula and these foci (Tri project centrally. The cell body regions are shaded light grey.

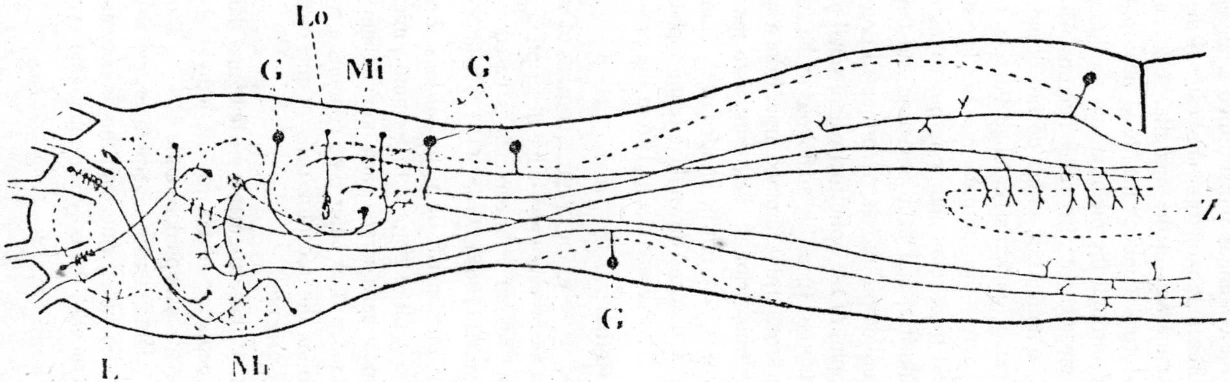

Fig. 10.9 : Horizontal section through the protocerebrum of the isopod crustacean *Ligidia occidentalis* with some of ist "main neuron tracts." L. lamina. ME. medulla. Large axons (G) project to protocerebrum and to contralateral optic loes. Z. central body. After Hansirom (1924) m Hansirom (1928). the illustration shows a level of resolution typical of Hanstroms neuroanatomical descroanatomical descriptions.

There are three neurophils lamina, medulla, and lobula distinguished with emphasis on the columnar and stratified arrangements of neurons. As in insects, photoreceptor axons are "implicator elements" of retinotopic columns in the lamina and in the medulla. Associated through-going columnar relay neurons connect the three neurophils and relay the relinolopic mosaic through the medulla into the lobula. Superimposed on the columnar mosaic are also relay neurons whose spacing is in proportion to the columns, but which do not occur in every column. Others, such as anaxonal (amacrine) or tangential relay neurons, contribute to planar networks across the columnar mosaic. The stratification of the neurophil is derived from relay neuron terminals, as well as from their collaterals and dendrites, and from amacrine and tangential Cells.

Finally, the retinotopic mosaic is projected onto the dendrites in higher order neurophil of the optic foci where small groups (or unique) higher order interneurons (possibly analogous to the descending neurons of insects) derive a structurally summed input from a specific population of lobula relay neurons prior to their entry into the optic nerve. The cerebral visual centres have not been adequately studied anatomically; nor have neurons that descend down the circumesophageal commissures to thoracic ganglia. These are omitted from the present account.

2. The Crayfish Retina

The compound retina of the crayfish consists of between 2,000 and 3,000 ommatidia. Our present knowledge description of the anatomy of the ommatidial and retinal arrangements is based on descriptions by *Parker* (1897), *Eguchi* (1965), *Krebs* (1972), and *Nassel* (1976a).

Each ommatidium consists of a corncal lens, a crystalline cone, a crystalline tract and eight receptor cells forming a microvillar rhabdom. Two types of pigment cells screen each ommatidium, one type distally and another proximally. The latter type contains a reflecting pigment. In addition, seven of the receptor cells of the ommatidium contain screening pigment granula.

The positions of these pigments in light-and dark-adapted states, allow different acceptance angles for the individual receptors. In the light-adapted state the acceptance angle is 4° ± 2° and in the dark-adapted, 24° ± 8°. Light entering several facets can reach one individual receptor in the light-adapted state and this is accentuated in the dark-adapted state. Classically, the crayfish compound eye is referred to as a superposition type of eye.

Two main classes of photoreceptors can be distinguished. The receptors R 1-R 7 constitute the first class and form the large proximal rhabdom which is spindleshaped and has orthogonal layers of microvilli. The other class comprises the R 8 cells. Each of these forms a small separate rhabdomere with horizontal microvilli situated distally in the ommatidium.

Cobalt and Golgi studies of R 1 R 8 receptor axons show that axons from R 1 R 7 terminate in the lamina is picxiform layer, wherer R 8 axons terminate in the medulla, further suggesting that the two morphological types represent two functional classes of receptors. The R 1 R 7 cells can be divided into two subclasses on the basis of their microvillar orientations. Those of R 1, R 2, R 5, and R 6 are oriented vertically, while those of R 3, R 4, and R 7 form horizontally arranged microvilli.

Rutherford and *Horridge* (1965) suggested that the two microvilli directions found in the lobster rhabdom comprise a mechanism for two-channel e-vector analysis. Differential sensitivity of retinular cells to the e-vector of polarized light was demonstrated by *Shaw* (1966, 1973). Found tow classes of cells sensitive to horizontal and vertical e-vectors, respectively, and other studies resolved polarization sensitivity rations of up to 1 : 9 and 1 : 12 (1970). The mechanisms for detection and analysis of polarized light are dealt with in detail by *Waterman* (1980). Eividence that information about the e-vector of polarized liglit is actually transferred to higher-order neurons is derived from recordings of neurons in the crab medulla externa and crayfish and stomatopod optic nerves.

Two visual pigments have been demonstrated in the crayfish retina. These are both rhodopsins imparting to the receptors maximum sensitivities at 565-570 nm and 425-445 nm, respectively.

Dichromatism has been suggested from intracellular recordings of photorcceptor cells. The two classes of colour receptors are present in each ommatidium, the majority of the cells were shown to be yellow-green sensitive, whereas only a small fraction (10%-19% of the total impalements) were violet sensitive.

Since the recorded cells were not intracellularily labelled, there is yet no correlation between morphological and physiological units that are violet or yellow-green sensitive, respectively. *Goldsmith* (1978), however, demonstrated with microspic-trophotometry that a pigment with an L_{max} in the shorter wavelengths is absent in the proximal (R1-R7) rhahdom. It is thus possible that R 1-R 7 are yellow-green sensitive and R 8 (which has not yet been analysed in this respect) is violet sensitive. Recordings from higher-order interneurons have so far shown convergent inputs from both the violet and yellow-green channels

3. Projection of the Retina into the Lamina

The pattern of retinal projections onto the lamina has recently been studied in the crayfish by means of cobalt uptake. By contrast, earlier attempts at tracing individual axon bundles by serial section analysis were hindered, inter alia, because of the vast distance between the retina and the lamina as well as the dense packing and heavy pigmentation of the receptor axons. *Srowe* (1977). however showed conclusively that in the crab *Leptograpsus* the receptor mosaic is projected retinotopically into the lamina. Serial sections from the retina to the lamina revealed that the axons from one ominatidium innervate one underlying columnar module (optic cartridge) and retain their neighbourhood relationships amongst themselves between the retina and lamina. the sampling raster of receptors is probably mapped homoioptically to deeper neuronal levels, as it is in insects.

Since so many other features of the crab visual system are similar or identical to those of the crayfish it could be assumed that the crayfish retina lamina projection has an identical relinotopic organization. Receptor axon penetrations through the retinal basement membrane permutate a map of the retina (below the retina) was based upon observations of short stretches of receptor axons. It was throught that initial crossover patterns were of a divergent and combinatorial nature. This is, however quite erroneous. The cited studies were incomplete in so far as the axons were not traced through the total distance between the retina and the lamina. Apparently, as is the case in *Leptograpsus* (*Srowe* 1977), groups of axons below the basement membrane, derived from three ommatidia, collect into single bundles. These then separate proximally above the lamina where retinotopic unpermutated, mappings of receptor axons are re-established.

Cobalt diffusion into the crayfish retina reveals sets of retinotopic axon bundles that project to the lamina, each accompanied by one long visual fibre axon from the R 8 photoreceptor of each ommatidium. Together, R 8 axons project retinotopically through the lamina and then across the first optic chiasma to the medulla where they map an anterioposterior inversion of the retinal mosaic.

4. Neural Arrangements in the Lamina

Compared to the homologous region in Diptcra, or, for example, Hymenoptera, the columnar and layered organization of the crayfish lamina is less precisely arranged but more orderly than, *e.g.*, the lamina of a nocturnal superposition lepidopteran. The structural "noise", or imprecision, can be related to variations of shapes of receptor axons and the structural variations amongst a class of interneurons. However, despite this, all species of interneurons anaxonal or relay monopolar cells are relatively simple to classify as distinct types according to characteristic features of their profiles. It is also apparent from electron microscopy studies that the connectivity patterns of receptor axons and interneurons are precise and seem to be reiterated in each cartridge. Parallel channels maintain the retinotopic mosaic, and lateral interactions between discrete sets of elements provide subsbrates for inhibition, excitation, or adaptation pools, amongst large domains of nerve cells.

The neural components, and some of their synaptic relations within and across the lamina's columnar array. Layering in crayfish nenrophil is more pronounced in deeper optic neurophil. However, it is also present in the lamina and serves as a structural paradigm for layered laminae of other Crustacea and many species of insects, particularly those suspected to have trichromatic vision.

5. Types of Neurons

Centripetal input to the optic lobes is provided by R1-R8 photoreceptor axons. The short photoreccptor axons (R1-R7) end at two levels in the lamina, indicating two main lamina strata, whereas axons from the 8th receptor cells project through the lamina, alongside the R 1-R 7 cells from the same ommatidium, and finally terminate superficially in the medulla. Ten species of neurons have so far been distinguished in the crayfish lamina. These are as follows: live types of monopolar neurons, two types of tangential T-neurons, one type of small field T-neuron, one type of centrifugal cell, and one type of amacrine cell. All neurons, bar the anaxonal amacrine cells, conneet the lamina with the medulla, or vice versa, according to their structural and functional polarity.

Interneurons can be characterized with respect to their dendritie or terminal domains in the lamina and in the medulla, as well as with respect to their cell body locations. Dendrites of relay neurons are restricted to either fo the two lamina strata or are distributed in both. Their fields have characteristic domains extending either to one single columnar unit (optic cartridge) or extending through a number of columns.

Monopolar neurons of the lamina of the lamina are derived from cell bodies distal to the neurophil, whereas the amacrine cells and

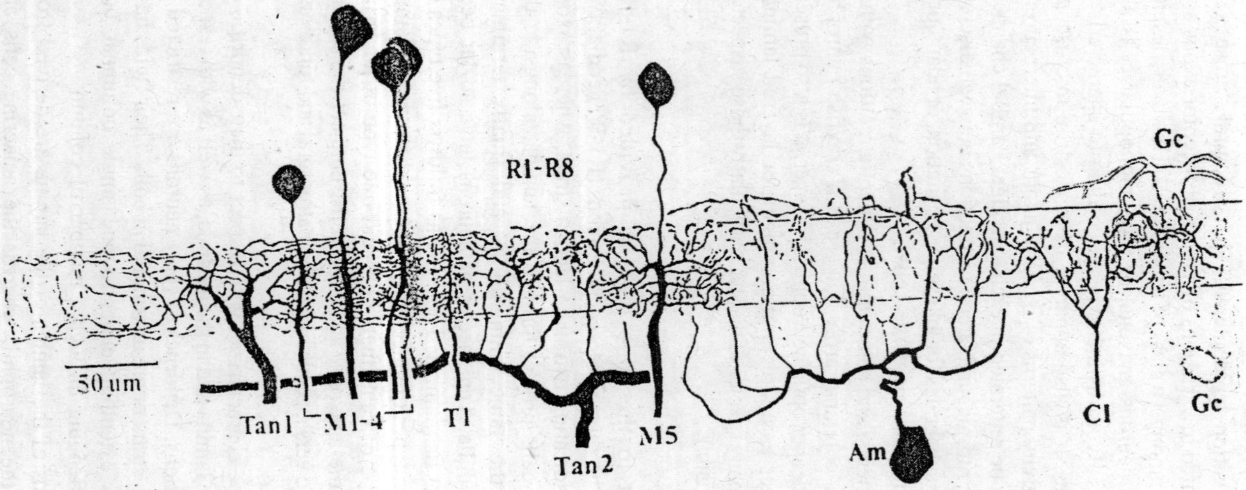

Fig. 10.10 : Diagram of lamina cells in the crayfish. Columnar elements invade every cartridge. These are the monopolars M 1-M 4 and the T 1 as well as receptor axons R 1-R 8. Two types of wide-field tangential neurons. Tan 1 and Tan 2. form dense arbors comprising large lamina and overlap freely. Cells with intermediate fields of arborizations include the monopolar cell M 5 and centrifugal neurons C 1. both of which are present as a proportion of the lamina columns. Large amacrine cells (Am) invade the lamina from below. Glial cells (Gc) from velate processes in the lamina plexiform layer.

T-neurons are derived from cell bodies beneath the lamina of proximal to the crossover of fibres in the first optic chiasma, respectively.

6. Columnar and Layered Organization of Lamina Cells

Seven receptor axon terminals define each cartridge, four in the distal and three in the proximal lamina stratum. As in the crab, retinotopic receptor projections bring ihe channels of one sampling unit (an ommatidium) to a single unit of organization in the neurophil.

Each cartridge comprises, then a set of seven receptor terminals (plus the en passant R 8) and a set of relay neurons, the monopolar cells. By analogy with insects these are termed "columnar" neurons. Axons of columnar (relay) neurons project alongside axons of long visual fibres (R 8) and transfer the retinotopic mosaic of the lamina into the outer surface of the medulla. Cobalt studies show that each R 8 axon shares a coherent bundle with the relevant set of relay axons, and there is no evidence of any lateral permutation of neighbourhood relationships.

(a) Monopolar Cells

Observations of cobalt-filled neuron populations. Together with Golgi preparations and electron microscopy, reveal that several types of monopolar cells are found reiterated in each cartridge. Among these are the monopolar neurons M1 M4, where by there is apair of M1 cells in each cartridge (Mla,b). M1 M4 are all small-field elements, with their dendritec arbors restricted to the parent cartridge. Their division into four categories in the lamina is based upon their overall *Gestalten* and on the distribution of their dendrites in this region. M1 and M2 are non-stratified, having dendrites in both lamina strata, whereas M3 and M4 are stratified, M3 with dendrites in the outer stratum and M 4 with dendrites exclusively in the inner stratum.

(b) T-Cells

Another columnar neuron, the small–field T–neuron (T1), has small–field processes disposed in both lamina strata. Each T–cell is derived froma cell body situated above the medulla and each gives rise to a small distal arborization in each medullary column. In this respect. Monopolar cell and small–field T–cells bear direct analogy with lamina relay cells (L1, L2, L3, L5) and medulla–to–lamina T–cells (T1) of dipterous insects. All species of relay neuorns can be identified by electron microscopy, and each has a characteristic set of connections in each cartridge.

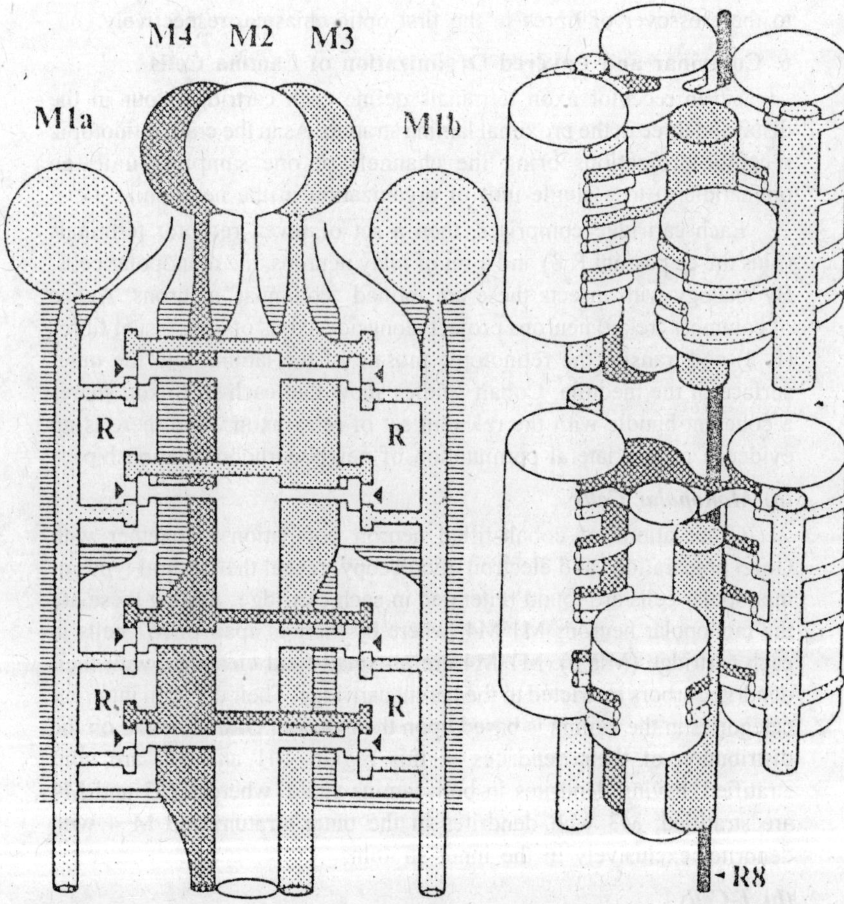

Fig. 10.11 : Schematic representation of the synaptic connections between photoreceptors (white profiles labelled R) and five columnar monopolar cells *M1a, M1b, and M2-M4* (*M2,* white; the rest, stippled) in the crayfish. The arrowwheads indicate the triad arrangements of synapses in the left hand figures. *M3, M2, M1a,* and *M1b* receive input in the distal lamina stratum (from four receptor terminals), whereas *M4, M2, M1a, and M1b* are postsynaptic to three terminals in the inner stratum. *M1a* and *M1b* divide the receptor output so that *M1* receive input from two distal and two proximal receptor endings and *M1b* from the two other distal and the remaining single proximal terminal. The long visual fibre R8 projects centrally through the cartridge and appears devoid of synaptic relationships in the lamina. The right-hand figure shows a dissected view of these elements.

(c) Resolution of Cell Assemblies

The remaining types of neurons (tangential cells, centrifugal cells, and amacrines) are regularly distributed. One to several cartridges. according to cobalt uptake experiments. However, their processes appear to participate in the mosaic organnization of the entrie neurophil and, most probably, each cell type has a prcise and reiterated relationship within each column. Cobalt studies have shown that each optic cartridge contains specializations from all types of lamina elements, as is the case in the lamina of many species of insect. The following observation serves as an example of this. Injection of "Co" direclty into crayfish neurophil has resolved populations of amacrines and, in other preparations, wide–field centrifugals or tangential elements. Within the range of the cobalt pool what is presumed to be the entire local population of any given cell is resolved. Observations of these "cell assemblies" reveal that even though single amacrine cells vary in their lateral domain, all amacrine cell together contribute to a homogeneous network or "isomorphic neural assembly" whereby each cartridge receives the same kind of distributions have been resolved for other wide-field nerve cells. The same kind of distributions have been resolved for other wide-field neurons that spread through several columnar units and which together comprise a regular palisade of neurons, one to every six or eight cartridges, with dendrite assemblies that invade every cartridge of the lamina.

Another class of neurons that comprise tha mosaic but are numerically fewer than the number of cartridges, are the wide-field tangential neurons. These elements have processes that extend over large portions of the projected retinotopic mosaic. Two types of such neurons have been distinguished; both contribute to regular and isomorphic networks of fibres.

(d) Tan 1

This species of tangential neuron cell (Tan 1) has long processes whose arborizations invade both the lamina strata in a bistratified rather than a diffuse pattern. The arrangement of these cells across the lamina make up a distal hexagonal network of processes which, deeper in the neurophil, is compressed to a rectilinear lattice. Adjacent Tan 1 cells overlap each other extensively. Their thich axons connect the lamina with the medulla where they also give rise to extensive arborization.

(e) Tan 2

The second type of tangential neuron (Tan 2) has large vertically arranged branches beneath the lamina from which proccsses ascend distally into the lamina's plexifrom layer. This neuron is connected to the medulla by a thick axon that gives rise to a large dendritic arborization.

(f) Centrifugal Cells

One type of centrifugal neuron has a regularly distribution amongst optic cartridges. By morphological definition this cell type in insects is classed as a centrifugal neuron derived from a unipolar cell body proximal to the medulla. Its ascending axon gives rise to a characteristic, layered pattern of collaterals or dendrites at several levels in the medulla, and thereafter its axon projects distally, across the first chiasma to arborize diffusely over several cartridge in the lamina.

(g) Amacrine Cells

One type of anaxonal or amacrine element has been found in the lamina and is derived from cell bodies that are located below the lamina's inner surface. Cell body fibres give rise to tangential branches form which numerous processes project through the plexifrom layer, finally giving rise to lateral branchlets at the distal surface of the lamina. There is considerable, overlap between adjacent amacrine neurons. Cobalt uptake by amacrines show this cell type to be spaced one to several cartridges.

7. Synaptic Arrangements in Cartridges and Stratification of the Lamina

(a) Synaptology

The lamina columns (optic cartridges) of the crayfish are less regular and less extensively separated by glial elemnts than is the case in several studied insects. It is, however, clear that seven photoreceptor endings at two strata define each columnar unit, with a synaptic output onto five monopolar neurons. This pattern is most likely identical throughout the complete set of cartridges in the lamina. Each cartridge also contains several other components derived from columnar and non-columnar neurons.

The shapes and arborization patterns of caryfish visual neurons are too imprecise to allow identification of dendrites by conventional electron microscopy alone. However combined Golgi and EM techniques prove a valuable tool for routine analysis. With this technique each cell is first identified by light microscopy and is then resectioned for electron microscopical observation. It was found that the five monopolar neurons (M1a, M1b, M2-M4) of each cartridge are characteristically wired up to specific receptor terminal combinations and thus comprise five distinct parallel channels that project to medulla relay cells of each central column.

Channel 1. M2 has numerous postsynaptic sites to all seven receptors of the cartridge. This neuron is thus capable of relaying from both levels of R1 R7 receptors and would be likcly to mix discrete channels that carry information about the *e-vector* of polarized light.

Channels 2, 3. M3 is postsynaptic to the shallow receptor terminals and M4 to the three deep ones.

Channels 4, 5. The two M1 neurons share the receptor terminations in the one receves input from two shallow and deep terminals, whereas the other derives an input from two shallow and one deep receptor terminal. Posssible separation of the receptor input might provide two e-vector channels however, the significance of five parallel pathways, all of which derive a first-order input, is not clear. The evidence is that the R1-R7 receptors have identical spectral sensitivities (*Goldsmith,* 1978), and colour-coded channels appear to be unlikely amongst these elcments.

In spite of some variations in the morphology of the individual lamina neurons, at least at the light macroscopical level, precise synaptic arrangements are manifested by the receptor-monopolar junctions. Characteristically, the receptor terminals from membrane evaginations (with a presynaptic dense bar and clear vesicles) adjacent to triads of postsynaptic profiles. The central profile in these postsynaptic triads is always a spine from M2, irrespective of its live.

In the distal lamina stratum, one of the lateral profiles is derived from M3 and the other from one of the M1 neurons. In the proximal stratum the one lateral profile is derived from M4 and from the M1s, respectively.

The lateral postsynaptic profiles of M1, M3, and M4 neurons form small knob-like invaginations into the receptor terminals. No specialized membranes, synaptic bars or vesicles have been found in or adjacent to these invaginations that could be suggestive of chemical synapses. The significance of these structures unknown.

Wide-field, T-cell and other lamina neurons have not yet been studied by electron microscopy in detail. It seems however, form combined Golgi and electron microscopical observations that neither M5 nor Tan 1 neurons are directly postsynaptic to receptor elements, and they most probably interact with sets of relay neurons in the lamina, as do L4 monopolar cells, tangential cells and amacrines in insects. It would not

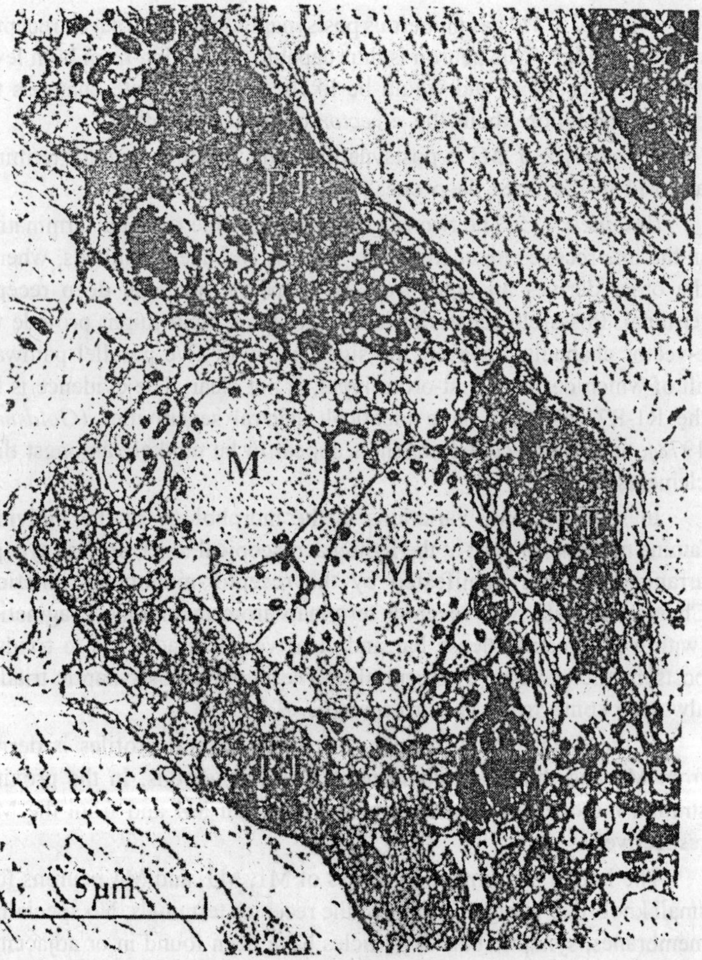

Fig. 10.12 : Electron micrograph of a Golgi preparation (Golgi-EM) showing a fortuitous resolution of two M1 neurons (M1a and M1b) which have been impregnated with silver chromate in the same cartridge of the crayfish lamina (black profiles). The micrograph originates from the lower part of the lamina where three photoreceptor terminals (RT) have their expanded endings. In the center of the cartridge of the crayfish lamina (lined by two tangential profiles) the axis fibres of other monopolars be seen, two of which are labelled M (These are probably M2 and M2 and M3. whereas M4 in this region is represented by diffuse arborizations rather than a single large profile).

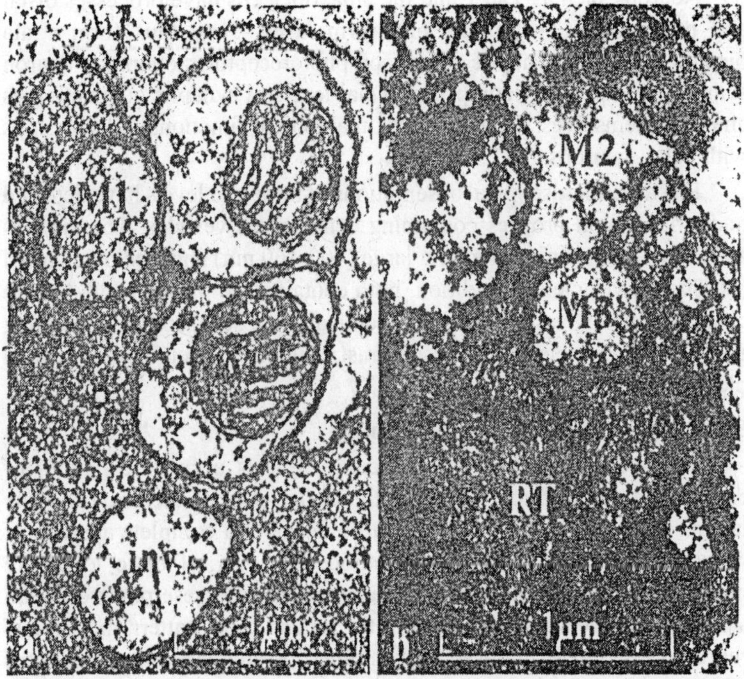

Fig. 10.13 : Electron micrograph of a photoreceptor-triad synapse form the distal lamina in the crayfish where the postsynaptic profiles are *M1, M2,* and *M3*. A small arrow-shaped presynaptic structure resides in the evaginated membrane area of the receptor terminal. A profile (ing) is also shown invaginating inot the receptor from M3 dendrite, bA Golgi-EM micrograph of a receptor terminal (*RT*) and a postsynaptic triad black profile with its invaginating spines, belongs to an M1 monopolar.

be surprising if some of the circuitry described from the fly, purportedly candidates for lateral inhibition and neural adaptation are also present in the crayfish; however, there is at present no physiological evidence for corresponding electro-physiological phenomena at the level of the lamina in Crustacea. If such circuits are present, neurons such as Tan 1, C1 and amacrine cells would also bear good comparison with wide-Field centrifugal neurons, C2, C3, centrifugal cells and a-proccsses in the lamina of the fly.

(b) Stratification

In addition to the two main lamina layers, defined by the receptor terminals and M3 M4 dendrites, there are additional strata which reflect the layers of synaptic specializations of other cell types.

Neurons such as Tan 1 and the amacrine cells have processes distal and proximal lo the lamina plexiform layer. They thus form strata outside the main synaptic region of photoreceptor-monopolar neuron synapses and possibly comprise special levels of synaptic connectivities amongst themselves, distal and proximal to their functional relationships with relay neurons.

Electron microscopical observations of these two levels have detected numerous neural profiles containing dense-cored vesicles. The profiles are of two types, one containing large (100-160 nm) and (the other small (60-80 nm) vesicles. In addition, both contain clear vesicles (20-40 nm) characteristically associated with chemical synapses. Profiles containing dense-cored vesicles are, by contrast, seldom detected in the main plexiform layer.

8. Aminergic Neurons in the Lamina

Formaldehyde-induced fluorescence reveals a specific pattern of catecholaminergic fluoi :scence in the optic lobes of the crayfish. The lamina normally displays a weak fluorescence or a complete absence of neurescerice with the standard method. However, if optic lobes are incubated in α-melhyl noradrenalin or dopamine. after reserpine treatment, a clear fluorescent pattern arises in the lamina. This pattern most likely represents two superimposed neuron types, by analogy with the patterns of fibres revealed after Golgi impregnation or cobalt injection. The most prominent pattern (which is occasionally revealed without inducement) appears to be derived from a small-field lamina T-neuron. linked to shallow arborizations in the medulla. The cell bodies of this neuron reside in the chiasmatic region between the two neurophils. The entire stucture of each cell much resembles the T1 neurons that link the lamina to the medulla and which have been observed as single stochiastically impregnated elements by the Golgi technique and as assemblies ol cells after cobalt uptake.

The second fluorescent cell type is characterized by the presence of large cell bodies below the lamina from which axis fibres invade the lamina to give rise to tangentially oriented processes, distal and proximal to the main lamina plexiform layer. The only neurons resolved in Golgi and cobalt preparations that match this pattern are the lamina amacrine neurons.

Profiles detected by electron microscopy, containing large dense-cored vesicles, partly coincide with these fluorescent patterns. Their tangentially arranged components reside distal and proximal to the plexiform layer. However, since the uptake of exogenous catecholamines is cytoplasmic, it is possible that neurons resolved by conventional electron microscopy, characterized by their contents of dense-cored vesicles, do not fully match the fluorescence-pattern, and it may be that some lamina neurons are still refractory to either Golgi methods or to cobalt-silver analysis.

9. Physiology of Lamina Neurons

Surprisingly few eleclrophysiological studies performed on crustacean optic lobes shed any light on the functional significance of a lamina neuron. The only recordings from neurons of this level are from the crab *Leptograpsus.* The crab lamina cells (unspecified and unmarked) exhibit the same kind of hyperpolarizing potentials as found in some insects, where modulations of light intensity cause a modulation of the amplitude of the interneuron membrane potential, rather than eliciting spikes.

10. Organization of thc Medulla

The medulla is divided into an outer and inner neurophil by a layer of thick tangential axons that separate the outer three quarters of its depth from the inner eighth These axons which comprise the serpentine layer, are destined for regions in the midbrain and contralateral optic lobes.

The distal three-quarters of the medulla [termed ME (o)] contain the terminals of M2 M4 lamina monopolar cells, the paired M2 terminals, the shallow endings of the long visual fibres (R8) and the arborization of lamina tangentials. Tan 1 and Tan 2. In addition, the majority of the dendrites of medulla columnar neurons (transmedullary neurons), as well as amacrine arbors, are observed to reside within the outer layers of the region.

The inner eighth of the medulla [ME (i)], beneath the serpentine layer, contains axon collaterals of some transmedullary neurons as well as dendrites from some cell types, such as T-neurons, that project to the lobula from exclusively this layer. Thus the basic principle organization is most reminiscent of the medulla in insects, particularly the diurnal Lepidoptera. Hymnoptera, and Diptera, almost similar to crustaceans.

The retinotopic mosaic is carried through the medulla by small-field transmedullary neurons which are arranged periodically (1:1) with respect to the columnar arrangements of long visual fibres and their accompanying sets of five monopolar cell endings and T1 cell terminals. Examples of types of transmedullary neurons are illustrated, as observed in Golgi

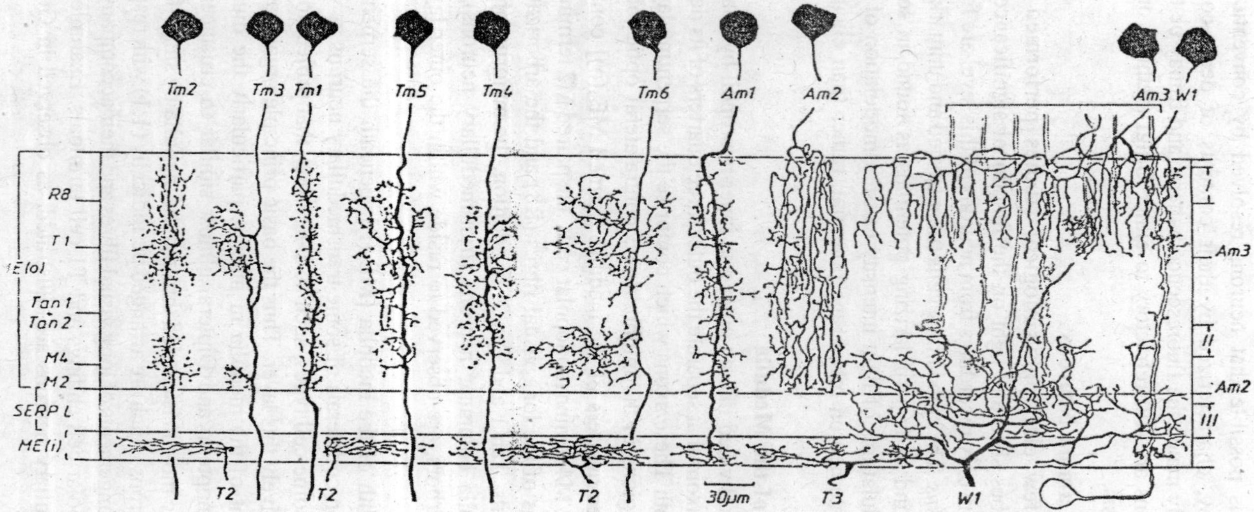

Fig. 10.14 : The medulla of the crayfish. On the left-hand side are indicated levels of terminations of some lamina elements as well as the main layers; ME(o), serpentine layer SERP.L. and ME(i). The amacrine Am2 and Am3 layers, as as well as three main tangential layers (I-III), are indicated to th e right. In this simplified diagram only examples of medulla relay cells are shown. Tinl-Tm6 are six characteristic types of transmedullary neurons classified according to their arborization patterns. Am1-Am3 are three types of amacrine cells. Wl is a type of wide-field non-columnar superpcriodic transmedullary cell. T2-T3 are T-neurons restricted to then Me is layer. Lastly, the input from the lamina is indicated by open profiles (inside bracket). From left to right these are M3, M4, Ml, M 2, Tl, R8, and C1, some of

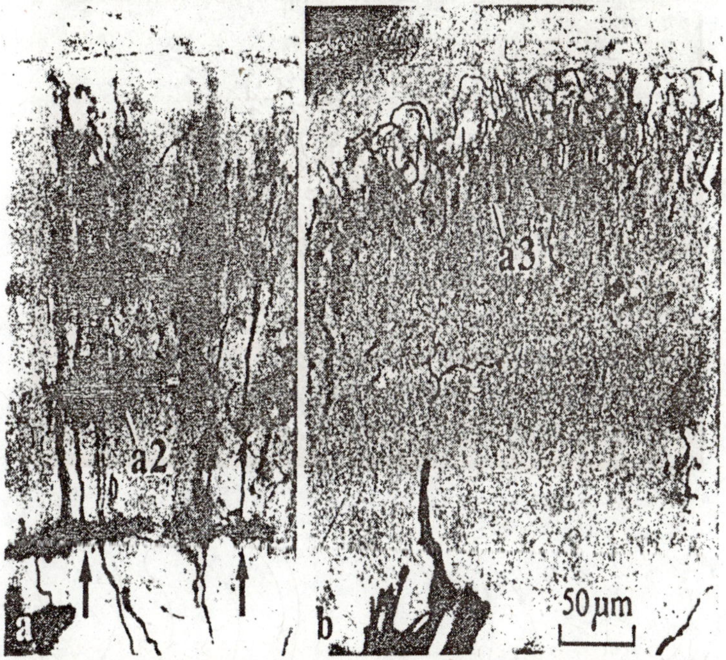

Fig. 10.15 : Three types of amacrines in the crayfish medulla seen in Gogli peparations: a Amacrines Am 1 and 2 from am 1 connects ME (o) and ME (i) (arrowed) where processes from some T-neurons are also seen. Am 2 extends through about two-thirds of the medulla depth. b the third kind of amacrine (AM3) which has a large-field arborization in the distal portion of the medulla.

preparations. These neurons are characterized by a perpendicular axis fibre one or several sets of dendritic arborizations in the ME (o), some types also having collaterals in ME (o) and axons projecting to the lobula via the second optic chiasma. The dendritic arbors of transmedullary neurons are disposed at several levels which bring them into apposition with arborization domains of other species of neurons and /or with one or more terminal specializations derived from the lamina. Despite their diffuse appearance these cells impart precision to layered structures of the neurophil as is shown by mass staining with coblt. Another class of neurons that most clearly contributes to the stratification of the medulla is the amacrine cell.

Characteristically, amacrines are either restricted to a single column or a specific domain of medulla columns and project to different depths

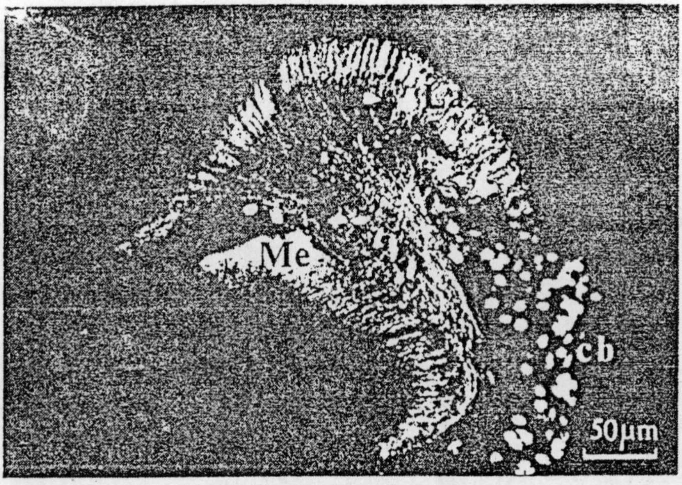

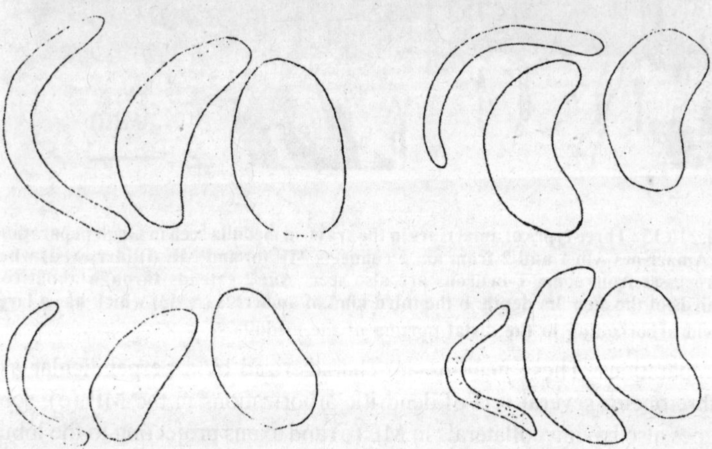

Fig 10.16: Horizontal section through the crayfish lamina (La) and medulla (Me) displaying catecholamine fluorescence after reserpine treatment followed by dopamine incubation. The neurons resolved are small-field columnar T-neurons connecting the lamina with the medulla. The cell bodies (cb) are stuated outside the chiasma between the two neurophils. The two proximal bands of fluorescence are too weak to be seen in this macrograph. (Elolsson, Nassfl. and Myhrbfrg. 1977). b Schematic drawing of catecholamine fluorescence patterns in the opitc lobes of four Crustacca. The neurophils are form left to right, lamina, medulla. Upper left. *Pandalus borealis* (decapod shrimp); upper right, *Neomysis* integer (mysid shrimp); lower left. *Astacus* astacus (crayfish); lower right. *Portterns depurator* (crab). (The different animals are not to scale.) After Elolsson and Khmm (1972) with permisson of R, Elofsson and Springer-verlag.

of the neurophil. Whether these merely provide and ciruitry within a stratum is not clear.

The remaining cell types resolved in the medulla have extensive abrorizations throughtout large portions of the projected retinotopic mosaic. These can be divided into three main classes (1) large cells that connect the medulla with the lobula; (2) cells that connect the medulla with the lobula and optic foci; and (3) cells that project directly protocerebral centers or contralateral optic lobes.

Cell types that constutute these classes either have diffuse arborization or form precise planar layers. Large cells of group 3 invade the serpentine layer and also have branches in two more distal strata as indicated in Fig. 10.14 The thickest axons within these three tangential layers have perpendicularly oriented secondary brances invading specific strata in the ME(o), suggesting functional relationships with columnar efferents from the lamina.

Catecholamine fluorescence is demonstrated in three layers of the medulla, a broad band distally and two thiner bands proximally, just above the serpentine layer. Only in the case of the distal band can the fluorophore be correlated with any single type of neuron, the lamina medulla small-field T–neuron, as mentioned previously.

11. The Lobula

Inputs from the medulla to the lobula comprise bundles of retinotopic columnar relay neurons and columnar T–neurons from ME(i). Postsynaptic columnar relay cells in the lobula seem, however, to be arranged as a subperiod of this lattice, thus coarsening the rectinotopic mosaic as is the case in insects. Accordingly, all cell types in the lobula which are arranged as columnar palisades have dendritic free domains which correspond to specifec groups of input columns projected from the medulla illustrate the denstrate the dendritic arbors of some columnar cells in either one of two main distal strata. These layers correspond to two main levels of input from the medulla. Two other levels of medulla terminals are distinguished more proximally in the lobula at levels of columnar neurons whose dendrites span large protions of the projected mosaic. Interspersed between the distal and proximal strata are diffusely arranged dendritic or terminal expansions of other cell types whose topological relationships seem to be concommitant with axons of lobula relay neurons rather than with primary input of medulla-derived efferents. Lobula stratification is also apparent in the distribution fo catecholamine fluorescence in five layers of the neurophil three of which appear to correspond to three of the lobula input layers.

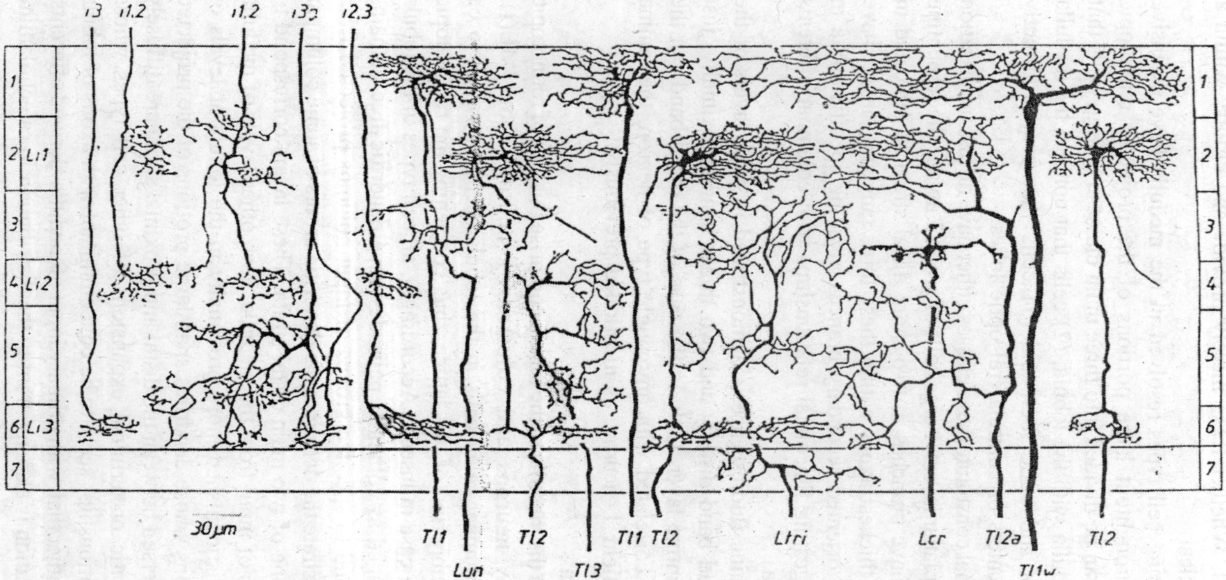

Fig. 10.17 : Diagram of elements in the crayfish lobula. The lobula has seven main strata (1-7), there of which receive input from the medulla (Li1-Li3). Input elements from the medulla are shown to the left (i3; i1,2; i3a;i2,3), most of which are derived from transmedullary neurons. Columnar elements have processes in either of two distal strata, TL1 elements at the distalmost and Tl2 at second level. T12 also has processes in stratum 6. A third kind of columnar element, Tl3, has processes in the lower half of the lobula. Lun. Liri. and Lun. Lcr are neurons that appear that appear to be non-columnar in structure.

Amacrine cells have not been seen in the crayfish lobula, though one type of large intrinsic T–neuron has been resolved which appears to join the outer layer of the lobula with its innermost layers, reminiscent of T–cells in undivided insect lobulae, such as, for example, in the bee. This region is also invaded by large-field tangential neurons comprising diffusely arborizing elements that connect the lobula.

STATOCYSTS

The statocysts are a pair of small, white, bead-like cuticular and spherical hollow sacs. They are said to be the organs of balance and position. One statocyst lies inside the precoxa of each antennule, attached to its dorsal wall. The minute statocystic aperture opens dorsally in the concave surface or depression of the precoxa and remains covered by a small fold of, integument. A small statocystic branch of the antennular nerve supplies the statocyst. On cutting a section the cavity of the statocyst is found full of minute sand particles surrounded by a number of elongated, delicate, receptor setae, which are attached to the inner wall of the statocyst in an oval outline. Each receptor seta is innervated by a small branch of the statocystic nerve. It consists of a swollen base and a long tapering shaft, which points towards the centre of the oval area. The shaft is bent in the middle and bears fine bristles beyond the bend.

The statocysts perceive the direction of the force of gravity and function as the organs of orientation and equilibrium. The sand particles function as statoliths.

Tactile Organs

There are in the form of some plumose setae, fringing the flattened portions of the appendages like the rami of the pleopods. Each tactile seta is a hollow cuticular outgrowth, supplied with, supplied with a nerve fibre, and consists of two segments. The basal segment or the shaft is slightly swollen and attached to the integument by a membrane. The distal segment or the blade gradually tapers and bears two rows of small barbs.

The elongated feelers of both antennae are also said to be tactile in function.

Olfactory Organs

The inner smaller branch of the outer feeler of each antennule bears a longitudinal groove containing containing numerous olfactory seta. Each seta consists of a basal segment or shaft, which is attached to the integument by a flexible membrane, and a distal segment or blade which is bluntly rounded. A small nerve fibre from the olfactory branch of the antennulary nerve innervates each seta.

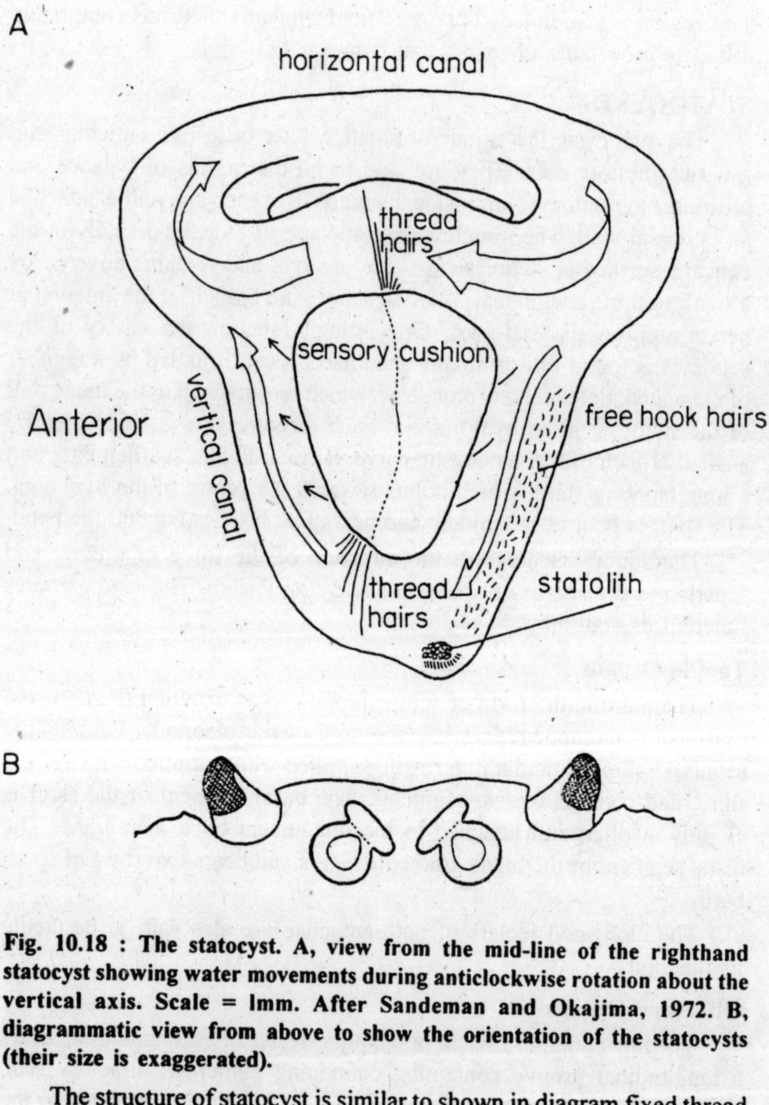

Fig. 10.18 : The statocyst. A, view from the mid-line of the righthand statocyst showing water movements during anticlockwise rotation about the vertical axis. Scale = lmm. After Sandeman and Okajima, 1972. B, diagrammatic view from above to show the orientation of the statocysts (their size is exaggerated).

The structure of statocyst is similar to shown in diagram fixed thread still bent), but the free hook hairs and the statolith hairs return rapidly to their normal positions and stop firing. This provides a mechanism for signalling 'off' which might be important to combat giddiness.

The function of the thread hairs and free hook hairs is probably primarily the detection of rotational movements. The statolith hairs, however, are too short to show much response to normal rotational movements and since these hairs support the statoliths, it seems likely that their function is the detection of gravity. The statoliths of a Mediterranean spider crab *Maia verrucosa* were described by Dikgraaf (1956) as a 'cluster of numerous miniscule stone particles. . . apparently identical to volcanic particles on the surrounding sea floor'. It seems probable that, like the statoliths of prawns and crayfish, crab statoliths are of external origin and are inserted by the crab just after molting while the cuticle is still soft and before the statocysts are sealed off from the outside. The statoliths lie on top of the statolith hairs at the bottom of the posterior ann of the vertical canal (Fig. 10.5) and any movement of the statoliths clue to gravity should be transmitted to the hairs space, since the iron filings respond to the magnetic field and not to the field of gravity.

Apart from the hooklikc sensory hairs which are connected to the statolith, most statocysts also contain threadlike hairs which register the turning movement of the animal, through movements of the water mass within the statocyst as do chose in our aural labyrinths. Apart from this, the degree of muscle tension is also regulated by the statocysts.

With regard to the chemical senses-especially the senses of smell and taste-in the Higher Crustacea there are large numbers of tubelike sense cells (aesthetascs) on the first antenna; they respond to substances dissolved in the water, and probably to odors in the air as well. Apart from these, organs of taste are also found at the ends of the walking legs and on the oral appendages.

REFERENCES

1. Handbokk of Sensory Physiology VII/6C. Vision is Invertebrates. Invertebrate Visuals Control and Behaviour II. Edited by *Hansgo Chem Autrum,* (1986), Springer Verlag Berlin Heidelberg Newyork.
2. Freshwater Crab, Diwan A.D. (2000); CIFE, (ICAR Publication) Versova, Mumbai.

11
ENDOCRINE GLAND AND HORMONES

Endocrine glands, the internal regulation in the crayfish or lobster is partly nervous but depends also on a variety of hormones. Activities affected by these hormones include dispersion and concentration of the pigment in chromatophores, migration of retinal pigment in the eyes, frequency of molting, deposition of the gastroliths, growth of the ovaries, amount of blood sugar, and rate of heart beat. Hormones influencing all these activities are secreted by the sinus glands, one of which is located in each eyestalk. In addition at least two chromatophorotropic hormones. are secreted by unidentified cells in certain parts of the central nervous system.

The sinus glands are extremely difficult to locate in preserved specimens. In a fresh specimen they are more easily found, because of a characteristic bluish color. If, in such a specimen, the exoskeleton, the hypodermis, and the dorsal retractor muscle are carefully removed from the dorsal surface of the eyestalk, as far out as the base of the eye. the sinus gland will then be seen, lying on the posterodorsal surface of the second and third optic ganglia. The gland is roughly stellate in shape, with blunt lobes. It can be seen more clearly after removal of the thin sheath which covers it and the optic ganglia. Another gland may then be revealed, named the organ, which has been suspected to be endocrine. In the crayfish this organ is very small and is located some distance proximal to the sinus gland. In the lobster the x-organ is larger, shaped like a bunch of grapes and located on the anterior surface of the optic ganglia, just opposite the sinus gland.

HORMONES

An animal has an integrated system, all part of which must be co-ordinated if the whole is to fuction satisfactorily. The nervous and endocrine system together performs this essential function of integration.

The endocrine system is made up of several glands each of which secretes on or more chemical substances. These substances, hormones, circulation in the body fluids and bring about reactions in other organs or organ systems of the body. The nature of the hormones, their mode of transport and their effects depend upon the reacting tissues as well as the chemical properties of hormone. Invertebrates, some of the most important, hormones are secreted by the pituitary a gland is located just below the brain, in many invertebrates, endocrine cells are found in most of the ganglia of the nerve chain. Special chemical hormones may be found among animals of nearly all phyla and classes and transported by the body fluids to all areas of the body.

Invertenbrate-Endocrine-System

Because invertebrate comprises 95 percent of the species in the animal kingdom. We might expect a correspondingly great diversity of invertebrate endocrine mechanism. While this expectation may ultimately be confirmed, our present knowledge of the endocrine systems of many invertebrates groups is still highly incomplete. Here it describes only a handful of relatively well studied examples of invertebrate endocrine system. The invertebrates systems are grouped in a seperate section for the convenience of the readers, Rather than because of any fundamental differences in organization or action between invertebrate and vertebrate systems, In fact, the basic principles presented in this chapter apply equally well to endocrine systems of vertebrates and invertebrates. The only general difference is that in most groups of invertebrates, unlike vertebrates, neurosecrefcary system are distinctly more prominent than non-neural endocrines glands, indeed in many invertebrate group, nonneural endocrine glands appear to be absent.

Hormonal System of Crustaceans

Crustaceans have a complex endocrine system, decapod crustacean such as crabs, shrimps, lobsters and cray fish have well developed endocrine systems that are second only to insects in complexity among invertebrates. We shall focus on two prominent neurosecretory system that (in some cases alongwith nonneural glands) control molting, reproduction, mineral and water balance, colour chances and heart function. *Neurosecretion.* The importance of specialized neurosecretory cells and organs has been recognized in invertebrates long before their importance was generally established in vertebrates. Neurosecretory cells are difficult to define and characterize so that they are easily and unequivocally recognized; the major approach for locating and pinpointing such cells still depends on the use of certain stains that have a special affinity for neurosecretory cells.

Neurosecretory organs consist of a group of neurons which are the source of secretion. These cells have nerve fibers, in which the secreted agent is transported, and which usually terminate in close association with a vascular structure. Here they form a nervohemal *organ*, where the secreted product is stored and released. Neurohemal organs are very widespread, but it is not certain that they are universally present in connection with all neurosecretory organs.

In all the more highly organized animal phyla neurosecretory systems are central to the control and operation of endocrine mechanisms. We have already seen the importance of neurosecretion in the control of vertebrate endocrine organs; among invertebrates neurosecreting seems to be even more important. In some insects, neurosecretion and neuro-hemal'organs have a central role in the control systems. Neurosecretory systems are also important in annelids and crustaceans, in which they are involved in control of reproduction. metabolism, molting, pigmentation, etc. Although, some of the endocrine glands of vertebrates as well as invertebrates are outside direct neuroendocrine control, the normal function of most or all glands ultimately depends on regulation from the nervous system, either directly through nerve pathways or througli hormonal mechanisms. It seems that control by neurosecretory mechanism is the general rule in all animals in which endocrine mechanisms play a major role.

Hormones and Endocrine Function

Endocrine function is best understood and has been most extensively studied in those invertebrates which are morphologically highly organized. There are two reasons for this situation; one is that a highly organized animal such as an insect needs more detailed control and integration than a less organized animal such as a sea anemone. The other reason that the more highly oiganized animals are better understood is that many functions are delegated to specialized organs. This permits experiments such as removal of organs, reimplanlation, extraction, and other procedures that are difficult or impossible in less highly organized animals.

Most invertebrate hormones are different from vertebrate hormones. both in chemical constitution and in their effects. Many vertebrate hormones which have profound effects on vertebrate growth, development of gonads. metabolic processes, and so on, have no effect whatsoever if introduced into invertebrate animals.

Endocrine function has been demonstrated in many invertebrates, but the mere fact that some extract has a physiological effect does not demonstrate that the extract contains a substance that has a normal as a hormone in this animal. More careful study is required and the necessary criteria include the demonstration of facts such as a normal occurrence of the substance in question, demonstration of where it originates, examination of the effects of removal of its source, or of the blocking of its secretion, the purification and isolation of the purification and eventually its synthesis, including the demonstration that the synthetic product is effective in similar concentrations to the natural product. As a result, detailed knowledge of invertebrate endocrinology is restricted to a relatively few forms.

The groups in which organs of internal secretion have been clearly demonstrated and are reasonable well understood are moiluscs (notably cephalapods), annelids crustaceans, insects, and tunicates. All these are highly organized animals where the complexity of function and the specialization of organs facilitate endocrine research, to particular, insects have been studied in great detail for reasons which include their easy accessibility, the ease of collecting breed and beeding them, their tolerance 10 drastic surgical procedures and their economic importance.

The X-organ Sinus Gland Neurosecretary System and Control of Molting

In decapod crustaceans a sizeable parts of the brain is located in the eyestalks, the major neurosecretary system in dicapods is in this eyestalk portion of brain. The X-organ is a cluster of all bodies of neurosecretory neurons within the ganglia of each eyestalk. These cell bodies send a bundle of axons to the sinus gland. The neurorenal organ also located in the eyestalked, The crustacean X-organ-sinus gland complex is analogous in many was to the systems in insects composed of the brain neurosecretary cells and corpus cardium the anatomical arrangement of the crustacean system, however, has advantages for experimentation, because eyestalk removal functionally removes the neurosecretary system, Neurohormones released at the sinus gland control several processes of which molting has been studied most extensively. Crustaceans like insects, must periodically molt the exoskeleton in order to grow.

Neurons of the X-organ sinus gland complex produce and secrete a molt inhibiting hormone. Removal of the eyestalks of a decapod thus induces an early molt. The molt inhibiting hormones acts on paved non-neural glands called Y-Organs, The Y-Organs sometimes termed ecdysal glands are analogous and perhaps homologous to insect prothoracic glands. They secretes ecdysteroid, principally ecdysone and 20 Hydrocyecdysons which directly control the various processes involved in molting, thus, the control of molting in crustaceans is strikingly parallel to that in insects with the major exception that in crustaceans the neurosecretory control of steroid secretion by the excitatary neuro-secretary control of the prothoracic glands in insects. Other harmones participate in control of molting besides molt-inhibiting hormone and ecdysteroids. For example, hormones from the sinus gland and elsewhere regulate molt related water balances and calcium.

The heavy deposition of calcium salts in the already rigid chitinous skeleton makes gradual growth of the animal impossible. Growth therefore occurs in definite spurts which take place during the time the exoskeleton is soft after each molt. These molts are a good example of hormonal processes, which are especially well developed in the crustaceans. Glands which excrete their compounds internally (endocrine glands), controlling the process, are partial developments of the nervous system and lie as neurosecretory cell islands at the base of the eyestalk in decapods (X-organ and sinus gland); their secretions control molting, chitin formation, and color change (chromatophorotropine). The non-neurosccretory gland, the Y-organ, found near the epidermis of the second maxillae or second antennae, also plays a part in the hormonal activity involved in molting, its secretions initiating the molt.

Finally, in 1954, endocrine glands which secrete a male hormone (androgen) were discovered in amphipods; afterward, they were also found in other Higher Crustacea. They lie at the end of the vas deferens; when they are transplanted into a female, it changes into a male; they therefore contain true sexual hormones. Too little is known of other hormonal secretions in the Higher Crustacea, but numerous other life processes (*e.g.,* the carbohydrate level in the blood, calcium uptake and water uptake of the tissues) are at least partially controlled by internal secretions.

Reproduction and Sex Determination

Hormones play an important role in sex determination in crustaceans. Young genetic males develop an androgenic gland associated with the genital tract. The androgenic gland which does not develop in genetic females, masculinizes the differentiating gonads into testis and also includes male secondary sexual characteristics. Implantation of an androgenic gland into a young genetic female causes transportation of the ovaries into testis. Thus, as in mammals the crustacean reproductive system becomes male if provided with an appropriate hormones signal and becomes female if this signal is absent.

In malacostracans hormones control llic differentiation of the primary and secondary sexual characters of the male. The gland responsible, the androgcnic gland, was first demonstrated in the amphipod *Orchestia gammarella*, where it lies in the haemocoel alongside the Vas deferens. A corresponding gland has been found in other amphipods, and in some decapods, including *Carcinus maenas*, and isopods. In females differentiation of the gonad is not under hormonal control, though, as in vertebrates, the ovary secretes a hormone which is responsible for the development of temporary characters associated with the incubation of the eggs. Secretion occurs during the period of deposition ol yolk in the ova, and controls the enlargement of the oostegites forming the brood pouch in isopods (they are reduced between breeding periods), the lengthening of the marginal setae on the oostegites in gammarids (in which otherwise they remain unchanged) and the development of egg-carrying setae on the pleopods of decapods. The deposition of yolk is itself controlled by a hormone secreted by an endocrine organ in the eyestalk known as the X-organ which limits egg production, and hence the appearance of secondary sexual characters, to intervals characteristic of each species. This recalls the role of the hypophysis in vertebrates.

In some amphipods, such as *Orchestia cavimana* the gonad, in early developmental stages, starts by containing both male and female cells; in males the presence of the androgenic gland suppresses further development of eggs, which remain abortive and fail to lay down yolk (*ovt*); in females absence of the gland prevents development of the primordial male cells. The presence of the androgenic gland in decapods appears to explain the feminization of male decapods parasitized by rhizocephalan cirripedes such as *Sacculina.* The gland is gradually destroyed by the parasite; spermatogenesis ceases and the external secondary sexual characters are just at a subsequent molt and replaced by more feminine ones. The suppression of yolk forniation in parasitized females inhibits the appearance of temporary sex characters related to incubation.

Chromatophores and Colour Change

Many crustaceans are capable of pronounced change in colour mediated by chromatophores. Each chromatophores is a highly branched cell within which pigment can be concentrated or dispersed, many crustaceans have several colour classes of chromatophores and can control them rather independently. The control of crustacean chromatophores is hormonal mediated by Chromatophorotropins from the eyestalks and other neurosecretory centres, Each colour class of chromatophres is under control of antagonistic concentrating hormone and black pigment dispersing hormone, similar hormones control the movement of screening pigments during light and dark adaptation of the eyes, the various pigments concentrating hormones are quite specific and independent in their actions but recent evidences suggests that a single hormone mediates light adaptation and pigments dispersal in all chromatophore classes.

Pericardial Organs and Cardioregulation

They are the second most prominent neurohemal organs of higher Crustacea. They lie in the dorsal venous cavity surroudning the heart and exert their clearest effects on heart function, this cell bodies of most of the neurosecretory neurons that have ending in pericardial organs lie in or near thoracic ganglia of the central nervous system. Their axons reach the pericardial organs through segmented nerves. Some cell bodies, however, are located in the pericardial organs are the biogenic amines octopamine, seratonin and dopamine, plus at least one peptide. All have excitatory effects on the heart and also facilitate neurally evoked contraction of peripheral skeleton muscles, the hormonal release of cardiostimulatory amines by the pericardial organs is analogous in certain ways to release of calcicholamines by the vertebrate adrenal medulla. It also serves to remind us that although, most known neurohormones are peptides, not all are.

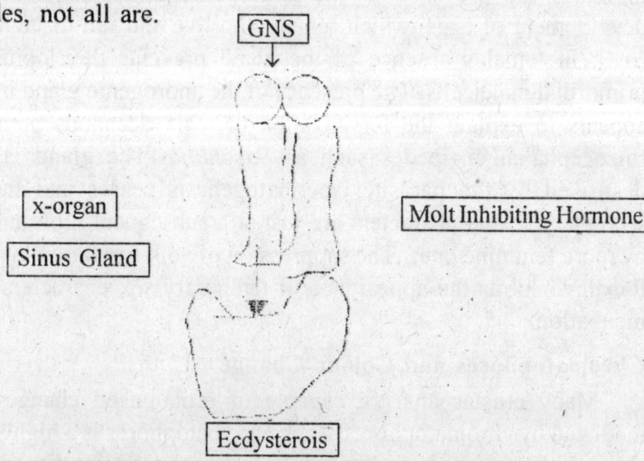

Endocrine Control of Molting in Crustaceans

Fig. 11.1

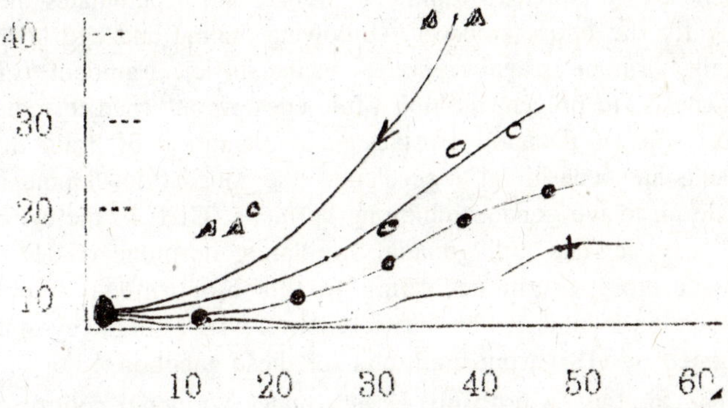

Fig. 11.2

Rate of increase weight of the shrimp length, weight index determined by dividing the wet weight of hte ovaries by the body length head.

Hormonal Pattern in Crustaceans

In its diversity in the crustacean hormone secretory cells are neurosecretory cell. These are specially in "medulla terminalis of its brain. There some of them constitutes the medulla triminalis and organ know as "sinus gland" is situated at the base of the eye stalk. The X organ produces a molt inhibiting hormone and probably also in some cases a molt accelerating hormone. The other hormones concerned in molting is secreted by Y-organ is found in other the antennal or the maxillary segment. and is analogus to ecdysial of insect. Removed of eye stalks in juvenile or non-breeding females leads to a rapid increase in the ovarian weight. through the stimulation of vitellogenesis.

The neurones cells in eyestalk complex secrete the gonado inhibiting hormone and when the eyestalk are removed surgically or when the activities of cell are surprised ovary development with yolk production is initiated in females while testiscular development occurs in males under natural condition the change in photoperiods seems to regulate the activity. A high titer of sinus gland extract also inhibits in molting in egg bearing females. In crayfish for example, the spring molt is post poned in females while they are earning developing eggs attached to their pleopods. If their sinus glands are removed they molt at the same time as the male or non-breeding female. The sinus gland hormone first inhibits vitellogenesis is as its secretion gradually under changing environment condition (photoperiods.)

The ovary increases rapidly in weight and accumulates the food supply for the future embryos, following mating and egg laying the eyestalk hormone is again secreted in the sufficient amount to inhibit villogenesis to prevent molting while eggs are attached to female. In several spp. of decapod crustaceans implantation of brain thoracic segments has been found to accelerate the gonadal development. This in addition to the gonado inhibiting hormone (GIH) of the eye stocks there may be also and gonadal stipulating hormone (GIH) of the neurosecretory origin according to this biohormonal hypothesis reproduction is inhibited by the secretion of GIH from the eyestalk and stimulated by GSH from brain and for those ganglion.

The molting is probably largely under hormonai control and is modified via central nervous system according to both internal and external environmental conditions. Molt initiation, however, is the only part of the control to be well understood. The X-organs-groups of neurosecretory cells situated in each eyestalk-produce a molt-inhibiting hormone which passes down the axons to the sinus glands. Removal of both eyestalks, or of the X-organ/sinus gland complex, results in molt initiation; reimplantation of the organ complex delays the molt. The target organs of the molt-inhibiting hormone are the Y-organs which are situated at the front of the body beneath the branchiostegites, one on either side. These, when released from inhibition, produce a molting hormone. The effect of the molting hormone is to initiate premolt stage D. Removal of both eyestalks and Y-organs does not result in molt initiation, but reimplantation of the organs alone leads to a molt. By stage D_2 molt initiation is irreversible; removal of Y-organs at this time does not delay the molt and neither is it accelerated by removal of the eyestalks. Thus the X- and Y-organs control the important transition from D_4 to D in the molt cycle. The control of the sequence of events from D_2 back to C_4 is less well understood. Each event up to C_4 may be triggered by the preceding event but it is likely that further hormonal control is involved.

Terminal anecdysis (C_4T) has been explained in terms of the X- and Y- organs by Carlisle (1957) who showed that there were different control mechanisms in *Maia* and *Carcin. s.* In *Maia*, C_4T is the result of degeneration of the Y-organs eyestalk removal in C_4T does not produce molting. In *Carcinus*, however, C_4T is achieved by constant release of molt-inhibiting hormone and removal of the eyestalks does produce molting.

This dual control of molt initiation is similar/to theJsituation found in insects except that in the latter group the release of molting hormone is stimulated by a neurosecretion instead of being inhibited. A further similarity, which in this case links all arthropods, is that the chemistry of the molting hormone-ecdysone is similar in all groups and hormone extracts from one group can be used to initiate molting in other groups. The crustacean molting hormone is called crustecdysone.

An important part of the hormonal control of molting is that, since the X-organs are part of the nervous system, the release of molt-inhibiting hormone is under nervous control. This provides a link between the molt cycle and the external and internal environments. The X-organs presumably stop secreting when sufficient metabolic reserves are present, when there are no reproductive duties such as egg carrying, and when external conditions are favourable. When conditions are unfavourable, however, the X-organs continue to secrete molt- inhibiting hormone. External conditions which tend to inhibit molting are constant light, low temperature and crowding. In temperate regions mature females tend to molt in the autumn; molting is inhibited by cold in the winter and by egg carrying during spring and summer. In mature males molting is more common in the early summer; many species only copulate when the female is soft, just after a molt, and thus the males need to be hard-shelled in the autumn.

Ecdysis

Although, of relatively short duration ecdysis is obviously one of the most important events in the molt cycle. Descriptions of ecdysis generally divide it into a passive phase followed by an active phase. The passive phase, which overlaps with D_4, is a time of swelling and splitting while the active phase involves muscular movements culminating in the withdrawal of the crab from the old skeleton. During both phases the crab is usually hidden away in a burrow or under a stone. It generally, avoids members of its own species since most crabs have cannibalistic tendencies.

In the passive phase water uptake commences and a split develops from posterior to antérior along the epimeral line of each branchiostegite (Fig. 11.3) and between the back of the carapace and the first abdominal segment. The back of the old carapaçe gradually rises up revealing the new carapace underneath. This generally appears rather wrinkled but brightly coloured. Uptake of'cwater and the consequent increase in haemocoel volume is no doubt responsible for these changes. The pericardial may be important here since they swell progressively during the passive phase and bulge from the branchial cavity under the new carapace. It is likely, however that their main function is to store fluid displaced from the limbs during the active phase. Towards the end,of the passive phase the epimeral split extends anteriorly almost to the mouth.

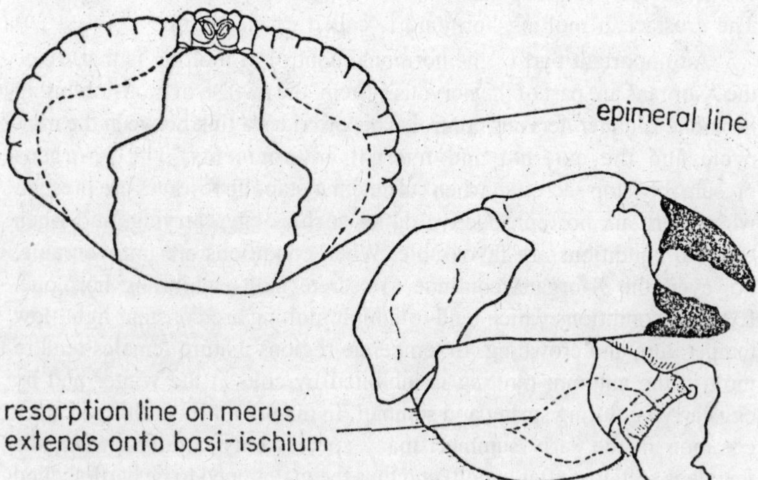

Fig. 11.3 : Ventral views of the carapace and chela of the edible crab *Cancer pagurus* showing reabsorption lines at which splitting occurs during ecdysis.

The active phase begins when the carapace has been raised like the lid of a box hinged at anterior and of animal to an angle of 30°. Contraction of muscles attached to leg bases commence and serve to loosen lins within old sketelon the outword sign of this activity is a rhythmic sereis of depressions appearing and disappearing on alternate sides of a new casapace (to which leg muscles are ultemately attached) in *Pachygrapsus* each contraction last about 10 seconds but in much larger snow crab *Chironocetes o-pilio* from northeast Canadian waters each contraction last about 55 seconds. This loosening activity exists for few minutes (*Pachygrapsus*) or a few hours (*Chironoecetes*). Eventually, however, beining at a posterior end, the limbs are gradually with own from old skeleton the pericardialsac which are maximally swollen during the early part of active phase now deflate as the fluid is redistributed to emerging limbs the gelatinous layer between old and new skeleton may lubricate withdrawl quickly as gills are withdrawn the respiratory mechanism is temporarily put out of action. The chelipods with their swollen distal ends present special problems but these are over come by premot breakdown of chelipod muscles and by selective reabsorption and splitting along fracture lines on basal podomeres which allow bulkly propodus to be withdrawn. All parts of old skeleton which have been reabsorbed are left behind with cast shell endophragmal skeleton, apomes, qastric mill etc.

Pigment Migrations

Decapod crustaceans, especially brachrurans, have provided much of the experimental material for the study of neurohormones in invertebrates and have provided definitive evidence of neuro endocrine cells in invertebrates and also of endocrine cells of nonnervous origin. The changeover of a light-adopted animal to a dark adopted one and vice-versa, takes place within an hour or two and in brought about by dispersion or concentratioin of the pigment granules in the chromatophores lying superficially in the hypoderimis or more deeply in underlying tissues. Typically, in decapods, there are three sets of retinal pigments contained in different cells enclosing each ommatidium. The proximal pigment, which is black, is in the retinular cells, the distal pigment, also black, is in chromatophores that surround the crystalline cone; and the light reflecting pigments lies in another set around the reticular cells; In palaemonectes all three moves outward or inward in response to differences in the intensity of incident light; *Astacus*, the reflecting pigment does not move, but remain always just above the basement membrane and in *Homarus* only the proximal pigment migrates, the distal and reflecting pigment remaining fixed in position. There in no nerve supply to either the eyestalk produced any kind of change in dissection and histological examination of the eyestalk revealed a thickened area on the dorso-lateral surface of the ganglia within them, the relay station for impulses originating in the retinular cells.

In decapods this area is flasked shaped, representing about one hundredth of the total stalk tissue, and endoses a hemolymph sinus the majority of them are clustered in a group called the X-organ in medulla terminalis ganglion; some are in the optic peduncle and elsewhere within the eyestalk and some in other parts of the central nervous system. In species without eyestalks there organs lie in comparable regions of the cephalic ganglia. Sinus gland is, therefore actually an isomer, for this structure is not intrinsically glandular but is a storage depot for products of the neuroendocrine cells, the name has, however, been retained for it, even though its true nature as a neurohemal organ is now fully established. It soon became evident that the sinus gland stores and discharges into the circulatory system a complex of hormones which are involved in wide range of physiological processes, including growth and molting, regeneration, reproduction and various aspects of metabolism as well as chromatophore and retinal pigment migrations and that the X-organ-sinus gland complex is a major part of a delicately balanced endocrine system comparable to that of the hypothalamic neurohypophysal system of vertebrates.

Molting

Molting in decapod crustaceans is also influenced by hormones discharged form the sinus glands. Molting is not simply the act of emerging from the old exoskeleton, but a continuous series of events that go on during most of the life of an individual and lead to the actual act of ecdysis. The molting cycles is divide into atleast four major stages. The first of thre is premolt or proecdysis, second is ecdysis, third is post molt and last in inter molt.

The first intimation of neuroendocrine control of molting was observed that the intermolt period was significantly shortened in animal deprived of the eye stalk. Later experiments on crabs have shown that removal of the sinus gland alone does not induced precautious molting, but removal of the X-organ in eyestalks animal prolongs the intermolt period. It is, therefore, postulated that the X-organ produces a molt inhibiting hormone that is stored in the sinus gland and released from it into the hemolymph in active concetration during postmolt and intermolt periods.

Molting is promoted by a hormone liberated from the Y-organ, small masses of glandular tissues that are situated ventrally and laterally to the eye stalk just above the junction of the branchiostegite with the cuticle of the lateral. body wall at the anterior end of the branchial chamber. If these are removed during intermolt or in vary early procedysis, the animal will not molt again, but if the Y-organ are excised when proecdysis is under way, the events of that molt proceed as usual Y-organ are necessary to trigger the molting process and that once this has begun, it will inevitably continue to its natural condition.

Water Balance

Water balance is also affected during proecdysis, and under natural condition, relatively large amounts are taken in through the forgut or other areas. With a terrestrial mode of life, the pericardial sacs are the principal areas of water storage. and may function also in the actual uptake of water. There are membranous pouches that project from the pericardial sinus into the branchial chamber.

There change in water content and in the conectration of materials in the hemolymph impose upon the animal the necessity for osmoregulation and so relate this vital physiological process to a hormonal control, at least during proecdysis.

Evidence for neuroendocrine involvement derived from ligation and injection experiments in cray fish *Procambarus clarkii* and the crab *Metapograpsis messor*. Ligation of the eyestalk leads to a rapid decrease in the osmotic concentration of the crabs hemolymph, since this fall in concentration can be some extent prevented by injection of eyestalk homogenates, a contribution to osmoregulation by substances in the eyestalk is implied. In postecdysis the external dimension of a newly molted animal increase greatly, due principally to its uptake of water, which now becomes distributed through the tissues causing expansion of the soft exoskeletion.

Reproduction

The reproductive organs of crustaceans are also sites of hormone production. The interaction between the molt inhibiting hormone, or the molting hormone, and the reproductive state of female crayfishes has already been noted. It has been found, in general, among the species investigated that removal of the eyestalk from immature females results in rapid and precocious development of being through the activity of a hormone developed in the X-organ. In the male crustaceans an androgenic gland has been identified as a solid strand of glandular cells located between the coxopodite of the last walking leg and the vas deferens, whose activity is coutrolled by centers in the eyestalk. If this gland is implanted in an immature female, her ovary develops as a testis and the secondary sex character of the male, most evident in the modification of the first abdominal appendage as a passage for sperm, are aquired by her in the cousse of the next two or three molts. These altered females do not, however, become functional males, for although, sperm duct are formed, they remain solid cords and never develop a lumen. On the other hand, males from which the and organic glands have been removed do not becomes females. If the eyestalk are removed from young males, the androgenic gland hypertrophies and testicular development is promoted. This effect is comparable to the testicular promotion of ovarian development in eyestalk less females and points to a similar gonad inhibiting factor revealed/released through the sinus gland.

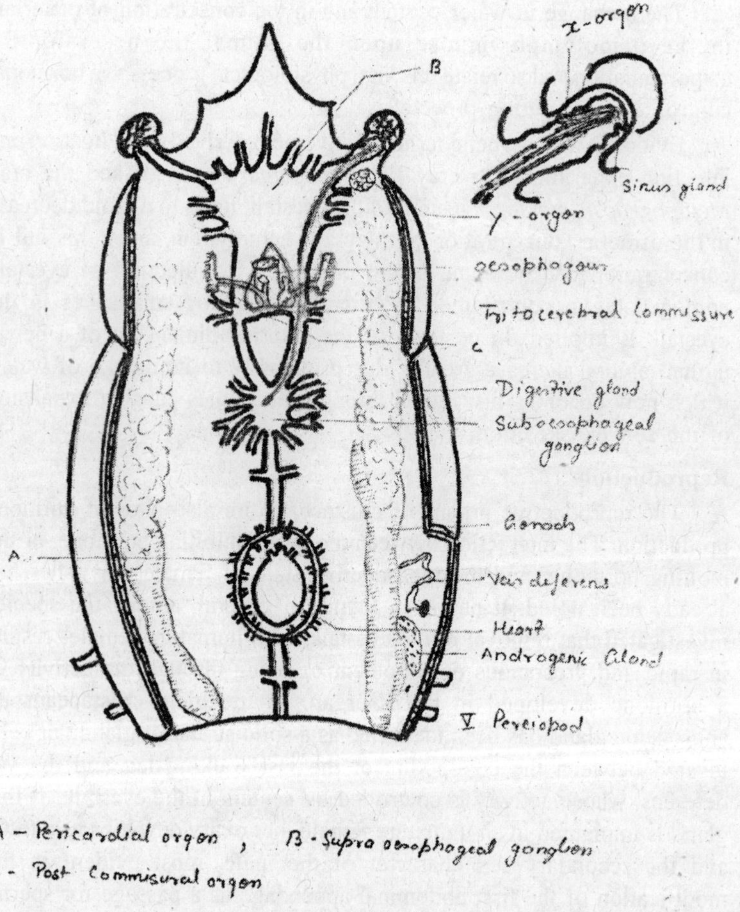

Fig. 11.4 : Hormonal System in Decapoda.

REFERENCES

1. Gardiner, Mary S. "The Biology of Invertebrates," MacGraw Hill Book Company, New York, 1972, pp. 700-717
2. Jordon, E. L. and Verma, P. C. "Invertebrate Zoology," Rastogi Publication Meerut.

12

REPRODUCTIVE SYSTEM

INTRODUCTION

The reproductive process concerns all aspects of the replacement of the present generation by its successor. Accordingly, we have to consider under this heading not only the various expressions of sexuality of the adult individual, but also courtship, copulation, egg laying, parental protection of eggs and young, development stage at emergence of the young, aspects of larval behaviour, and any other matters which may have some bearing on the proportions of offspring which survive to rich maturity. This can best be done separately under these various headings. First however, let us designate what is likely to be the aboriginal reproductive pattern in the Crustacea, from which other patterns have arisen as deviations. In some cases these deviations supply a major feature in the adaptive radiation of the phylum and class. In other cases deviations from the norm may be assembled to indicate probable progressive steps in an evolutionary process, *e.g.,* the protection of the eggs and early developmental stages. Yet other deviations may be minor end points in adaptive radiation which possibly solve a functional problem for one particular species occupying one particular habitat.

It is suggested as a basis for discussion that the sexes were separate, that there were no accessory sexual organ and no special genital ducts, and that the gametes were discharged via the nephridia into the sea water. The eggs were numerous, small, and lacking in yolk. Cleavage and gastrulation occurred rapidly while drifting in the sea, and the organism hatched as a planktonic trochophore larva. The parental generation was solely concerned with the emission of large numbers of gametes for which no protection was provided.

Special Provision for the Fertilised Ova

Crustacea in which the ova and the spermatozoa were discharged in clouds into the sea water.

The crab, is dioecious. The two sexes can easily be distinguished:

(1) The abdomen in case of male is much narrower than in female.

(2) There are only two pairs of pleopods in the male whereas in the female there are four pairs.

(3) The genital opening in the coxa of the last pair of pereiopods while that of the female is on the sternum of the sixth thoracic somite.

Male Reproductive System

The male genital system consists of two testicular masses which are medially interconnected by a commissure so that the whole system appears to be "H" shaped structure. Each testicular mass is a convoluted tubular organ with many lobes situated just dorsal to the digestive gland and under the hypodermis. It extends from the middle region to anterior and posterior sides. It passes laterally to the stomach and above; the midgut it turns inward and joins the similar portion from the other side to form a bridge or commissure behind the pyloric foregut. The portion of the testis posterior to the commissure is tightly bound by connective tissue. It also binds all the loops of the testis together. There is some variation in the dimension of the testis. Throughout its anterior portion the testis measures about 1.5 to 3.0 mm and towards the middle portion where the convolutions are tight it measures 4.0 mm. in dimension. The vas deferens arises from the posterior end of the testis and its lumen is connected with that of the seminiferous duct. These are paired long convoluted tubes passing backward from the testis to the posterior of the thorax. Each opens to the exterior on the coxopodite of the last pereiopod.

Histology of the Testis

The transverse section of the testis shows that it is composed of numerous follicles which are closely packed together. Each follicle has a thin wall enclosing a large central cavity and this cavity is continuous with that of the vas deferencs. In the early stages of maturation the cells of the follicle are not clearly differentiated and are very small. But in matured testis it is difficult to distinguish the follicles as the cavities of the follicles become obliterated by the growth of the spermatoblast. The outermost is the connective tissue layer and the inner one is the germinal epithelial layer. Each cell of the follicle becomes spermatoblast and increases greatly in size. Each spermatoblast give rise

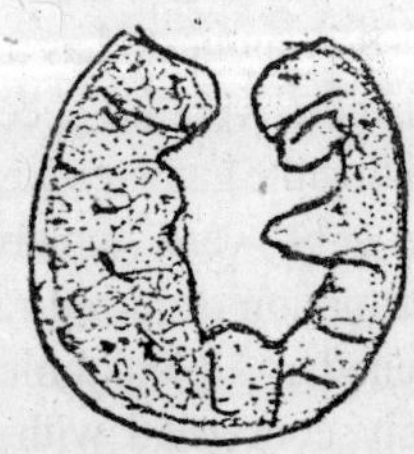

Fig. 12.1.

to large number of spermatozoa. The spermatozoa consists of dark central portion containing a nucleus and outer clear membrane. When spermatozoa mature they break and carried into the vas deferens. There they are collected together in small groups and each group becomes surrounded by a capsule to form spermatophore. In the mature crab, the vas deferens appears to be whitish due to the presence of spermatophores whereas in immature one the vas deferens in almost transparent.

Vas Deferens

The vas deferens originates from the posterior end of the testis. It passes through the thoracic cavity and pereiopodial musculature of the eight thoracic somite where it ends as a more genital opening on the coxa of the fifth pereiopod. The vas deferens is divisible into three different portion's, anterior vas deferens (AVD), median vas deferens (MVD) an posterior vas deferens (PVD). Each portion has got different function.

Anterior Vas Deferens (AVD)

This is the first part of the vas deferens which arises from the posterior end of the testis just near the commissure. It a narrow tube-like structure tightly coiled much convoluted and is bound together as one mass. Sperm are evacuated from the testis through this part of the vas deferens. The AVD measures about 217-326μ in diameter, The AVD appears to be white because of the presence of sperm which are being formed into spermatophores.

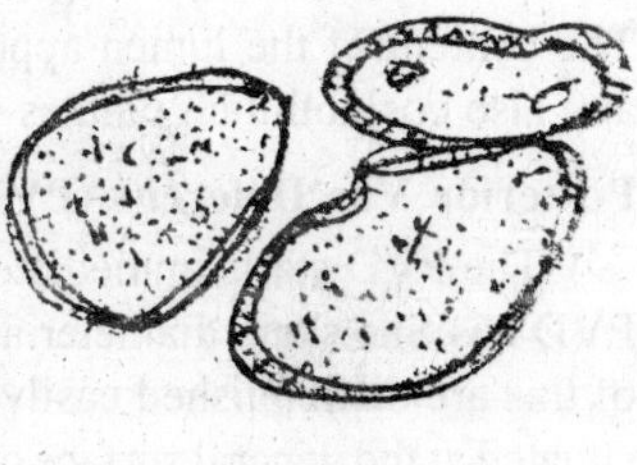

Fig. 12.2.

The histological section of the AVD reveals that there is a single layer of columnar epithelial cells which line the lumen. These cells are situated on the basement membranes. The fold of the AVD may be divided into three to five branches. The epithelial cells which lie on the folds are highly columnar with spindle shaped nuclei. The epithelium is covered on its outer side by circular muscle fibres. On outer side of the duct there is a thick sheath of connective tissue. The central part of the lumen is filled up with number of sperm. The spaces between the loops are occupied by the blood vessels and connective tissue fibres.

The second portion of the AVD is longer which has many coils and is bound by connective tissue into a compact mass. It is difficult to separate the coils because the tube is so delicate that it always ruptures.

The diameter of the second portion of the AVD is more than the first portion and this indicates that further the same tube becomes the MVD by increasing its diameter.

Median Vas Deferens (MVD)

This is the median portion of the vas deferens which is closely coiled. It is opaque white tube which consists of major coils having more diameter and minor coils having less diameter. The major coils are situated at the anterior arid ventrolateral portion of the pericardium. The posterior two thirds portion of the MVD forms the minor coils which lie ventral to the pericardium. At the beginning the diameter of the MVD is 434μ after which it slowly increases to 860μ, the length of the MVD is found to bc approximately 60 to 80 mm.

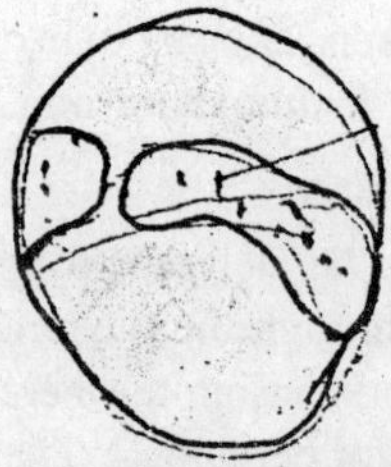

Fig. 12.3.

The epithelial layer lining the MVD is composed of columnar cells which are multinucleate with basally arranged spherical clusters of nuclei. The content of the lumen appears to be filled with the spermatophores and also eosinothilic granules which give the white colour to the MVD.

Posterior Vas Deferens (PVD)

The PVD starts immediately after the minor coils of the MVD. The PVD has She same diameter as the-minor coils of the MVD. The loops of this are distinguished easily by their colourless appearance. They are situated at the veneral surface of the pericardium are bound by connective tissue. The coils of the PVD run alongwith the digestive gland lateral to the intestine. At the posterior end of the pericardium just near the sella turcica the PVD changes its direction and passes anteriorly through the musculature of the eighth thoracic somite. The duct emerges as a penile papilla on the medial surface of the coxa of the last walking leg. The PVD can be divided into two portions: the first part extends from the minor coils of MVD up to the ejaculatory duct and other one is the main ejaculatory duct. The diameter of the PVD in the beginning is 717.0μ and slowly it decreases a it passes downward. The PVD is the longest portion of the vas deferens and its length including ejaculatory duct is approximately 80 to 100 mm.

The histology of PVD is very similar to that of MVD except the epithelial cells lining the lumen are smaller in size. The ejaculatory portion of the PVD shows epithelial lining of low columnar cells with centrally placed nuclei and are provided with cilia.

Penis

The penis is the last portion of the ejaculatory duct it is white tube which lies ventrally in the proximal foramen of the first pleopod of the matured crab. It measures 434.0μ in diameter. The T.S. of the penis shows that it is lined by a single layer of epithelial cells. The cells are culumnar and are provided with cell. The lumen in occupied by typholosolar folds. The typholosolar folds are lined by secretory epithelial cells.

Accessory Structures

Incase of male there are two abdominal appendages which are situated on the first and second abdominal somites and they act as copulatory organs. The male genital opening is situated on the ventral side of the coxopodite of the last perioped. As the thoracic sterna are deeply concave in male the locking apparatus here is well developed and the abdomen is closely applied to it. The two pairs of abdominal appendages together from a single copulatory organ. The first appendage is tubular and into this tube the posterior rod-like appendages is inserted during copulation. Such copulatory organ is introduced into one of the vulvae of the female during copulation. As the genital papillae are very small to reach the vulvae, the abdominal appendages get modified to from a sexual organ. Spermatophores are poured into the tube from the genital papilla and the rod-like a second appendage helps in forcing the male sexual products into the spermatheca of the female.

Androgenic Gland

The occurrence and physiological function of the androgenic gland was firs discovered by Chamiaux-Cotton 1954 in amphipod, *Orckestia ammarellus*. Subsequently it was described in various other *malacostracans*. The androgenic gland of *Bathytelphusa cunicularis* is located between the muscles of the coxopodite of the last waling leg and is attached to the subterminal region of the vas deferens. The gland is made up of solid strand of cells folded beck on itself forming a

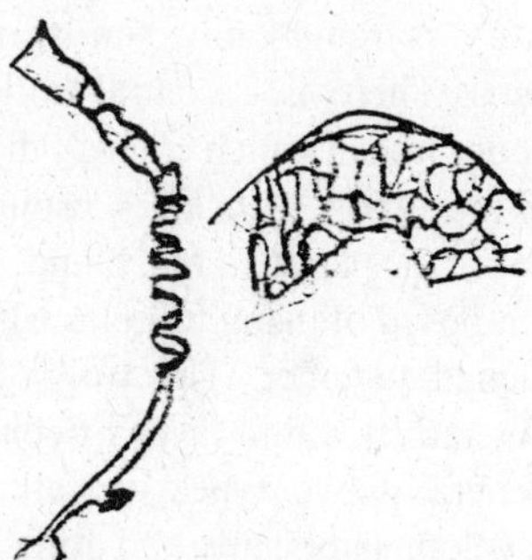

Fig. 12.4.

compact mass. It measures about 2-3 mm. in length. The cells of the androgenic gland are isodiametric and show histological evidence of secretion. The nuclei are round and situated at the periphery. Then plasm is vaculated *Carcinus.* The androgenic gland is of the historically type in histology. *Charniux-cotton* had demonstrated the role and function of this structure in determining the primary and secondary male sexual characters in Crustaceans. The androgenic gland is a unique instance of in endocrine sex gland separate from the testis.

Female Reproductive System

The ovaries are paired structures and occupy similar position as the testes in the male reproductive system. Each ovary is divided into two unequal lobes. The lobes are triangular in shape measuring about 3-4 mm. in diameter and are situated on the dorsal surface of the gland. As in the male, here also the paired ovaries are connected together at the posterior end by a commissure behind the pyloric foregut which forms 'H' shaped structure. The most anterior part of the ovary is found to be extended up to the base of the anteriolateral spines. The ovary curves medially at lateral spines and runs lateral to the stomach. Posterior tot he commissure the ovary passes ventral to the pericardium and further it extends up to the extreme posterior end of the thorax. At this region it gives rise to large sac called as spermatheca or seminal receptacle. In young crab the spermatheca is very small and difficult to trace out. On the ventral side spermatheca is continued into a short tube called as oviduct which means by the exterior by means of vulva on the stemum of the sixth thoracic somite.

Histology of Ovary

The general histological pattern of the ovary is remarkably similar to that of the testis. There is a central hollow shaft with empty lumen which extends the whole length of the ovary with laces projecting outward from this shaft. This gland is seen to be composed of many follicles which are closely packed together. The wall of the ovary is covered by a thin layer of connective tissue.

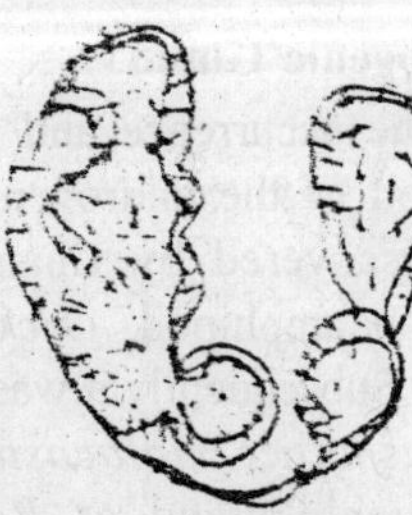

Fig. 12.5.

Generally two types of cells are present in the lobes of the ovary developing oocytes and follicular cells. The developing oocytes have large rounded nuclei with prominent nuclesioli. The cytoplasm of the oocyte is compact and contain many granules without any cacuoles. The cell membrane can be seen dearly in the developed oocyte. Each follicular cell has a thin oval nucleus. The outermost layer of the ovary is the germinal epithelium or the generative zone from which oocytes develop. These oocytes extend further from the gererative zone along the wall of the central shaft and then into the lobes. These oocytes are of various sizes but they have similar nuclei as those of the advanced oocytes within the lobes. The histological structure is essentially the same throughout the length of the ovary and its commissure.

Spermatheca and Oviduct

The spermatheca and the oviduct join with the ovary at the posterior end of the thorax. The oviduct opens to the exterior on the somite of sixth thoracic. The spermatheca measures 10 mm. in height and 8 mm. in width, its gretest development occurs at the time of copulation. It is somewhat roughly oval in shape. The opening between the ovary and spermatheca is located at the level of the sepatation of dorsal and ventral portions of the spermatheca. The ventral part of the spermatheca farrows to form a small oviduct which passes through the pereiopodial musculature of the sixth thoracic segment. The diameter of the oviduct is about 978.0 μ. The oviduct opens outside by two soft ovals in the sternal region. Generally these ovals are 2 to 3 mm. wide.

Histology

In TS it is seen that the wall of the spermatheca consists of two stratified layers. The outer layer consists of columnar epithelial cells. External to the outer stratified layer there is a compact fibrous coating. The fibrous coating possibly contains muscular elements which merge with a fibrous parenchymatous tissue. Extending from this outer stratified layer to the lumen there is a layer of inner stratified muscle. In this series from the compact-layer towards the lumen, the cells are in various stages of degeberation. Many of these cells are displaced and loose in the lumen. The lumen is occupied by large amorphous mass which is granular.

The wall of the oviduct also consists of a layer of columbar epithelial cells which are similar to that of the spermatheca. The whole duct is filled with circular and longitudinal striated muscle layers.

Accessory Structures

In female, the abdomen is broad and all the somites are freely movable. Four pairs of abdominal appendages or pleopods function as an accessory reproductive structure. These appendages are located on the second to fifth abdominal segments. Each appendage is biramus and consists of a basal protopodite from which arise the endopodite and exopodite. Except the first pair of appendage all the three pairs are similar in structure. The exopodite in these three pairs are modified into a elongate fan-like structure having large number of setae on its anteric and posterior borders. The endopodite, on the other hand, is segmented and nears cluster of long setae on the distal borders of each segment. The long setae of the endopodites are the setae to which the ova are attached after ovulation.

The anterior levator originates in the endophragmal skeleton and its long apodeme makes two connections with the dorsal rim of the basi-ischium. One is its tough flexible insertion onto the cuticular plug which secures the breakage-plane; the oilier is an articulation, proximal to the insertion, between the hard cuticle of the apodeme head and a hard projection from the rim of the basi-ischium anterior to the cuticular plug. During normal levation the pull of the muscle is transmitted to the basiischium through this articulation and not through the insertion to the cuticular plug. It is, of course, important that force should not be applied to the plug since this risks autotomy. Both anterior and posterior levator muscles help in this respect sinct particular parts of both muscles, during contraction, keep the abutting faces of the articulation pressed together. By preventing disarticulation, and the consequent transfer of stress to the plug, they act together as a safety catch. The posterior levator is a small muscle originating within the coxa and its apodeme inserts next to that of the anterior levator on the rim of the basi-ischium posterior to the plug. The posterior levator apodeme is set at an angle to that of the anterior levator and the heads of both apodemes are connected; this allows the contraction of the posterior levator to influence the operation of the anterior levator–acting as a safety catch in normal levation and springing the catch during autotomy. Catch-springing occurs as a result of contraction of a special part of the posterior levator which causes the head of the apodeme to rotate and to nudge the head of the anterior levator apodeme. The nudge disarticulates the apodeme and transfers stress to the plug.

The neural side of autotomy complements the mechanical side. Precise control is required and this is supplied by the CNS and the cuticular stress receptor organs. During autotomy a special motor unit is recruited and produces a particularly strong contraction in the anterior levator muscle. The resultant stress is transmitted, as in normal levation, through the articulation. The limb levates as far as possible but the stress is maintained and strains the articulation; the safety catch part of the posterior levator prevents disarticulation at this stage. The strain in the articulation produces mechanical deformation of the cuticular stress receptor which lies just anterior to the cuticular plug. When the strain is sufficiently great the stress receptor stimulates the CNS which responds by inhibiting the safety catch part of the posterior levator and stimulating that part which produces disarticulation. The strong force is suddenly transferred to the plug subjecting it to a strain-shock. This, combined with the elastic recoil of the adjacent articulation, is sufficient to shear the plug and the limb breaks off. It is likely that the sudden imposition of the various stresses is important since there is evidence that materials resist sudden strains less well than gradually applied ones (McVean and Findlay, 1976).

Crabs autotomizc limbs which are badly damaged or seized by an attacker. The mangrove crab *Goniopsis sometimes* responds to a human attacker by pinching hard with a chela and then autotomizing the chelipcd. The chela continues to pinch after autotomy and in one's haste to detach it one often lets the crab escape. Loss of body fluids following autotomy is prevented by an internal partition which remains behind as a seal over the stump. This is soon supplemented by clotted blood which accumulates to form a scab.

Autotomized limbs can only be replaced by molting. Loss of limbs stimulates the onset of the next molt but may decrease the size increment since resources are diverted to the regenerating limbs. Regeneration starts soon after autotomy by the formation of a small limb bud (Adiyodi, 1972). The bud develops during intermoult (stage C) into a tiny limb folded up inside a flexible sack which projects from the stump. During prcmoull (stage D) the bud grows rapidly and when the crab moults its length is about a quarter of the carapace width. Immediately after the moult the regenerated limb unfolds and expands but is still somewhat smaller than the other limbs. It regains its full size after the next moult. In the case of the chelipeds the form of the limb may change during regeneration. If a crab with dimorphic chelae (*e.g. Pilumiius*) loses its major cheliped it usually replaces it at the next moult by changing the morphology of the original minor to that of a major. The change is generally from righthandcd (commonest) to lefthanded crabs. Probably the major chelipcd is more important in feeding than the minor and this strategy results in it being replaced most rapidly. In male *Uca* in which the minor chela is most important in feeding (and is narrower anyway than most limb buds) the opposite situation obtains: a lost major is replaced by a regenerated major on the same side (Crane, 1975).

Molting

Drach showed that ecdysis or casting of the old exoskeleton was only a part, and in terms of duration a very small part, of the full moult cycle. Ecdysis used to be thought of as a regular physical necessity which disrupted normal life for a short time. We now know that physical and physiological changes associated with molting and growth take. place throughout the interval between ecdyses. Ecdysis itself is simply one of the more obvious changes. The realization of this constant state of change led to a classification of stages in the moult cycle. This originated with Drach (1939) but has since been modified by a variety of authors. In the following classification the characters of the exoskeleton protein and lipid, and chitin and protein respectively. Glucose derived from glycogen is polymerized to form chitin. This glucose may come from the glycogen which accumulates in the epidermis, or from the blood glycogen store. The glucose content of lobster blood increases by 35 per cent between C_4 and D_4. Glycogen may also be derived from lipids in the hepatopancreas but this source is probably more important during endocuticle secretion during A and B.

Cuticular resportion is the most important change to take place during the later sub-stages of D. The old cuticle is broken down by the action of chitinase. The first part to go is the membranous layer which degenerates into a gelatinous layer. Exoskeleton resorption proceeds to different degrees in different species and generally more of the organic content than the mineral content is resorbed, for instance, *Carcinus resorbs* 79 per cent of the organic content but only 18 per cent of the inorganic content from the carapace. A considerable amount of the resorbed minerals may be excreted in the urine.

At ecdysis the large volume of sea water which is absorbed (mainly through the gut wall) disturbs the normal ionic balance of the haemolymph. Magnesium in particular is high and this may help to inhibit movement during A_1.

The major depletion of hepatopancreas reserves takes place during A and B. The exoskeleton is too soft to permit feeding and the crab must draw on its reserves to synthesize the endocuticle. This is the main layer of the exoskeleton and is composed of heavily calcified chitin. Much glycogen is required for chitin synthesis and probably also as an energy source during this time of high metabolic activity. This glycogen is derived from the breakdown of hepatopancreas lipids. Mineral reserves are depleted since the endocuticle is mineralized as it is secreted. However, a substantial part of the required inorganic ions are absorbed directly from the surrounding sea water.

When feeding starts again in late B_2 or C_1 the reserves are fully depleted. However, the crab has grown only in the sense that it has increased in size. Its exoskeleton is still thin and its tissues do not fill the space inside. A crab in this state, however large, is not a useful addition to any menu; the hepatopancreas is depleted and the muscles are atrophied and watery. The food intake in the first half of C is therefore utilized in tissue growth and only when this has caught up with the earlier exoskeleton growth does the crab set about accumulating its reserves once more.

Control

The moult cycle is probably largely under hormonal control and is modified via the central nervous system according to both internal and external environmental conditions. Moult initiation, however, is the only part of the control to be well understood. The X-organs—groups of neurosecretory cells situated in each eyestalk produce a moult-inhibiting hormone which passes down the axons to the sinus glands. Removal of both eyestalks, or of the X-organ/sinus gland complex, results in moult initiation; reimplantation of the organ complex delays the moult. The target organs of the moult-inhibiting hormone are the Y-organs which are situated at the front of the body beneath the branchiostegites, one on either side. These, when released from inhibition, produce a molting hormone. The effect of the molting hormone is to initiate premoult stage D. Removal of both eyestalks and Y-organs does not result in moult initiation, but reimplantation of the Y-organs alone leads to a moult. By stage D_2 moult initiation is irreversible; removal of Y-organs at this time does not delay (lie moult and neither is it accelerated by removal of the eyestalks. Thus the X- and Y-organs control the important transition from C_4 to D in the moult cycle. The control of the sequence of events from D_2 back to C_4 is less well understood. Each event up to C_4 may be triggered by the preceding event but it is likely that further hormonal control is involved.

Terminal anecdysis (C_4T) has been explained in terms of the X- and Y-organs by Carlisle (1957) wlio showed that there were different control mechanisms in *Maia* and *Carcinus*. In *Maia* C_4T is the result of degeneration of the Y-organs; eyestalk removal in C_4T does not produce molting. In Curcinus, however, C_4T is achieved by constant release of moult-inhibiting hormone and removal of the eyestalks does produce molting.

This dual control of moult initiation is similar to the situation found in insects except that in the latter group the release of molting hormone is stimulated by a neurosecretion instead of being inhibited. A further similarity, which in this case links all arthropods, is that the chemistry of the molting hormone—ecdysone—is similar in all groups and hormone extracts from one group can be used to initiate molting in other groups (Krishnakumaran and Schneiderman, 1970). The crustacean molting hormone is called crustecdysone.

An important part of the hormonal control of molting is that, since the X-organs are part of the nervous system, the release of moult-inhibiting hormone is under nervous control. This provides a link between the moult cycle and the external and internal environments. The X-organs presumably stop secreting when sufficient metabolic reserves are present, when there are no reproductive duties such as egg carrying and when external conditions are favourable. When conditions are unfavourable, however, the X-organs continue to secrete moultinhibiting hormone. External conditions which tend to inhibit molting are constant light, low temperature and crowding. In temperate regions mature females tend to moult in the autumn; molting is inhibited by cold in the winter and by egg carrying during spring and summer. In mature males molting is more common in the early summer; many species only copulate when the female is soft, just after a moult, and thus the males need to be hard-shelled in the autumn.

Ecdysis

Although of relatively short duration ecdysis is obviously one of the most important events in the moult cycle. Descriptions of ecdysis generally divide it into a passive phase followed by an active phase. The passive phase, which overlaps with D_4, is a time of swelling and splitting while the active phase involves muscular movements culminating in the withdrawal of the crab from the old skeleton. During both phases the crab is usually hidden away in a burrow or under a stone. It generally avoids members of its own species since most crabs have cannibalistic tendencies.

In the passive phase water uptake commences and a split develops from posterior to anterior along the epimeral line of each branchiostegite and between the back of the carapace and the first abdominal segment. The back of the old carapace gradually rises up revealing the new carapace underneath. This generally appears rather wrinkled but brightly coloured. Uptake of water and the consequent increase in haemocoel volume is no doubt responsible for these changes. The pericardial sacs may be important here since they swell progressively during the passive phase and bulge from the branchial cavity under the new carapace (Watson, 1971). It is likely, however, that their main function is to store fluid displaced from the limbs during the active phase (Rao, 1968). Towards the end of the passive phase the epimeral split extends anteriorly almost to the mouth.

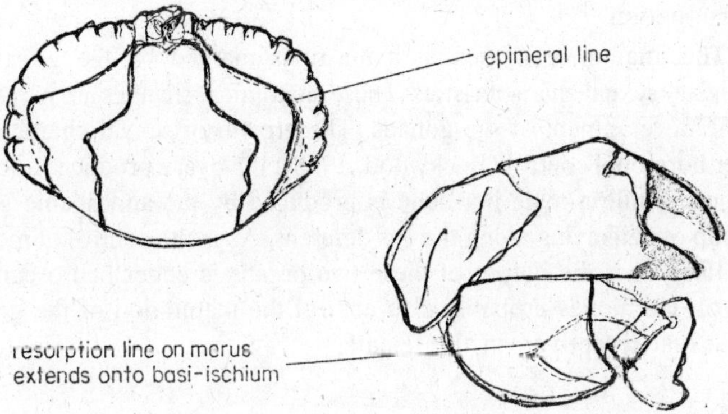

Fig. 12.6 : Ventral views of the carapace and chela of tlie edible crab *Cancer pagurus* showing resorption lines at which splitting occurs during ecdysis.

The active pliase begins when the carapace has been raised 'like the lid of a box hinged at the anterior end of the animal' (Hiatt, 1948) to an angle of about 30°. Contractions of muscles attached to the leg bases commence and serve to loosen the limbs within the old skeleton. The outward sign of this activity is a rhythmic series of depressions appearing and disappearing on alternate sides of the new carapace (to which the leg muscles are ultimately attached). In *Pachygrapsus* each contraction lasts about 10 seconds (Hiatt, 1948) but in the much larger snow crab *Chionoecetes* opilio from northeast Canadian waters, each contraction lasts about 55 seconds (Watson, 1971). This loosening activity persists for a few minutes (Pachygrapsus) or a few hours (Chionoecetes). Eventually, however, beginning at the posterior end, the limbs are gradually withdrawn from the old skeleton. The pericardial sacs which are maximally swollen during the early part of the active phase now deflate as the fluid is redistributed to the emerging limbs. The crab then withdraws quite rapidly from the old skeleton, emerging backwards. The gelatinous layer between the old and new skeletons may lubricate withdrawal which has to be quick since as the gills are withdrawn the respiratory mechanism is temporarily put out of action. The chelipeds with their swollen distal ends present special problems but these are overcome by premoult breakdown of the cheliped muscle (Skinner, 1966) and by selective resorption and splitting along fracture lines on the basal podomeres which allow the bulky propodus to be withdrawn. All parts of the old skeleton which have not been resorbed are left behind with the cast shell endophragmal skeleton, apodemcs, gastric mill, etc.

Reproduction

The attainment of sexual maturity is marked by the growth of secondary sexual characteristics. These anatomical changes are paralleled by the development of the gonads. The growth of sexual characters is under hormonal control (Lockwood, 1968); the ovary produces a female hormone while a male hormone is produced by the androgenic gland, a group of cells lying along the vas deferens. As in the control of molting it is likely that the output of these two organs is under neurosecretory control. Hormones probably also control the maturation of the gonads and sexual receptivity in the female.

Eggs

Females which become sexually receptive when hard shelled generally lay their eggs shortly after copulation. However, those which copulate just after molting are in a depleted state and do not generally ovuiate for some time. Sperm is stored by the female in the spermathecae; it can be retained through a moult and remains viable for a long time. Egg carrying is seasonal in crabs from temperate regions and tends to occur in spring and summer. In mature female *Cancer pagurus* ecdysis and copulation occur in September but egg laying does not take place until December of the following year, 14 months later. The eggs are fertilized as they are laid since they pass through a spermatheca on their way out. New laid eggs are yolky and orange or khaki in colour; they adhere to the hairs on the pleopods and are retained and brooded in a mass under the broad abdomen of the female. Such females are described as ovigerous or berried. In *Cancer* the eggs hatch in July having been brooded for six months; a second batch may be laid the following December before the next ecdysis and copulation.

In tropical crabs breeding is often continuous throughout the year but may show a lunar rhythm in which egg hatching coincides with full or new moon. In the mangrove tree crab *Aratus pisoni* the time from laying to hatching is only 16 days; more than one egg batch may be laid between consecutive ecdyses and an average of about six batches is laid per year. In terms of egg batches per year and the length of time taken for the eggs to develop, *Cancer* (large, temperate) and *Aratus* (small, tropical) probably represent two extremes.

The number of eggs in a batch is generally very large and varies with the size of the crab. *Aratus* of 16 mm carapace width lays about 5000 eggs and a 36 mm *Pachygrapsus* lays about 48000 eggs. Large crabs like *Cancer pagurus* lay up to three million eggs when full grown. The size of a newlaid egg, however, is very small; they are generally slightly oval in shape with a maximum diameter of between 0-25 and 0-35 mm. The eggs swell as they develop so that by the time they are ready to hatch they are roughly double their new-laid volume. During development the colour of the egg changes through brown to grey as the yolk is used up and the outline of the embryo becomes visible. The eyes and pigment spots appear first, followed by the outlines of the abdomen and cephalothorax.

When the eggs hatch the larvae are released from beneath the abdomen of the female. Generally the female assists this process by rafting the abdomen to and fro and by digging into the egg mass with the chelae. In *Aratus* the female climbs down a mangrove root into the water and rapidly vibrates the abdomen back and forth releasing a cloud of larvae. In the northwest Atlantic spider crab *Libinia emarginata* unusual egg hatching behaviour involving the male has been described. Male *Libinia* can detect the presence of a female about to release larvae. They become very aggressive and fight for the possession of such a female. When a male has captured the female he positions her behind him with the backs of their carapaces touching and clasps his last pair of walking legs beneath her abdomen. Holding her in this 'obstetric' position he backs into a protected area and while the female releases the larvae by fanning her abdomen he drives away any other crabs that approach. After the larvae are liberated the female cleans the empty egg cases from her pleopods and the male then releases her.

Larvae

The larval life of a crab is generally spent in the plankton. It may last from one to several months and during this time the larva moults several times before settling on the sea bed and molting to the first crab stage. There are two larval forms, the zoea and the megalopa (or, in hermit crabs, the glaucothoe). There are between 2 and 6 (usually 4) zoeal stages and these are generally followed by one megalopa or glaucothoe stage.

Structure and Development

The zoea larva hatches in an embryonic form as a protozoa prezoea but this quickly moults to the first zoea. These larvae are tiny shrimp-like animals, they have a full complement of functional head appendages but only the first two thoracic appendages, the 1st and 2nd maxillipeds, are developed and functional; the rest of the appendages are absent or present only as limb buds. The head appendages all appear to function as they do in the adult but the maxillipeds are used for swimming: the exopods bear long plumose setae and the zoea beats these against the water to achieve fairly fast but jerky progress.

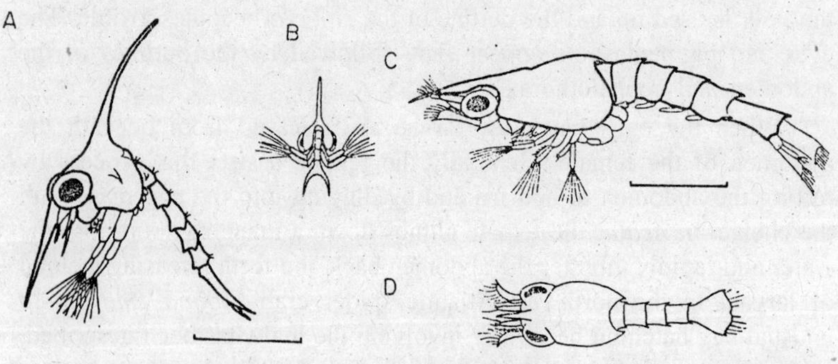

Fig. 12.7 : Zoea larvae. A, side view of the third zoea of the swimming crab *Macropipus puber.* B. posterior view of the first zoea of *Carcinus maenas* C, side view of the fifth zoea of the West Indian land hermit crab *Coenobita clypeatus.* D, dorsal view of the first zoea of the hermit crab Pagunis *bernhardits.* A and B after Lebour, 1928: C after Provenzano, 1962: D after MacDonald a at.. 1957. Scales 1 mm.

Figure shows that hermit crab zoeae are more shrimp-like than brachyuran zoeae; this difference may reflect the greater degree of evolutionary change in the Brachyura away from the common ancestor Further differences lie in the development of spines on the carapace and in the shape of the telson. In typical brachyuran zoeae the carapace bears a ventrally pointing rostal spine and a tall dorsal spine; lateral spines are also frequent. In the Anomura spine development is more variable. Hermit crab larvae and stone crab larvae (Lithodidae) usually have short spines and often only the rostral spine is developed; the zoeae of porcelain crabs and of squat lobsters have rather long carapace spines. However, in contrast to brachyuran larvae, anomuran larval spines are always orientated antero-posteriorly and contribute to the shrimp-like shape. The telson in brachyuran zoeae is usually forked whereas in anomuran zoeae it is often spade shaped and forms a tail fan. A list of the differences between brachyuran and anomuran zoeae was given by Williamson (1965).

At each moult the zoea grows and develops by adding setae to the existing appendages and adding appendages to the posterior half of the body. Addition of swimming setae to the exopods of the maxillipeds occurs in quite a regular fashion in brachyuran zoeae and can be used to 'stage' larvae. First zoeae almost always have 4 swimming setae on each maxilliped, this increases to 6 in the second zoea, usually 8 in the third zoea, and 9 or 10 in the fourth zoea. The development of other appendages in brachyuran zoeae is generally restricted to limb and pleopod buds. These usually appear first in the penultimate zoeal stage and by the last zoea (or metazoea, usually the fourth) they may bear a few hairs but are still non-functional. In contrast, in anomuran zoeae two other functional appendages are developed: 3rd maxillipeds used in swimming appear in the second zoea and functional uropods forming a wide tail fan with the telson are developed by the third zoea.

The megalopas and glaucothoes resemble the adult forms more closely than do the zoeae; indeed, they are often referred to as post- larval stages. There is a reduction in carapace armature and a complete set of functional appendages including chelipeds and walking legs is present. Locomotion is still largely by swimming but this is achieved by the pleopods borne on the extended abdomen and not by the maxillipeds. An interesting difference, apart from shape, between the megalopa and the glaucothoe is the presence in the latter of well developed uropods forming a tail fan with the telson; in the megalopa the appendages of the last abdominal segment take the form of pleopods and there is no tail fan.

Swimming

Zoeae usually swim backwards propelled by the anteriorly directed beating of the maxillipeds. Brachyuran zoeae swim head downwards with the dorsal spine foremost and the abdomen trailing.. Anomuran zoeae, however, generally swim telson first with the abdomen extended and usually with the dorsal side uppermost. The direction of swimming, therefore, is along the axis of the spines in both types of larvae. The speed of swimming is generally l-2cm/s. Reversals or escape movements which occur on collision are effected in anomuran zoeae by tail-flips using the tail fan. In brachyuran zoeae, which lack a tail fan, there is no clear escape movement. When the larvae stop swimming they sink and the rate of sinking depends on the development of spines. A narcotized zoea of the porcelain crab *Porcellana sank* in a horizontal posture at a rate of about 0-3cm/s; *Porcellana* zoeae are characterized by the presence of enormously long carapace spines extending posteriorly and anteriorly. Narcotized brachyuran zoeae, however, with their globular carapaces and shorter spines, sank head first at a rate of about l-2cm/s. Thus, one function of long spines may be to aid flotation and, indeed, anomuran larvae possessing long spines (Porcellanidae and Galatheidae) frequently rest in a horizontal position hardly sinking at all. Resting is rarely observed in brachyuran zoeae which remain in the same place by swimming constantly upwards. It is possible that short-spined zocae gain the advantage of manoeuvrability; *Porcellana* larvae had difficulty in turning but could change direction more quickly when the rostal spine was cut off.

The direction of swimming seems to involve responses to both light and gravity. Zoeae swim towards a light source and can perceive both direction and intensity of light. In the dark the normal swimming orientation is maintained indicating perception of gravity. However, in brachyuran zoeae control is lost in the dark (but not in the light) if the carapace spines are cut off. Probably the spines bear hairs sensitive to water movement. These might distinguish relative water movements due to sinking and thus 'up' and 'down' could be perceived. Similarly differential water movements on either side of the spines during swimming could give directional information. In the light, visual cues could supplement or replace the information from the spines.

The megalopa and glaucothoe swim forwards with the dorsal side uppermost. The propulsive force is produced by the pleopods which beat backwards. In the megalopa during swimming the first two pairs of walking legs are tucked up underneath the body while the last two pairs are held above the carapace, the 3rd pair being extended forwards over the eyes. This disposition of the legs helps to streamline the animal. The megalopas and glaucothoes are the settlement stages and after a time swimming in the plankton they adopt a benthic existence. The glaucothoes take up residence in tiny empty gastropod shells while the megalopas walk about with their abdomens flexed under their bodies.

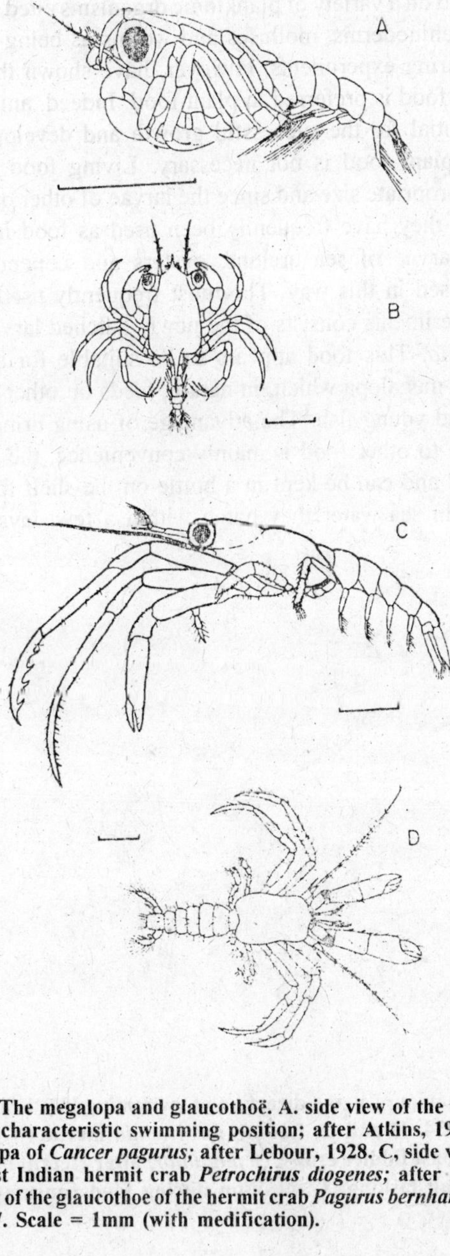

Fig. 12.8 : The megalopa and glaucothoë. A. side view of the megalopa of *Carcinus maenas* in characteristic swimming position; after Atkins, 1954. B. dorsal view of the megalopa of *Cancer pagurus;* after Lebour, 1928. C, side view of the glaucothoe of the West Indian hermit crab *Petrochirus diogenes;* after Provenzano, 1968. D, dorsal view of the glaucothoe of the hermit crab *Pagurus bernhardus;* after MacDonald *et al.*, 1957. Scale = 1mm (with medification).

Feeding

Zoea larvae feed on a variety of planktonic organisms cited. Diatoms and the larvae of echinoderms, molluscs and worms as being common food in nature. Rearing experiments, however, have shown that living and moving animal food is preferred to plant food. Indeed, animal food appears to be essential for the successful growth and development of crab larvae while plant food is not necessary. Living food must, of course, be of an appropriate size and since the larvae of other organisms fit this requirement they have frequently been used as food in rearing experiments. The larvae of sea urchins, oysters and copepods have successfully been used in this way. The most frequently used food in modern rearing experiments consists of the newly hatched larvae of the brine shrimp *Artemia.* This food appears to be suitable for all larval stages including the megalopa which, in nature, feeds on other decapod larvae, copepods and young fish. The advantage of using brine shrimp larvae in preference to other food is mainly convenience; the eggs are remarkably resistant and can be kept in a bottle on the shelf for a long time; when placed in sea water they hatch within a few days.

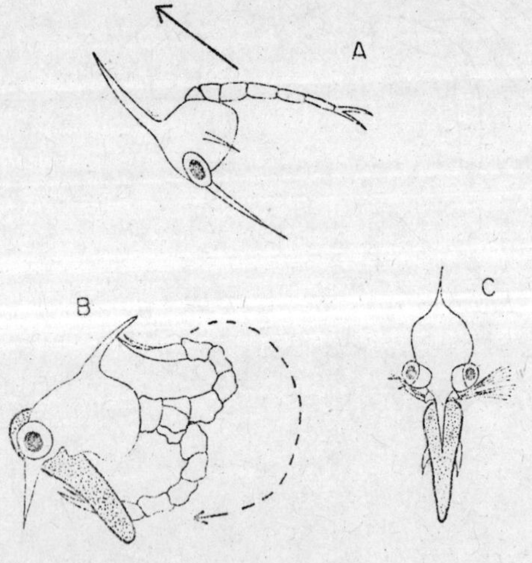

Fig. 12.9 : Larval behaviour. A. the direction of swimming in a brachvuran zoea; after Foxon, 1934. II and C, feeding on *Artemia* larvae by the zoea of an American east coast fiddler crab *Uca pugilator;* after Hcrrnkind. 1968. B, food capture by tail lashing; C. position during food consumption.

It appears that food is taken by individual acts of capture. This has been described in *Uca* zoeae: the prey, *Artemia* larvae, were captured by lashing the abdomen backwards and forwards. An individual *Artemia* larva is pinned between the forked telson and the rostral spine of the zoea and adjusted until the anterior end is in contact with the mandibles, it is then consumed. Tail-lashing was stimulated by the presence of *Artemia* larvae and can, therefore, be regarded as a behaviour particularly associated with feeding. Objects other than *Artemia* larvae which were captured by tail-lashing (other zoeae, sea urchin larvae) were rejected immediately indicating some sensory discrimination in feeding.

Megalopa larvae seize their prey with their chelae which often have hooked tips to prevent escape. Some glauchothoes capture prey with their long 1st and 2nd pairs of walking legs which stretch out anteriorly during swimming; in these forms the legs may be armed with teeth with which to grasp and hold the prey.

Effects of Temperature, Salinity and Starvation

Temperature, salinity and the availability of food are three environmental variables which have received study during laboratory rearing experiments with crab larvae. Like most animals, crab larvae are able to survive under a range of environmental conditions; however, for each species there is generally an optimum area in the range of each variable within which survival, development, growth, etc. are at a maximum. These variables do not, however, act independently on survival; they are all part of the same environment and they tend to interact such that combinations of sub-optimal conditions usually, but not always, reduce survival to a greater degree than a single sub-optimal condition would do.

Within the temperature tolerance range, low temperature lends to slow development and prolongs the larval life. In the natural environment this probably leads to increased mortality from the chance causes of predation, disease, etc. A long larval life, however, should also lead to increased dispersal which, as argued below, may be advantageous. Conversely, high temperatures give rapid development and may lead to increased survival and less dispersal. In the case of low salinity it has been shown that the later larval stages are more tolerant than are the early larvae. This is a necessity for estuarine crabs in which the later larval stages must search out the habitat of the adults. The effects of starvation on crab larvae have shown up a partial independence of molting and growth. When not enough food is available the larvae may continue to moult but may not grow and the larval life is prolonged. This may result in extra zoeal stages since it appears that metamorphosis to the megalopa or glaucothoe depends on the attainment of a certain bulk.

Combinations of temperature and low food show up the interactions of sub-optimal conditions. At low temperature, low food further prolongs the larval life. At high temperatures, with their attendant high metabolic rates, low food may produce considerable mortality from starvation.

Reduction in Larval Life

In various groups of crabs there is a reduction in the number of larval stages or in the length of the larval life. In spider crabs, for instance, only two zoeal stages are found. This may lead, in tropical species, to a very short larval life of less than two weeks. In cooler regions, however, a larval life of about one month is more common even with only two zoeae.

Complete suppression of the larval life is very rare in marine crabs; it occurs in a few genera of the Dromiidae and Maiidae. In the fresh water Potamidae, however, direct development is the rule. The eggs are large and yolky and relatively few in number: a female potamid of 100mm carapace width might carry about 200 eggs. There are no free larval stages and the eggs hatch directly into young crabs. Hatchlings of the polamid Kingsleyu are about 6mm in carapace width; they are called 'marsupial young' since they remain under their mother's abdomen for a time before venturing out on their own. Some fresh water grapsids also have reduced larval lives. One of these is the bromeliad crab *Meteipauliis clepressus* which lives in the water trapped in the leaf axis of epiphytic bromcliads in the mountain forests of Jamaica. This crab lays relatively few large yolky eggs—about 100, each with a diameter of 1-5 mm, laid by a female of 18 m carapace width. There are two non-feeding zoeal stages lasting, together, for three days, and a megalopa lasting for six days. The suppression of the larval life in fresh water crabs has presumably evolved because fresh water is unsuitable for the dispersal of larvae. The same applies to land, but the land crabs of the family Gecarcinidae and the land hermit crabs have retained a full larval life. Land crabs, when the eggs are ready to hatch, may migrate considerable distances at night to liberate their larvae in the sea.

Function of Larvae

Rough calculations of survival to various stages in the life cycle of the mangrove tree crab *Aratus* gave 0-041 per cent survival from egg to newly settled crab and 17 per cent survival from newly settled crab to an adult of 18mm carapace width. Thus, considerable mortality occurs during the larval life and this is almost certainly true for all marine animals with planktonic larvae. It is likely, therefore, that there is some particular advantage to be gained by releasing the young as an enormous number of tiny larvae rather than as a small number of young crabs. The existence of the latter state in potamids, although in this case an adaptation to fresh water life, shows that crabs are capable of evolving this system.

One advantage that is gained by species with a long planktonic larval life is dispersal. Larvae which drift in tidal and ocean currents are likely to settle some considerable distance away from their parents and may establish new populations in areas in which their species did not previously occur. Similarly, if a local population were to be wiped out by a temporary catastrophe-such as a very cold winter-rapid recolonization from the plankton could occur. A difficulty attendant on dispersed it that a proportion of surviving larvae maybe transported to unfavourable environment and may not survive long after setting. Delayed metamorphosis occur till favourable conditions. The ability of megalopa and glaucothoe to survive both in plankton and on the sea bed indicates that some selection of environment before metamorphosis may also occur in crab. The huge mortality that occurs during larval life is compensated for by vast numbers of larvae produced the relatively long length of life in most crab species also help to reduce the danger.

Crayfish Reproductive System

To display the gonad it is necessary to remove the heart and the tough, membranous floor of the pericardial sinus. In the female A care is necessary not to damage the thin-walled oviducts which lie immediately below the pericardial floor and which may be confused with the transverse bands of fibers in that floor.

In the crayfish there is a single, Y-shaped gonad, very small in immature specimens, becoming much larger at maturity, particularly in the female. The paired lobes of the gonad extend forwards on either side of the stomach, while the unpaired lobe may extend back a short distance into the abdomen. In a male the testis often resembles the hepatopancreas (q.v.) in its whitish color and soft consistency, and thus, rather easily becomes indistinguishable if at all squashed during dissection. In some species of crayfishes the three distal extremities of the testis have the form of little, bulbous swellings, of which the posterior one sometimes is found twisted below the intestine.

In the lobster the ovary is H-shaped, the short cross bar being placed just in front of the pericardial sinus. The paired longitudinal parts may extend two-thirds the length of the animal.*

The testis of the lobster also is H-shaped, or the cross bar may be lacking so that there are separate paired testes.

The oviducts are paired and start below the pericardial sinus, each passing outwards to the body wall, inside of which it then runs downwards to open on the inner aspect of the coxa of the third pereiopod. Each duct is wide and ribbon-like as it leaves the ovary, but quickly tapers to a small cylindrical tube. Often the ducts are overlooked because they are thin-walled and transparent. But at the time of egg-laying the walls actually become quite glandular, contributing an albuminous or mucilaginous secretion.

The vasa deferentia are paired and leave the gonad at the same level as do the oviducts. In the crayfish each Vas deferens is a small, coiled, relatively thick-walled tube, often pinkish in color, which opens at the tip of a flexible papilla on the inner surface of the coxa of the fifth pereiopod. Usually, the coils are especially numerous in the upper part of the duct, and care is needed not. to be deceived as to the point where the duct leaves the testis. The walls of the duct secrete a gelatinous material which surrounds the sperm as it passes out. In the lobster the duct first runs outwards from the testis to the body wall, then curves ventrally and thickens along its dorsoposterior margin. This thickened part of the duct is glandular and coils in a S-shaped fashion. It is followed by a small swelling, representing a sphincter, and this in turn by a thinner

---

* In fresh lobsters the color of the ovary provides interesting evidence regarding the reproductive history of the individual. For 4 or 5 years the ovary slowly matures, going through color phases of bright yellow, flesh color, salmon, light olive green, and a, rich dark green. Copulation then occurs (immediately after a molt), and 1 or 2 months later the eggs are laid. The ovary now is grayish while with green flecks (the remains of eggs that failed to be discharged). After 5 or 6 weeks the ovary turns light green and the flecks become, orange. The eggs which were laid are retained on the pleopods for about 11 months, at the end of which time the larvae hatch. Shortly after that there is another molt and copulation again occurs. Meanwhile the shade of green of the ovary steadily darkens, until a second batch of eggs is laid, 2 years after the first, laying. The ovary is again grayish white for a few weeks, but this time has both orange flecks and green flecks (remains of ova which failed to be discharged at the first and second layings respectively).

ductus ejaculatorius. Microscopic examination of some of the contents of the vas deferens may reveal some of the peculiar spermatozoa, small rounded cells, from the periphery of each of which radiate a number of long, fine, curved processes.

A seminal receptacle is present between the bases of the fourth and fifth pairs of perciopods in the female of the lobster and of certain crayfish genera. This is not an internal structure but is simply a small pocket in the exoskeleton. Frequently, it is termed the annulus. It is thought to have evolved independently in the lobster and in crayfishes respectively.

During copulation the male holds the female on her back by means of his chelipods and other periopods aided in some crayfishes by "the spiny projections on certain of the ischia. The copulatory pleopods, which normally lie forwards tightly pressed against the ventral surface of the thorax, are lowered to about a 45° angle, and then are held in that position by one of the fifth pereiopods carefully crossed under the body. The flexible papillae, on which the vasa deferentia open, fit into the grooves of the first pleopods, and issuing sperm, surrounded by a gelatinous secretion, flows along these grooves. If there is a seminal receptacle, the tips of the pleopods are inserted into this and the sperm passes in, being securely retained there by the gelatinous secretion which hardens into a plug. Where there is no seminal receptacle, the sperm is deposited in packets, termed spermatophores, on various parts of the female's body, but particularly between the bases of the last two pairs of pereiopod. In crayfishes copulation occurs when both male and female have completely hardened exoskeletons, but in the lobster copulation always occurs shortly after a female molt, when her exoskeleton is soft or at most of a leathery consistency. In many species of crayfishes the first pleopods of the mature males alternate between two forms at successive molts, only one of these forms being adapted for copulation.

Egg-laying occurs quite some time after copulation: 10-45 days later in some crayfishes without any seminal receptacle, a few days to several months later in crayfishes with a seminal receptacle, and up to a year or more later in the lobster. The sperm remain viable all this time, and indeed have been found to survive at least 2 years in the receptacle of the lobster. During egg-laying the female lies on her back, with the abdomen curled forwards so tightly that the space in which the pleopods occur is closed except for an opening in front. The eggs emerge from the oviducts in a mucilaginous secretion and are carried back into the temporary brood chamber by a current of water caused by the beating of the second and perhaps other pairs of pleopods. The ova are fertilized by sperm released by an unknown mechanism from the receptacle or from spermatophorcs, and are attached to the pleopods by a secretion which hardens into a cement. After some weeks or months, during which the developing embryos are continually supplied with fresh water by movements of the pleopods, the young hatch, as schizopod larvae in the case of the lobster, or as miniatures of the adult in the case of a crayfish. Newly hatched crayfish remain attached to the maternal pleopods for a further short period of time. Molting of ovigerous females is specially delayed until after the release of the young.

Reproduction

The mature of reprocuction in lobsters is extremly varried. Copulation generally occurs head to head and belly to belly, but sperm deposition may be internal. Eggs are carried from 1-11 months cemented to pelepods under the tail and fecundity varieds from few 100 to two thousand egg /♀ some palmonids spawn more than once a year. Given with in a species size at the several naturally varies greatly with geographic location.

Secondary Sexual Characters

In the *nephropidae* the sexes can easily be distinghuished externallly by defferences in the first *pair of pleopods,* the presence of *seminal reciptical* in the female and the size defferences in the claws and abdomen of mature animals the first pleopods of the male are *selender* and *rigid* while the first pleopods of female are divilev to the other but *greally reduced* in size. First two pairs of pleopods in (♂)and (♀) nephros shows that *claws* of *adult* (♂)are considerably larger than those of females this diapnerphrume is probably the evolutionary result of inter male, comptetion for females as has been hypothetised for fresh water crayfish. The abdomen of adult females broods than that of males bulge to abdomen segments which are display outward to accomodate the large number of eggs. Female clawed loboster have non plumes (Ovigerous) settle on the endopods of the pleopods and on the external ridges of abdomen the spines on the sternal bars of the abdomen are sharp in males and blunt in females.

Internal Reproduction Organ

Female : The ovaries confirm to the general plecopod plan being paired cylimdrical rods dorsal and lateral to the alimentary canal united *by transveres* bridge underneath the heart they extend from anterior end of stomach at about the level cyes. The insertion of abdomen of beyond and a thin development of the ovaries have very general plan a shortduct connect to *the ovary with genital* apeture the coxa of 3rd leg. The immature vary is *creamy white in colour* which changes to green in *Homarus and Nephrose nervigicées to* organge and latered in *gases and Penulirus,* the royal blue in *Metanephros andomanicous.*

Males : The testes are similar species in all *i.e.,* paired white tubes connected by a transverse. Commissure in the middle beneath the heart they extend from the anterior end of the foregut to the posterior margin of the cephalothoxas. The convulated vasa deferentia arise mid may on the posterior hornes of the testies and rim to the gonopore. On the coxae of the fifth legs the Vas defrens can be sub divided into proximal portion which recives the sperm from the testes and conducts it to an enlarged glandular section this second section secrets a gelationous matrix that evelops the sperm the terminal ejaculation done by a sphincter muscle forces the spermatophores out of gonopore the spermatophores stored in ejaculatory region until copulation attched to exterior of distal portion of the Vas deferens is androgenic gland. The secretion of this gland androgenic hormone controls the expression of primary and secondary sexval character in male lobsters.

Reproduction cycle : In many but not all species of lobsters mating must occur shortly after the female molts. During the soft shelled condition this imposses a cyclic nature to reproduction behaviour of the female. The female carries the eggs externally and cann't molt during the time with out loss of eggs. The molt cycle and ovarian cycle are totally intervened males, on the other hand apperar to able to copulate succesfully at any time of the year and spermatogenesis occurs continuously after maturity in *Nephrops norvegicus* and *Jasus.*

Most of the sping lobster been eggs and are said to be in berry from once or twice in a yera *Homarus* spawns only every other year after reaching maturity and *Nephrops* spawns every year the molting hormone. MIH and MH they are responsible for additionally seasonal factor, temperature, photoperiod are implicated the pigment from reabsorping ovaries colour the blood a light green some times when the lobsters are held in very adverse condition the reabsoption occur at a later time in the cycle and heamolymph turns very dark green in a so called black lobster.

Maturity : Males sexual variety has not been reliabaly assess by external characteristics. The presence of mature one spermatozoa in testes is a good indicator of physiological maturity. In sping lobster the sudden lengthning of second and third walking legs has been suggested as reliable indicator of maturity. External characteristics other than presence of use to determine maturity in females. The abdominal increases with the onset of the maturity. the things are associated with change in the environment, the temperature etc.

Mating and Fertilization : In nephropidae female mate only in soft shelled state after moltings in *Homarus americanas* and possibly in other clawed lobster. It appears that a sexpheromone is released by the female at or before molting. Copulation occurs shortly after molting when the female is soft and pair may remain together for several days afterward. A report on intermolt (hard shelled) mating in *H. americanas* is surprising in *Penulirus* the female is not usually soft shelled when mated although there is a partial syhchrony of molting and mating. In *Panulirus* and *Palinurus* the spermatophore is placed externally on the sternal plate of female. They have maturity is gelatinous and relatively short lived. In semination the physical implantation of sperm packet or on in the female must be distinguished from fertilization. Since the two may be widely seperated in time. Just prior to female uses the modified 5th legs to open the spermatophores and release the eggs flow from oviduct over the newly open sperm mass they are fertilized in a brood chamber form between the extended uropods at the function of the telson and inner uropod. They are apperently drawn into the brood chamber by current by exopodite of the posterior pleeorod which beat systematically forwards the telson. In sleeper lobster Sylleridae the fertilization of eggs is usually external but Lyons (1970) suggested that it is internal in syllarids and *Syllarus*. No extenal spermatophoric mass was detected.

Feeundity, Incubation and Matching : Fecundity is generaly estimated as number of egges carries externally it varies between species and among the size classes. *Homarus americanas* 5000–11,5000 *Jasus* 3,85,000–1,940,000. In general spiny lobster shorter carring time more eggs and more broods/year than clawed lobster. The larval stages of spiny lobster last much longer and are thus more liable to natural mortality and including predation egg loss other the incubation period affects the size of brood hatched. The relative reproductive contribution of very large female was deminished by small proportion that they comprise of population although the breed to four times per season and carry more eggs than their similar counterpart.

Incubation depend upon water quality, temperature, the nephropid lobster have long period for egg (6-11 month) and short (3-6 week) larval stage while just reverse is true for spiny and sleeper lobster. The typical development period for *H. americanas, H. gamarus* is (9-10 months) in eggs (3-6 weak) in larval stage in *Nephrons* incubation period 6-10 month larval duration. A spiny lobster from deep cold water long incubation period of atleast and months while warmer water species *P. cyznus* carry eggs for 3-6 weeks. In contrast to clawed lobster spiny lobster and sleeper lobster larval remain or plankton for the 3-11 months.

The macrura might be roughly divided into prawn like forms, active swimmers, with a relative this shell and the more heavily armoured lobsters and their kindered which crawl on bottom of the sea. Of the prawn like forms one large group is tribe penacidea, which comprises most of the larger larger prawns of tropical seas. These may generally be recognized by having the first three pairs of legs provided with pincers claw (Chelae) and nearly alike in size. Many of the larger species of *Penaeus* which are common in the warmer seas and may reach the dimensions of small lobsters and are used for food. This fishery is of considerable importance especially in India, Japan and the Southern United States. *Penaeus caramote* is a common mediterranean species highly esteemed for the table.

Palaemon (Prawn) is found in Indian rivers which have sufficient lime salts for their exoskeleton the body is bluish-green in colour, it is elongate and carapace of cephalothorax and an abdomen which is often turned downwards. Cephalothorax is some what cylindrical and shows no joints, the abdomen is compressed laterally and is six jointed. The body is covered with a stout cuticle which forms an exoskeleton, it is hardened by deposits of $CaCO_3$ and has various pigments.

Reproductive System

Prawn is dioecious as are all Arthropoda with a few exceptions. The male is larger and has bigger second pair of chelate legs than in the female.

Male Organs

Their is a pair of testes lying below the heart above the hepatopancreas. The two testes are jointed together anteriorly; they are made of many lobules. From each testis arises a convoluted Vas deferens which passes downwards and dilates terminally into a seminal vesicle in which bundles of spermatozoa are stored. The vas deferens opens bya gential aperarture on the arthrodial membrane on the fifth walking leg. The endopodite of each second pleopod has an appendix masculine by whch spermatophores are transferred to the female.

Female Organs

There is a pair of ovaries below the heart and above the hepato-pancreas. Two ovaries have their ends closely approximated. From each ovary arises an oviduct having a broad funnel at its commencement. The oviduct runs downwards and opens by a gential aperature on the coxa of third walking leg.

Gonads are hollow enclosing remains of coelom, they are continous with their ducts because of the reduction of coelom; the vasdeferens and oviduct are coelomo ducts.

A glutinous substance is secreted by tegumental glands of female which makes the legs stick to the pleopods. The appendices interna of two sides get interlocked for carrying eggs by the female . The eggs are large with much yolk forming a central core, the yolk is surrounded by peripheral protoplasm; such eggs are called centrolecithal. In copulation a male puts the female down on her back and deposits spermatophores on the ventral surface of the female where fertilizaion occurs.

Development of Prawn

The development of prawn resembles that of *Astacus*. The eggs are big with a large amount of central yolk and a layer of protoplasm all around, and egg is centrolecithal; it has a surrounding egg membrane, the nucleus lies in the middle of the yolk with a small amount of cytoplasm around it. Segmentation of the zygote is meroblastic, only the peipheral cytoplasm being divided. The nucleus of the zygote divides but the cytoplasm does not, so that an uncleaved egg is formed which has a large number of syncitial nuclei in the center, gradually these nuclei migrate to the periphery where each nucleus gets surrounded by some cytoplasm and cell membranes but does not extend into the yolk. Thus a morula is formed which is multinucleate. The peripheral layer of cells called blastoderm enclosed the yolk. The yolk gets segmented forming radiating yolk pyramids, but soon these fuse into one mass of yolk. Thus a *solid blastula* is formed, its blasotocoel being filled with yolk. Future endoderm ectoderm and mesoderm can be distinguished in the *blastoderm*. The blastoderm cells increase on one side and become columnar, they from a ventral plate. The ventral plate acquires five thickenings, there are two head lobes from which eyes and brain will develop, two thoracio-abdominal rudiments, and one endoderm disc.

The blastoderm cells invaginate or sink in the endoderm disc to from *archenteron* cavity having an external, opening or blastophore, thus a gastrula is formed, its blastoderm is the ectoderm, cells linning the *archenteron* are the endodermal cells, the space between ectoderm and endoderm is filled with yolk. The endoderm buds off cells near the blastophore which from the mesoderm lying between the ectoderm and endoderm. The blastophore closes to shut off the *archenteron.*

The thoraxico-abdominal rudiments unite and anterior to these three pairs of ectodermal elevations arise into which mesoderm grows to form rudiments of antennules, antennae and mandibles. This embryo is now known as in a nauplius stage which is still an egg like.

Depressions of the ectoderm from the *stomodaeum* and *proctodaeum* and their openings are the mouth and the anus. The *stomodaeum* and *proetodaeum* opens after a time into the *archenteron* to form the enteric canal. Endoderm cells of the enteric canal grow radially outwards ingesting the yolk, till they become long columns touching the ectoderm. Thoracicoabdominal rudiments grow into a forked process which bends forward lying below the egg. Eyes stalks and a labrum are formed, a carapace appears around the thoracic region. Paired maxillulae, maxillae, three pairs of maxillipedes and five pairs of walking legs are formed. Cephalothoracic region becomes large and globular but the abdomen is small and bent under the cephalothorax, rudiments of abdominal appendages and a telson are formed. The embryo hatches from the egg but clings to the mother by chelae of first pair of walking legs and also by several threads secreted along the telson. The time taken upto hatching of the embryo is about six weeks.

The germ layers begun to form the organs, the ectoderm forms the epidermis, covering gills, nervous system, sensory parts of eyes and statocysts and the lining of *stomodaeum* and *proctodaeum*, endoderm forms the epithelium of the midgut and hepatopancreas; the mesoderm gives rise to muscles, connecting tissue, vascular system, gonads and parts of excretory organs . All this time yolk is progressively used up. The embryo molts ones to become independent of the mother.then by successive molts it assumes the adult form and size. Molting is under the control of a hormone produced by sinus gland located in each eye stalk, the hormone has an inhibitory effect on molting.

The female lays eggs 6 to 20 hours after mating and one whole batch of eggs is laid in 20 minutes. The laid eggs are held with pleopods and are considered to be fertilized at the time of extrusion. Unmated ripe females also lays eggs within 24 hours after the molt, but eggs are not fertilized and drop off in 2 or 3 days. The incubation period is 19 days at room temperature of 26°C to 28°C. The eggs change colour from yellow to stale grey as development proceeds.

Larvae pass through eleven molts but only eight morphological stages can be recognized distinctly. The last larval stage emerges 30 to 40 days after hatching. The growth rate is fairly even during the three larval stages, but later becomes irregular. Juveniles emerge when the last stage larvae metamorphosis. They cease their pelagic life, sink down and settle on the bottom as crawlers then cling to vegetation and submerged objects. They feed on small worms, crusaceans, insects, larvae as well as on organic matter of animal and plant origin. They rapidly molt every 5 to 10 days and attain a length of 5 cm. in 12 months.

In nature juveniles remain in brackish water for one or two weeks and then shortly migrate upstream into water of less salinity. When two months old they can swim against the current. They climb over bunds or dams 2 to 3 feet high if water drops over the dam (bunds). During rainy season they migrate through temporary streamlets into ponds, paddy fields.

They are found to cover 200 Km by migration

Regeneration

Crustacea especially young forms, have the power to regenerate lost parts at subsequent moltings. Sometime a different appendage replaces a lost one *e.g.,* removal of a defective eye may be replaced by an antenna, this regeneration of a new part is called *heteromorphosis.*

The gonads of arthropods arise in development from coelomic pouches continuous with ducts which are coelomoducts. Typically the sexes are separate in crustaceans, though in one class, the Cirripedia, hermaphroditism is general and it also occurs sporadically among the Malacostraca notostracan and branchiopods. The gonads are usually paired and spread through several segments of the body. In malacostracans and some entomostracans male and female ducts open on different segments. This suggests that there may have been a primitive segmental arrangement and that the retention of different members of the series is associated with different functional needs in the two sexes. In the male the there anterior tubules become testes, two become supporting structures and the last, together with the main duct, becomes the vas deferens, with extends to the last thoracic segment of the adult and joins its partner from the other side to give a single genital papilla. In the female the elongated tube becomes the ovary and one tubule becomes the oviduct which opens on the sixth thoracic segment; the others become suspensory.

The sexes are usually distinguished by secondary characters associated with the transfer of sperm and the protection of eggs and

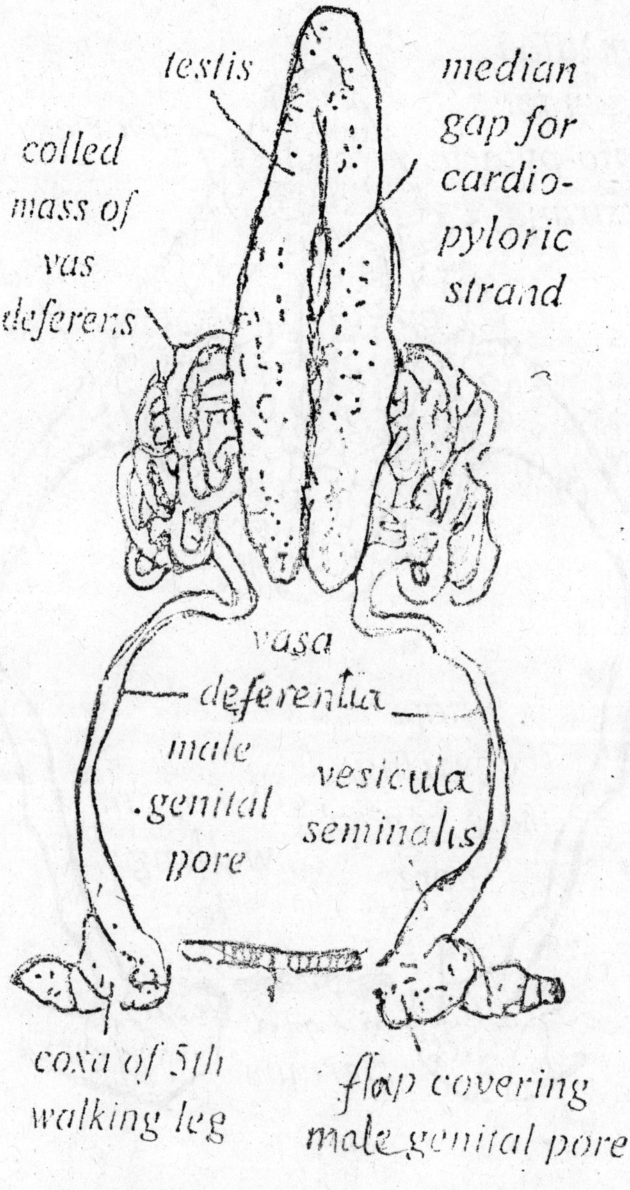

Fig. 12.10 : Male gentital organs of a prawn.

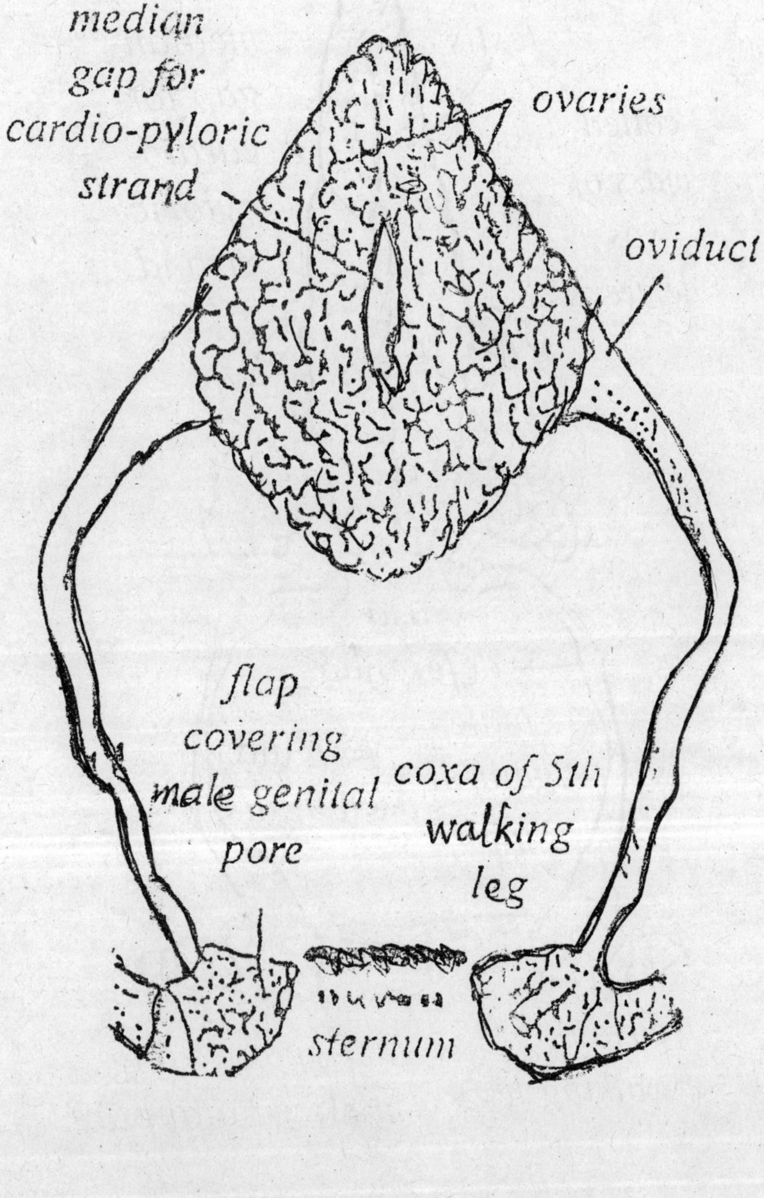

Fig. 12.11 : Female gentital organs of a prawn.

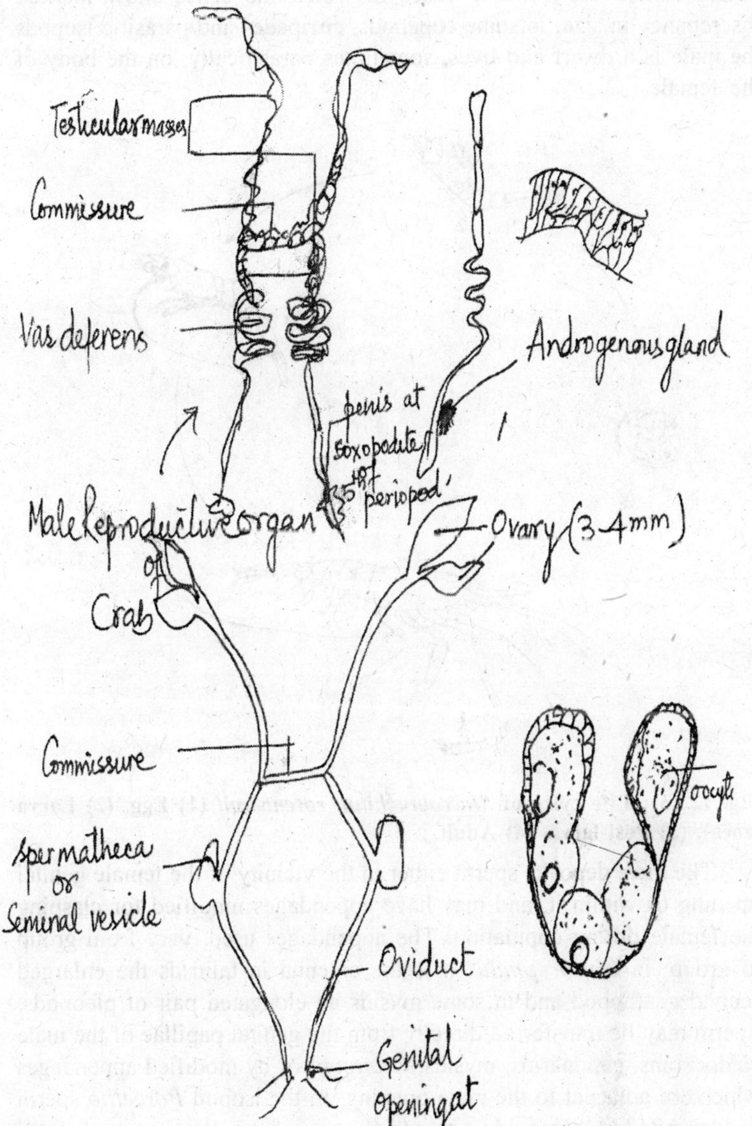

Fig. 12.12 : Female Reproductive organs of Crab.

young. There are extreme cases in which the sexes show marked discrepancy in size: in some copepods, cirripedes and parasitic isopods the male is a dwarf and lives, sometimes parasitically, on the body of the female.

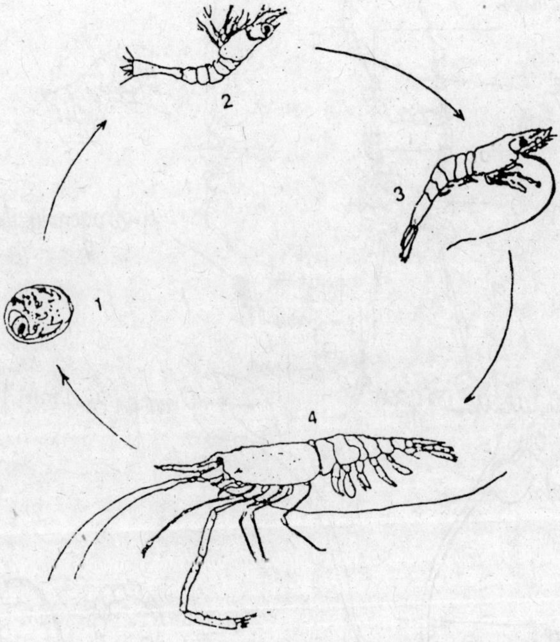

Fig. 12.13 : Life cylce of *Macrobrachium rosenbergil* (1) Egg. (2) Larva (zoea). (3) Post-larva. (4) Adult.

The male deposits sperm either in the vicinity of the female genital opening or within it, and may have appendages modified for clasping the female during copulation. The appendages used, vary from group to group: in *Chirocephalus* it is the antenna in talitrids the enlarged second gnathopod and in some mysids an elongated pair of pleopods. Sperm may lie transferred directly from the genital papillae of the male cladocerans, gammarids, mysids, *Crangon*) or by modified appendages which are adjacent to the male opening. In the isopod *Porcellio* sperm are deposited in each oviduct by one of a pair of copulatory styles formed from the endopods of the second pleopods. Each is grooved longitudinally and the genital papilla projects into the base of the groove to discharge seminal fluid.

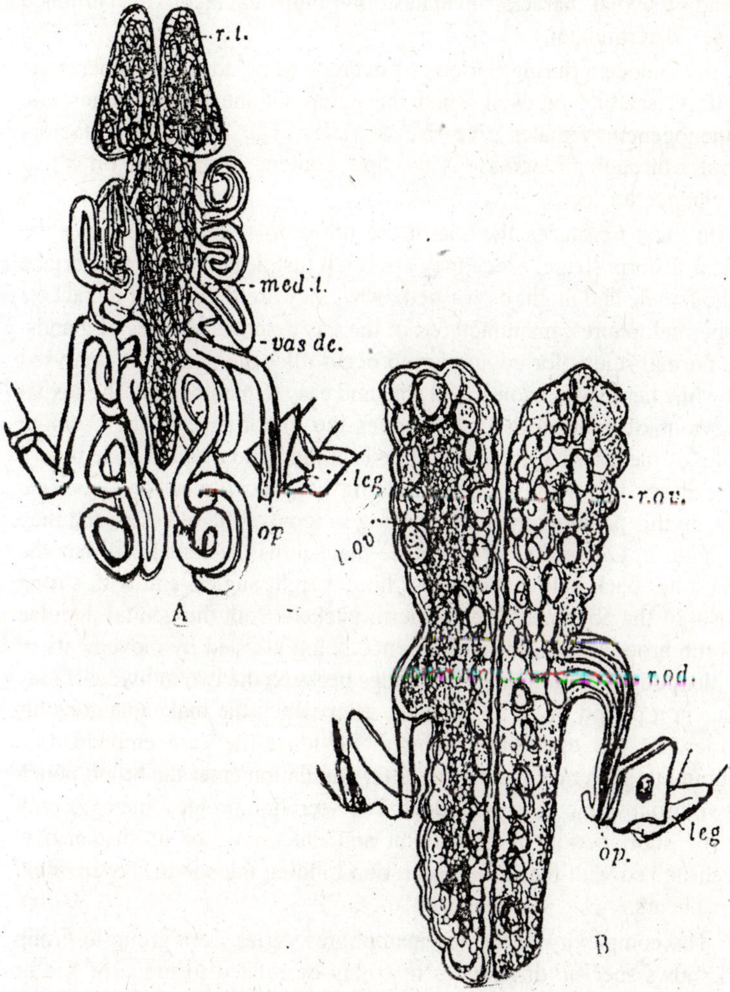

Fig. 12.14 : A, Male reproductive organs of *Astacus fluviatilis.* From Howes. r.t right anterior lobe of testis; *med.t.* median posterior lobe of testis vas de. vas deferens; op external opening of vas deferens; leg, right fourth ambulatory leg on which the vas deferens opens. B, Female reproductive organs of *Astacus fluviatilis.* From Howes. r.od. right oviduct; the left oviduct is shown partly opened; r,ov right lobe of ovary; left lobe of ovary with the upper half removed to show the ovarian cavity, which is the remains of the coelom and into which the ripe ova drop; op. external opening of oviduct; leg, right second ambulatory leg on which the oviduct opens.

In malacostracans hormones are responsible for primary and secondary sexual characters in males. Environmental factors also influnce the sex determinator.

In Cladocera during periods of overcrowding adverse temperature, and food scarcity or even when the nature of the food changes, the parthenogenetic females give rise to males. The last of these factors operates through an increase in the lipid content. of the algae on which the cladoceran feeds.

In most Crustacea the sperm are not motile and do not have the typical filiform shape. Since they are often just deposited on the surface of the female and might be washed away, they are transferred in packets of spermalophores, manufactured in the vas deferens by special glands The normal rule is for copulation to occur after the female has ecdysed and while her exoskeleton is still so it and easy to manipulate. The sexual behaviour of gammarids demonstrates the importance of the accurate timing of these two events. The male is attracted to a ripe female before she ecdyses and swims around holding her, sometimes for days. She molts in this position, the male helping to remove the old skin, and they copulate. In *Gammarus duebeni* the male turns the female so that she lies on her back below and across him at right angles, and with strong thrusts of the abdomen directs sperm packets from the genital papillae into the brood pouch, the transference being assisted by movements of his pleopods. One to four such matings between the two individuals may occur in a period of 10-30 minutes, after which the male abandons his partner. As the eggs pass along the oviduct they are embedded in albuminous secretion and shortly after copulation enter the brood pouch in two groups, one from each duct. The secretion protects the eggs only for the short period of maturation and the timing of its dispersal is synchronized with that of the secretion binding the sperm. Fertilization then occurs.

The complexity of the spermatophores varies from group to group and shows specific differences that may be related to mode of life or to devices for the discharge of the sperm. In the planktonic copepod *Calanus* the spermatophore is an elongated sac nearly 0.5 mm long and at copulation one end is attached near the oviducal opening on the genital segment of the female, adhering by its sticky outer wall. The sperm pass into the spermathecae (spth), two sacs which lie, one on either side, just within the median genital papilla. The spermatozoa are stored there until they meet ova approaching the female opening (ovd) to be discharged.

The spermatophores of decapods are simple sacs or rod-like structures.

In some (hermit crabs) they are stalked and become attached in large numbers to the pleopods of the female. The stalk often fastens them to a basal ribbon (r) which acts as a conveyor belt from their site of manufacture in the spiral coils of the vas deferens. It keeps them together as they pass through the genital aperture to the tube formed by the apposition of the first and second pleopods which deposits them on to the female. The spermatophores also remain external to the female in macrurans. In the lobster (*Homarus*) the gelatinous transparent rods up to 25 mm long are directed by the copulatory appendages into a pouch between the bases of the last walking legs of the female. They thus lie across the route taken by the ova passing from the genital opening to the pleopods. In *Astacus* the spermatophores are shorter rods deposited on the pleopods and ventral surface of the abdomen.

Copulation in crabs results in sperm being transferred directly into a spermatheca on each oviduct, where they are stored. Consequently, the spermatophore has lost much of its protective importance. In *Carcinus maenas* the copulating pair lie sternum to sternum with the abdomen unflexed, the female under the male who grips her with the anterior pereiopods. The right and left genital papillac of the male are inserted into the basal grooves of the first abdominal appendage (abd.app I) and the appendages enter the oviducts. Sausage-shaped spermatophores, freed from the genital openings, are then pushed into the female by the piston-like action of the second abdominal appendage (abd.app 2). The sperm may remain in the spermalheca for months, sealed in by a plug of secretion provided by the male, before fertilizing the ova.

The spermatozoa of decapods have a remarkably complex structure which varies from species to species. Their complexity is linked with their immobility. The simplest, found in natant decapods, have the nucleus surrounded by undifferentiated protoplasm prolonged into a single spine comparable to the tail of a motile sperm. In reptant forms the spermatocyte gives rise to a capsule which explodes to deposit the male nucleus in the cytoplasm of the egg, in which fertilization occurs in the hermit crab *Diogenes pugilator*. The spermatozoon (similar to that of the loboster) comprises a capsule, beneath this a central undifferentiated mass of protoplasm from which arise three long spines (or anchorage (sp), and then the nucleus (n). The wall of the capsule consists of a rigid outer membrane (out.mem), an extensible inner one (inn.mem) and fluid between. It encloses a central canal slim by an operculum at one end (op), and enclosing at the other an acrosome (ac) comparable to a firing-pin and having *frail* protoplasmic connections with the operculum (proto).

The sperm attaches itself to the egg by means of the spines, with the tip of the capsule towards the egg membrane (mem). The operculum breaks, severing its connection with the firing-pin. With this release of pressure the inner membrane expands, folding over the outer, by way of the opening made by breakage of the operculum and it presses on the surface of the egg, rupturing the membrane. The capsule becomes bell-shaped as the canal opens out and then the firing-pin penetrates the cytoplasm of the egg, dragging the nucleus behind it, the spines being carried passively behind the nucleus. Later all parts except the nucleus are resorbed. In crabs the sperm nucleus is in the form of a cupule molded to the base of the capsule, so that the two are intimately associated; in sperm which are transferred directly to the oviduct, spines are frequently reduced.

In contrast to such specialized arrangements for ensuring successful fertilization is the condition in barnacles. In these sessile forms the sperm are filiform and motile and transferred from one individual to the next by a long flexible penis here are no spermatophores. The penis is placed accurately into the mantle cavity of the partner where the spermatozoa are discharged and the eggs fertilized. To do this it may more than double its length.

The eggs may pass from the oviducts into the sea but are more usually retained on the b dy, either entangled in a special secretion or enclosed in a pouch, during some or all of their developmental stages. Those which are freed, predominantly in marine planktonic species (*e.g., Calanus*, most euphausiids, penaeid prawns), are relatively smaller and more numerous. Membranes covering the eggs may be produced solely by the oviduct, which characteristically secretes a thin transparent covering of chitin around each (indicating its ectodermal origin). Additional secretions from the duct differ considerably from group to group or may be absent; they afford an amount of protection varying with the requirements of the young and the method of incubation. Gammarids protect the young in a brood pouch where each completes its development within a transparent membrane. In copepods an oviducal secretion cements the eggs together to form an ovisac-really just a mass of eggs-which hangs from the genital segment and retains the young until they have reached the nauplius stage. The most lasting covering is found in branchiopods (*e.g., Chirocephalus, Artemia, Triops*) in which the protection of the eggs during periods-perhaps years of prolonged desiccation is essential to the survival of the species. *Chirocephalus diaphanus*, an inhabitant of temporary freshwater pools, has a median

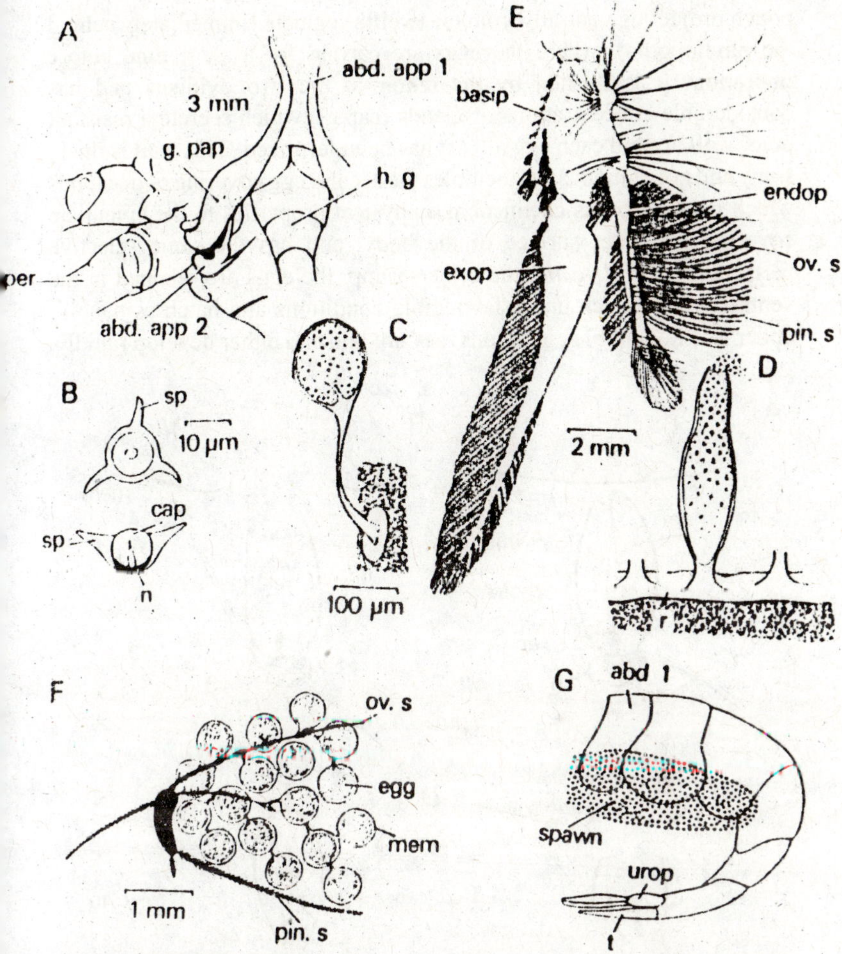

Fig.12.15 : ***Carcinus maenas*; A, copulatory appendages of male as seen on underside of abdomen. B, spermatozoon in (upper figure) apical and (lower figure) lateral view. C, *Diogenes figilator*, spermatophore. D, *Eupagurus bernhardus*, spermatophore discharging. *Crangon vulgaris*: E, first right pleopod of female in breeding dress; F, transverse section of endopodite of this pleopod to show eggs attached to ovigcrous setae; G, lateral view of abdomen to show that the spawn mass does not interfere with its flexure during backward escape movements, (C, after Bloch; D, after Jackson; E, F, G, after Yonge) abd, abdominal segment; abd.app i, first abdominal appendage; abd.app 2, second abdominal appendage, its tip inserted in cavity of first; basip, hasipodite; cap, capsule; endop, endopoditc; exop, exopoditc; g.pap, genital papilla in cavity of first abdominal appendage; h.g, region of find-gut; mem, egg membrane: n, nucleus; ov.s, ovigerous setae; per, last pcreiopod; pin.s, pinnate setae; r, ribbon to which spcrmatophores are attached; sp, spine; t, telson; urop, uropod.**

pouch projecting ventrally from the twelfth segment (immediately behind the phyllopodia), where the eggs are carried for a short time before liberation. It is formed by the union of the two oviducts and has considerable concentrations of glands (cap.gl) which secrete a resistant outer wall around each egg after it has been fertilized. This wall is thick, spiny and not chitinous. It resembles the second egg covering of decapods which is allied by its chemical and physical properties to the epicuticle covering the outer surface of the body, and has the same selective permeability. In *Artemia*, the brine shrimp, the eggs are retained in the ventral brood pouch under favourable conditions and hatch as nauplii, under less favourable conditions they are laid and either develop rapidly.

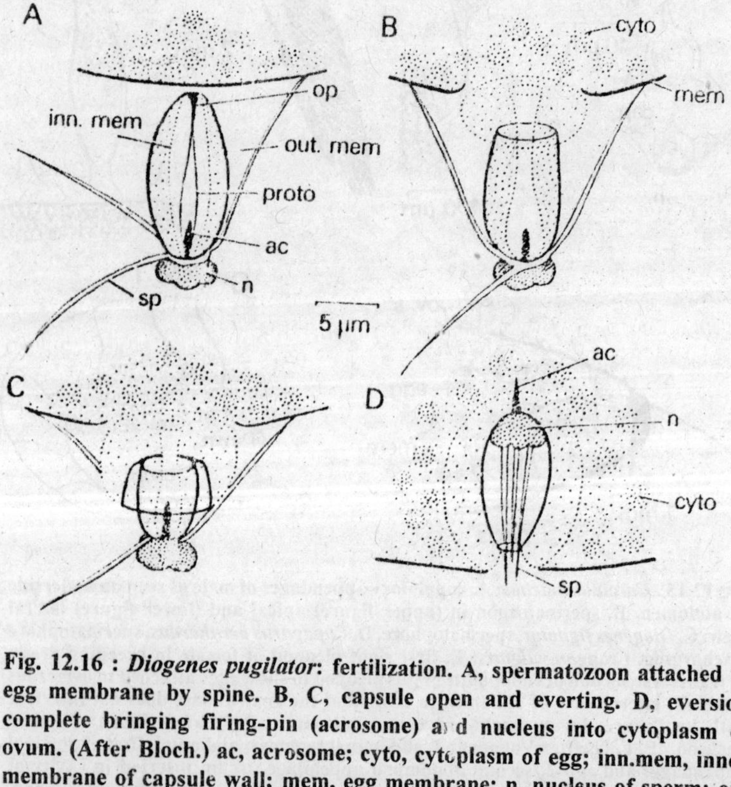

Fig. 12.16 : *Diogenes pugilator*: fertilization. A, spermatozoon attached to egg membrane by spine. B, C, capsule open and everting. D, eversion complete bringing firing-pin (acrosome) a) d nucleus into cytoplasm of ovum. (After Bloch.) ac, acrosome; cyto, cytcplasm of egg; inn.mem, inner membrane of capsule wall; mem, egg membrane; n, nucleus of sperm; op, operculum; out. mem, outer membrane of capsule wall; proto, protoplasmic connection between acrosome and operculum; sp, spine. .

Marine planktonic species (*e.g., Calanus*, most euphausiids, penacid prawns), are relatively smaller and more *numerus*. Membranes covering the eggs may be produced solely by the oviduct, which characteristically secretes a thin transparent covering of chitin around each (indicating its ectodermal origin). Additional secretions from the duct differ considerably from group to group or may be absent; they afford an amount of protection varying with the requirements of the young and the method of incubation. Gammarids protect the young in a brood pouch where each completes its development within a transparent membrane. In copepods an oviducal secretion cements the eggs together to form an ovisac—really just a mass of eggs-which hangs from the genital segment and retains the young until they have reached the nauplius stage. The most lasting covering is found in brainchiopods (*e.g., Chirocephalus, Artemia, Triops*) in which the protection of the eggs during period perhaps years of prolonged desiccation is essential to the survival of the species. *Chirocephalus diaphanus*, an inhabitant of temporary freshwater pools, has a median pouch projecting ventrally from the twelfth segment (immediately behind the phyllopodia), where the eggs are carried for a short time before liberation. It is formed by the union of the two oviducts and has considerable concentrations of glands (cap.gl) which secrete a resistant outer wall around each egg or lie quiescent. *Apus* has a long series of phyllopodia extending beyond the genital segment, and a median egg sac would interfere, with their functioning. Each oviduct has its own formed from the two exits of the phyllopodium of the genital segment (IIth thoracic) and in no way interfering with the functioning of the other parts of the appendage. Copulation occurs after ecdysis and within a few hours a batch of eggs is extruded into each brood sac. They are freed immediately before the next ecdysis and their sticky surfaces anchor them to plants or stones.

Similarly, in higher crustaceans parts of appendages not vital to other activities of the animal are used to carry the eggs. The pleopods of female decapods support the relatively enormous volume of a single spawn mass beneath the abdomen until the young have readied an advanced stage of development and are ready to hatch as larvae, or even as miniatures of the adult (*Astacus*). This may take several months. Indeed the exploitation of some habitats by the more advanced crustaceans is largely due to the retention of the developing eggs and the protection afforded them by the integumental secretion. It comes from cement glands related to special ovigerous setae, and not only forms the outer egg membrane but also the stalk by which the eggs are attached to the seta. The placing of the spawn in decapods varies with the mode of life and the use to which the abdomen is put. In species in which it is an important locomotor organ, as in *Crangon, Leander* and *Palaemonectes*, its rapid flexing for escape movements would be impeded by a spawn mass extending along its length, and eggs are not carried beyond the 4th pair of pleopods, and are, moreover, concentrated on the protopodites of the pleopods and carried close to the body. The position of the spawn in the shrimp (*Crangon vulgaris*) is still, further, influenced by the burrowing habit of this species. The egg mass is conspicuously long and shallow and is attached to unbranched setae arising from the protopodites of the last two pairs of thoracic appendages, as well as those of the first four pairs of pleopods and from the endopodites of the first pleopods. This leaves the exopodites of the abdominal appendages free to carry out burrowing activities.

Only the thorax is concerned with locomotion in crabs and the appendages of the reduced abdomen are concerned solely with reproduction. The abdomen shows marked sexual dimorphism. In the female it is broad, all the segments are movable and it is not so closely applied to the thorax as the abdomen of the male. The first pair of pleopods at the angle between thorax and forwardly directed abdomen is reduced, but the endopodites of pleopods 2-5 bear ovigerous setae. They retain the eggs in the space between posterior tliorax and upturned abdomen, and the exopodites embrace the mass externally; in a large *Cancer* this may comprise 3×10^6 eggs. The abdomen of the hermit crab, *Eupagurus*, is protected by a coiled gastropod shell and both rami of the pleopods carry eggs. The asymmetry and configuration of the shell are reflected in the development of the pleopods, which are present only on the left side, and in the fact that no eggs are attached to the last pleopod in the narrower apical region of the shell.

In most reptant decapods female secondary sexual characters, such as the broad abdomen with deep sides and the ovigerous setae with their special glands, are acquired at maturity and appear to persist. In natant forms a number of these characters are seasonal. In shrimps and prawns (*Crangon, Leander, Palaemonetes*) the setae appear only when the female molts into the egg-carrying condition. Other setae associated with the breeding dress may function only for a brief period during the spawning process. Some of these appear on the posterior sterna of the thorax and guide the eggs from the genital openings to the place of attachment of the spermatophores and towards the pleopods. The abdomen is turned forward beneath the thorax to form a chamber into which the eggs are shed and other special setae temporarily close it at the sides to prevent their escape. The fertilized eggs are propelled into it by setae on the first pleopod and rotated in the cementing fluid which, owing to its low surface tension, flows evenly over each, binding all to the ovigerous setae.

Some hermit and true crabs, abundant in the tropics, have the gill chambers adapted for breathing air and in this respect have gone far towards becoming land dwellers. They live far from the sea but must return at intervals since, as in other decapods, the eggs hatch as larvae. The only crustaceans which (an live and reproduce entirely on land are members of the Oniscoidea, a suborder of the Isopoda. In these, as in other peracaridans, the young are retained by the female until they have the form of the adult. The pouch in which they develop lies between the ventral surface of the thorax and thin-walled plates, the epipodites or *oostegites* attached to the inner surface of the coxopodites of some thoracic legs. It receives the openings of the oviducts (ovd). Movement of the plates allows interchange between the medium within the pouch and that outside, and in mysids their pulsation maintains a flow of water over the developing young which supplies oxygen and carries away waste. In terrestrial isopods the eggs develop in a watery medium and are by no means *cleidoic*. The system of capillary channels over the surface of the animal by which external water is conducted to the respiratory surface of the pleopods also serves the brood pouch. The developing embryos may recei e nourishment from the parent by way of four vascular folds of tissue, the cotyledons, which, in *Oniscus* and *Porcellio*, project into the brood pouch from the ventral body wall. Cotyledons and oostegites are discarded at the molt which follows liberation of the young and the mother reverts to the nonbreeding dress. However, if another batch of eggs has begun development, new cotyledons and oostegites appear with the new integument.

In some Crustacea, part of the cavity enclosed by the carapace is used as a brood chamber in direct communication with the oviducts and away from interference by the feeding and locomotor activities of the limbs. The lower part of the mantle cavity of a *Barnacle* accommodates several hundred shelled eggs which are agglutinated by secretion from the distal part of the oviduct. They hatch as nauplii. A rather more circumscribed part of the cavity is used to accommodate the eggs of water fleas such as *Daphnia* and *Simocephalus*. Again it is dorsal, between body wall and carapace, but it is closed posteriorly by two strong processes projecting into it from the trunk. Movements of the trunk draw these away from the opening, and allow a circulation of Water through the pouch. Thin-shelled, parthenogenetic eggs, laid during favourable conditions, complete development in 2-3 days and are liberated. At other times the eggs must be fertilized before they will develop. Only two are produced, a thick resistant shell is secreted around them and they are cast off when the female molts. At thin molt a saddle-shaped, dorsal area of the carapace, the epithelium, becomes especially thick, and separates from the rest to form a case around the resting eggs. There are some cladocerans, such as *Moina,* which have a more intimate association between embryo and parent: the brood pouch is closed and the embryos are nourished by secretion passed into it from the parent.

The life cycle of crustaceans agrees with that of other arthropods in being made up of instars, separated by ecdyses or molts. Most species are marine and their early stages are usually planktonic larvae, spreading the species and growing up at the same time. The more primitive groups, the branchiopods and free-living copepods, have the more primitive life histories in which larva and adult have much the same habitat and lead much the same sort of life, and transition from one to the other is gradual. In more advanced and specialized groups larva and adult frequently pursue a different mode of life, for the adult may be benthic, as in crabs, sessile, as in barnacles, or parasitic, as in some cirripedes, copepods and isopods. The transformation from early larva to adult then involves changes in body format certain stages which are sudden and radical and constitute a metamorphosis as striking as that of an insect.

The embryo either hatches as a nauplius or, if hatching is delayed, a naupliar stage is frequently recognizable during development. The nauplius larva is characteristic of the class, occurring in both Conchostraca and Malacostraca and suggesting the monophylactic origin of a very varied assemblage of forms. It resembles the trochophore of annelids in consisting essentially of cephalic and anal regions in juxtaposition. There are three pairs of jointed appendages which will become the first three pairs in the adult; a large labium (labr) overhangs the mouth (m) and dorsally there is a cuticular shield (car) through which can be seen a simple median eye (e). The antennules (ant') are mainly balancing organs, the antennae (ant") the principal locomotor organs, whilst these and the mandibles (md) may share the function of brood collecting and have special spines and processes for this activity. Each molt introduces additional somites and limbs which develop the characteristics of the adult, differentiation proceeding backwards along the trunk. This along the trunk. This pattern of development is best displayed by the branchiopods in which changes between one instar and the next are of more or less equal importance, and so small that the development may be called continuous. In the Copepoda, however, changes in body form, both external and internal, are more pronounced after certain ecdyses. In *Calanus* the first six stages, all naupliar, have only three functional pairs of appendages, the masticatory bristles on the last joints of the antennae are not well developed until the third stage, for the first two do not feed; at the fifth, maxillules are present and at the sixth, maxillae. The next molt brings greater transformations. The larva then has maxillipeds and two pairs of biramus swimming feet on the thorax, and a hind region from which all remaining somites of thorax and abdomen will later differentiate; there are also several important internal advances towards adult organization. This stage is known as the first *copepodid* and by a series for instars (six in *Calanus*) leads without further abrupt transition into the adult.

In barnacles also there are two planktonic forms which constitute the initial part of the life history but differences between them are more striking than between those of *Calanus*. This nauplius, as in all cirripeds, is distinguished by anteriolateral spines on the dorsal shield. After a series of six naupliar stages which show steady increase in size in the number of setae on the appendages and, at the sixth stage, the appearance of rudimentary maxillules, the next ecdysis gives a cypris larva so named because of its superficial resemblance to the ostracod *Cypris*. The carapace (edge) is bivalved, with an adductor muscle (add.mu) drawing two halves together so that it encloses the whole body; through it can be seen a pair of compound eyes (comp) as well as the nauplius one (e). The large antennae, mandibles and caudal spine of the nauplius are reduced and six pairs of biramus thoracic appendages (th. app) function in swimming. The antennules (ant') enlarge, project from the front of the carapace, and each develops a sucker-like disk bearing the opening of cement glands (gl) which help in attaching the larva when it ssettles. Development to this stage, with a body from essentially similar to that of the adult, takes about three weeks. The *Cypris* swims for a time and then attaches itself by the antennules to a stone or rock in the vicinity of *Barnacles* of the same species. Metamorphosis into the sessile adult involves a reorientation of the body and changes in the proportions of its parts: the thorax comes to lie across the long axis of the larval carapace, which is lost and replaced by another enveloping the body and reinforced by calcareous plates; the toracic appendages lengthen and their setose coverting is elaborated. In this life history metamorphosis takes place in two steps, the first (nauplius-*Cypris*) leading to the adult organization, the second (*Cypris* settling) orientating this to suit the definitive mode of life.

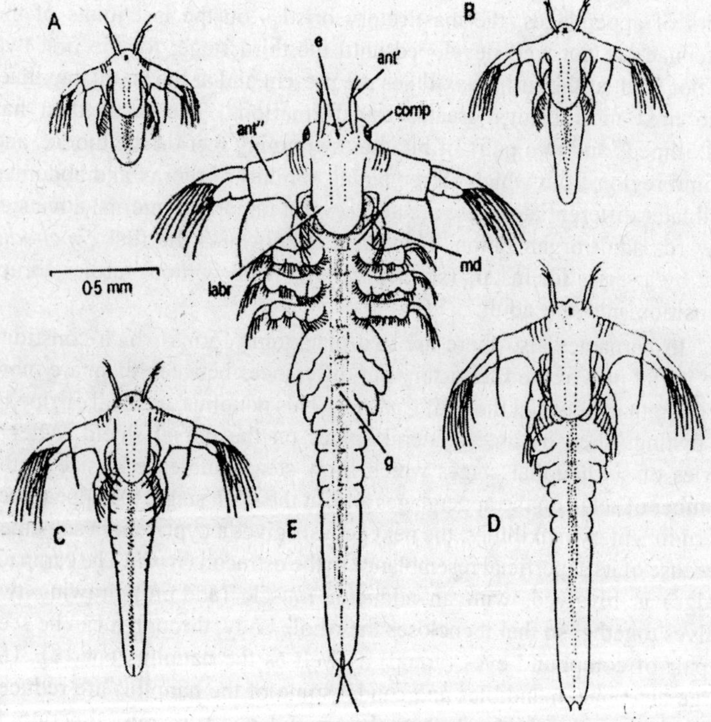

Fig. 12.17 : *Artemia salina*: larval stages showing progressive increase in the number of segments and appendages. (After Sars.) ant', antennule; ant", antenna; comp, compound eye; e, median eye; g, gut; labr, labrum; md, mandible.

Essentially the same kind of life cycle is exhibited by many higher malacostracans with benthic adults. Since in them, however, the nauplius is represented by an embryonic phase the start of free life is at a higher level of organization. This is the zoea, the larval stage of the prawns, shrimps and crabs. It already shows certain malacostracan features. There is a cephalothorax somewhat laterally compressed and covered by a carapace (car) fused with the thoracic terga, and a relatively large abdomen, with five (later six) distinct segments, ending in a telson (t).

There are lateral compound eyes (comp) and sometimes a nauplius eye. All the head appendages (ant', ant", mx', mx") and the first tow (crabs) or three (prawns) pairs of thoracic appendages are developed, but none is posterior to these; the rest of the thorax is still rudimentary with segments barely differentiated. The carapace is produced into spines: a rostral spine (r.sp) is of frequent occurrence, together with one dorsal (d.sp) and a lateral spine (lat.sp) on each side in crabs (though rarely in other decapods), or with only a pair of posterior lateral spines which are long in squat lobsters (*Golathea, Porcellana*) and short in hermit crabs. The larva swims with the thoracic appendages (mxp) which will become the maxillipeds of the adult; they are biramus, the rami being long and setose. The abdomen assists in swimming and is rhythmically flexed beneath the thorax and then stretched out behind. In some macrurous forms, like the common shrimp *Crangon*, the last pair of abdominal limbs, the uropods, develops precociously and forms an effective tail fan with the telson. Certain structures of the larva-spines on the carapace, large forked telson, elongated maxillipeds-fulfil larval needs only and are later modified or lost. The larva lives in the surface water feeding on minute planktonic organisms and growing in size and complexity with repeated castings of its skin; the differentiation of its thoracic segments and appendages occurs in a regular anterioposterior succession.

The common edible crab *Cancer pagiirus* has five zoeal stages each lasting a few days and distinguished by an increasing number of setae on the swimming legs as the larva grows. Rudimentary pleopods appear at the third stage when rudiments of the thoracic appendages are obvious. At the fifth molt the zoea metamorphosis to a form known as a megalopa transitional between zoea and adult, which lasts for about twelve days. This post larval stage, charateristic of all brachyuran crabs, exhibits most of the external features of the adult: the cephalo-thorax is depressed; the maxillipeds are solely mouth parts and no longer rotatory; the four posterior pairs of thoracic appendages are walking legs held close to the body whilst the animal swims, and the great chelae have developed claws; the long spines on the carapace have disappeared, though there is still a short rostral spine and backwardly projecting dorsal spine; the abdomen is relatively large and projects horizontally behind the carapace, exposing the pleopods (pleop) which are the functional swimming organs and bear long swimming setae. The megalopa, although still planktonic, tends to sink through the water, eventually reaching the benthic habitat of the adult where it secures its position by strong spines (sp) projecting

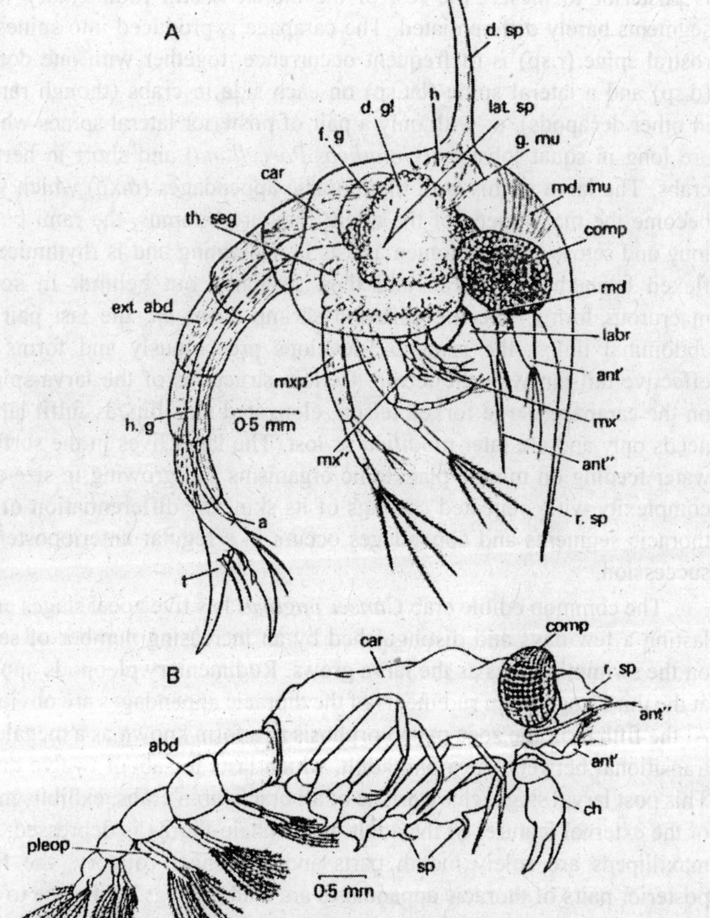

Fig. 12.18 : A, *Carcinus maenas*: first zoea. B, *Pilumnus heitellus:* megalopa. a, anus; abd, abdomen; ant, antennule; ant'', antenna; car, carapace; ch, chelipod, comp, compound eye; d.gl, digestive gland; d.sp, dorsal spine; spene, ext.abd. extensor muscles of abdomen; f.g, fore gut, g,mu, muscle of gastric mill; h, g, hind-gut; babr, labrum; labrum; lat,sp, lateral spine; mx' maxillule; mx'', md, mandible, md.mu, mandibular muscle; mxp, maxilliped; pleop, uniramus pleopod; r.sp, spines, t, telson; the.seg, thoracic segment developing.

from the basal joints of the chelae and three anterior pairs of pereiopods as well as by the claws cf these appendages.

The life cycle of sonic natant eucarids (euphausiids and penacid Prawns) includes both the basic crustacean larva, the nauplius, and the malacostracan larva, the zoea, followed by a post-larval stage with abdominal propulsion which merges into the adult. Larvae and adults live in the same habitat so there is no marked metamorphosis. The egg has sufficient yolk to make it unnecessary for the nauplius to feed and this stage consequently lacks mouth and food-collecting processes on the antennae and mandibles, which are solely swimming appendages. In the prawn the developmental stages are well marked by changes in the type of swimming organs, for whereas these are at first cephalic appendages, they are thoracic in the zoea and abdominal in the post-larva. The penaeid zoea has all the thoracic segments distinct and equally developed, bearing appendages with swimming exopods. This is in contrast to the cuphausiids (except *Euphausia superba*) and higher decapods, like the crab already described, in which thoracic segments and appendages are much delayed in their appearance. In respect of its thoracic structure the penaeid zoea resembles the later larval stages of the *Caricican* shrimps and prawns, and also the form which hatches from the egg in the *reptant* macrurans such as *Nephrops* and *Homarus*. It may therefore he regarded as the type of zoea characteristic of the macrurans, sometimes referred to as the "schizopod" or "mysis" stage on account of its biramus thoracic legs, both names which, unfortunately, imply false relationships. In the post-larval stage the pleopods become, setose and functional and exopods disappear from the legs.

The second cleavage took place at 00.30 hrs. The cleavage continucd and at 01:30 hrs. The blastula stage was oberved. (1 hr upto the blastula stage)

Fig. 12.19 : (d) Cleavages start, (e) Cleavage continued, (f) Blastula stage.

An embryonic membrane was clearly visible during the blastula stage.

– Gastrulation started at 02.15 hrs. and continued up to 02.55 hrs. At 04.50 hrs the embryonal mass become consticted laterally and the appendages started differentiating

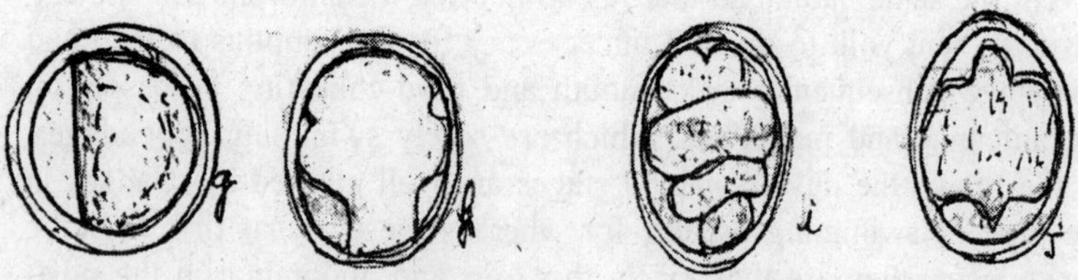

Fig. 12.20 : (g) Gastrula, (h) embryonic mass develop, (i) appendages start differentiating, (j).

– By 07.45 hrs all the 3 nauplear appendages could be seen as lateral thickenings which become tipped with short spine like setge by 09.45 hrs.

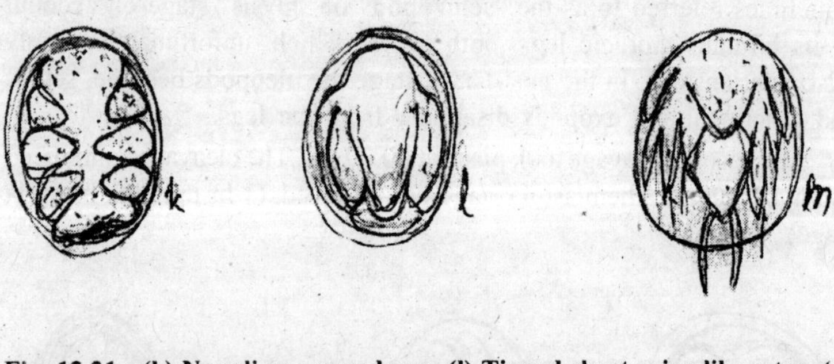

Fig. 12.21 : (k) Naupli or appendages, (l) Tipped short spine like setae, (m) 3-Appendages fully formed.

At 13.00 hrs the 3 appendages were fully formed space inside setae. The embryo occupied the entire space inside the egg and the movements were restricted to sudden jerks of appendages. The furcal setac first pierced the egg membrans and the nauplius wriggled but of the egg 16 to 17 hrs. after the eggs were spawned.

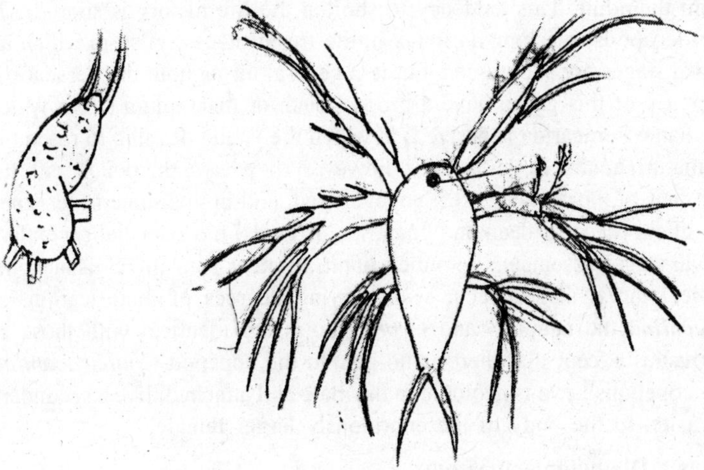

Fig. 12.22 : Lateral view of Nauplius, Nauplius.

Post Larval Development of *Penaeus* Spp.

The post larval are long and selender. Rostrum long with 1 or 2 dorsal spines, supra orbital and hepatic spines are present on carapace, postero dorsal and postero lateral spines present. On 5th and 6th abdominal segment. Telson with 8 pairs of setae. *Scaphocerite long and narrow, broader anteriorly than proximally.* A 2 flagellum 2-3 segemented. Mandibular palp, cylindrical with distal segment smaller than that of same width as proximal segment are visible.

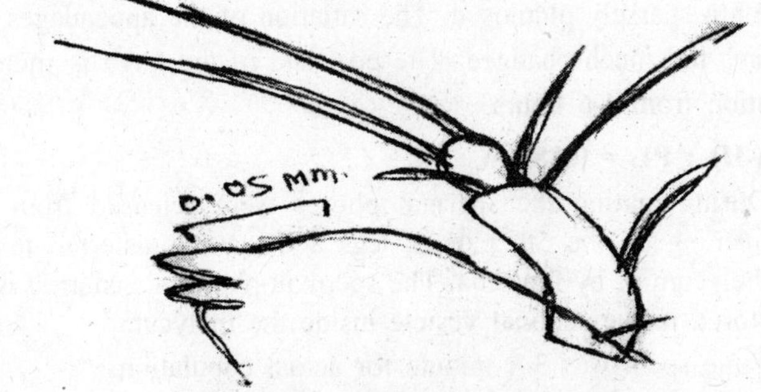

Fig. 12.23 : Post larva-*Penaeus*.

In a number of malacostracans the embryonic period ol development is extended and the young hatch in a form which does not differ essentially from the adult. This tendency to shorten the life history is sporadic in the decapods, reaching a climax in the freshwater crayfish in which all larval stages are suppressed, but it is general throughout the Peracarida. Females of this group have a brood pouch or marsupium (from which the name Peracarida is derived) in which the young develop to resemble miniature adults. In the isopods, however, they leave the pouch with the last pair of thoracic legs still undeveloped, and in this same order some members (decacaridean and gnathiids, isopods) have special secondary larval stages associated with their habit. In other parasitic Crustacea the typical larval stages occur with varying degrees of modification. In *Sacculina* the nauplius and *Cypris* stages are identical with those of barnacles except that there is no gut. In the copepod *Chondracanthus* the copepods larva is neotonic in the male and attached, like a secondary parasite, to the body of the enormously larger female.

Eggs : Diameter = 0.35mm

The eggs were demersal, spherical and brownish in appearance. There was a clear space between the perivitelline membrane and the egg capsule. The eggs hatched is about 18 to 24 hours.

Nauplius-I : PL = 0.22mm

Pyriform, brownish, with smooth carapace, three pair of appendages, an anterioventral eye and two small spires at the port. The 1st antenna is longer than the rest of the appendages. 2nd antenna is biramus, but unsegmented the setae at this stage appear unplumosed. Duration of this stage is from 4-6 hours.

Nauplius-II : TL = 0.25mm

The only apparent change at this stage is that all the setae are sparsely plumosed. The satiation of the appendages remains not much changed. The post end of the larva is squarer. Duration from 4-6 hours.

Nauplius-III : PL = 0.28mm

- During mating the spermatophores were released from the opening on the 5th pair of legs and were transferred to the thelycum of by Petasma. The spermatophores transferred were stored in the seminal vesicle inside the thelycum.
- Time spent was 3-4 minute for actual copulation.

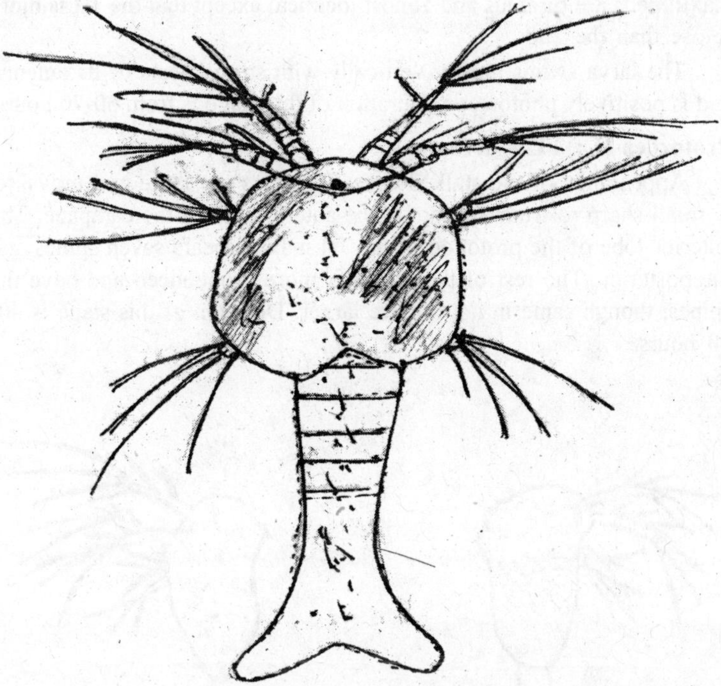

Fig. 12.24 : Protozoea larva-I.

- After mating the whitish mass of spermatophores can be seen through the plates of thelycum in females. This is used to confirm mating in hatchery.

Spawning

- Took place in early night hours (8-12 midnight)
- Eggs release continuously in the water column.
- While eggs are extruded the spermatophores are also extruded from the seminal vesicles but into the sea water.
- Fertilization is external (in sea water).
- Lunar periodicity is also observed.
- In *P. indicus* at the time of mating (impregnation) the ovary of female is not fully mature. There is a time gap between mating and spawning.

Developed third one the 1st antenna is three segmented the two maxillipeds are biramus and almost identical except that the 1st is more setose than the 2nd.

The larva swims mostly vertically with strong beats of its antenna and is positively phototactic. Duration of this stage is from 60-70 hours.

Protozoea-II : TL = 1.75mm

Appearance of the stalked eyes. The eyes are large and movable. A small sharp rostrum appears at the anterior end of the carapace. The anterior lobe of the protopod of the 1st maxilla bears seven spines and the posterior. The rest of the telson is more pronounced and have the spines. though same in no. but are larger. Duration of this stage is 40-60 hours.

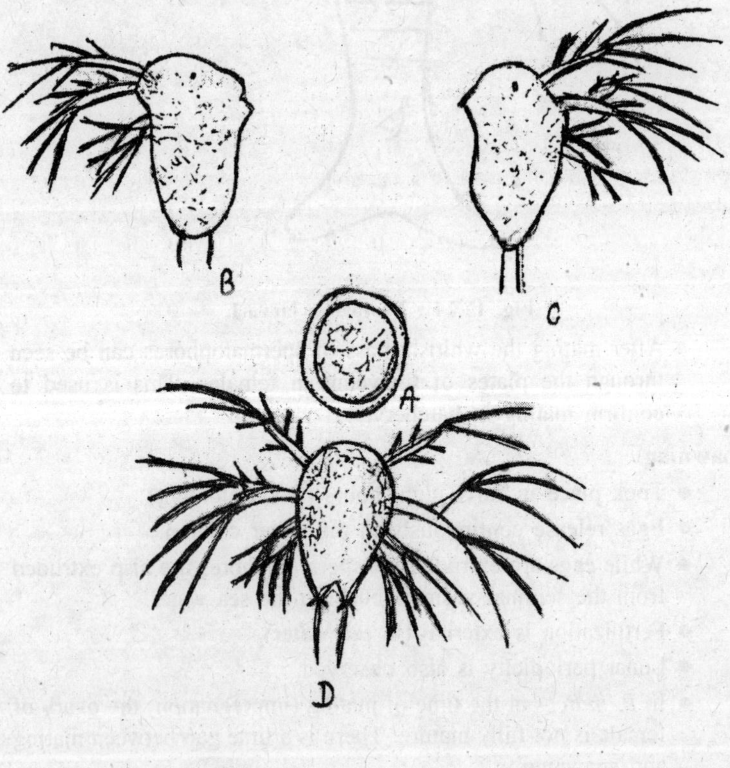

Fig. 12.25 : A; Egg B; Nauplius-I, C; Nauplius II, D; Nauplius II.

Protozoea-III: TL = 2.13mm

Major development of this stage is that the abdominal segments bears sharp dorsal spines on the 1st 5 segments, biramus uropod. The telson bears biramus uropods of which only the outer one is setose, bearing 5 setae the fork of the telson bears same number of spines as in the previous stage.

Commercial penacid prawns breed in the inshore area of the sea.

P. monodon	50-70 meter depth zone
M. dabsoni	Within 25 meter depth zone
P. stylifera	
M. monoceros	50-60 meter depth zone
P. indicus	
M. affinis	Very shallow coastal areas sometimes in estuarine mouth regions

Table 12.1 : Sexual characteristics of male and female *M. rosenbergii.*

Character	Male	Female
Size	Males grow much larger with proportionately larger head than female.	Females are smaller in size with a small head.
Claws	Second pereoped occur as very long, robust, spinous claws or chelipeds.	Much smaller and slender claws than males.
Gonopore	A flap covered gonopore on the arthroidal membrane between the coxae of each 5th pereopod and sternum. Gonopore are visible at 5.9 CL.	Gonopore are located as oval apertures on the medial surface of the coxae of the 3rd pereiopod and are covered with a thin membrane. Gonopore are visible at 7.6 mm CL.
Appendix masculina	It is present at the junction between the appendices internae and the endopods of the 2nd pair of pleopods (Fig.2)	Appendix masculina is absent. Only appendices internae is present.
Abdominal pleura	In matured males the vertical surface of the first abdominal pleura becomes slightly pointed centrally. chamber at 20 mm CL.	In mature females the 1-3 abdominal pleura elongated and widen to form a brood
Reproductive setae	Absent	Special reproductive setae appear on the thorax and pleopods of maturing females
Androgenic and	A solid strand of cells is attached to the subterminal region of vas deferns.	Absent
Clean Lengh & Weight at maturation	Length 175mm Weight 35g	Length 150mm Weight 25g

Mating and Spawning

The mating and spawning behaviour described here are based mainly on the observation of Hudinaga (1942) on P. *japonicus*.

Mating

- Took place during night
- One male followed by other
- In few minutes female moulted and becomes "Soft' premating molting, (not a periodical normal molting)

74 mm : *M. monoceros*

72 mm : *M. affinis*

54 mm : *M. dobsoni.*

The maturity can be identified by observing the size of the ovary, keeping the prawn above a bright light source.

130.2 mm : *P. indicus*

64.1mm : *M. dobsoni*

88.6 mm : *M. affinis*

100 mm : *M. brevicomis*

M-Macrobrachium, P-Penaeus

SPAWNING SEASON

Most of the commercial penacid prawns shows protracted breeding season extending throughout the year. However, the peak spawning periods vary from species to species, place to place and year to year usually two spawning peaks coinciding with the on set of the south east monsoon and north east monsoon.

SPAWNING GROUND

Five Maturity Stage

Stage-I

Testis lobes are not develop or spermatozoa are not formed/generative cells and not having the tubular portion upto vas deferens.

Stage-II

Testis lobes are fairly developed/generative cells becomes prominent in all the tubular portions.

Stage-III

Testis lobes are well developed both generative cells and well formed spermatozoa present in tubular form

Stage-IV

Testis lobes well developed and spermatozoa can be found in the distal portion of the vas deferens. The terminal ampoules ready for release. This is fully mature stage.

Stage-V

Testis lobes contain only spermatozoa and not other stages of spermatogenesis.

Size at First Sexual Maturity

Male : The size at 1st sexual maturity of males on the basis effusion of two halves of petasmal endopodite is found to be:

102 mm : *Penaeus indicus*

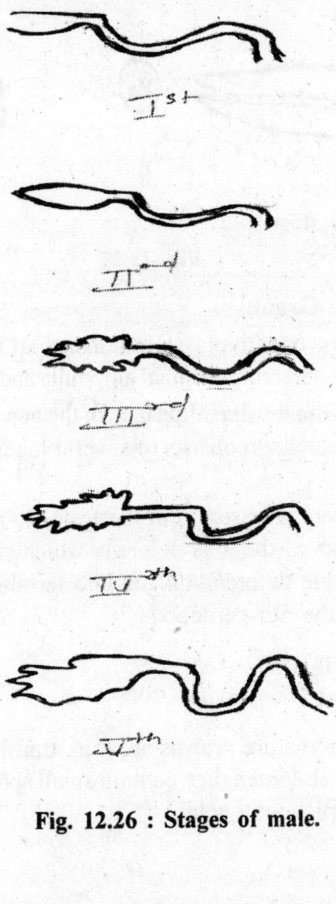

Fig. 12.26 : Stages of male.

Late Maturing Stage

The ovary is light green and visible through exoskeleton. The anterior and middle lobes are fully developed the maturing ova are opaque due to accumulation of yolk.

Mature Stage

The ovary is dark green and clearly visible through exoskeleton. The ova are larger than in the proceeding stage and the peripheral region becomes transparent.

Spent Recovery Stage

After the extrusion of eggs, the gonad reverts almost immediately to the immature condition. This stage is, therefore, distinguishable from that of the immature only from the size of the prawn.

Late maturity stage **Mature stage**

Fig. 12.27

Male Reproductive Organs

- The male reproductive system consists of a pair of testes, vas efferens, vas deferens; terminal ampoulla and a pair of gonopores.
- The testes lie on the dorsal surface of the hepatopancreas, beneath the heart. Each half consists of several lobes extending over the pancreas.
- The Vas efferens arises from the testes lobes and join into the proximal part of the Vas-deferens which runs downwards and backwards, the deferens bulges into terminal ampoules before it opens in the 5th pereopods.

MATURITY STAGES

1. Immature Stage

The ovaries of immature prawns are thin, translucent, unpigmented and confirmed to the abdomen they contain small spherical ova and clear cytoplasm and conspicuous nuclei.

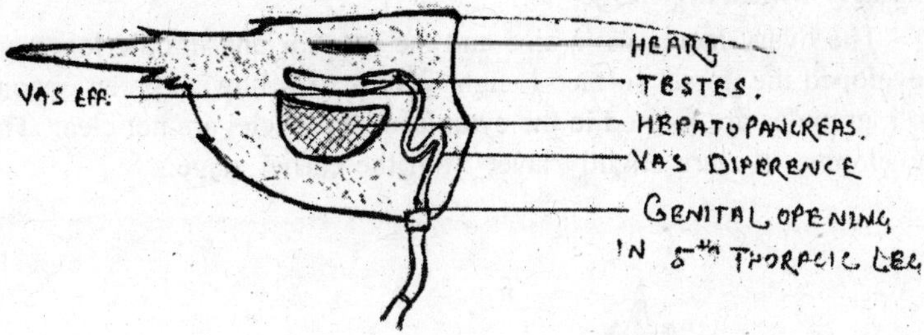
HEART
TESTES.
VAS EFF.
HEPATOPANCREAS.
VAS DIFERENCE
GENITAL OPENING
IN 5th THORACIC LEG

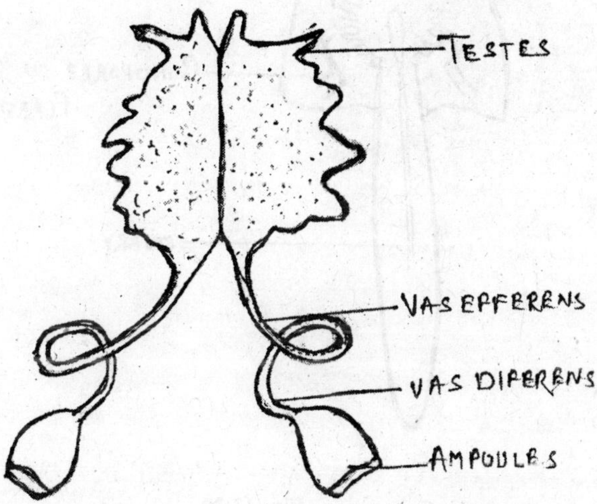
TESTES
VAS EFFERENS
VAS DIFERENS
AMPOULES

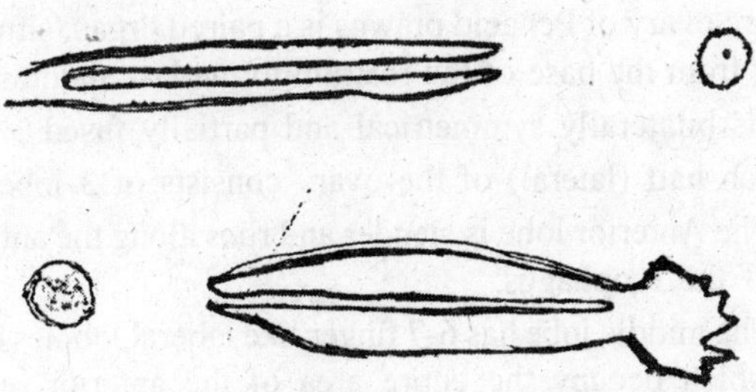

Fig. 12.28.

2. Early Maturing Stage

The ovary increases in size and the anterior and middle lobes are developed the dorsal surface is light yellow to yellowish green opaque yolk granules are formed in the cytoplasm and nuclei are not clear. The developing ova are slightly layer from the earlier stage.

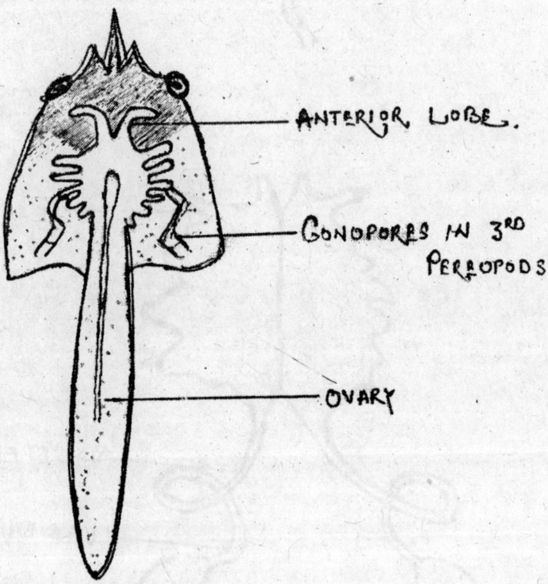

Fig. 12.29.

Female Reproductive Organs

Mature ovary of Penaeid prawns is a paired organ; situated dorsally, extending from the base of the rostrum to the last abdominal segments.

- It is bilaterally sytnmetrical and partially fused.
- Each half (lateral) of the ovary consists of 3 lobes.

(i) The Anterior lobe is slender and runs along the anterior portions of the esophagus.

(ii) The middle lobe has 6-7 finger !ike laberal lobules or projections which occupy the entire area of the anterior region and the posterior border of the cephalothorax.

(iii) The posterior lobe extends the length of the abdomen

- The two halves of the ovary are united by two commissure, one at the base of the tip of the posterior lobes in the 6th abdominal segments.
- The oviduct seen from the tips of the lobules (middle lobe) on either side to the external gonopores on the 3rd pereiopods.

Body of the larva becomes rounded anteriorily and elongated posteriorly. The 1st antenna shows unchanged satiation. The mandibles remain unchanged. Duration from 4-6 hours.

Nauplius-IV : PL = 0.31mm

Frontal organs are quite evident at this stage and buds of thoracic appendages are formed. The exopodite of the 2nd antenna shows six segments and seven setae. The mandibles remain unchanged. Duration from 4-6 hours.

Nauplius-V : PL =- 0.35mm

The caudal furcii bear seven spines each. The 1st antenna shows five basal annulations with three ventral, three dorsal and twice terminal setae. The segmentation of the exopod varies from 7-11 and bears 8 long setae and one short terminal spine. Duration from 4-6 hours.

Protozoea-I

The 1st protozoea has a clear transparent body with an oval carapace. 6 segmented thorax and an unsegmented post bifurcate abdomen. The larva has two pairs of antennae, two pairs of maxillae and two pairs of fully formed maxillipeds and an under,

Knowledge of reproductive biology is essential for, selective breeding, large scale, production of seeds in hatcheries.

Reproductive capacity in relation to the environment in captivity can be increased. Sexes and sexual difference can be marked easily.

Sexes are separate generally females larger than males, endopodite of Ist pair of pleopods modified into copulatory organ.

"Petasma" or "Andricum"

In "Thelycum" sperm storing organ is situated in between the last tow pairs of thoracic legs (seminal vesicles inside thelycum). In consists of an anterior plate a posterior plate and two lateral plates.

The opening of the genital ducts: Situated on the base of the coxa of 3rd pair of pereiopods and on the last (5th) pair in males.

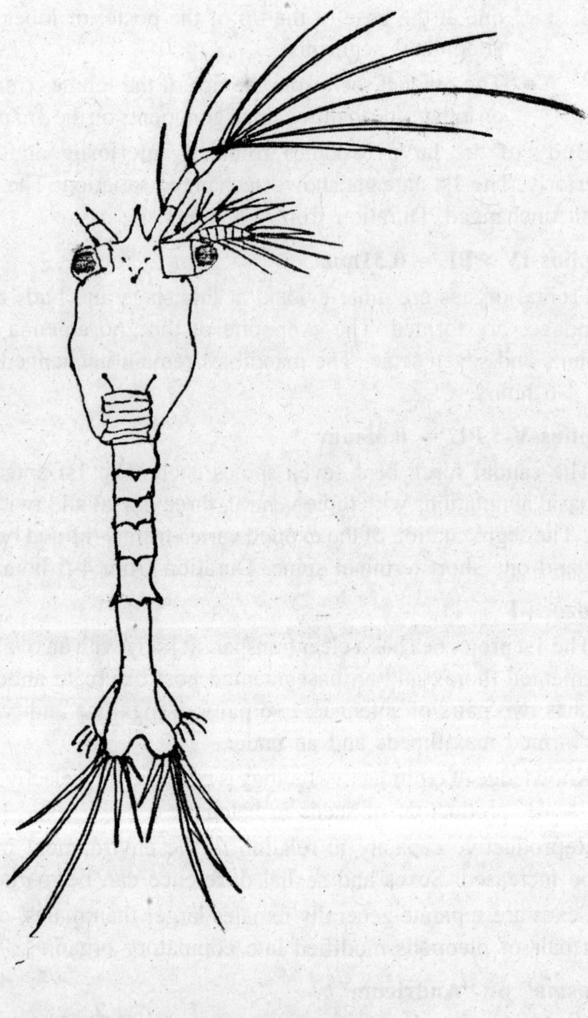

Fig. 12.30 : Protozoea.

Mysis-I : TL = 2.25mm

The Rostrum is sharp and projects beyond the tip of the eye. Sharp orbital and pterygostomial spines occur on the anteteriolateral margin of carapace. The 1st antenna is three segmented with the distal segment giving out two branches the inner one of which bears setae and the outer one long short. The basal segment bears strong ventro median spine and is swollen at the base. The telson is deeply cleft and bears a pair of lateral and 6 pairs of post spines. Duration of this stage is 40-50 hours.

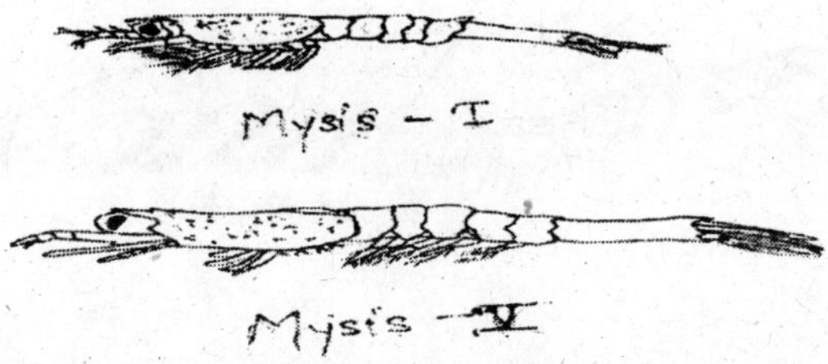

Fig. 12.31 : Mysis.

Mysis-III : TL = 3.20mm

Major development of this stage is the occurrence of hepatic spine and two segmented plcopods. The Rostrum no longer projects beyond the tip of the eye. The telson is longer but not changed. Duration of this stage ranges from 30-40 hours, then changed into Mysis-IV and Mysis-V respectively.

Mysis-IV : TL = 3.30mm

No change in external morphology. (30-40 hours)

Mysis-V : TL = 3.42mm

The pleopods become setose and the 6th abdominal segment loses the posteriolateral spines. Duration of this stage is 30-40. hours.

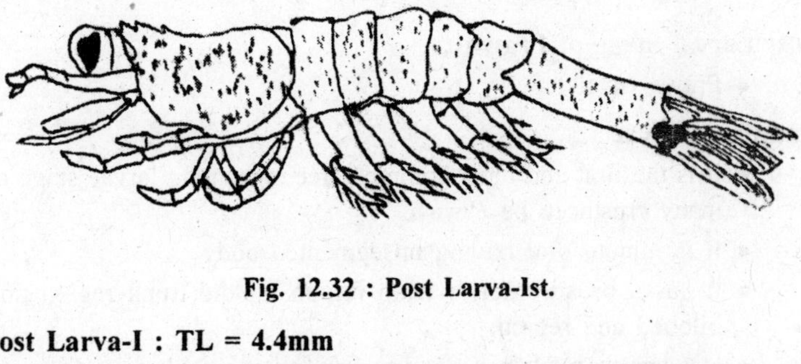

Fig. 12.32 : Post Larva-Ist.

Post Larva-I : TL = 4.4mm

The most apparent change at this stage is in the mode of the swimming. The postlarva makes full use of the pleopods for propulsion and swims extremely swiftly. It can also walk on the bottom and sides of the container. The pleopods are still uniramus the distal segment bears 10 setae while the basal bears three.

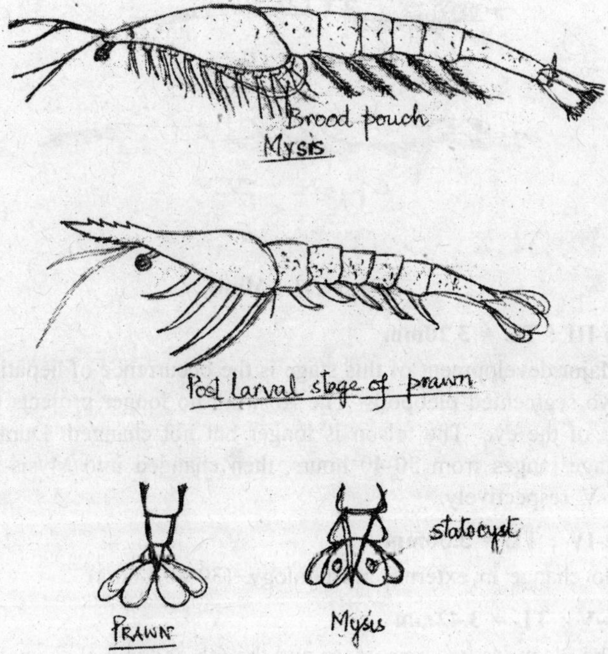

Fig. 12.33

Post Larval Stage of Prawn

- Prawn Mysis.

Nauplius Larvae

- It is the first and most common free swimming larvae stage of many crustacca i.e *Prawn.*
- It is minute oral having unsegmented body.
- It has a broad anterior head region middle trunk region and bilobed and region.
- Head be are median eye and a pair of antennules bearing terminal setae.

- Nauplices under goes a series of molts to pass through several intermediate larval forms before it reaches the adult stage.

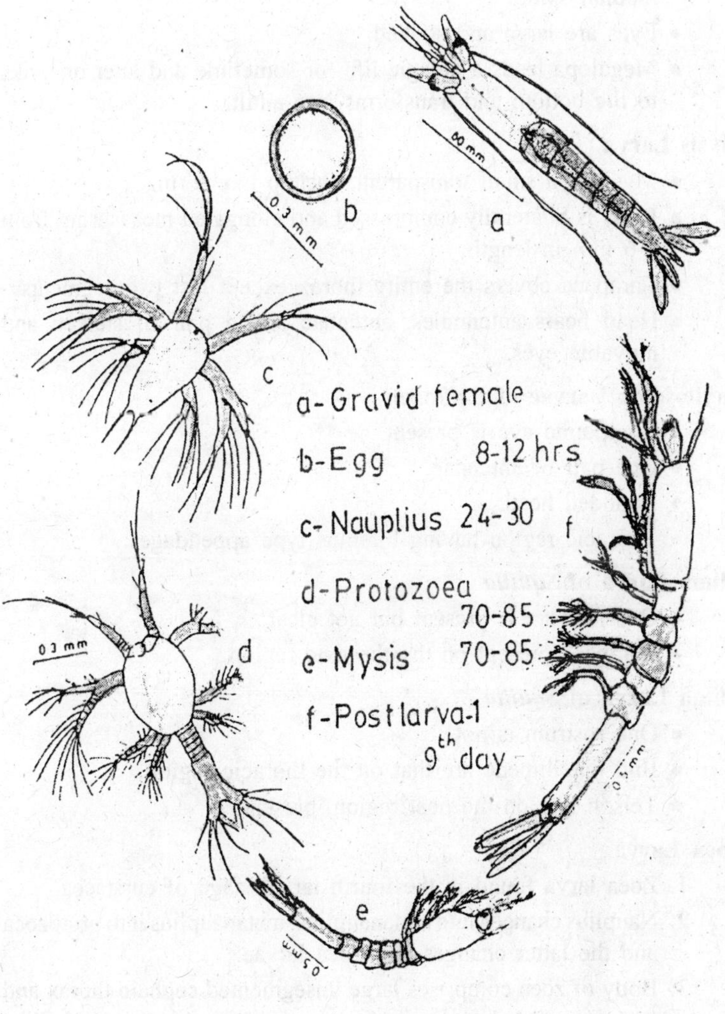

Fig. 12.34 : Life cyle of prawn.

Megalopa Larva

- Megalopa larva is the larva of crab and it develops from zoea larva through successive molts.
- It has a broad and crab-like cephalothorax bearing an anterior median spine.
- Eyes are large and stalked.
- Megalopa leads a pelagic life for sometime and later on sinks to the bottom and transforms into adult.

Mysis Larva

- Mysis is a small transparent, shrimp like form.
- Body is bilaterally compressed and elongated meaasuring from 2-6 mm in length.
- Carapace covers the entire thorax except last two segments.
- Head bears antennules, antennac and a pair of stalked and movable eyes.

Phyllosoma Larvae of *Palinurus*

- Compound eye is present.
- One pair of antennae.
- Rounded head.
- Thoracic region having biramus type appendages.

Alima Larva of *Squilla*

- One rostrum is present but not clear.
- Carapace present on the thoraxic region.

Alima Larva of *Squilla*

- One rostrum is not.
- IInd maxillipede are that on the thoracic region.
- Telson that on the post region, present.

Zoea Larva

1. Zoea larva found in the fourth larval stage of curstacea.
2. Nauplius changes into metanauplius, metanauplius into protozoca and the latter changes into zoea larvae.
3. Body of zoea comprises large unsegmented cephalo thorax and long segmented abdomen.
4. Carapace of cephalothorax is produced into long spines of which one is rostral, a median dorsal and two lateral.

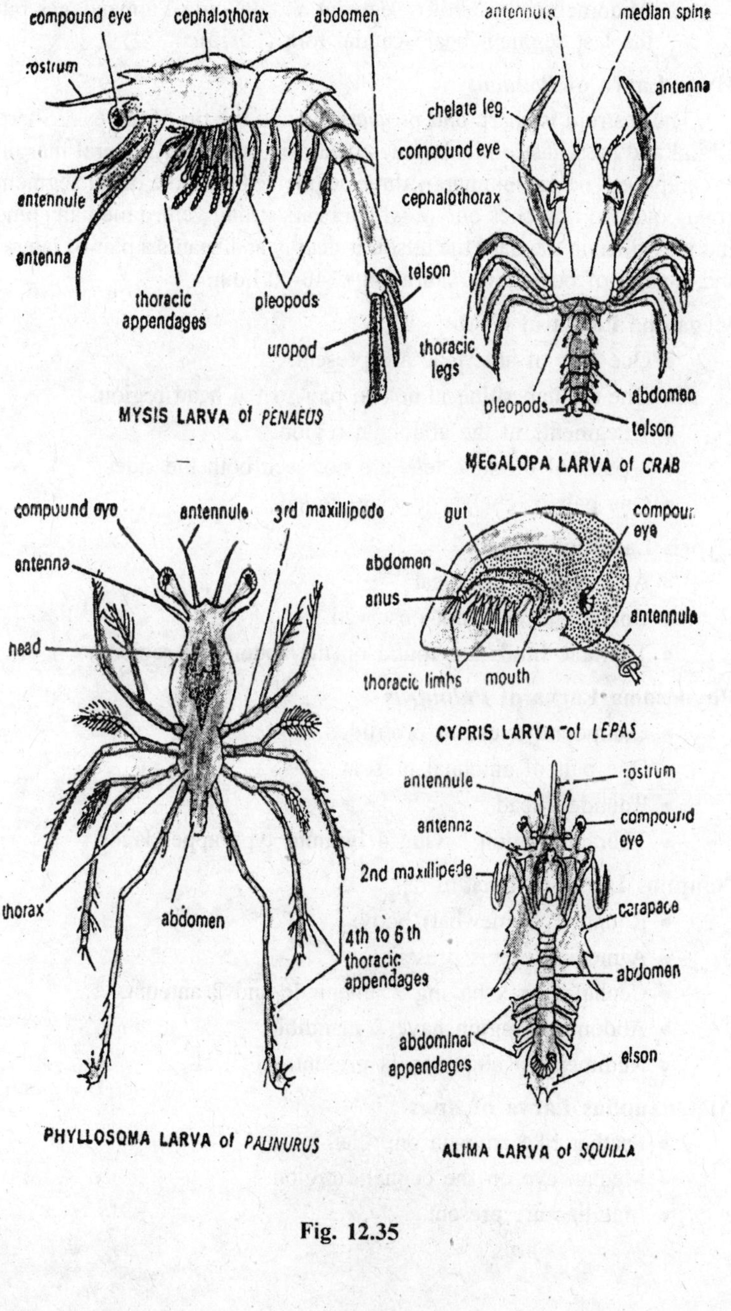
compound eye
cephalothorax
abdomen
rostrum
antennule
antenna
thoracic appendages
pleopods
telson
uropod
MYSIS LARVA of PENAEUS
antennule
median spine
antenna
chelate leg
compound eye
cephalothorax
thoracic legs
abdomen
pleopods
telson
MEGALOPA LARVA of CRAB
compound eye
antennule
3rd maxillipede
antenna
head
thorax
abdomen
4th to 6th thoracic appendages
PHYLLOSOMA LARVA of PALINURUS
gut
eye
abdomen
anus
antennule
thoracic limbs
mouth
CYPRIS LARVA of LEPAS
antennule
rostrum
antenna
compound eye
2nd maxillipede
carapace
abdomen
abdominal appendages
ALIMA LARVA of SQUILLA

Fig. 12.35

5. Compound eyes are paired and movable.
6. Abdomen consists of 6 segment and devoid of appandages but the last segment bears caudal fork (Furca).

Mysis Larva of *Penaeus*

The rostrum is sharp and projects beyond the tip of the eye. Sharp orbital and pterygastrominal spines occur on the antero lateral margin of carapace. The 1st antenna is three segmented with the distal segment giving out two branches one of which bears strong ventro median spine and is swollen at the base the telson is deeply and beams a pair of lateral and 6 pairs of post spins. Duration = 40-50 hours.

Megalopa Larva of Crab

- One pair of antermae are present.
- One median spine is not an pair to the head region.
- 7 segments in the abdomen region.
- 4 pairs of thoracic legs are not seen both the sides.
- One pair of chelate legs are found.

Cypris Larva of *Lepus*

- Antennulc is not clear.
- Compound eyes not observed.
- Thoracic limbs are found on the abdominal region.

Phyllosoma Larva of *Palinurus*

- Compound eyes are protruded.
- One pair of antennal present.
- Rounded head.
- Thoracic region having 4 biramus type appendages.

Nauplius Larva of *Cyclops*

- Rounded (somewhat) body.
- A median eye is present.
- Cephalothorax having 2 antennule and 2 antenna.
- Abdominal region have 2 mandible.
- Rudimentary abdomen is present.

Metanauplius Larva of *Apus*

- Feather like antenna one that.
- Median eye on the cephalic region.
- Maxillae are present.

- Maxillipeds are that on abdominal region.
- Candal fork, is present on the caudal region of the larval.

Protozoea Larva of *Euphausia*

- Two antennule are present.
- Biramus antenna are that on cephalic region.
- Rudiments of maxillae are observed.
- Mandibles present.
- Uropod and telson found posterior.

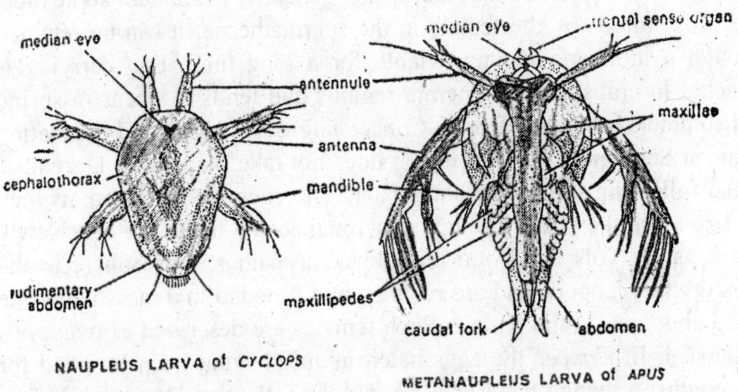

Fig. 12.36.

Zoea Larva of Crab

- Median dorsal spine is posterior to head.
- One rostrum is present.
- Ist and IInd maxillipedes mandible, maxilla, are that cephalo thoraxic region.
- Eyes are promiment.
- Caudal furca having 2 cirri.
- 5-7 abdominal segments.

Female which become sexualhy receptive when hard shelled generally lay their eggs short after copulation those which copulate just after molting are in depended state and do not generally ovulate for some time. Sperm is stored by the female in the spermathecae; it can be retained through a molt and remains viable for a long time. Egg carrying is seasonal in crabs from temperate regions and tends to occur in spring and summer. In mature female *Cancer pagurus* ecdysis and copulation occur in September but egg laying does not take place until December of the following year, 14 months later. The eggs are fertilized .as they are laid since they pass through a spermatheca on their way out. Newly laid eggs are yolky and orange or khaki in-colour; they adhere to the hairs on the pleopods and are retained and brooded in a mass under the broad abdomen of the female. Such females are described as ovigerous or berried. In *Cancer* the eggs hatch in July having been brooded for six months; a second batch may be laid the following December before the next ecdysis and copulation.

In tropical crabs breeding is often continuous throughout the year but may show a lunar rhythm in which egg hatching coincides with full or new moon. In the mangrove tree crab *Aratus pisoni* the time from laying to hatching is only 16 days; more than one egg batch may be laid between consecutive ecdyses and an average of about six batches is laid per year. In terms of egg batches per year and the length of time taken for the eggs to develop, *Cancer* (large, temperate) and *Aratus* (small, tropical) probably represent two extremes.

The number of eggs in a batch is generally very large and varies with the size of the crab. *Aratus* of 16mm carapace width lays about 5000 eggs and a 36mm *Pachygrapsus* lays about 48000 eggs. Large crabs like *Cancer pagurus* lay up to three million eggs when full grown. The size of a newlaid egg, however, is very small; they are generally slightly oval in shape with a maximum diameter of between 0.25 and 0.35 mm. The eggs swell as they develop so that by the time they are ready to hatch they are roughly double their new-laid volume. During development the colour of the egg changes through brown to grey as the yolk is used up and the outline ot the embryo becomes visible. The eyes and pigment spots appear first, followed by the outlines of the abdomen and cephalothorax.

When the eggs hatch the larvae are released from beneath the abdomen of the female. Generally the female assists this process by shafting the abdomen to and fro and by digging into the egg mass with the chelae. In *Aratus* the female climbs down a mangrove root into the water and rapidly vibrates the abdomen back and forth releasing a cloud of larvae. In the northwest Atlantic spider crab *Libinia emarginata* unusual egg hatching behaviour involving the male has been described, Male *Libinia* can detect the presence of a female about to release larvae. They become very aggressive and fight for the possession of such a female. When a male has captured the female he positions her behind him with the backs of their carapaces touching and clasps his last pair of walking legs beneath her abdomen. Holding her in this 'obstetric' position he backs into a protected area and while the female releases the larvae by fanning her abdomen he drives away any other crabs that approach. After the larvae are liberated the female cleans the empty egg cases from her pleopods and the male then releases her.

Larvae

The larval life of a crab is generally spent in the plankton. It may last from one to several months and during this time the larva molts several times before settling on the sea bed and molting to the first crab stage. There are two larval forms, the zoea and the megalopa (or, in hermit crabs, the glaucothoe). There are between 2 and 6 (usually 4) zoeal stages and these are generally followed by one megalopa or glaucothoe stage.

Structure and Development

The zoea larva hatches in an embryonic form as a protozoea but this quickly molts to the first zoea. These larvae are tiny shrimp-like animals they have a full complement of functional head appendages but only the first two thoracic appendages, the 1st and 2nd maxillipeds, are developed and functional; the rest of the appendages are absent or present only as limb buds. The head appendages all appear to function as they do in the adult but the maxillipeds are used for swimming: the exopods bear long plumose setae and the zoea beats these against the water to achieve fairly fast but jerky progress.

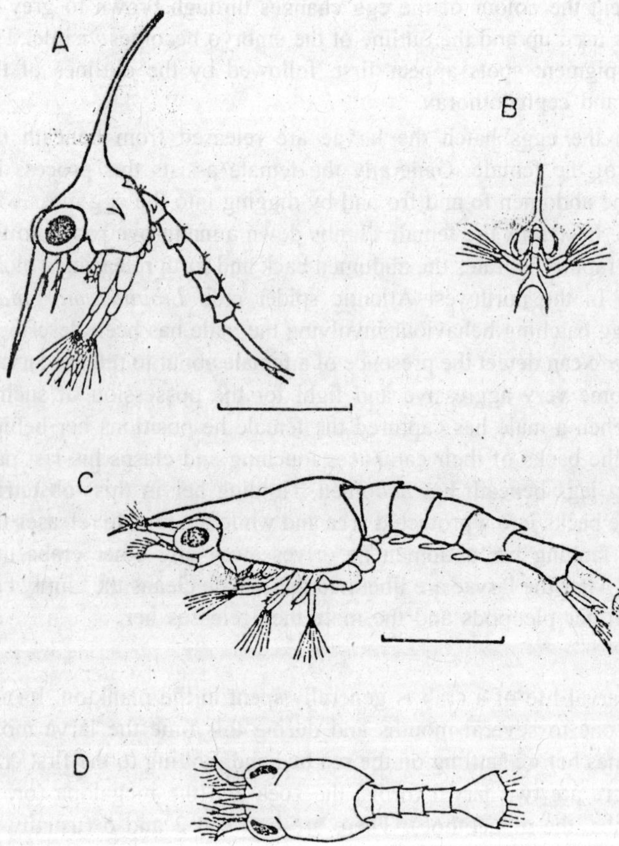

Fig. 12.37 : Zoea larvae. A, side view of the third zoea of the swimming crab *Macropipus puher.* B, posterior view of the first zoea of *Carcinus maenas* C, side view of the fifth zoea of the West Indian land hermit crab *Coenobita clypeatus.* D, dorsal view of the first zoea of the hermit crab *Pagiirns bernhardus.* A and B after Lebour, 1928; C after Provenzano. 1962; D after MacDonald *et al.*, 1957. Scales = 1mm.

Fig. shows that hermit crab zoeae are more shrimp-like than brachyuran zoeae; this difference may reflect the greater degree of evolutionary change in the brachyura awa/ from the common ancestor with the abdomen extended and usually with the dorsal side uppermost. The direction of swimming, therefore, is along the axis of the spines in both types of larvae. The speed of swimming is generally 1-2cm/s.

Reversals or escape movements a twitch occur on collision. Further differences lie in the development of spines on the carapace and in the shape of the telson. In.typical brachyuran zoeae the carapace bears a ventrally pointing rostral spine and a tall dorsal spine; lateral spines are also frequent. In the Anomura spine development is more variable. Hermit crab larvae and stone crab larvae (Lithodidac) usually have short spines and often only the rostral spine is developed; the zoeae of porcelain crabs and of squat lobsters have rather long carapace spines. However, in contrast fo brachyuran larvae, anomuran larval spines are always orientated antero-posteriorly and contribute to the shrimp-like shape. The telson in brachyuran zoeae is usually forked whereas in anomuran zoeae it is often spade shaped and forms a tail fan. A list of the differences between brachyuran and anomuran zoeae was given by Williamson (1965).

At each molt the zoea grows and develops by adding setae to the existing appendages and adding appendages to the posterior half of the body. Addition of swimming setae to the exopods of the maxillipeds occurs in quite a regular fashion in brachyuran zoeae and can be used to 'stage' larvae. First zoeae almost always have 4 swimming setae on each maxilliped, this increases to 6 in the second zoea, usually 8 in the third zoea, and 9 or 10 in the fourth zoea. The development of other appendages in brachyuran zoeae is generally restricted to limb and pleopod buds. These usually appear first in the penultimate zoeal stage and by the last zoea (or metazoea, usually the fourth) they may bear a few hairs but are still non-functional. In contrast, in anomuran zoea two other functional appendages are developed: 3rd maxillipeds used in swimming appear in the second zoea and functional uropods forming a wide tail fan with the telson are developed by the third zoea.

The megalopas and glaucothoes resemble the adult forms more closely than do the zoeae; indeed, they are often referred to as post-larval stages. There is a reduction in carapace armature and a complete set of functional appendages including chelipods and walking legs is present. Locomotion is still largely by swimming but this is achieved by the pleopods borne on the extended abdomen and not by the maxillipeds. An interesting difference, apart from shape, between the megalopa and the glaucothoe is the presence in the latter of well developed uropods forming a tail fan with the telson; in the megalopa the appendages of the last abdominal segment take the form of pleopods and there is no tail fan.

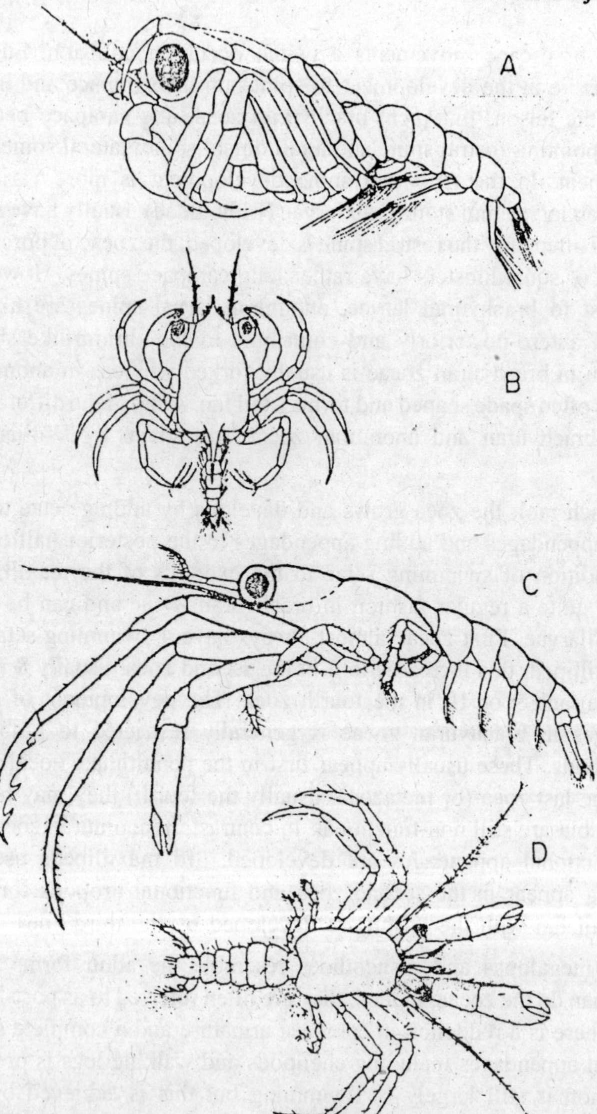

Fig. 12.38 : The megalopa and glaucothoe. A, side view of the megalopa of *Carcinus maenas* in characteristic swimming position; after Atkins, 1954. B, dorsal view of the megalopa of *Cancer pagurus;* after Lebour, 1928. C. side view of the glaucothoe of the West Indian hermit crab *Petrochirus diogenes*; after Proven/ano, 1968. D. dorsal view of the glaucolhoe of the hermit crab *Pagurus bernhardus*; after MacDonald *et al.*, 1957. Scale = 1mm.

Swimming

Zoeae usually swim backwards propelled by the anteriorly directed beating of the maxillipeds. Brachyuran zoeae swim head downwards with the dorsal spine foremost and the abdomen trailing. Anomuran zoeae, however, generally swim telson first anomuran zoeae by tail-flips using the tail fan. In brachyuran zoeae, which lack a tail fan, there is no clear escape movement. When the larvae stop swimming they sink and the rate of sinking depends on the development of spines. A narcotized zoea of the porcelain crab *Porcellana* sank in a horizontal posture at a rate of about 0.3cm/s; *Porcellana* zoeae are characterized by the presence of enormously long carapace spines extending posteriorly and anteriorly. Narcotized brachyuran zoeae, however, with their globular carapaces and shorter spines, sank head first at a rate of about 1-2cm/s. Thus one function of long spines may be to aid flotation and, indeed, anomuran larvae possessing long spines (Porcellanidae and Galatheidae) frequently rest in a horizontal position hardly sinking at all. Resting is rarely observed in brachyuran zoeae which remain in the same place by swimming constantly upwards. It is possible that short-spined zoeae gain the advantage of manoeuvrability; *Porcellana* larvae had difficulty in turning but could change direction more quickly when the rostal spine was cut off.

The direction of swimming seems to involve responses to both light and gravity. Zoeae swim towards a light source and can perceive both direction and intensity of light. In the dark the normal swimming orientation is maintained indicating perception of gravity. However, in brachyuran zoeae control is lost in the dark (but not in the light) if the carapace spines are cut off. Probably the spines bear hairs sensitive to water movement. These might distinguish relative water movements due to sinking and thus 'up' and 'down' could be perceived. Similarly differential water movements on either side of the spines during swimming could give directional information. In the light, visual cues could supplement or replace the information from the spines.

The megalopa and glaucothoe swim forwards with the dorsal side uppermost. The propulsive force is produced by the pleopods which beat backwards. In the megalopa during swimming the first two pairs of walking legs are tucked up underneath the body while the last two pairs are held above the carapace, the 3rd pair being extended forwards over. This disposition of the legs helps to streamline the animal. The megalopas and glaucothoes are the settlement stages and after a time swimming in the plankton they adopt a benthic existence. The glaucothoes take up residence in tiny empty gastropod shells while the megalopas walk about with their abdomens flexed under their bodies.

Feeding

Zoea larvae feed on a variety of planktonic organisms. Diatoms and the larvae of echinoderms, molluscs and worms as being common food in nature. Rearing experiments, however, have shown that living and moving animal food is preferred to plant food. Indeed, animal food appears to be essential for the successful growth and development of crab larvae while plant food is not necessary. Living food must, of course, be of an appropriate size and since the larvae of other organisms fit this requirement they have frequently been used as food in rearing experiments. The larvae of sea urchins, oysters and copepods have successfully been used in this way. The most frequently used food in modern rearing experiments consists of the newly hatched larvae of the

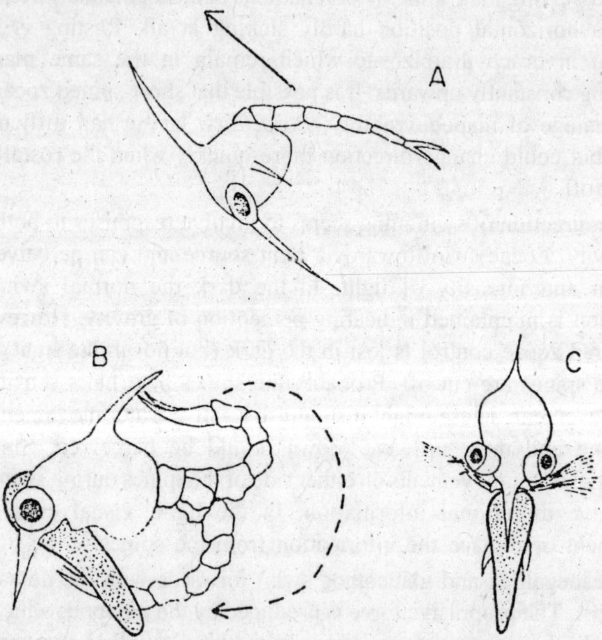

Fig. 12.39 : Larval behaviour. A, the direction of swimming in a brachyuran zoea; after Foxon, 1934. B and C, feeding on *Artemia* larvae by the zoea of an American east coast fiddler crab *Uca pugilator;* after Herrnkind, 1968. B, food capture by tail lashing; C, position during food consumption.

brine shrimp *Artemia*. This food appears to be suitable for all larval stages including the megalopa which, in nature, feeds on other decapod larvae, copepods and young fish. The advantage of using brine shrimp larvae in preference to other food is mainly convenience: the eggs are remarkably resistant and can be kept in a bottle on the shelf for a long time; when placed in sea water they hatch within a few days.

It appears that food is taken by individual acts of capture. This has been described in *Uca zoeae*: the prey, *Artemia* larvae, were captured by lashing the abdomen backwards and forwards. An individual *Artemia* larva is pinned between the forked telson and the rostral spine of the zoea and adjusted until the anterior end is in contact with the mandibles it is then consumed. Tail-lashing was stimulated by the presence of *Artemia* larvae and can, therefore, be regarded as a behaviour particularly associated with feeding. Objects other than *Artemia* larvae which were captured by tail-lashing (other zoeae, sea urchin larvae) were rejected immediately indicating some sensory discrimination in feeding.

Megalopa larvae seize their prey with their chelae which often have hooked tips to prevent escape. Some glauchothoes capture prey with their long 1st and 2nd pairs of walking legs which stretch out anteriorly during swimming; in these forms the legs may be armed with teeth with which to grasp and hold the prey.

Effects of Temperature, Salinity and Starvation

Temperature, salinity and the availability of food are three environmental variables which have received study during laboratory rearing experiments with crab larvae. Like most animals, crab larvae are able to survive under a range of environmental conditions; however, for each species there is generally an optimum area in the range of each variable within which survival, development, growth, etc. are at a maximum. These variables do not, however, act independently on survival; they are all part of the same environment and they tend to interact such that combinations of sub-optimal conditions usually, but not always, reduce survival to a greater degree than a single sub-optimal condition would do.

Within the temperature tolerance range, low temperature tends to slow development and prolongs the larval life. In the natural environment this probably leads to increased mortality from the chance causes of predation, disease, etc. A long larval life, however, should also lead to increased dispersal which, as argued below, may be advantageous. Conversely, high temperatures give rapid development and may lead to increased survival and less dispersal. In the case of low salinity it has been shown that the later larval stages are more tolerant than are the early larvae. This is a necessity for estuarine crabs in which the later larval stages must search out the habitat of the adults. The effects of starvation on crab larvae have shown up a partial independence of molting and growth. When not enough food is available the larvae may continue to molt but may not grow and the larval life is prolonged. This may result in extra zoeal stages since it appears that metamorphosis to the megalopa or glaucothoe depends on the attainment of a certain bulk.

Combinations of temperature and low food show up the interactions of sub-optimal conditions. At high temperature, low food further prolongs the larval life. At high temperatures, with their attendant high metabolic rates low food may produce considerable mortality from starvation.

Reduction in Larval Life

In various groups of crabs there is a reduction in the number of larval stages or in the length of the larval life. In spider crabs, for instance, only two zoeal stages are found. This may lead, in tropical species, to a very short larval life of less than two weeks. In cooler regions, however, a larval life of about one month is more common even with only two zoeae.

Complete suppression of the larval life is very rare in marine crabs; it occurs in a few genera of the Dromiidae and Maiidae. In the fresh water Potamidae, however, direct development is the rule. The eggs are large and yolky and relatively few in number: a female potamid of 100mm carapace width might carry about 200 eggs. There are no free larval stages and the eggs hatch directly into young crabs. Hatchlings of the potamid *Kingsleya* are about 6mm in carapace width; they are called 'marsupial young' since they remain under their mother's abdomen for a time before venturing out on their own. Some fresh water grapsids also have reduced larval lives. One of these is the bromeliad crab *Metapaulus depressus* which lives in the water trapped in the leaf axils ofepiphytic bromcliads in the mountain forests of Jamaica (Hartnoll, 1964). This crab lays relatively few large yolky eggs—about 100, each with a diameter of 1-5 mm, laid by a female of 18m carapace width. There are two non-feeding zoeal stages lasting, together, for three days, and a megalopa lasting for six days. The supression of the larval life in fresh water crabs has presumably evolved because fresh water is unsuitable for the dispersal of larvae. The same applies to land, but the land crabs of the family Gecarcinidae and the land hermit crabs have retained a full larval life. Land crabs, when the eggs are ready to hatch, may migrate considerable distances at night to liberate their larvae in the sea.

Functisaa of Larvae

'Rough calculations of survival to various stages in the life cycle of the mangrove tree crab *Aratus* gave 0.041 per cent survival from egg to newdy settled crab and 17 per cent survival from newly settled crab to an adult of 18mm carapace width. Thus, considerable mortality occurs during the larval life and this is almost certainly true for all marine animals with planktonic larvae. It is likely, therefore, that there is some particular advantage to be gained by releasing the young as an enormous number of tiny larvae rather than as a small number of young crabs. The existence of the latter state in potamids, although in this case an adaptation to fresh water life, shows that crabs are capable of evolving this system.

One advantage that is gained by species with a long planktonic larval life is dispersal. Larvae which drift in tidal and ocean currents are likely to settle some considerable distance away from their parents and may establish new populations in areas in which their species did not previously occur. Similarly, if a local population were to be wiped out by a temporary catastrophe such as a very cold winter-rapid recolonization from the plankton could occur. A difficulty attendant on dispersal is that a proportion of the surviving larvae may be transported to unfavourable environments and may not survive long after settling. This danger is to some extent avoided in many groups with planktonic larvae by delaying metamorphosis until the right environment is present. The ability of the megalopa and the glaucothoe to survive both in the plankton and on the sea bed indicates that some selection of the environment before metamorphosis may also occur in crabs.

A further possible advantage of a reproductive strategy which involves planktonic larvae derives from the fact that the vast majority of the primary production in the sea takes place in the plankton. This primary production, directly or indirectly, provides most of the food for-the animals on the sea bed: the plankton is a net exporter of food whereas the sea bed communities are importers. In general, therefore, food must be more abundant in the plankton than elsewhere in the sea. Planktonic larvae may be able to exploit this abundance and grow more quickly than would be possible in other environments. There are, of course, exceptions to the generalization that food is more abundant in the plankton. The sea bed, being two dimensional, may accumulate food, as detritus

Table 12.2 : Distinguishing characters of the postlarval penaeids

	P. indicus (Fig. 1,a-d)	P. mergurensis (Fig. 1,e-h)	P. monodon (Fig. 1, 1-1)	P. semisulcatus (Fig. 2, a-d)	P. Japomicus (Fig. 2, e-h)	M. dobsoni (Fig. 2i-l)	M. monoceros (Fig. 2, m-p)	M. affinis (Fig. 1, m-p)
1. Chromatophores								
i Telson	present in distal only	present from base half to distal end	present from base to distal end	Present from base to distal end	present from base to distal end	two prominetn to distal end	present from base to distral end	present from base to distal end
ii. Inner uropod ramus	may have a minute one, but usually absent	3 to 4 on median aspect.	a row of chromatophores along median aspect.	numerous, on distal half only	one near base on median aspect	one prominent in the middle	2-3 present	one present at distal end
iii. Outer uropod ramus	absent	absent	3 to 5 in the middle on median aspect	numerous on distal half only	absent	one prominent in the middle	asent	one present at distal end
iv. 6th abdominal segment								
a. Anteriolateral chromatophores	present	present	absent	present	absent	absent	present	present
b. Ventral chromatophores	6	5 to 6 more prominent than in P. indicus	13 to 18	9 to 10	10 to 11	single	4 to 5	2
c. Dorsal chromatophores	absent	absent	absent	absent	absent	usually absent	single	single

2. Colour of chro matophores in live specimens	red	red	maroon	brown	crimson	brown	reddish brown	browr
3. Posteriomedian dorsal spines on 5th abdominal segment	absent after two rostral spine stage	present up to 5 rostral spine stage	absent after two rostral spine stage	present up to stage	present up to 4 rostral spine spine stage	absent 4 rostral	absent	absent
4. Rostrum	reaches end of eye	reaches end of eye	reaches end of eye	long, surpasses end of eye	short, reaching only half length of eye	very short blunt	very short pointed	very short pointed
5. Spinules on dorsal carina of 6th abdominal segment	absent	absent	absent	absent	present all along length of carina	2 to 3 may be present near posterior end	1 to 2 may be preset near near posteri-or end	1 to 2 may be present near posterior end
6. Antennal spine on carapace	absent	absent	absent	absent	prominent	present	present	absent

*The inner end outer uropods of the earlier postiarvae of P. indicus in the sea (up to 3 rostral spine stagel have dhromatophores in the distal half. These chromatophores disappear in the postlarave with 4 rostral spines i.e. the stage at which they nomrally enter the brackishwater regions.

LARVAL CRUSTACEANS

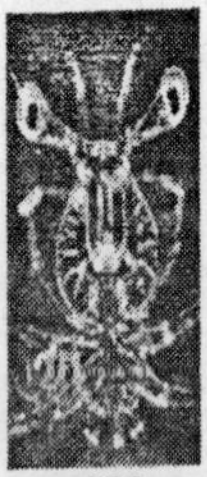

a035 Slipper Lobster Phyllosoma

a033 Slipper Lobster Phyllosoma on Medusa

a017 Slipper Lobster Phyllosoma

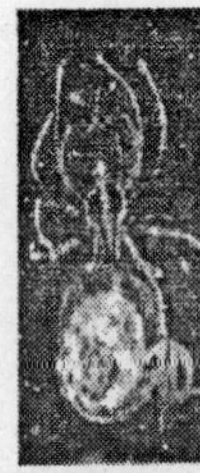

a030 Slipper Lobster Phyllosoma on Medusa

a001 Slipper Lobster Phyllosoma on Medusa

a023 Slipper Lobster Phyllosema on Medusa

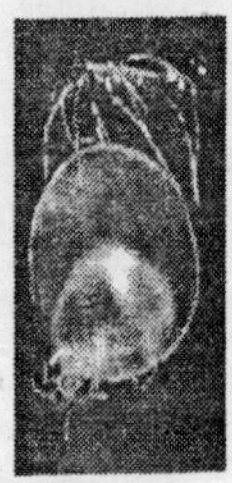

a029 Slipper Lobster Phyllosoma on Medusa

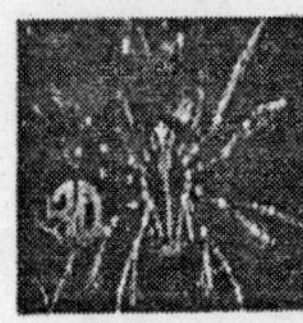

a028 Slipper Lobster Phyllosoma

a026 Slipper Lobster

a027 Slipper Lobster

a016 Porcelain Crab Larva

a034 Squat Lobster

Fig. 12.40 : Downloaded from internet website (Contd..).

a004 Crustacean

a007 Mantis Shrimp antizoea

a006 Mantis Shrimp

a005 Mantis Shrimp

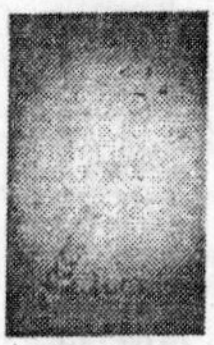
a008 Mantis shrimp

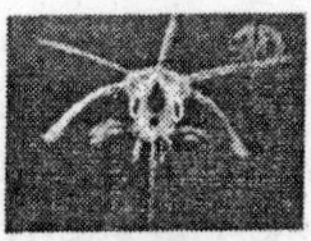
a002 Crab Zoea

a011 Swimming Crab Zoea

a040 Crab Zoea

a015 Flat Crab Megalopa

a038 Flat Crab Megalopa

a043 Flat Crab Megalopa

a022 Flat Crab Megalopa

a021 Crustacean

a020 Crustacean

a037 Crustacean

a014 Shrimp Eretmocaris

a018 Sergistid Shrimp

a013 Sergestid Shrimp

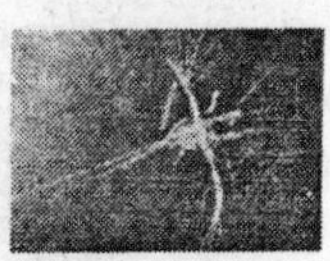

Fig. 12.41 : Downloaded from internet website.

and as resident organisms, to an extent that the local abundance of food exceeds that in the three dimensional plankton above it. However, as the excess of production over consumption in the plankton should result in less competition for food, and less efficiency may be required to exploit the food than is necessary in an environment like the sea bed where production (imports) is more nearly balanced by consumption.

The huge mortality that occurs during the larval life is compensated for by the vast numbers of larvae produced and is, therefore, relatively unimportant. Some danger, however, lies in the variability in this mortality. Good years and bad years for larval survival are well known from fisheries studies (although their causes are not so well understood) and species with planktonic larvae must be able to survive about a year as well as to take advantage of a.good year. Dispersal helps in this context since the year may not be bad over the whole range of the species and settlement of larvae hatched in other areas may still occur. The relatively long length of life in most crab species also helps to reduce the danger of bad years; it is mainly the larvae which die, the adults survive to breed again.

P. semiculcatus: Larval Development

The diameter of the egg ranges from 0.27 to 0.32mm, with a mean of 0.30mm: the smaller sizes are newly spawned eggs with previtelline space almost absent and were rare in plankton. Advanced eggs however, are larger in size and possess a narrow previtelline.

Nauplius I

Total length (TL) 0.32-0.35 mm, mean 0.34 mm. The greatest body width is 0.15-0.17mm with a mean of 0.16mm. The pear-shaped nauplius possesses a ventrally projecting "labrum and bears three appendages. Antennule A1 is uniramus and bears three long setae and one minute terminal seta together with two ventrolateral setae, the proximal being shorter. Antenna A2 is biramus, the endopod baring two long terminal setae and three short ventrolateral setae; the exposed unsegmented and bears three shrot ventrolateral setae, the outer two of which are slightly shorter, and two long ventrolateral setae. The posterior portion of the body is rounded and bears a single pair of caudal setae. An ocellus, which is present in all nauplius substages, is located in the anteriomedian region. The mandible is biramus with exposed and endopod each bearing three long terminal setae which persist in all subsequent naupliar substages. The body is smooth when viewed laterally.

Nauplius V

TL 0.39-0.41 mm, mean 0.43 mm, and mean body width 0.19 mm. The body is elongated and pear-shaped with the anterior protion being wider, and the posterior portion having a pronounced cleft with two rounded furcal processes each bearing five setae. The outline of a developing carapace can be seen on the dorsal surpace of the body, frontal organs and rudiments of ventral appendages are-present. A1 1s segmented and bears two "long setae and one short seta terminally, and two long distal setae and one short proximal seta are present on the inner margin. The A2 endopod bears three long terminal setae and three shorter setae on the inner margin. The exopod is divided into ten segments, and bears four terminal setae with the outer seta being shorter; four long and one shrot proximat setae are present on the inner margin.

Protozoea I

Carapace round and well developed, CL 0.42-0.48 mm, mean 0.46 mm. A pair of compound eyes seen through the carapace, are present although the naupliar eye persists. The posterior portion of the body is divided into a six-segmented thorax and an unsegmented abdomen connected with the telson, which is deeply notched and possesses seven spines on each lobs. The outer and inner most spines are shorter than the remaining spines. A1 is divided into three segments with the basal segment faintly divided into five subsegments. The distal segment bears two long and three short terminal setae, and one. short ventrolateral seta. The middle segment is longer with one short mid-ventrolateral seta. One short distal ventrolateral seta is present on the basal segment. A2 is biramus, endopod is two segmented with five terminal setae, four of which are very long. Two shorter setae are present at the junction between the first and the second segment. One ventrolateral seta is present on the second segment and one at the junction between the second segment and the protopod. The endopod setae are usually presented as 1 + 1 + 1 + 5T (T = terminal). This setation formula persists throughout the subsequent protozoeal substages. The exopod of A2 is ten-segmented having eleven setae on the inner and distal margin, and two on the outer margin.

Protozoed II

The carapace is elongated ranging in length from 0.64 to 0.73 mm with a mean of 0.71mm. The presence of compound stalked eyes, rostrum, a pair of bifuracted supraorbital spines, and segmented abdomen, are the most apparent modifications distinguishing this stage from the

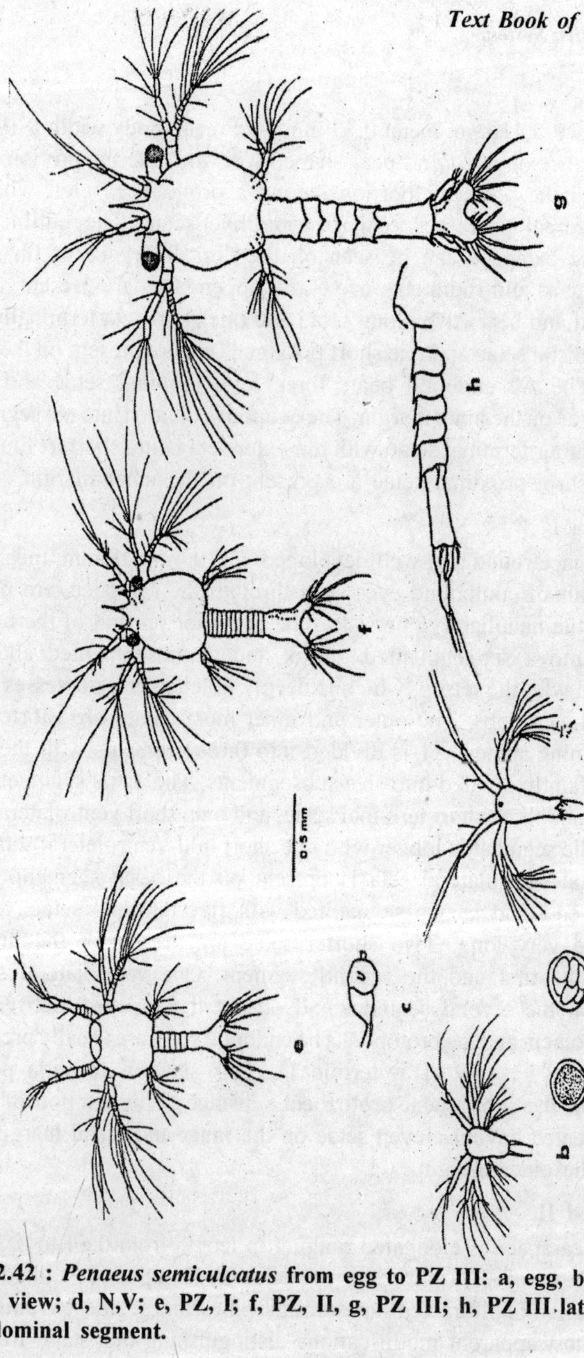

Fig. 12.42 : ***Penaeus semiculcatus*** **from egg to PZ III: a, egg, b, NI; cNI lateral view; d, N,V; e, PZ, I; f, PZ, II, g, PZ III; h, PZ III lateral view of abdominal segment.**

preceding substage. The segmentation of A1 and A2 remains as described in PZ I. Only one short seta is added on the outer margin of A1. The setation of A2 and telson spines remains unchanged.

Protozoea III

The carapace is elongated, CL 0.76-0.90 mm with a mean of 0.86 mm. The major difference between this substage and the preceding one are the presence of abdominal spines, the biramus uropodos, simple supraorbital spines, and the absence of the five subsegments of the A1 basal segment. A1 bears four long terminal setae plus three short setae; two short setae are present on the inner margin of the middle and basal segments. These segmentation and setation of A2 remain as in PZ I and 'II. The abdomen is six-segmented, the sixth segment being three times longer than any other individuai segment. Distinct dorsomedian spines are present on out abdominal segments except the sixth. The fifth and sixth segments each bear a pair of posteriolateral spines. A pair of ventrolateral spines and a single ventromedian spine are also present on the sixth segment. The exopod of the uropod possesses three short setae, and the endopod one only. An additional pair of spines are added to the caudal furca.

Mysis I

The carapace length (CL) range is 0.95-1.15 mm with a mean on 1.00 mm. The difference between this substage and the preceding PZ III are the appearance of functional pereiopods with long setose exopods; the exopods of A2 which are flattened to form the antennal blades; and the distal segment of A1 which is branched. One pair of hepatic and pterygostomial spines are present on the carapace. The rostrum is long and extends beyond the eye tip, one pair of supraorbital spines persists. A1 is three-sigmented, the distal segment has two terminal branches, the external one of which is longer, and both branches possess several terminal setae. The mid-segment is the same length as the distal segment and bears two long setae at the juction with the distal segment and one of the inner margin. The proximal segment is as long as the other tow segments confined and hears a prominent ventromedian spine and a rudiment of the statocyst. Two setae are present above the statocyst. Numerous small setae are present at the junctions of the three antenaal segments. The A2 endopod is rodlike, without setae, and about half the length of the exopod. The exopod is broadened having ten terminal and lateral setae. The dorsomedian spine on abdominal segments IV, V and

VI persists. The pairs of posteroventrat spines on segments V and VI are still present along with a single ventromedian spine on segment VI. The tetson bears seven pairs of terminal and subterminal spines plus one pair of lateral spines making a total of nine pairs, which is characteristic of all *Penaeus* species. The telson spine formula (9 + 9) persists in all the subsequent mysis substages and early postlarval stages. The uropod is biramus, the exopod possesses ten setae on its terminal and inner margins and bears a prominent posterolateral spine and a posterioventral spine on the protopod of the uropod. These spines persist in all subsequent substages of the mysis. The endopod of the uropod is unsegmented and bears seven terminal and marginal setae.

Mysis II (M II)

Carapace length ranging from 1.21 mm to 1.32 mm with a mean of 1.3 mm. The major differences between M I and M II are the presence of unsegmented pleopods in the latter substage, together with a segmented outer branch to the At distal segment. The rostrum extends beyond the eye and supraorbital, hepatic, and pterygostomial spines are present. Chelae on the first three pereiopods are well developed and each bears a single terminal seta. Spination of the abdominal segments remains as before. One seta is added to the inner branch of the distal segment of A1 and the outer branch is now two-segmented,. Five additional setae are added to the exopod of A2 and the endopod now bears one terminal seta. Telson spines remain unchanged, the exopod and endopod of the uropod each possess six additional setae.

Mysis III

The carapace length ranges from 1.46 mm to 1.53 mm with a mean of 1.50 mm. The main changes between this subsequent and the preceding substage are the presence of two segmented pleopods and a two - three segmented endopod of A2. A dorsorostral spine is now present on the carapace. The inner branch of A1 is obscurely two-segmented in smaller M III and distictty divided into two segments in larger M III. Both branches are equal in length; the statocyst at the base of the basal segment is well developed. The A2 endopod is two-segmented and bears two terminal setae. Abdominal and telsor spines persist as before.

Postlarva 1.44 mm CL

The postlarval stage differes from the last mysis substage in the presence of functional pleopods and the reduction of the exopods of the perciopods in early postlarvae, and their complete disappearance in advanced postlarvae. The rostrum now extends beyond the eye and bears four dorsal spines and fourteen short dorsomaginal setae. Supraorbital and pterygos tomial spines are present at CL 1.10 mm and both disappear at CL over 1.10 mm, although, hepatic spines persist. The dorsomedian and posterolateral spines on the fifth and sixth abdominal segments, and the ventrolateral spines on the sixth segment are still present. The outer branch of the distal segment of A1 is four-segmented, and the inner branch three-segmented. The A2 flagellum (endopod) is shorter than the antennal blade (exopod) and is seven-segmented. Nine pairs of telson spines persist, the medioposterior margin of the let son is now pointed. Five greenish chromatophores are present on the posterior dorsal.surface of the telson, two on the posteromedian surface of the exopod of the uropod, and six on the posteriomedian surface of the endopod. Six similar chromatophores are seen on the posterotateral and ventrolateral surfaces of the sixth abdominal segment, and one on the posterioventral surface fo the fifth segment.

Postlarval stages (1.10-2.6 mm CL) rostral and telson spination. The change in carapace length in relation to the eye stalk length and the increase in the number of dorsal and ventral spines on the rostrum, together with .the reduction in the number of telson spines as the postlarva gradually metamorphoses into the juvenile stage which resembles the adult shrimp. In the plankton late myses have been observed with two dorsorostral spines and nine pairs of telson spines. Postlarvae of 1.10 mm CL have an additional dorsorostral spine, nine pairs of telson spines, one pair of supraorbital spines, one pair of pterygostomial and hepatic spines, the latter persist in subsequent stages. The telson cleft is slightly shallower than in M III. At CL 1.44 mm, one more dorsorostral spine is added, and pterygostomial and supraorbital spines vanish. Telson spines remain as 9 + 9. At CL 1.87 mm, two additional spines are added on the rostrum, one dorsally and one ventrally. A rudiment of a developing antennal spines appears, but telson spines remain as 9 + 9. At CL 2.2mm two more rostral spines are added on the ventral and the dorsal margins making a total of seven dorsal spines and three ventral spines (7 + 3), as is usually observed in the adult shrimp. Telson spines remain as 9 + 9 with the posterior margin more pointed than in earlier postlarval instars. At CL 2.5mm and 2.6 the rostral spines remain as before, but the telson spines are reduced to eight pairs at CL 2.6. At this stage specimens may be regarded as juvenile shrimp.

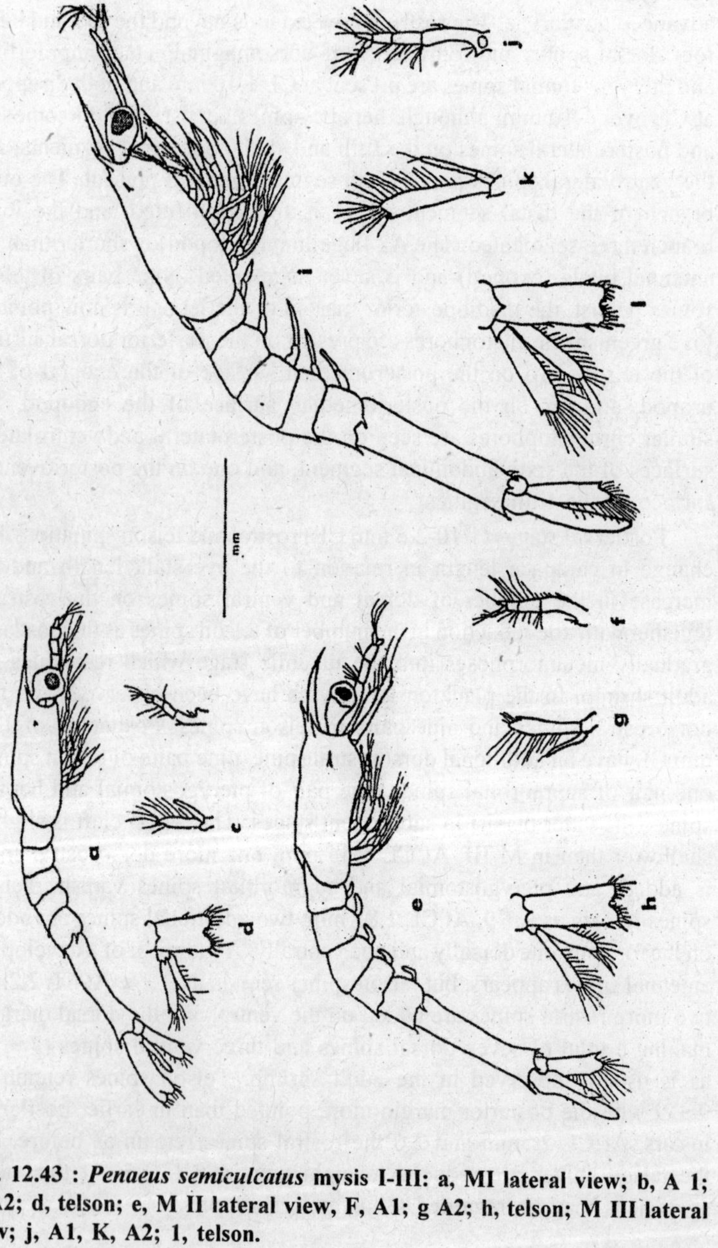

Fig. 12.43 : *Penaeus semiculcatus* mysis I-III: a, MI lateral view; b, A 1; c, A2; d, telson; e, M II lateral view, F, A1; g A2; h, telson; M III lateral view; j, A1, K, A2; 1, telson.

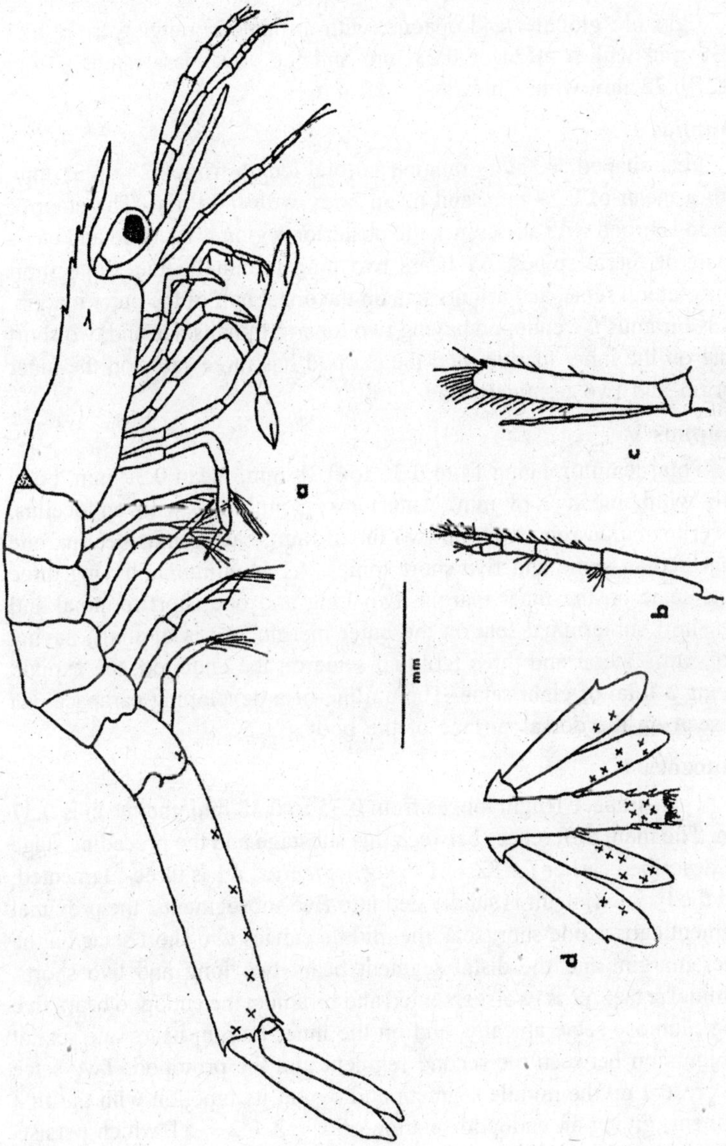

Fig. 12.44 : *Penaeus semiculcatus* postlarva 1.44 mm CL: a, lateral view; b, A1, c, A2; d, telson.

Metapenaeus sp.: Larval development

Eggs are globular and opaque, with a diameter ranging from 0.27 -0.30 mm with a mean of 0.28 mm and the yolk mass ranging from 0.12 -0.22 mm with a mean of 0.20 mm.

Nauplius I

Pear-shaped and stout ranging in total length from 0.28 to .31 mm, with a mean of 0.29 mm, and mean body with 0.09 mm. The anterior region is broad with an ocellus, the posterior region is rounded and bears a pair of furcal spines. A1 bears two long terminal setae, three short ventrolateral setae, and a short seta on the distal half of the outer margin. A2 is biramus the endopod having two long terminal setae and two short setae on the inner margin, and the exopod has three setae on the inner margin and two terminal setae.

Nauplius V

Total length ranging from 0.36 to 0.38 mm, mean 0.37 mm, body with width mean 0.18 mm. Anterior region broad with an ocellus, posterior region rounded with two furcal rami, each ramus bearing one long median spine and five short spines. A1 is uniramus having three short setae on the inner margin, two long and one short terminal and one short subterminal seta on the outer margin. A2 is biramus, having two ventrotaterat and three terminal setae on the endopod, the exopod having a total of eight setae. The outline of a developing carapace can be seen on the dorsal surface of the body.

Protozoea I

The carapace length ranges from 0.35 to 0.38 mm, the mean is 0.37 mm. The main differences between this substage and the preceding stage are described earlier for PZ I of *P. semiculcatus*. A1 is three-segmented, and the basal segment is subdivided into five subsegments; the proximal segment carries one short seta, the middle carries two short setae on the inner margin, and the distal segment bears two long and two short'. terminal setae. A2 is two-segmented and biramus; the endopod bears five long plumose setae apically, and on the inner margin bears one seta at the junction between the second segment and the protopod. Two setae are present on the middle segment and two at its junction with the first segment, giving an endopod setation of 1 + 2 + 2 + 5T which persists in the subsequent substages and characterises the protozoea stages of *Metapenaeus*. Exopod of A2, 10 segmented bearing six ventrotateral, two dorsohiteral,. and three terminal setae, five distinct thoracic segments are present. The abdomen is unsegmented with a round medial bulge, which disappears in the subsequent substages, two rami are present posteriorly producing a deep median deft; each ramus is armed with seven pairs of spines which persist in subsequent substages.

Protozoea II

The CL ranges from 0.42 to 0.49 mm with a mean of 0.46 mm. Rostrum compound eyes, and a simple pair of supraorbital spines are present.

The anteriolateral region of the carapace between the base of the rostrum and the supraorbital spine is concave. Five thoracic and six abdominal segments are present the latter being longer. Telson spines remain unchanged. A1 is three-segmented, five subsegments of the basal segment persist, and four terminal, one subterminal and three ventrolateral setae, are present. A2 is biramus; segments and setae remain unchanged.

Protozoea III

The CL ranges from 0.53 to 0.55 mm, mean 0.54 mm. Supraorbitat spines are prominent, dorsomedian spines are present on the first five abdominal segments, the fifth segment carries a pair of postcrolateral spines, and the sixth a pair of ventrolateral spines. Uropods are biramus and the telson deeply cleft bearing seven pairs of spines. Subsegments If A1 are now fused, the distal segment carries five terminal, three ventrolateral, and one dorsolateral setae. Setation and segmentation of A2 remains unchanged.

Mysis I

The CL ranges from 0.59 to 0.65 mm, the mean is 0.61 mm. The rostrum almost reaches the tip of the eye, but has no dorsal spines, the supraorbiral.spines disappears, a pair of antennal and pterygostomial spines is present. Dorsomedian spines on abdominal segments V and VI persist Pereiopods are present and functional, the first three chelate. A1 is three-segmented, the distal segment having two rudimentary subequal branches bearing four terminal setae, three dorsolateral, and one at the junction with the middle segment. The middle segment carries two dorsolateral, and two ventrolateral setae at the junction with the basal segment. The basal segment is longer than the other segments combined, and bears a prominent ventromedian spine together with a developing statocyst. The endopod of A2 is unsegmented, and is about-half the length of the exopod. The exopod is blade-like with five terminal and thee ventrolateral setae, a posteriolateral spine is absent. Tetson deeply notched bearing-six pairs of terminal and subterminal spines in addition to a pair of lateral spines. The uropod is biramus, the endopod carrying nine ventrolateral and terminal setae and the exopod one posteroventral and one posterolateral spine plus 13 setae.

Mysis II

The CL ranges from 0.66 to 0.72 mm, the mean is 0.69 mm. The rostrum has one dorsal spine and a small hepatic spine; rudiments of pleopods are present. Carapace and abdominal spines are unchanged. A1 is three-segmented with a total of eighteen to twenty setae, the distal segment has two unsegmented branches, the outer branch slightly longer, and both bearing three terminal setae. Statocyst and ventromedian spine persist, a basal spine is now present with three short setae just above it. The A2 exopod is unsegmented and bears a distolateral spine and fifteen setae. A spine has been added to the terminus of the protopod. The endopod is unsegmented, reaching the middle of the exopod, and carrying one terminal and one subterminal seta. The telson has a shallow cleft and seven pairs of spines, exopod of uropod with a posteriolateral spine and 11 terminal and ventrolateral setae; a spine has been added to the end of the protopod. The endopod is unsegmented, and bears eight ventrolateral and terminal setae.

Mysis III

The CL ranges from 0.68 to 0.91 mm, the mean is 0.82 mm. The rostrum is reduced in length and bears two dorsal spines; carapace spines remain unchanged. The first three pereioppds have well developed cheale; pleopods are two-segmented; dorsomedian spines of abdomen persist; an additional ventromedian spine (anal) is present on the sixth abdominal segment. A1 has numerous setae, the distal segment with two unsegmented branches, the outer longer and with five to six setae, the inner branch with three setae. Ventromedian and basal spines and statocyst persist. The A2 exopod has 18 setae, the distolateral and basal spines persist, the endopod is two-segmented and bears four setae.

***Metapenaeus spp.* postlarva, 1.44 mm CL**

Rostrum, short reaching middle eye, five dorsorostral spines, three to four ventrorostral, and six dorsorostral setae are present. Hepatic and antennal spines persist. Pterygostomial and dorsomedian spines on abdominal segment V are absent. Abdominal segment VI bears one dorsomedian and one ventromedian spine. The posteroventral corner of segment VI is pointed, and four short setae are present just above the dorsomedian spine. Pleopods have nine to thirteen setae. A1 is three-segmented and the distal segment has two branches. The outer branch is two-segmented with four setac; the inner branch is three-segmented with numerous setae, the proximal segment is twice as long as the middle and distal segments combined. Ventromedian and basal spines and a well developed statocyst are present with four short setae just above the statocyst. A long apical seta is present on the basal segment. The A2 endopod Uantennal ftagellum) is longer than the exopod and eleven-segmented; the exopod has 30 ventrolateral and terminal setae and a distinct distolateral spine. A smaller spine is also present on the end of the protopod. Telson spines are reduced to six pairs. In early postlarval stages of Metapenaeus spp. the seven pairs of spines seen during the mysis substages remain unchanged.

Metapenaeus sp. **: Postlarval Development**

Metapenaeus affinis postlarva 1.5 mm CL.

The rostrum is short, reaching the middle of the eye, devoid of ventral spines, but with five dorsal spines. Well developed antennal and hepatic spines are present. A band of several large and small greenish chromatophores are present on the mid-carapacc surface, one chromatophore is present near the orbital margin, and one near the dorsal surface of the posterior margin. Two small chromatophores are also seen on the hepatic region of the carapace. This pattern of chromatophores is also present on advanced postlarval stages of this species up to 3.0 mm CL. Two pairs of lateral spines and three pairs of terminal spines are present on the telson, the mid-terminal pair is longer than the all other telson spines.

Metapenaeus affinis **postlarva, 2.5 mm CL**

Rostrum reaches beyond middle of the eye and bears six dorsal spines, antennal and hepatic spines persist and an orbital process begins to appear. Telson bears two pairs of lateral spines and a iarge posteriomedian spine.

Metapenaeus affinis **postlarva, 2.95 mm CL**

Rostrum reaches slightly beyond eye tip, is slightly curved downwards, and bears eight dorsal spines. The telson has a large posteromedian spine and four smaller lateral spines. In adult *Metapenaeus* telson spines become numerous and microscopic. At CL 4.00 mm an additional dorsal spine is added on the rostrum to make a total of nine. At CL 6.00 mm there are 10 dorsal spines, and this number remains unchanged during subsequent growth.

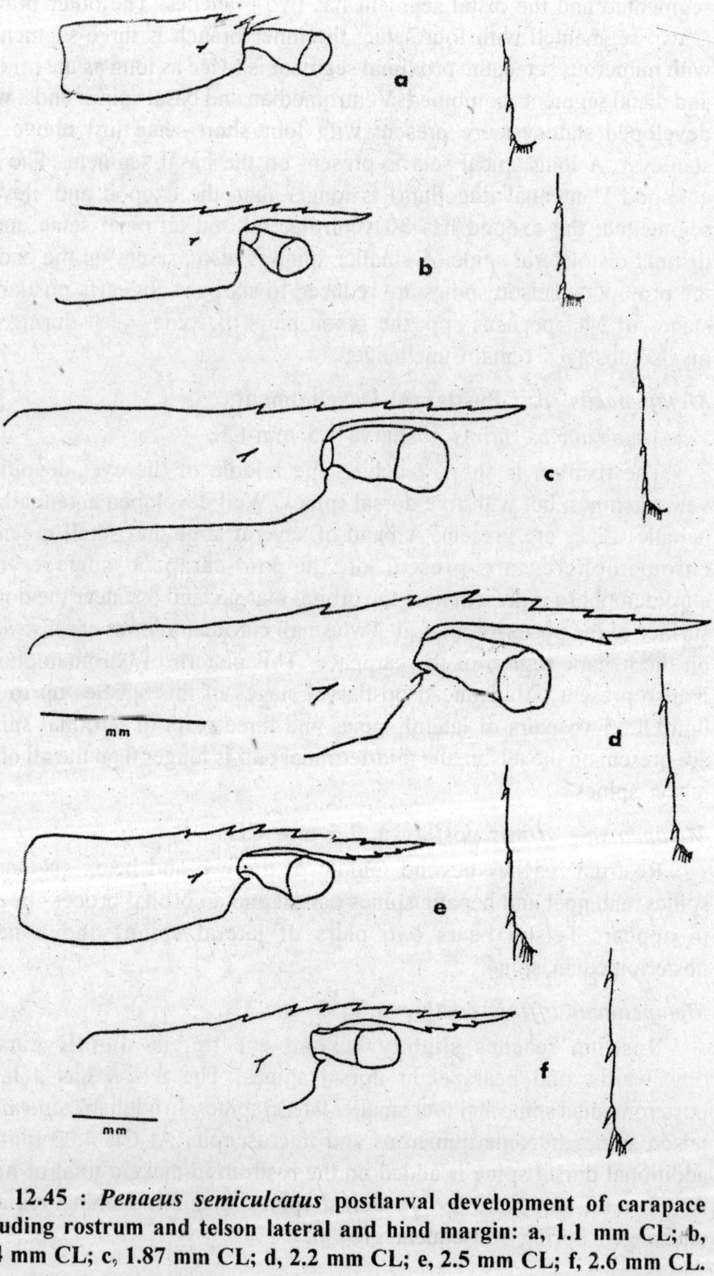

Fig. 12.45 : *Penaeus semiculcatus* postlarval development of carapace including rostrum and telson lateral and hind margin: a, 1.1 mm CL; b, 1.44 mm CL; c, 1.87 mm CL; d, 2.2 mm CL; e, 2.5 mm CL; f, 2.6 mm CL.

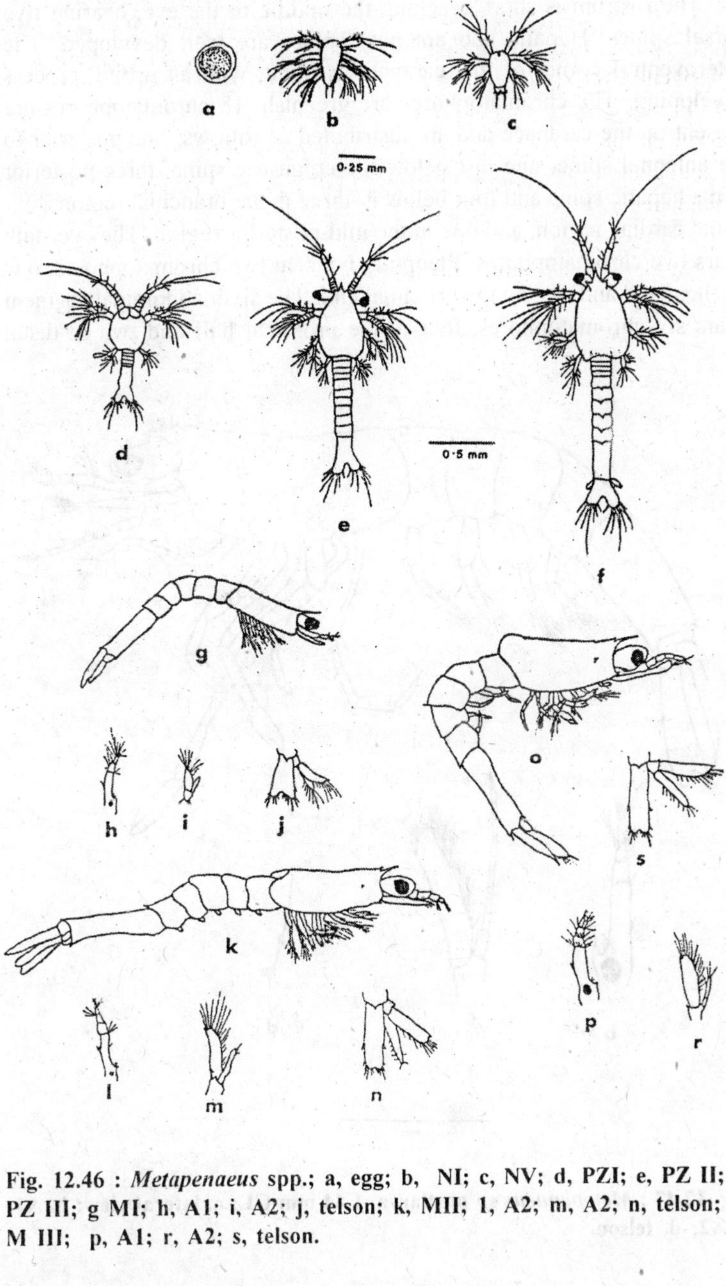

Fig. 12.46 : *Metapenaeus* spp.; a, egg; b, NI; c, NV; d, PZI; e, PZ II; f, PZ III; g MI; h, A1; i, A2; j, telson; k, MII; 1, A2; m, A2; n, telson; o, M III; p, A1; r, A2; s, telson.

Metapenaeus stebbingi **postlarva, 1.82. mm CL**

The rostrum is short, reaching the middle of the eye, bearing five dorsal spines. Hepatic and antennal spines are well developed. The anteroventral corner of the. carapace is blunt, with an orbital process developing. The chromatophores are greenish. 18 chromatophores are present on the carapace and are distributed as follows: one posterior to the antennal spine, one just below the epigastric spine, three posterior to the hepatic spine and four below it, three in the branchial region, four in the cardiac region, and one in the mid-posterior region. The eye stalk bears two chromatophores. Pleopods 1-5 beat two chromatophores, one on the.distal and one on the proximal end. The sixth abdominal segment bears six chromatophores, four in the proximat half and two in distal

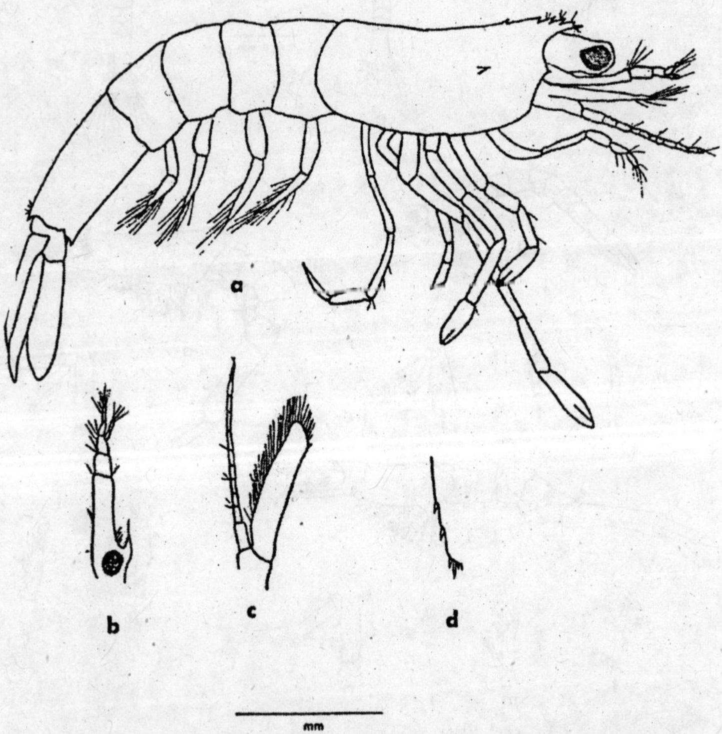

Fig. 12.47 : *Metapenaeus sp* postlarva, 1.44 mm CL; a, lateral view; b, A1; c, A2; d, telson.

half. The telson has four vertically located chromatophores on the distal half and two on the middle surface. The protopod of the uropod, with one only, expod, with five, and endopod with four. This pattern of chromatophore distribution characteristics postlarvae ranging in CL from 1.55 to 3.00 mm. Telson spines at these sizes do not differ from those of *Metapenaeus affinis* postlarvae of the same size range, but the rostrum in *Metapenaeus stebbingi* does not extend beyond the eye, even at 3.00mm CL.

Metapenaeus stebbingi **juvenile, 8.57 mm CL**

The rostrum is long, reaching beyond the eye and bearing ten dorsal spines. The distance between first rostral spine and rostral tip is very short. Hepatic, antennal, and orbital spines are present, the last being less conspicuous. Numerous greenish chromatophores are present over the entire body. The chromatophore distribution on the carapace, sixth abdominal segment, and tetson. Numerous minute chromatophores are present in the posterior region of the exopod of the uropod. These chromatophores give a greenish colour to the exopod. The chromatophore distribution described above remains almost unchanged up to the adult size.

***Parapenaeopsis stylifera*:** **Larval Development**

Egg diameter ranges from 0.30 to 0.42 mm, the mean is 0.35 mm the yolk and the embryonic mass diameter ranges from 0.21 to 0.27 mm with a mean of 0.23 mm. All eggs collected from plankton had a large previtelline space, and a blue sheen was observed around the egg chorion. Most eggs collected from the plankton were at an advanced stage.

Nauplins I

Total length ranges from 0.27 to 0.30 mm with a mean of 0.28 mm, the body width is Q.13-0.15 mm with a mean of 0.14 mm. The body is pear-shaped and fragile, an oceltus is present in the anterior region and the posterior end is narrower without a cleft. Two distinct long furcat spines are present. The three pairs of appendages are also fragile. At bears two long and one short terminal setae, three ventrolateral setae of almost equal size and one dorsoventral seta. A2 is biramus, with endopod bearing two,short ventrolateral and two longer terminal setae, and exopod bearing two ventrolateral and three terminal setae. The mandible (not seen in the figure) is biramus, exopod and endopod each bear three terminal setae.

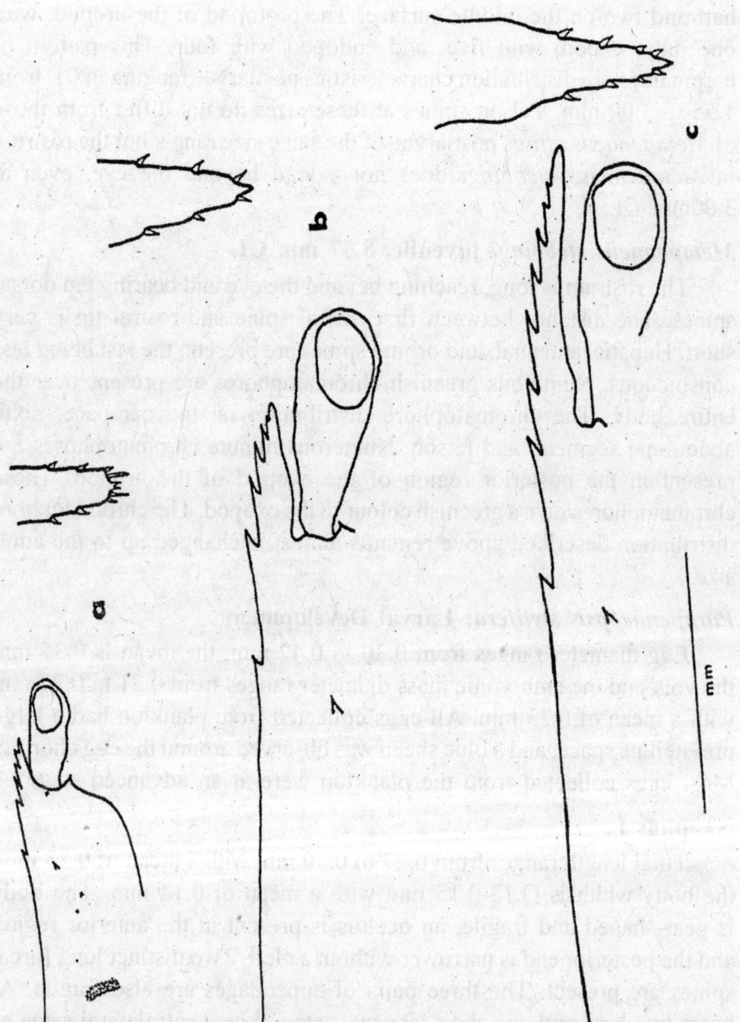

Fig. 12.48 : ***Metapenaeus affinis*** **postlarval development of carapace and rostrum together with telson: a, CL 1.50 mm; b, CL 2.5 mm; c, 2.95 mm.**

Nauplius V

TL ranges from 0.32 to 0.41 mm with a mean of 0.36 mm, BW is 0.14-0.16 mm with a mean of 0.15 mm. Body and appendages are fragile, a small ocellus is present anteriority, the posterior region is bifuraate with six pairs of spines, the medioterminat pair being distinctly longer than the others. The outline of a developing carapace may be seen. At is two-segmented, the second segment being faintly subdivided into five subsegments. A long ventrolateral seta is present at the junction between the first and second segment. This seta is oftern seen pinting towards the anterior region of the body, and occasionally over apping the opposing seta. The other two ventrolateral setae are shor three terminal and one dorsotateral setae are also present. A2 1s biramus, the endopod has two long termianl and three ventrolateral setae, the exopod is divided into five segments with seven long ventrolateral and terminal setae and one short terminal setae. When viewed laterally nauplius stages of *P. stylifera* have a dorsal protuberance.

Protozoea I

CL ranges from 0.26 to 0.30 mm, the mean is 0.28 mm. The carapace is round and glabrous, with a pair of compound eyes showing through with the naupliar oceltus persists. The abdomen is unsegmented, the tetson is deeply deft with seven pairs of terminal and subterminal spines, the medioterminal pair of which are longer than the others. A1 is three-segmented , the distal segment bearing one ventrolateral and five long terminal setae, one of which is characteristically longer than the others and persists in the subsequent substages. The middle segment bears one ventrolateral and one dorsoventral seta at the junction with the proximat segment. The proximal segment is subdivided into four and bears one short ventrolateral seta. A2 is biramus, and the endopod is three-segmented, with two pairs of long ventrolateral setae, four long and one short terminal setae. No setae are found at the junction between the second segment and the protopod. Both number and position of endopod setae are characteristic of the protozoeal stages of a *Parapenaeopsis species*. The A2 exopod is.ten-segmented, bearing 10 ventrolateral and terminal setae.

Protozoea II

CL ranges from 0.40 to 0.49 mm, the mean is 0.43 mm. The carapace is round and devoid of supraorbital spines. Compound stalked eyes, a small rostrum, and a segmented abdomen characterise this substage.

Telson spines remain unchaged. A1 is three-segmented and th subdivisions of the proximal segment persist. Two setae are added to the outer margin of the distal segment and one seta is added to the middle segment. Segments and setae of A2 remain unchanged.

Protozoea III

CL ranges from 0.49 to 0.54 mm, the mean changes in this substage are the presence of a fourth segment in A1, abdominal spines, the biramus uropod and an additional pair of spines on the telson. The carapace is devoid of supraorbitat spines. Dorsomedian spines are present on abdominal segments IV, although the dorsomedian spine of the first segment is minute. The fifth segment bears a pair of posterolateral spines and the sixth segment a pair of ventrolateral spines. The exopod of the uropod bears five terminal setae. Setae are absent form the endopod.

Mysis I

CL ranges from 0.57 to 0.68 mm, the mean is 0.64 mm. The main difference between mysis I and PZ III are as follows: presence of well developed pereiopods, the first three with chelae, dorsomedian spines on abdominal segments I-IV absent, a dorsomedian spine and a ventromedian spine now present on segment VI, posteriolateral pair of spines on abdominal segment V absent. The carapace is elongated but the rostrum is short not reaching the end of the eye; a pair of antennal and ptergostomial spines are present supraorbital and hepatic spine are absent. The telson is deeply notched with eight pairs of spines which persist in the subsequent substages. A1 is three-segmented, with the distal segment bearing two unsegmented branches, the inner branch longer than the outer branch and bearing three terminal setae, outer branch only one. Two lateroventral and one terminal spine present on the distal segment. The middle segment bears three setae at the junction with the distal segment. The proximal segment bears four setae at the junction with the middle segment, and a ventrolateral seta. A prominent ventromedian spine is also present, but a statocyst is absent. The basal spine rudiment is present with three short setae above it. The exopod of A2 is flattened bearing four terminal and ventrolateral setae. The endopod reaches one third of the length of the exopod, and bears one terminal seta. The distolateral spine at the end of the protopod are minute. The endopod of the uropod is shorter than the exopod and bears four small setae, the exopod bears only two setae and a small distolateral spine.

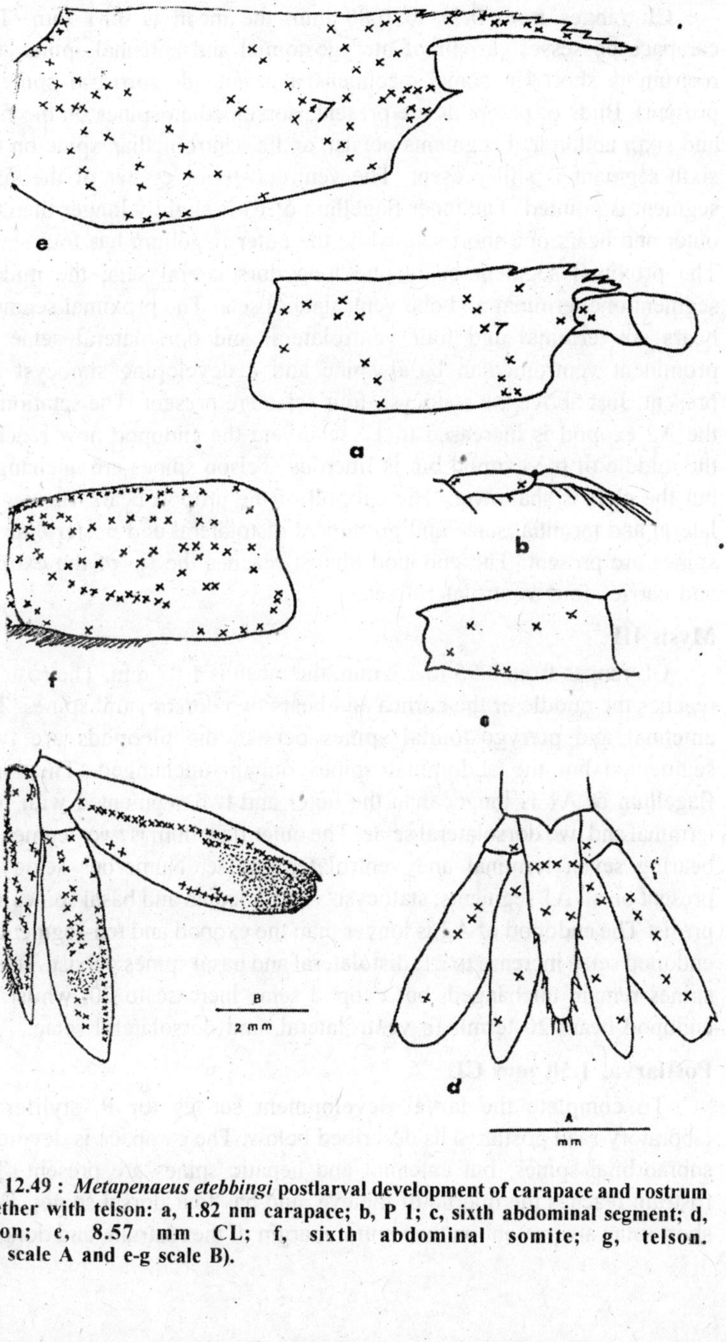

Fig. 12.49 : ***Metapenaeus etebbingi*** **postlarval development of carapace and rostrum together with telson: a, 1.82 nm carapace; b, P 1; c, sixth abdominal segment; d, telson; e, 8.57 mm CL; f, sixth abdominal somite; g, telson. (a-d scale A and e-g scale B).**

Mysis II

CL ranges from 0.71 to 0.88 mm, the mean is 0.81 mm. The carapace possesses a pair of pterygostomial and antennal spines, the rostrum is short (in some specimens a minute dorsorostral spine is present). Buds of pleopods are present, dorsomedian spines on the fifth and sixth abdominal segments persist, and a ventromedian spine on the sixth segment is still present. The ventroposterior corner of the sixth segment is pointed. The inner flagellum of At is slightly longer than the outer and bears one short seta while the outer flagellum has four setae. The proximal segment bears one long dorsolateral seta, the middle segment one terminal and one ventrolateral seta. The proximal segment bears six terminal and four ventrolateral and dorsolateral setae. A prominent ventromedian basal spine and a developing statocyst are present. Just above the statocyst, four setae are present. The setation of the A2 exopod is increased to 17 setae and the endopod now reaches the middle of the exopod but is fiberous. Telson spines are unchanged but the cleft is shallower. The exopod of the uropod bears 10 ventro-lateral and terminal setae and prominent distolateral and posteriolateral spines are present. The endopod almost reaches the tip of the exopod and carries four ventrolateral setae.

Mysis III

CL ranges from 0.95 to 1.2 mm, the mean is 1.07 mm. The rostrum reaches the middle of the cornea and bears two dorsorostral spines. The antennal and pterygostomial spines persist, the pleopods are two-segmented but the abdominal spines remain unchanged. The inner flagellum of A1 is longer than the outer and two segmented with four terminal and two dorsolateral setae. The outer flagellum is two-segmented bearing seven terminal and ventrolateral setae. Numerous setae are present on all A1 segments; statocyst, ventromedian and basal spines still presist The endopod of A2 is longer than the exopod and ten-segmented, endopod setae increase to 21, distolateral and basal spines persist. Telson spines remain unchanged, but exopod setae increase to 20, whilst the endopod bears 20 terminal, ventrolateral, and dorsolateral setae.

Postlarva, 1.50 mm CL

To complete the larval development ser-ies for P. stylifera a laboratory-read postlarva as described below. The carapace is devoid of supraorbital spines, but antennal and hepatic spines are present. The rostrum reaches the middle of the eye, and has four dorsal spines. Four short setae are present on the ventral margin of the rostrum, and dorsally there are two posterior to the second spine, three between the penultimate and the epigastric spine, and three posterior to the epigastric spine. A dorsomedian spine is absent but the posterioventral corner of the sixth abdominal segment is pointed. Six setae are present on the ventral margin of the fifth and sixth segments. A1 is three-segmented with two unequal branches, the- inner branch being longer with five segments, the outer having only three segments. A well developed statocyst and basal projection are also present, together with two long terminal, and six shorter ventrolateral and dorsolateral setae on the basal segment. The exopod of A2 is broad with 37 setae, the endopod is much longer and divided into 20 segments. The telson bears three terminal spines, one long posteriolateral pair, four small posteriomaginal pairs and one long posteriomedian spine; on each side of the latter spine, two long plumose setae are present, exceeding the length of the posteriolateral spine. Two shorter setae are present between the posteriolateral and posteriomarginal spine.

Postlarva 8.7 mm CL

The rostrum is very long and upcurved, reaching beyond the last segment of A1 and bearing seven dorsal spines, the first of which reaches to the middle segment of A1. The distal half of the rostrum is devoid of spines. Numerous short setae are present just anterior to the first dorsal spine and these continue on the dorsal margin of all spines with the exception of the penultimate spine. A prominent antennal spine is present together with a small orbital spine. The telson carries three pairs of small spines laterally and a large posterior spine. Numerous long and short setae are also present on the lateral margins.

***Metapenaeopsis stridulans*: postlarval stage 2.2 mm CL**

The rostrum is short, reaching slightly beyond the cornea of the eye, carrying seven dorsal spines. One short seta is present anterior to the first rostral spine, two on its dorsal margin, one on the dorsal margin of the second rostral spine, and three on the third. Seventeen setae of different lengths are present on the ventral margin of the carapace. Conspicuous hepatic, antennal, and pterygostomial spines are present, and the orbital portion of the carapace is conves, the eye is very large and has a short stalk.

A1 is three-segmented, the distal segment having two branches; the inner branch is longer and has seven segments with four short terminal setae and thirteen ventrolateral and dorsolateral setae. A1 distal segment carries 10 ventrolateral and seven dorsolateral setae. The middle segment

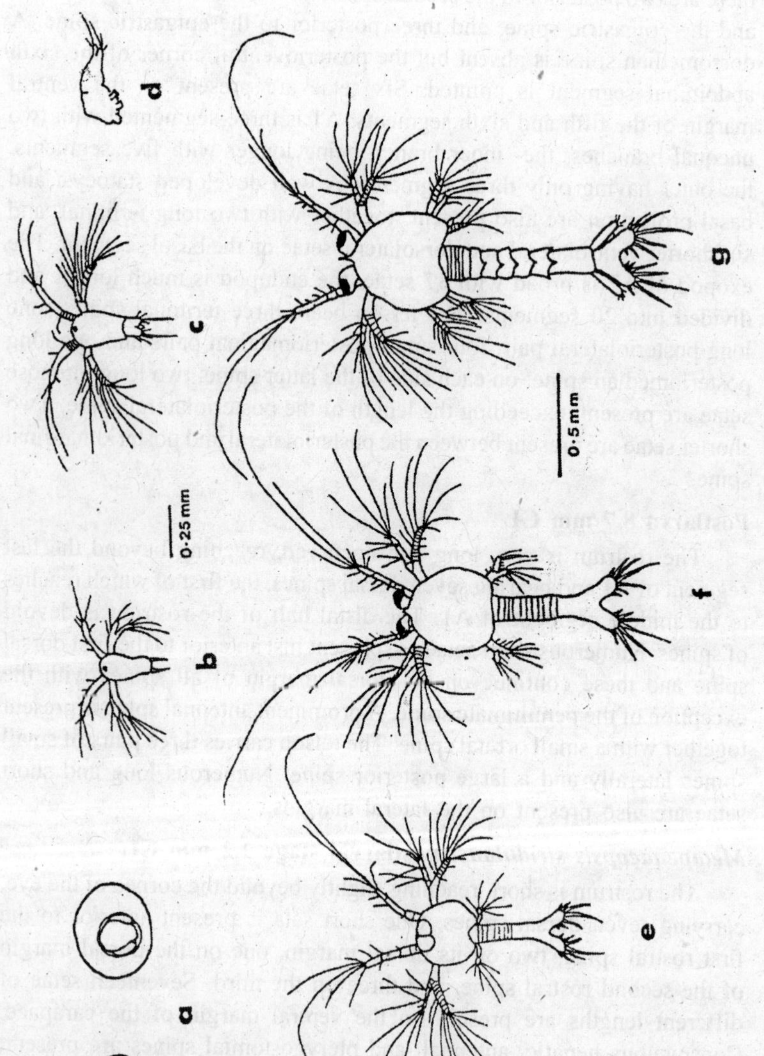

Fig. 12.50 : *Parapenaeopsis stylifera:* a, egg; b, NI, c, NV, lateral view; e, ZI; f, Z II; g, z III, (a-d first scale and e-g second scale).

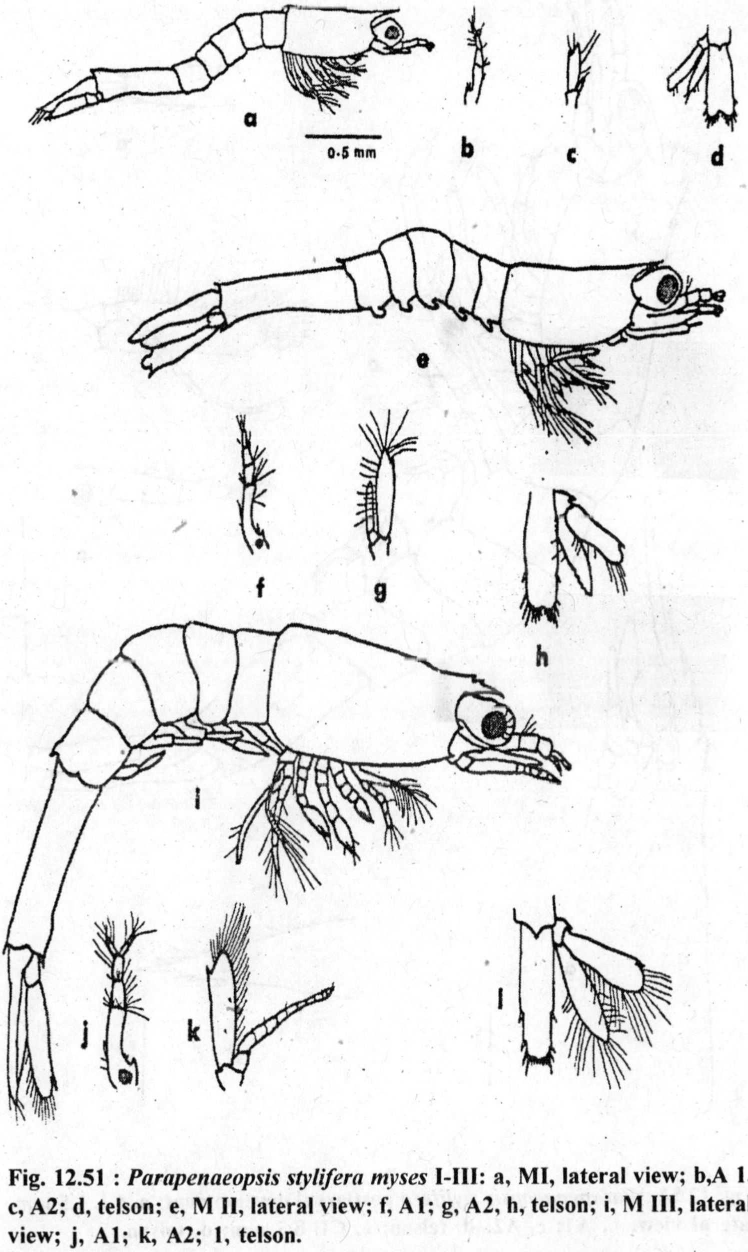

Fig. 12.51 : ***Parapenaeopsis stylifera myses*** **I-III: a, MI, lateral view; b,A 1, c, A2; d, telson; e, M II, lateral view; f, A1; g, A2, h, telson; i, M III, lateral view; j, A1; k, A2; 1, telson.**

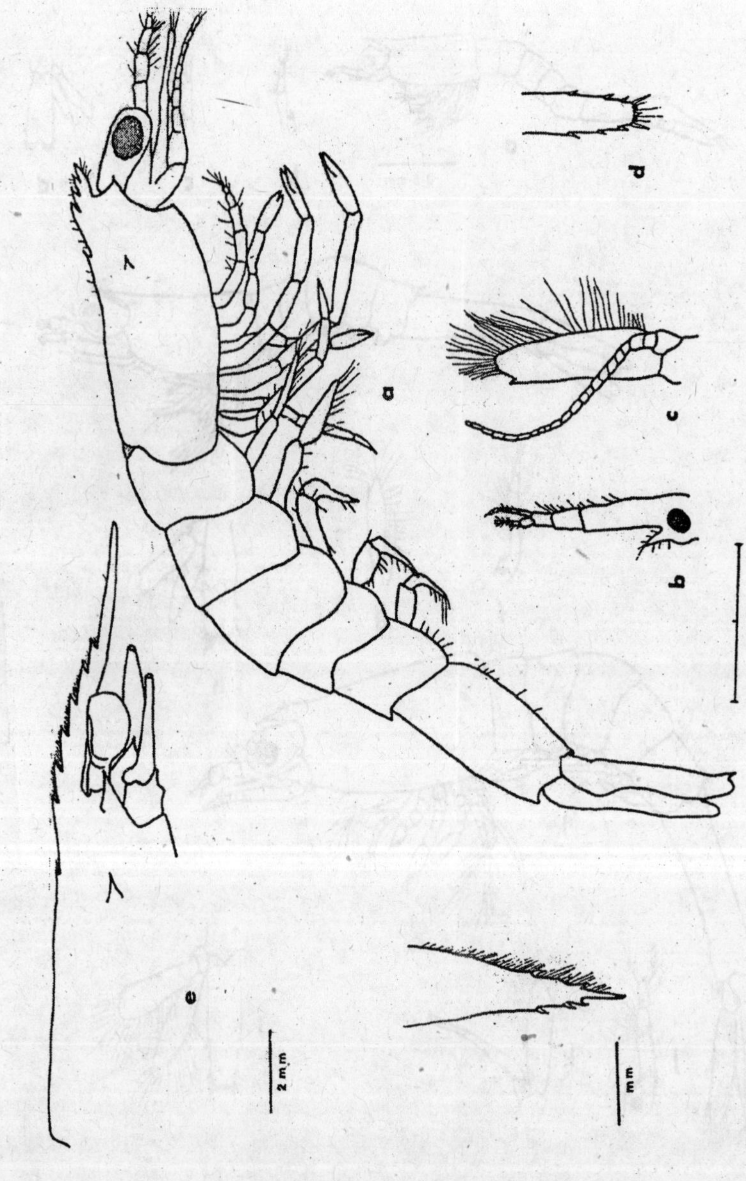

Fig. 12.52 : ***Parapenaeopsis stylifera*** **postlarval development: a, CL 1.5 mm, lateral view; b, A1; c, A2; d, telson; e, CL 8.7 mm; d, telson.**

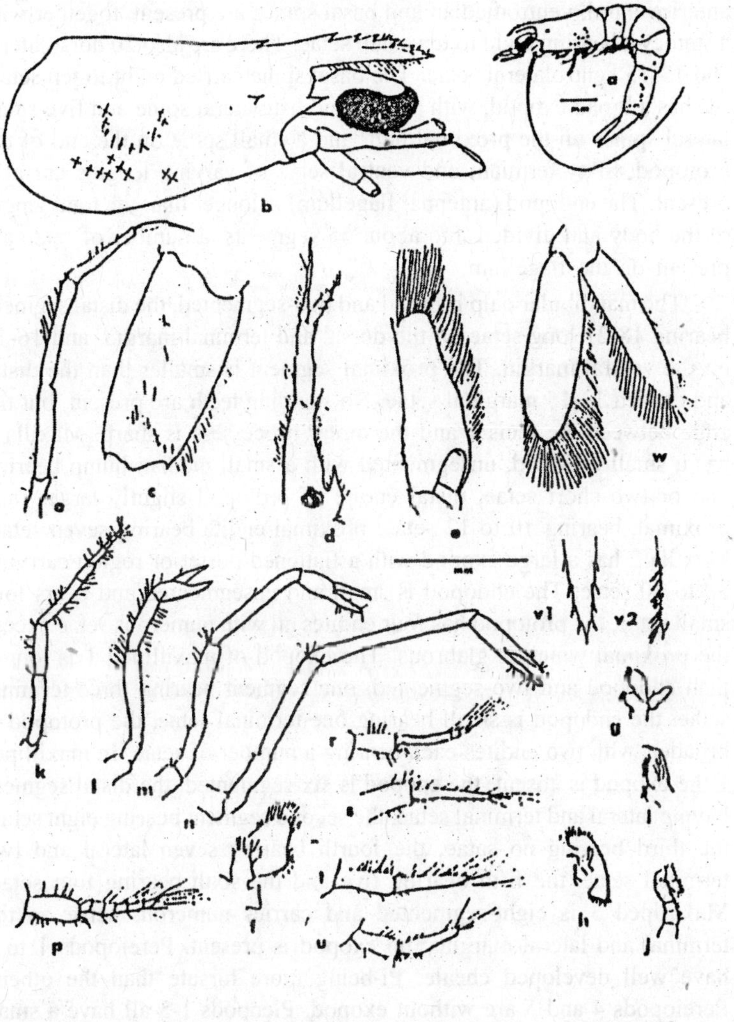

Fig.12.53 : *Metapenaeopsis stridulans* postlarval development 2.2.mm. a, lareval view; b, chromatophore distribution on carapace; c, chromatophore distribution on the sixth abdominal segment; d A 1; e, A 2; f, mandible; g, maxilla; h, maxilla; i, maxillipod; j, maxillipod; k, maxillipod; 1, pereiopod; m, pereiopod; n, pereiopod; o, pereiopod; p, pteopod; r, pleopod; s, pleopod; t, pleopod; t, pleopod; u, pleopod; v1, telson spines; v2, telson setae; w, uropod.

carries 10 setae on each margin and the proximal segment is twice as long as the distal and middle segments combined. Conspicuous anterioventral, ventromedian and basal spines are present, together with a statocyst bearing eight to ten small setae. There are also 20 dorsolateral and 13-15 ventrolateral setae. The basal spine carried eight to ten setae. A2 has a broad exopod, with prominent distolateral spine and five small lateral spines on the proximal half, and a small spine on the end of the protopod. Sixty termiani and ventral setae of varying lengths are also present. The endopod (antennal flagellum) is longer than the total length of the body and divided into about 55 segments; a number of setae are present on the flagellum.

The mandibular palp is broad and two-segmented, the distal segment bearing 18-21 long setae on the dorsal and terminal margin, and 16-17 on the ventral margin. The proximal segment is smaller than the distal and carried 13-15 marginal setae. No standing teeth are present, but the edge between the incisor and the molar processess is sharp. Maxilla 1 has a small endopod, unsegmented with a small median hump bearing one or two short setae, distal endite of protopod slightly larger than proximal, bearing 10 to 12 setae; proximal endite bearing seven setae. Maxilla 2 has a large exopod with a flattened posterior region carrying 38 to 40 setae. The endopod is small and unsegmented and bears four small setae, the protopod has four endites all with numerous setae except the proximal which is glabrous. The exopod of maxilliped 1 is longer than endopod and two-segmented, one segment bearing three terminal setae: the endopod is small bearing one terminal setae, the protopod is broader, with two endites each bearing a number of setae. In maxillipes 2 the exopod is absent, the enopod is six-segmented, the distal segment having lateral and terminal setae, the second segment bearing eight setae, the third bearing no satae, the fourth bearing seven lateral and two terminal setae, the fifth bearing five and the sixth bearing four setae. Maxitliped 3 is eight-segmented and carries numerous setae on the terminal and lateral margins; no exopod is present. Pereiopods 1 to 3 have well developed cheale. Pi-being more hirsute than the others. Pereiopods 4 and 5 are without exopod. Pleopods 1-5 all have a small bud-like exopod except P 2: P 1-5 are two-segmented with a broad basal segment. The telson has four pairs of hump bearing one or two short setae, distal endite of protopod slightly larger than proximal, bearing 10 to 12 setae; proximal endite bearing seven setae. Maxilla 2 has a large exopod with a flattened posterior region carrying 38 to 40 setae. The endopod is small and unsegmented and bears four small setae, the protopod has four endites all with numerous setae except the proximal which is glabrous. The exopod of maxilliped 1 is longer than endopod and two-segmented, one segment bearing three terminal setae: the endopod is small bearing one terminal setae, the protopod is broader, with two endites each bearing a number of setae. In maxillipes 2 the exopod is absent, the enopod is six-segmented, the distal segment having lateral and terminal setae, the second segment bearing eight setae, the third bearing no satae, the fourth bearing seven lateral and two terminal setae, the fifth bearing five and the sixth bearing four setae. Maxitliped 3 is eight-segmented and carries numerous setae on the terminal and lateral margins; no exopod is present. Pereiopods 1 to 3 have well developed cheale, P1 being more hirsute than the others. Pereiopods 4 and 5 are without exopod. Pleopods 1-5 all have a small bud-like exopod except P 2: P 1 - 5 are two-segmented with a broad basal segment. The telson has four pairs of spines and a large terminal spine. Lateral plumose setae are also present on the tetson with one pair between the first and second telsonic spine, three pairs between second and third, two pairs between third and fourth, and two pairs between the fourth and terminal spines. The exopod of the uropod carries 49 long terminal and ventrolataral plumcs setae and nine short seate on-the dorsal margin, a posterotateral spine and a small spine on the-hind margin of the protopod are also present. The endopod is shorter than the exopod, and bears 58 plumose setae which decrese in length anteriorly.

Bluish chromatophores are usually found over the entire body,but their distribution on the carapace and the sixth abdominalsegment is of taxonomic importance. Eleven of these chromatophores form a crescent shape on the .branchial region of the carapace, nine chromatophores are found below the crescent. On the sixth abdominal segment, there are eleven chromatophores, seven of which form a line beginning from the posteroventral corner extending in a wide are to the ventral margin.

Most penaeid shrimp eggs are found, in plankton during their advanced stages. Night plankton samples may contain the earliest developmental stages of penae.id eggs. Assigning these eggs to genera of species is a difficult task, which has been completely avoided by most authors. This is especially true in those areas where species from a number of genera are equally dominant. The same problem is encountered when attempts'are made to identify nauplii.

In the Arabian Gulf, however, it has been possible to identify planktonic penaeid eggs and nauplius to species. This has been due to the differences in the time of spawning exhibited by dominant species, which also tend to belong to different genera. The spawning of *P. semiculcatus* occurs before that of *Metapenaeus spp.* or *P. stylifera.*

Eggs of Kuwaiti penaeid shrimp are of.different sizes, with the eggs P. semiculcatus being larger than those of *Metapenaeus sp.*, which have a very narrow previtelline space, but smaller than those of *P. stylifera* which have a much wider previtelline space. Similar observations have also been made elsewhere.

Similarly, Penaeus naupliar stages may be isolated easily from those of other genera as they lack a dorsal protuberance. In this investigation *P. semiculcatus nauplii* appear to lack the dorsal protuberance, whilst nauplii of *Metapenaeus sp.* and *P. stylifera* possess one. The relative length of the ventrolateral setae on the A1 in N and NV described in this investigation provide another characteristic feature by which the nauptii of *P. semiculcatus, Metapenaeus sp.*, and *P. stylifera* may be separated. In *P. semiculcatus* nauplii substages I and V there are three pairs of ventrolateral setae, the most distal of which is longer than the others. In *Metapenaeus spp.*, they are equal at N1, at later stages the proximal seta becomes longer than those in the middle, and those in the middle longer than the proximal. In *P. stylifera* N1, the proximal seta is longer than the other two and at Nil-NIV, the proximal setae become much longer than the other two and point anteriorlly, occasionally overlapping the opposing setae. It is reported the presence of a small posterior denticle or spine in N1 of *P. semiculcatus*; this spine has also been observed on the N1 of other *Penaeus* and *Metapenaeus* species. A posteriomedian spine is absent in *P. stylifera.* In this investigation a posterio- median spine on N1 of all species was not seen, but it would seem to be an important diagnostic character which could be used to distinguish the N1 of *Metapenaeus* and *Penaeus* species from those of *Parapenaeopsis*.

Protozoea I—III belonging to *P. semiculcatus, Metapenaeus* and *P. stytifera*, can be separated using the following characters. The A2 endopod setation formula for *P. semiculcatus* is 1 + 1 + 2 + 5T (T = terminal), for *Metapenaeus spp.*, it is 1 + 2 + 2 + 5T and for *P. stylifera* it is 0 + 2 + 2 + 5T. *M. affinis*, which is very common in Kuwaiti waters, has been described as having an endopod setation of 1 + 2 + 3 + 5T. In the present of work, however,- this pattern was not found in *Metapenaeus* protozoea recovered from the plankton. The A1 proximal segment at PZ I - II substages was divided into five subsegments in *P. semiculcatus* and Metapenaeus spp. In *P. stylifera*, however, four subsegments only were present. Rao (1973) reported five subsegments, whereas Muthu *et al.*, (1978b) reported four subsegments for PZ II of this species. Both authors seem to disagree with Haq and Hassan (1975) who reported three subsegments for PZ I, and four for PZ II of *Parapenaeopsis* species inhabiting Pakistan waters. The discrepancies encountered here may be due to different geographical varieties of the same species in different parts of the same species in different parts of the world.

Supraorbital spines are completely absent in *P. stylifera* protozoeal substages, but present in *P. semiculcatus* from PZ II persisting up to early planktonic postlarval stages (CL 1.10 mm). Seven pairs of telson spines are found in PZ I and II of all species and an additional pair is added in the PZ III substage of *P. semiculcatus* and *P. stylifera*, but not of *Metapenaeus* spp. The spines on abdominal segments may also be used to separate protozoea of different species. Dorsomedian spines are present on abdominal segments I-V in PZ III of all species described. Gurny (1972), however, reported the presence of dorsomedian spines on segment II-V only for *M stebbingi. P. semiculcatus* have one pair of ventromedian spines on the sixth abdominal segment, which has also been reported..in other *Penaeus* species. Devarajan *et al.*, (1978), however, overlooked the presence of the ventrolateral and ventromedian spines in PZ III of *P. semiculcatus* but reported their presence in MI.

The following differences between substages of the mysis stage of the three shrimp species may be used to isolate each species. The rostrum extends beyond the eye in all mysis substages of *P. semiculcatus*, but not *Metapenaeus*. In this species the rostrum is reduced in M II and M III, agreeing with,the observations made by Muthu *et. al*. (1978a) for *M. affinis*. The rostrum in *P. stylifera* mysis stages does not reach the eye. A pair os supraorbitat spines are present in mysis stages of *P. semiculcatus* and absent in all myses of the other two species. Hepatic spines, however are absent only in the mysis stages of *P. stylifera*, whereas an antennal spine is only absent in the substages of P. semiculcatus. The endopod of A2 is shorter than the exopod of M III of *P. semiculcatus* and *Metapenaeus* sp. but in M III of *P. stylifera* it is much longer. Telson spination is different for each species; in *P. semiculcatus* it is 9 + 9, in *P. stylifera* it is 8 + 8 and in *Metapenaeus* sp. it is 7 + 7. In addition the telson cleft is deep in M I-III of *P. semiculcatus* and *Metapenaeus* sp., but is shallow in M II and completely basent in M III of *P. stylifera*. These observations follow the general characteristic generic features shown by species of *Penaeus, Metapenaeus* and *Parapenaeopsis*.

The rostrum in nearly postlarval stages of *Penaeus semiculcatus* projects beyond the eye and bears three dorsal spines. In subsequent stages the number of spines increases up to seven dorsal and thee ventral. In contrast the early postlarval stages of M. affinis, *M. stebbingi, P. stylifera* and *M. stridulans*, have a short rostrum that does not bear ventral spines. The length of the rostrum, however, increases in advanced stages (CL over 2.95 mm) and it becomes armed with more than seven sorsat spines in all these species, excepting *P. stylifera* (CL 8.7 mm). The rostrum of this species bears seven dorsal spines, with the distal one located half way along the rostrum. Supraorbital spines are present only in the early postlarval stages of *P. semiculcatus* (CL 0.98-1.10 mm). Pterygostomial spines, however, vanish in advanced postlarval stages of *P. semiculcatus, M. affinis, M. stebbingi,* and *P. stylifera* (CL 1.5 mm and 8.7 mm), but are present in early postlarval stages (CL 1.10 mm) of *P. semisulcatus*. A large pterygostomial spine is seen in a 1.50 mm CL postlarva of *M. stridulans*. The antennal spine it well developed in early and advanced postlarval stages of all species except *P. semiculcatus* postlarvae until they reach CL 1.87 mm. The inner branch of A1 is three-segmented in early postlarvae of *Metapenaeus* sp., four-segmented in *P. semiculcatus*, five in *P. stylifera* and seven in *M. stridulans* (CL 2.20mm).

The A1 proximal segment in all species has a well developed basal and ventromedian spine, but in the *M. stridulans* postlarva, a conspicuous anterioventral spine is also present. This anteroventral spine is a characteristic feature of the genus *Metapenaeopsis*. The only species of *Metapendeopsis* recorded in Kuwaiti waters is *M. stridulans* and thus, the postlarva observed during present investigations is almost certainly *M. stridulans*. The endopod of A2 is shorter than the exopod in the early postlarval stages of *P. semiculcatus, Metapenaeus* sp., but longer in *P. stylifera* and *M. stridulans*. The latter species may be further identified as *M. stridulans* as the A2 in M II and completely basent In M III of *P. stytifera*. These observations follow the general characteristic generic features shown by species of *Penaeus, Metapenaeus* and *Parapenaeopsis*.

The rostrum in nearly postlarval stages of *Penaeus semiculcatus* projects beyond the eye and bears three dorsal spines. In subsequent stages the number of spines increases up to seven dorsal and thee ventral. In contrast the early posttarval stages of *M. affinis, M. stebbingi, P. stylifera* and *M. stridulans*, have a short rostrum that does not bear ventral spines. The length of the rostrum, however, increases in advanced stages (CL over 2.95 mm) and it becomes armed with more than seven sorsal spines in all these species, excepting *P. stylifera* (CL 8.7mm). The rostrum of this species bears seven dorsal spines, with the distal one located half way along the rostrum. Supraorbital spines are present only in the early postlarval stages of *P. semiculcatus* (CL 0.98-1.10mm). Pterygostomial spines, however, vanish in advanced postlarval stages of *P. semiculcatus, M. affinis, M. stebbingi,* and *P. stylifera* (CL 1.5mm and 8.7 mm), but are present in early postlarval stages (CL 1.10mm) of *P. semiculcatus*; A large pterygostomial spine is seen in a 1.50mm CL postlarva of M. stridulans. The antenna! spine is well developed in early and advanced postlarval stages of al l species except *P. semiculcatus* postlarvae until they reach CL 1.87mm. The inner branch of A1 is three-segmented in early postlarvae of *Metapenaeus* sp., four-segmented in *P. semiculcatus*, five in *P. stylifera* and seven in *M. stridulans* (CL 2.20 mm).

The At proximal segment in all species has a well developed basal and ventromedian spine, but in the *M. stridulans* postlarva, a conspicuous anteroventral spine is also present. This anteroventral spine is a characteristic feature of the genus *Metapenaeopsis*. The only species of *Metapendeopsis* recorded in Kuwaiti waters is *M. stridulans* and thus, the postlarva observed during present investigations is almost certainty *M. stridulans*. The endopod of A2 is shorter than the exopod in the early postlarval stages of *P. semiculcatus, Metapenaeus* sp., but longer in *P. stylifera* and *M. stridulans*. The latter species may be further identified as *M. stridulans* as the A2 exopod, in addition to distolateral and basal spines, bears five spines proximately. Telson spine formulae in the early postlarvae are 7 + 7 for *Metapenaeus* sp., 8 + 1 + 8 for *P. stylifera* 9 + 9 for *P. semiculcatus*, 4 + 1 + 4 for *M. stridulans* (CL 2.2mm). Telson spine formulae change drastically in later postlarval stages.

Chromatophore number and pattern of distribution on the carapace, abdominal segment VI, and telson were found to be useful in separating the postlarval stages of *P. semiculcatus, M. stebbingi, M. affinis* and *M. stridulans*. Chromatophore pattern of distribution, however, are not a reliable taxonomic character in preserved animals and should be avoided.

From the above it is evident that characteristic differences are present in the planktonic and benthic larval stages of penaeid shrimps inhabiting Kuwait waters. These provide useful criteria for differentiating the larval stages under consideration, and a key based on these characters is provided for the first time.

Eggs of *P. stylifera* have a wide previtelline space and a diameter range of 0.30-0.42mm (mean 0.35mm), which allows them to be isolated from eggs of *P. semisulactus* and *Metapenaeus* spp. The eggs of the latter species have narrower previtelline speces, although, their size range is similar. Thus, without the presence of the .subsequent stages in the plankton it may be difficult to separate them. Muthu *et al*. (1978a), however, measured the previtelline space and found that *Penaeus* species have a previtelline space of about 15 mm in width, 20-30 urn for *Metapenaeus*, except *Metapenaeus dobsoni* (Miers), and 60 urn for *Parapenaeopsis* species.

Nauplius

1. Dorsal protuberance present 2

   Porsal protuberance absent,

   posterlomedian spine present, ... *P. semiculcatus*

2. Body and appendages fragile, proximal seta of A1 is longer than the two anterior setae in N II-VI and points posteriorly, posteriomedian spine absent .... *P. stylifera*

   Body stout, distal ventrotaterat seta of A1 usually longer than the two posterior setae .... *Metapenaeus spp,*

Protozoea

1. Supraorbital spines absent 2
2. Endopod setation 0 + 2 + 2 + 5T .... *P. stylifera*

   Supraorbital spines present in

   protozoea II-III 3

3. Simple Supraorbital spine in protozoea II and III, A2 endopod setation 1 + 2 + 2 + 5T, telson spines in protozoea I-III 7 + 7 .... *Metapenaeus spp,*

   Bifurcated supraoribital spines in protozoea II and simple in protozoea III, telson spines 7 + 7 in protozoea I - II and 8 + 8 in protozoea III .... *P. semiculcatus*

CRAB LARVA

This is the larva of a marine shrimp (many freshwater and terrestrial crustaceans skip the larval stage). As you can see, these larvae bear little resemblance to the adult. The compound eyes are the most prominant feature, suggesting that vision is very important at this stage. Mechanosensory (vibration or hearing) and chemosensory (smell) setae sprout from the various appendages, giving them the senses they need to survive during this free swimming stage in the ocean. Eventually, this larva could have metamorphosed into the adult form to take its chances on a coral reef or the sea floor. But I (accidently) caught it first.

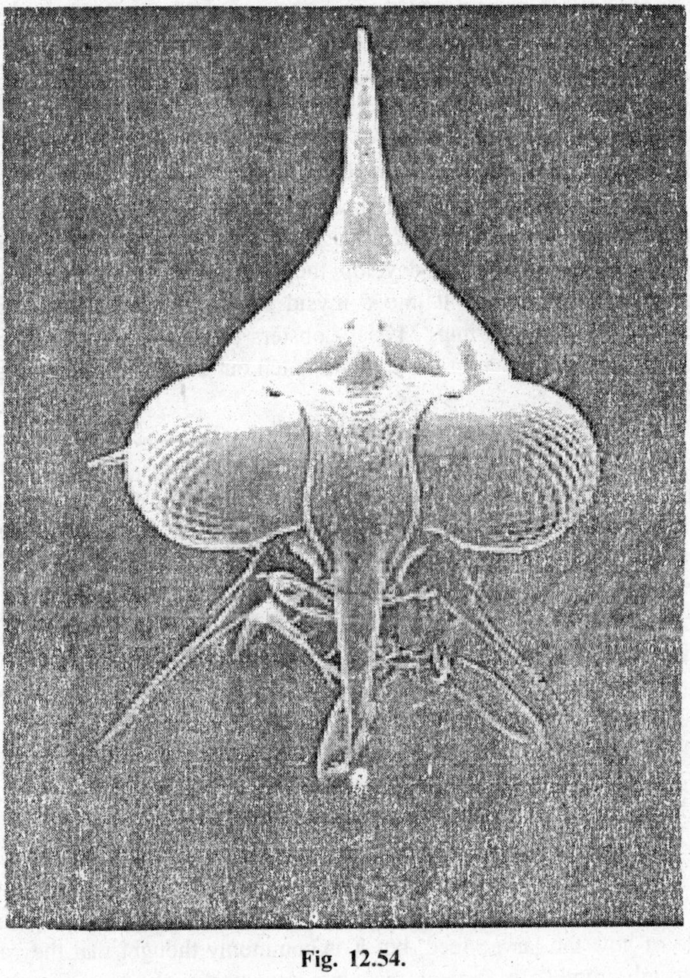

Fig. 12.54.

LARVAE AND POSTLARVAE

Planktonic larval life begins as the female lobster raises her abdomen and vigorously fans the pleopods to which the prelarvae are attached. This fanning breaks the cement attachment to the prelarvae and propels them upwards toward the water surface. Usually, the female lobsters releases only a small batch of prelarvae over a period of several nights to several weeks, possibly to prevent the larvae from being consumed all at once by various fish predators. When the attachment is broken, the prelarvae are released and they shed the cuticle binding their legs, in a true molt, to enter into the first larval stage. Lobster larvae are more developmentally advanced than are the larvae of other crustaceans. The minute lobster larvae are freed from their mother's pleopods, they are ready to begin feeding, while most other crustacean larvae hatch in non-feeding stages, or naupliar stages (from the Greek, nauplius meaning sailor). These naupliar stages subsist on remaining yolk reserves from the egg before molting into a protozoeal (from the Greek, proto; meaning first and zoon, meaning animal) stage. During the protozoeal stages, these types of crustaceans develop their digestive tract and feeding apparatus. They then molt into a mysid stage, named after the non-decapod swimming shrimp, Mysis. Lobsters undergo the naupliar and protozoeal stages while in the egg and hatch out directly into the mysid stage.

The first stage larva is about eight millimeters (one-sixth of an inch) long, completely transparent, with large dark eyestalks. Because of its transparency, all of its internal organs can be seen through its shell, or carapace. These larvae are typically concentrated at or near the water surface during the night and, depending upon the intensity of sunlight, may be near the surface or submerged to greater depths during daylight hours. This movement up and down through the water is estimated to be on the order of fifteen to thirty meters (45 to 90 feet) each day and is due to the response of the larvae towards light, which changes during and between each stage, so that they are either attracted to or repelled by light. Because other planktonic organisms, both animal and plant, perform similar vertical migrations, the lobster is always in the presence of a rich food supply. We know little of how the larvae feed, but it is commonly thought that they are raptorial (from Latin, *raptus* meaning "to seize"), chasing after their intended prey and seizing them with their legs or mouthpart.

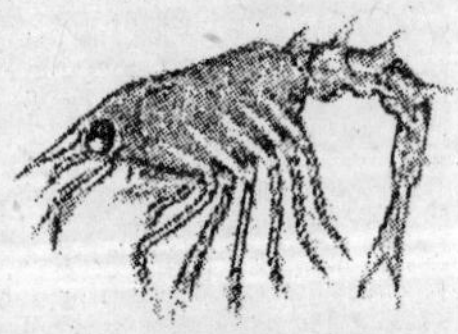

Fig. 12.55.

Movement is achieved by the larvae beating oar-like outer branches, or exopodites, of their thoracic appendages (both legs and maxillipeds). Beating these exopodites results in an upwardly directed movement, while cessation of the beating results in sinking. Backward movements can be achieved by a rapid flexion of the abdomen, but these flexions are not as effective as in later, benthic (bottom-dwelling) stages due to the relative size of the abdomen. The swimming of the larvae seems haphazard, with frequent somersaults and spiral swimming paths. Some of this undirected movement may be due to currents and waves, but it may also be related to the absence of a functional statocyst (gravity organ) until after metamorphosis.

Second stage larvae differ somewhat from first stage, in that they are about twice as large (one-third of an inch) have large pigment cells (called chromatophores), bear pleopod buds upon their abdominal segments, and have functional claws on their first three legs. The mandibles are also more developed and they bear more sensory hairs on all appendages. But they are still transparent and have large, dark eyestalks. These larvae behave similarly to the first stage larvae.

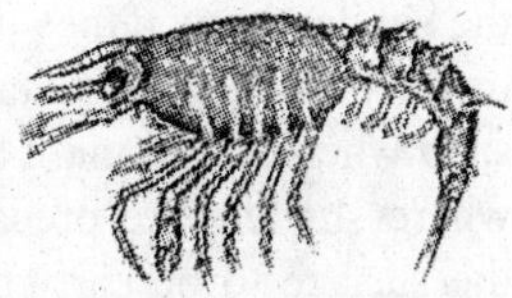

Fig. 12.56.

Third stage larvae are larger still, being about one centimeter (half an inch) in length. They have even more and larger pigment cells, the pleopods are nearly fully formed, but still not functional, uropods appear on either side of the telson, and their first pair of chelae are much larger. The eyestalks are longer and more slender than in the previous stages, the mandibles are thicker and bear more prominent teeth, and they have even more sensory hairs. But they, too, are still transparent. However, their behavior differs in that they are less light sensitive and so are found in the upper waters.

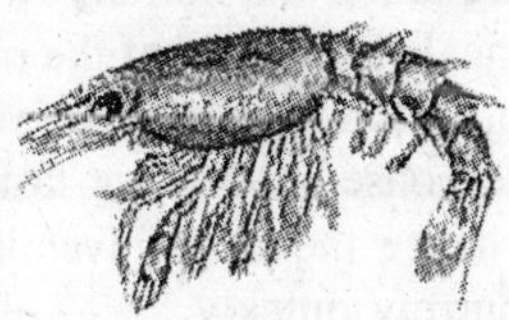

Fig. 12.57.

Depending on the water temperature, the larvae may take from eleven to over fifty days to molt through the three larval stages and into the metamorphic postlarval stage. During this time, which is usually between the months of June through August, the larvae are in waters with prevailing winds directed onshore. By the time they are preparing for metamorphosis, they are inshore. The metamorphic molt results in a postlarva that looks like a miniature lobster and is pigmented with a brownish-red color. It has lost the oar-like exopodites on all of its legs, retaining them only on the maxillipeds, and swims by beating the pleopods of the abdomen. This swimming is well-directed and strong enough that the postlarvae can swim against some currents. But, typically the postlarvae orient to the currents, which again will bring them closer inshore to find a suitable settlement habitat. The postlarvae are strongly light seeking at the beginning of this stage; however, as molting into the fifth stage draws near, light becomes repellent, and they begin to seek shelter in the benthic, or bottom, environment. It is currently thought that the postlarvae make repeated dives down to the bottom to examine the habitat there; if they judge it to be unsuitable or only marginally suitable (as in the case of sand where they cannot build a shelter, or mud where shelters are precarious), they return to the surface to be carried elsewhere and to try again. The postlarvae are capable of delaying the molt into the fifth stage for a period of time and will do so in the laboratory if they are not provided with a suitable settlement habitat.

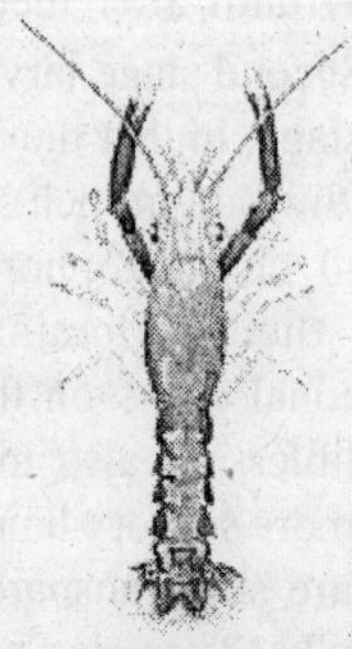

Fig. 12.58.

This transition from planktonic to benthic realm is the most dangerous period in their life. Only a small percent of lobsters actually survive the transition from the top to the bottom. Waiting for them on the bottom are crabs and bottom-dwelling fish ready to snatch them up for a quick meal. The claws of the postlarvae are small and useless for warding off predators. Their only defense is to flee by tail flipping, but this escape response often is not fast enough to get away from many fish. The only chance the early juveniles have of surviving is if they can get into a burrow quickly.

JUVENILES

The juvenile stage of the lobster's life starts when it metamorphoses into the basic form of the adult and settles to the bottom and lasts until the lobster reaches sexual maturity. Laboratory experiments have demonstrated that postlarval and juvenile lobsters can burrow in many types of substrate, but until recently, it was unknown which substrates offered the best protection against predators. Using a combination of both laboratory and field techniques, researchers found that cobble substrates, made up of

Fig. 12.59.

rocks of varying sizes (common in many rocky seashore areas), protected the newly-settled and juvenile lobsters against most predators. Peat from salt marshes, eelgrass, and mud substrates all offered some degree of protection, but not as great as that of cobble. Underwater research techniques, particularly that of suction sampling which employs a large tube that acts as a vacuum cleaner, have confirmed laboratory observations and have shown that newly settled lobsters are most abundant in *cobble habitats*.

The larger, more mobile juveniles, starting from when they reach a size of about four inches, are not really different from adults. They eat rhe same kind of prey, live in the same areas, and, in general, have all of the same behavior as sexually mature lobsters. Newly settled juveniles, however, are different from adults in all of the above attributes. There are also morphological differences between these younger and older juveniles: while the older juveniles have a heavy crusher claw and quicker cutter claw, the younger juvenile's claws are both small and of the same dimensions. Even the muscle fiber bundles within each claw, are identical but these fiber bundles change slowly over the first year of benthic life, gradually becoming mostly slow, tonic fibers in the crusher claw and fast, phasic fibers in the cutter claw.

Because of their vulnerability to predators, the recently settled juvenile lobsters are in a potential bind. In order to grow and thus become less vulnerable, they need to find food which would suggest that they need to forage. But long-term laboratory observations in "ant farm" aquaria suggested that these small juveniles never left their shelters. Similar SCUBA observations also suggested that these small juveniles might be shelter bound for nearly their entire first year of life. Lobsters measuring 20 mm in carapace length ("CL") or greater (three or more inches long) were the only small lobsters ever observed out of their shelters.

Fortuitously, the "ant farm" laboratory observations showed that the small lobsters spent a significant proportion of their time pleopod fanning which created a strong current through their shelter. The water within this current often carried planktonic organisms in it, and the lobsters would move their mouthparts, legs, and claws as if they were attempting to catch the plankton. Subsequent videoanalysis of these currents and the movements of the mouthparts and legs confirmed that these lobsters were indeed capturing the plankton with a combination of raptorial techniques carried over from the larval stage days and suspension feeding techniques used by other crustaceans. Feeding experiments demonstrated that postlarval and early juvenile lobsters could survive and grow rapidly on planktonic diets up to a size of 12 mm CL (about two inches long). If the diet were supplemented by organisms found within their shelters, such as worms and amphipods, it is possible that they could remain in their shelter for their entire first year and grow to a size of about 20 mm CL. Field experiments using differently sized tethered (or leashed) lobsters have demonstrated that predation pressures are greatly reduced for lobsters above 15 mm CL. Interestingly enough, this is also the same size at which the development of the two claws is complete. Thus, instead of going out for dinner, early juveniles eat in-and that is how they stay alive to become adults.

As mentioned, at about 20 mm CL juvenile lobsters begin to emerge from the shelters and begin short, near shelter foraging bouts. As they increase in size, the area foraged becomes larger and larger. The lobsters also move offshore with increasing size. Their larger, functionally separated claws allow these larger juvenile to crush mollusks, urchins, and all of the same foods as the sexually mature adults. These juveniles also alter their behavior, so that instead of responding to threats by escape maneuvers, they aggressively display with their claws. While fights-to-the-death between lobsters are rare, they do exhibit a progressive sequence of display behaviors, which, if an opponent refuses to withdraw, can escalate into a full-blown attack where antennae, claws, legs, or even eyestalks can be ripped off. These fights have extremely interesting consequences for the combatants. Laboratory-staged fights between evenly sized juvenile lobsters have demonstrated that the winner of a fight is truly a winner and the loser is a true loser. If a previously-fought lobster is then placed into another staged fight, its behavior is dependent upon its status from the previous fight. In other words, previous losers lose or refuse to fight with a new opponent. In the case where losers are pitted against other losers, no displays or fights occur. Winners again win, unless they are pitted against another winner in that case, the fights escalate and the lobsters cause extensive damage to each other. Usually, the dominant or subordinate status only lasts until the animal molts. Then, if a subordinate molts first, survives, and becomes larger than the dominant, it will often become the new dominant animal. It is currently believed that these seemingly long-term changes in status and behavior are determined by the level of two hormones known as serotonin and octopamine. Winners have higher blood levels of serotonin, while losers have higher levels of octopamine. If a lobster who has previously lost a fight is injected with serotonin, it becomes extremely belligerent and will attack lobsters much larger than itself. Even after being severely damaged (having claws and legs ripped from their joints), these serotonin-injected lobsters will still continue to fight. Studies such as these imply that aggressive behavior may have a hormonal component, and the lobster is currently being used as a model to better understand human aggression.

Development in Ancestral Decapods

The splitting of the Decapoda into the Pleocyemara and Dendrobranchiata by Burkenroad (1963) was based primarily on the care of the eggs (or lack of it). The Pleocyemata, as the name implies, are those which carry their eggs on the pleuron, while the Dendrobranchiata are those which do not (and are further distinguished by the form of their gills). There can be little doubt that the shedding of the eggs and development through a number of free-living naupliar stages is more primitive than the retention of the eggs and the occurrence of a nauplius-like embryonic phase. In these s respects, then, the development of the Dendrobranchiata is more primitive than that of the *Pleocyemata*. Within the zoeal phase, however, it seems that the gradual development from stage to stage seen in the *Pleocyemata* may well reflect the ancestral condition more closely than the marked metamorphosis which separates the early (protozoeal) stages from the later (mysid) stages in the Dendrobranchiata This is not to deny that most of the morphological zoeal features of the Dendrobranchiata are more primitive than those of the *Pleocyemata*, but it is suggested that the marked subdivision of the zoeal phase in the dendrobranchiate line came after the *pleocyemata* line had diverged.

Support for this view comes from the Euphausiacea and Amphionidacea, which are included with the Decapoda in the Eucarida. The early and late zoeal stages of the Euphausiacea are designated calyptopis and furcilia, respectively, but the distinction between them is based entirely on the form of the anterior carapace. All other features develop gradually, as do all features in the Amphionidacea.

The first three zoeal stages show certain very consistent features throughout the Eucarida, and these must have developed before the group became subdivided into its extant orders. Thus the paired eyes are sessile in stage 1 and stalked subsequently, and uropods first emerge in stage III. If uropods are defined as appendages of the sixth abdominal somite, which, with the telson, form a tail-fan, and if all phases of development are brought into consideration, then the occurrence of such appendages is by no means restricted to the Eucarida. Uropods are found, for example, in the Stomatopoda, the Anaspidacea and some Isopoda. It is, however, only in the Eucarida that they initially develop within the envelope of the telson, emerge as functional appendages, usually with setae, and appear before the pleopods. In some decar.ods with direct or abbreviated development pleopods may appear in stage I or 11, before the uropods, which, conservatively, still emerge in stage III. The Eubrachyura have no uropods in any developmental stage, but the occurrence of such appendages in the larvae of more primitive brachyurans, such as the Raninidae, shows this to be a secondary lons. Taking into account all these variations, it may be said that, in all groups of the Eucarida which hatch as nauplii, uropods make their first appearance in the third postnaupliar stage, and in all other eucaridan groups in which uropods occur these appendages make their first appearance in the third stage from hatching (excluding the prezoea).

The early appearance of uropods, before the pleopods, applies not only to the Decapoda, Amphionidacea and Euphausiancea but also to the Mysidacea. The sequence of development therefore strongly supports the proposal of Watling (1981, 1983), based on other criteria, that the Mysidacea should not be grouped with other forms with thoracic oostegites but should be included in the Eucarida. The closest living relatives of the mysidaceans are not the euphausiids, as the old grouping 'Schizopoda' implied, but zoea larvae. The resemblance of mysids to the similarly named mvsis larvae of the Penaeidae is more than superficial, but they show even more similarity to caridean and amphionidacean zoeas. This is not to suggest that the mysidaceans evolved from either of these groups. Rather, it is envisaged that the decapodan, amphionidacean, mysidacean and euphausiacean lines diverged independently from early eucaridan stock. Different forms of egg protection developed in the first three of these lines, leading to loss of the free naupl.ius and retention of verv similar zoea larvae. The mysidaceans have *neotenously* retained many zoeal features, including the form of most of the appendages, in the juvenile and adult.

Briefly, the zoeal phase of the ancestral Decapoda (and of the ancestral Eucarida) is seen as a series of at least nine stages (instars). The first three of these were probably similar to the protozoeal stages of modern Penaeidae and the last three similar to mysid stages of the same family. The additional ancestral stages provided a fairly gradual transition in the middle of the phase, in which successive biramous

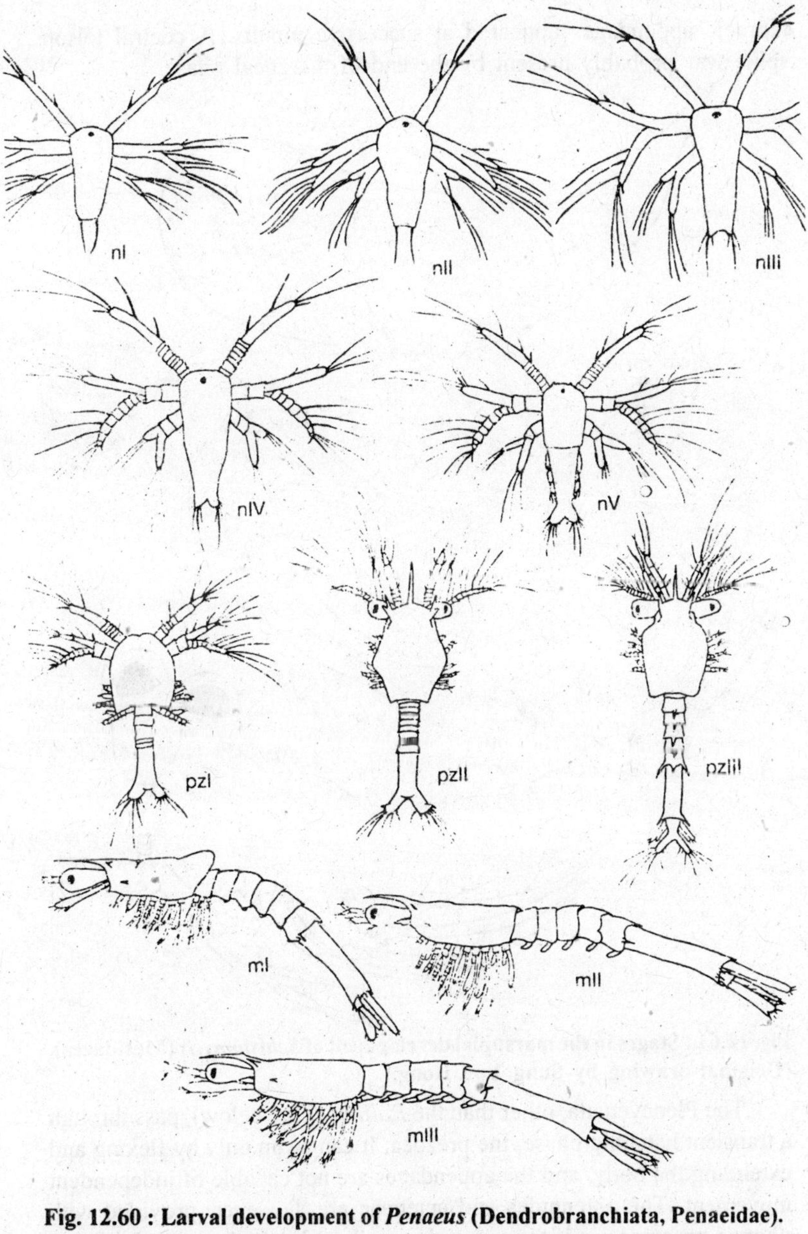

Fig. 12.60 : Larval development of *Penaeus* (Dendrobranchiata, Penaeidae). m = larva, n = nauplius, pz = protozoea; roman numerals denote stages. (adapted from Dobkin 1961).

thoracic appendages appeared at successive moults. A central telson spine was probably present by the end of the zoeal phase.

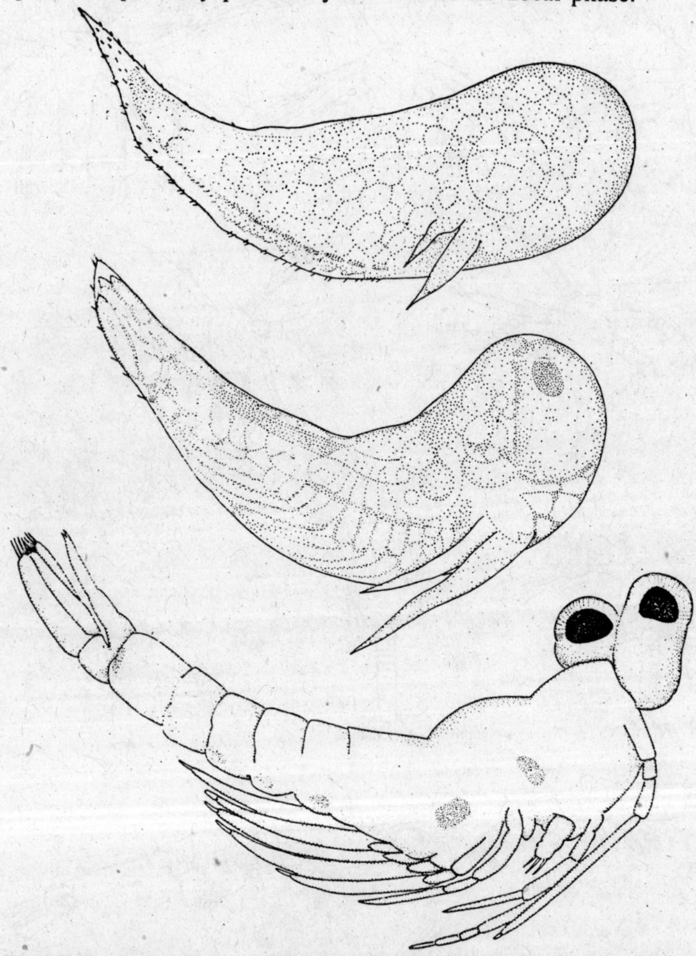

Fig. 12.61 : Stages in the marsupial development of *Schistomysis* (Mysidacea). (Original drawing by Sung Yun Hong.)

The Pleocyemata, other than the *Palinura* (see below), pass through a transient hatching phase, the prezoea. It can swim only by flexing and extending the body, and the appendages are not capable of independent movement. The antennules and antennae are, however, provided with plumed processes as if for swimming, and the bilobed telson also bears similar plumed processes. This prelarva shows many resemblances to a penaeid protozoea and seems to represent an ancestral zoeal form of the *Pleocyemata* fully consistent with that proposed above.

The Palinura

Phyllosoma larvae of the Palinuroidea, with their flattened form and reduced carapace, look very different from ervoneicus larvae of the *Eryonoidea*, in which the carapace is large and inflated. The two types of larvae, however, have one very important point in common: their uropods develop gradually, in synchrony with the pleopods, and bear no setae in the zoeal stages. This marks off the *Palinura* from the whole of the rest of the Eucarida, in which the sudden and precocious development of uropods was noted above. Other palinuran larval characters which contrast with those of the rest of the Eucarida concern the eyes and maxillae. The eyes are vestigial in all stages of the *Eryonoidea*, but in the Palinuroidea they are already stalked and movable in phyllosoma I, although the stalk is generally longer in phyllosoma II. The eyes are sessile in zoea I in all other Eucarida. The maxilla is completely without feeding endites in phyllosomas and in the corresponding stages of *eryoneicus* larvae (*i.e.*, those with natatory thoracic exopods), but the only other eucaridan zoeas which lack such endites are lecithotrophic forms, in which all the mouth-parts are vestigial.

There are no published observations on hatching in the Polychelidae, the only extant representatives of the Eryonoidea. The Palinuroidea, however, pass through a short-lived naupliosoma stage which resembles the prezoea of other Pleocyemata in having well-developed feathery antennules and antennae. It differs from a prezoea in that it swims with these appendages, much as does a nauplius, and it also differs in the development of the thorax and abdomen. The postnaupliar regions of a naupliosoma consist merely of the developing phyllosoma and show no naupliosomal features. The total lack of a naupliosomal telson contrasts particularly with the conspicuous bilobed plumed prezoeal telson. Thus, while the prelarval form of other Pleocycmata appears to be derived from a protozoea, that of the *Palinura* appears to he derived from a nauplius.

Fossil phyllosomas from the Upper Jurassic (Polz 1971) show that this type of larva has changed little in the last 250 million years.

Phyllosomas and early eryoneicus larvae are zoeas in the broad sense of larvae which swim with their thoracic exopods, but they have little else in common with the zoeas of other Decapoda. It seems impossible to derive either of the palinuran larval forms from any of the larvae in the assumed ancestral zoeal phase of the Eucarida, and the naupliosoma

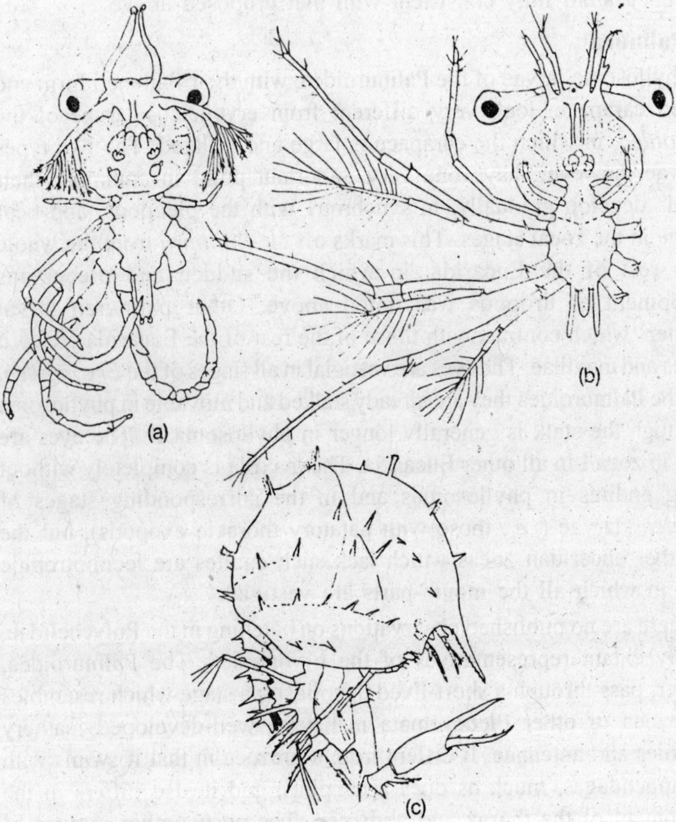

Fig. 12.62 : Larvae of *Palinura*: (a) naupliosoma and (b) phyllosoma III of *Jasus* (Palinuroidea); (c) eryoneicus larva of *Polycheles* (Eryonoidea), pleopods and uropods shown black, (a) and (b) redrawn from Lesser (1974); (c) redrawn from Selbie (1974, after modification).

fails to provide any evidence of such a zoea in the phylogeny of the Palinuroidea. Clearly, from an assessment of larval characters, the Palinura do not belong in the Decapoda nor even in the Eucarida. They are undeniably decapodous, in having five pairs of legs, and pleocyematous, in carrying their eggs under the abdomen, but, it is suggested, these characters might have evolved in parallel in the decapodan and palinuran lines of descent. Schram (1974) has pointed out resemblances between the *Palinura* and the fossil Pygocephalomorphs, morpha, but he ascribed these to convergence. The pygocephalomorphs seem to have shown variation in the number of maxillipeds and in the method of egg-care: some had oostegites but others had not. Perhaps the Pygocephalomorpha gave rise to the *Palinura*, and it is the resemblances between the *Palinura* and the Decapoda which are convergent.

Radiation Within the Pleocyemata

While the larvae of the *Palinura* cannot be derived from the ancestral zoeal phase postulated above, the zoea larvae of all the other Pleocvemata can, and gives a suggested sequence of branching in the evolution of this group.

It is debatable whether the production of fewer and larger eggs preceded egg retention in the early Pleocyemata or followed it, but the two processes are obviously closely associated from a functional point of view and were almost certainly also closely associated in time. Larger eggs, with more yolk, almost automatically produced longer embryonic development, and led to the nauplius being transferred from a larval to an embryonic phase. The protozoeal stages of the Dendrobranchiata permit a progressive transition from the naupliar method of swimming, with the antennae, to the zoeal method, with the thoracic exopods. With the elimination of the nauplius as a free larva there was no need for this transition, and very soon in the evolution of the Pleocyemata the antennae must have lost all natatory function. Segmentation of the antennal scale is associated with its use in swimming and vestigial segmentation of the distal part of the scale must have been a feature of the early zoeal stages of the ancestral Pleocyemata. Traces of this distal segmentation in the early stages are retained by some members of the Caridea and Stenopodidea and by one of the Thalassinidea but nct by any known member of the other groups.

Other trends which appear to have operated independently and probably irreversibly in several lines are progressive fusion of the endopodal segments of the maxillule and maxilla and progressive reduction in the number of thoracic exopods which become functional by the end of the zoeal phase. The number of thoracic exopods can vary from species to species within a few caridean genera, such as *Pandalus* (Pandalidae), but is usually a consistent character throughout families or larger groups. The number of legs bearing exopods is usually constant within families of the Thalassinidea and within some families and nearly all genera of the Caridea. In addition to the functional exopods indicated in the Table, non-setose exopods occur on maxilliped 3 in the Raninoidea and on legs 2-4 in *Dromia* (Dromioidea).

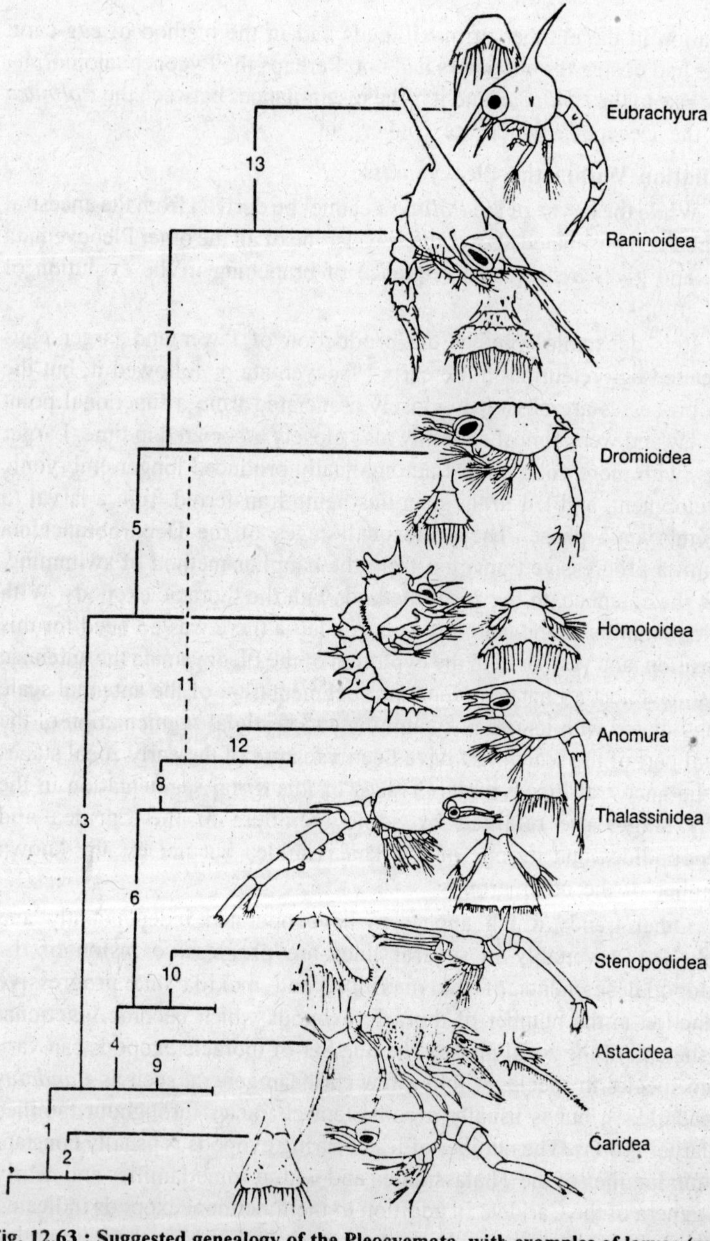

Fig. 12.63 : Suggested genealogy of the Pleocyemata, with examples of larvae (zoea III, lateral and telson, dorsal). Branch 11, leading to the Dromioidea, refers to larvae only. For further explanation see text.

Table 12.1 : Number of thoracic exopods with functional exopods in the final zoeal stage in different groups.

	maxillipeds	legs
Eubrachyura	2	0
Raninoidea	2	0
Dromioitlea	3	lord
Homoloidea	3	0
Anomura	3	0
ThalassiniJca	3	var
Stenopodidea	3	3
Nephropoidea	var	5
Caridea	3	var

var = variable number.

Abbreviations and occasionally extensions of larval life have evolved independently in many different lines. Within larger groups, such as the Caridea, Anomura or Brachyura, the number of stages can vary widely, and considerable variation can be found even between different species of the same caridean genus. In other genera, however, both within and without the Caridea, the number of zoeal stages has tended to stabilize, and it is somerimes consistent throughout large families or superfamilies. Thus there are nearly always four zoeal stages in the Paguridae (Anomura) and nearly always two in the Majoidea (Eubrachyura). In the Porcellanidae (Anomura) there are only two zoeal stages in eight of the nine known genera, but the genus *Petrocheles* is clearly the most primitive not only in its morphological larval features but also in passing through five zoeal stages.

The telson of ancestral decapods probably went through changes during the zoeal phase similar to those shown by many modern Penaeidae and Sergesridae. Thus, in these ancestral forms, the terminal processes were probably all in the form of plumed setae in the early zoeal stages, but many were replaced by spines later in the phase. This sequence has been retained by the Caridea, but in all other groups of Pleocyemata (branch 1) the first (outermost) pair of processes is spiniform from zoea I. The hair-like form of the second pair of processes, found in zoeas of the Stenopodidea, Thalassinidea, Anomura and Dromioidea, is envisaged as an inheritance from a single group (branch 4), while in the ancestors of the Brachyura (branch 5) tliese processes tend to be short spines from zoea I or they may fail to appear. The peculiar position of the Dromioidea, shown as arising from both these branches, are discussed below.

There was probably an ancestral tendency for the fourth pair of telson processes to be longer than the others in all stages and to develop into spines in later stages. When tliese spines are large, the telson takes on the form of a two-pronged fork, seen in such distantly related forms as *Cerataspides* include some factor which gives it precedence in development over the megalops or later phases so hybrid eggs would hatch as a no muran zoeas. Cytoplasmic substances play a regulaty role by signalling expression of specitic geves at the proper time. The maternal nuclear membrane and cytoplasm would favour the expression of maternal genes so all post zoeal phases in development correspond to a primiture brachyuran. All extant dromioidea have among their an cestors at least one of them hybrids resulting from cross fertilization Modern speces without zoeal development are closely related to those who supressed zoea larval for second time in their evolutinary history (Dendrobranchiata), some Galatheidae (Anomura) and the primitive bra chyuran family Raninidae. In this latter family the fourth process (.in each side is already a stout spine in zoea I, although it becomes progressively larger in succeeding stages and processes 1ù3 may become smaller. In most Eubrachyura the two-pronged fork, based on the fourth pair of processes, is already fully formed at hatching, one or more of processes 1-3 may be absent and there is rarely any appreciable further modification of these processes during zoeal development. The Eubrachyura is the only Eucaridan group in which uropods never appear in anv phase, and this, as stated earlier, must be regarded as a secondary loss. The fixed form of the telson in brachyuran zoeas may have some hearing on the absence of uropods but it cannot be a direct cause. The telson (although very different from that of the Brachyura) changes little during the zoeal phase in the Nephropoidea, but uropods still appear in zoea III.

The eubrachyuran zoea was probably derived from a form resembling the larva of a modern homoloid by reduction in the number of carapace spines and by changing the orientation of the rostrum. Intermediate conditions are shown by some Raninoidea (Williamson 1976; Rice 1980).

In ancestral forms, a central telson spine was probably not present in the early zoeal stages but made its ontogenetic appearance later in this phase, as it does in many modern Penaeidae. Its presence in all zoeal stages of modern Nephropoidea is related to their abbreviated development: lobsters of this group hatch as late zoeas, apart from the condition of the eyes and uropods. The central telson spine has been brought forward in development in several families of the Thalassinidea, while in other groups its appearance is delayed until the juvenile phase or the spine may be eliminated entirely.

The Dromioidea

The adults of the Dromioidea are crab-like but their zoea larvae are essentially anomuran. Larvae are known for at least 10 species of Dromiidae in six genera and for one species of Dynomenidae. The larvae show the true affinities of the group and, therefore, the adults must be crablike anomurans, comparable with the Lithodidae and Porcellanidae. Students of adult dromioids, however, are adamant that the group consists of primitive brachyurans and claim that the endophragmal skeleton and genital ducts are essentially brachyuran, not anomuran. These adult features indicate that the group might have evolved from homoloid-like ancestors but the larval features are not consistent with such an ancestry. Adult dromioids are primitive brachyurans but larval dromioids are anomurans, and to postulate that, at some point in its evolution, the group acquired the larvae of another group.

The possibility that a larval form, or the penes prescribing it, mav he transferred from one evolutionary lineage to another does not appear to have heen considered seriously in the past, and if the anomalies of the Dromioidea provided tlie onlv potential example then serious consideration would probably continue to be withheld. In other invertebrate groups, however, there are many examples of similar larvae and dissimilar adults which lend themselves to such an explanation. Comparable postulated transfers of genes prescribing larval form-can, for example, explain why radially symmetrical adult echinoderms have bilaterally symmetrical larvae, why the echinoderms and hemichordates show marked affinities as larvae and virtually none as adults and whv the method of metamorphosis is totally different in the two phyla, why echinoids and ophiuroids are so similar as larvae but so different as adults, why the adults and larvae of some echinoids and ophiuroids point to quite different groupings at the level of family or genus, and why phyla so widely different as Mollusca, Annelida, Echiura and Sipuncula all have representatives with very similar trochophore larvae cross-fertilization can result in the insertion of a new larval phase in a life-history, but it seems unlikely to lead to the substitution of one larval form for another. Thus, it is suggested that the members of the stem group of the Dromioidea probably had larvae wit the characters of primitive brachyurans, but it is not considered probable that this larva) form was directly supplanted by that of a primitive anomuran. It seems much more likely that cross-fertilization took place between a male anomuran with a zoea larva and a female dromioid which hatched as a megalopa. Many modern dromioids have no zoeal phase, and it is only necessary to suppose that this tendency towards direct development evolved fairly early in the history of the group to produce the following suggested sequence of events:

The earliest members of the dromioid line were primitive brachyurans, both as adults and larvae, and they gave rise to the ancestors of the modern Raninoidea and Eubrachyura. After the raninoideubrachyuran line (branch 7) had become established, the stock which retained the adult characters of the Dromioidea developed a tendency towards direct development, and a little later in evolutionary time all its surviving descendants were without zoea larvae but hatched as a megalopa.

The eggs of one of these early dromioids were then fertilized with the sperm of an early anomuran (branch 11). The genetic recipe for a zoeal phase must two characters differed occasionally from existing tables of xanthid megalopal characters (e.g. Andryszak & Gore 1981) that were derived from the same literature sources. Other characters are useless because they do not vary across taxa. Examples are the number of setae on the proximal segment of the mandibular palp, which is (with one exception) always 0, and the number of spines on the carapace (character 3), which is also 0 with only two clearly unrelated exceptions. However, sufficient information is available from the meristic data to suggest that certain patterns exist. Meristic counts tor 14 such characters were broken down into scores of 0, 1, or 2 for use in a phenetic analysis. As an example, the number of setae on the uropod (character 27) is almost always 9 or less in species with Group I larvae, bur always 10 or more in Group IV and most of Group III; this was scored as 1 and 0, respectively, for the phenetic analysis. Data for the original 28 characters are available upon request from the author.

Because megalopal characters were sought that might. differentiate between previously recognized zoeal groups, the methodology is obviously somewhat flawed in that it involves a degree of circularity and a subjective decision of which characters to employ. Certain taxa, for which over half of the scored characters are missing from the Tables, were deleted from the analysis. Two species, *Eurypanopeus abhreviatus* and *Eriphia gonagra*, came to my attention too late to be included. The resulting data matrix, consisting of 43 taxa and 16 characters (three of which were combined, so that a total of 14 characters was used was analysed for phonetic relationships by UPGMA (unweighted pair-group method using arithmetic averages) employing Euclidean distance and using the MINT computer package (Rohlf 1971). Specimens examined with scanning electron microscopy (SEM) were prepared following guidelines given by Felgenhauer (1987) but with 100% ethanol instead of amyi acetate used as the transitional fluid.

RESULTS

Descritions

The following descriptions are arranged according to the zoeal groupings of Rice (1980) and Martin (1984). it should not be assumed that the megalopas are readily divided into the same groups as the zoeas.

Group I Megalopas

Carapace usually quadrate, with or without acute anteriolateral projections (16 species with, 13 without), usually without produced acute rostrum, and without spines (except Hong's (1977) description of *Cycloxanthops truncatus*, in which the spine figured is in the same location as a commonly occurring mid-dorsal bump, and probably corresponds to the dorsal carapaces pine of the zoea. This spine sometimes persists in megalopas that die during the moult from fourth zoea; Hong's (1977) megalopa may be showing this abnormality.) Ventral ramus of amennule with 4 setae, 3 terminal and 1 subterminal; rarely with fewer terminal setae, occasionally with additional subterminal dorsal seta. Antenna with 7 to 11 (rarely 12) total segments. Mandibular palp 2-segmented (sometimes described with 3 segments, apparently a misinterpretation of segmentation by the author), no setae on proximal segment(s) (except for *Eurypanopeus abbreviates* with 1 seta) and 3 to 11 (usually 7 to 10) stout spines or setae on distal segment. Maxillule usually with 2 or 3 setae, one of which is basal, on proximal segment of endopod, and 2 to 6 (most commonly 4 or 6, arranged 2+2 or 2+2+2) setae on distal segment of endopod. Maxilla with 0, 1 or 2 non-plumose setae on endopod, or 6 to 8 setae in dumps of 2 to 3. Bilobed basal endite of maxilla with 4 to 9 setae on proximal lobe and 5 to 10 (12 in *Dyspanopeus texana*) setae on distal lobe. Scaphognathite fringed with 25 (*Rhithropanopeus*) to 58 (*Xantho incisus*) setae, usually 35 to 50. Epipods of maxillipeds 1 to 3 variable, with 1 to 13 (usually 4 to 7), 0 to 8 (usually 5 or fewer), and 6 to 21 (usually 11 to 13) long smooth setae; respectively. Basal distally-protruding enlargement of epipod of first maxilliped usually with only one such seta. Cheliped usually with strong recurved hook on ischium (most exceptions are from illustrations, so possibly this feature was overlooked; Kurata (unpublished) noted the spine does not exist in *Eurypanopeus depressus*, although Negreiros-Fransozo (1984) records it for E. abbreviatus). Spination of dactylus of periopods rarely documented; where described or illusrated, usually with 3 spines on ventral border, rarely 1 (*Rhithropanopeus barrisii*), 2 (*Eurypanopeus depressus*) or 4 (*Xantho incisus*). Dactylus of fifth pereiopod usually devoid of spines (with only *Panopeus occidentalis* and *P. americanus* with 1 and 3 spines, respectively) and usually with 3 long serrate setae (rarely 0 to 2). Spination of ischium of pereiopods 2-5 documented for too few species to allow generalizations. First and third pleopods with 11 (rarely fewer) to 19 setae on distal segment of exopod and 2 to 3 liouks on endopod (4 in *Xantho incisus*). Uropods with 5 to 10 setae, rarely fewer (3 to 4 in *Rhithropanopeus barrisii*) or more (11 in *Etisus laevimanus*, 12 in *Xantho incisus*). Telson variable in form, nearly circular or sharply truncate on posterior border.

Group II Megalopas

Carapace as described for Group I, with or without sharp anteriolateral projections and usually without produced rostrum; small sharp carapace spines appear only in *Pilumnus novaezealandiae* on anteriolateral dorsal surface. Ventral ramus of antennule usually with 5 or more setae, 1 or 2 subterminal. Antenna with 9 to 15 (usually 10) total segments. Mandibular palp similar to above description, with 9 to 15 spines or setae on distal segment. Endopod of maxillule with 1 or 2 setae on proximal segment (4 in *Pilumnus novaezealandiae*) and 2 to 6 setae on distal segment, at least 1 subterminal. Setation of endopod and basal endites of maxilla as described for Group 1. Scaphognathite fringed with 28 to 48 (usually 30 to 45) plumose setae. Setation of first to third maxillipedal epipods 4 to 12 (but 26 in *Pilumnus novaezealandiae*), 0 or 1 (17 in *P. novaezealandiae*), and 7 to 13 (19 in *P. novaezealandiae*). Basal protrusion of epipod of first maxilliped with 1 long seta. Cheliped either without large recurved ischial spine, or with more than one; condition seen in Group I larvae (single large recurved spine) not found. Spination of ventral border of dactylus of pereiopods 2 to 5 varies from 0 to 3. Fifth pereiopod with 0 to 3 long serrate setae, most commonly 0 or 1 rather than 3 as in Group I. Where documented, ischial segments of pereiopods 2 to 5 bear spines, with *Echinoecus pentagonus* the only exception. First pleopod with 10 to 15 (usually 13 to 15) setae on exopod and usually 2 to 3 hooks on endopod (4 in *Pilumnus novaezealandiae*, 6 in *P. lumpinus*). Third pleopod with 10 to 14 (usually 11 to 13, but 9 in

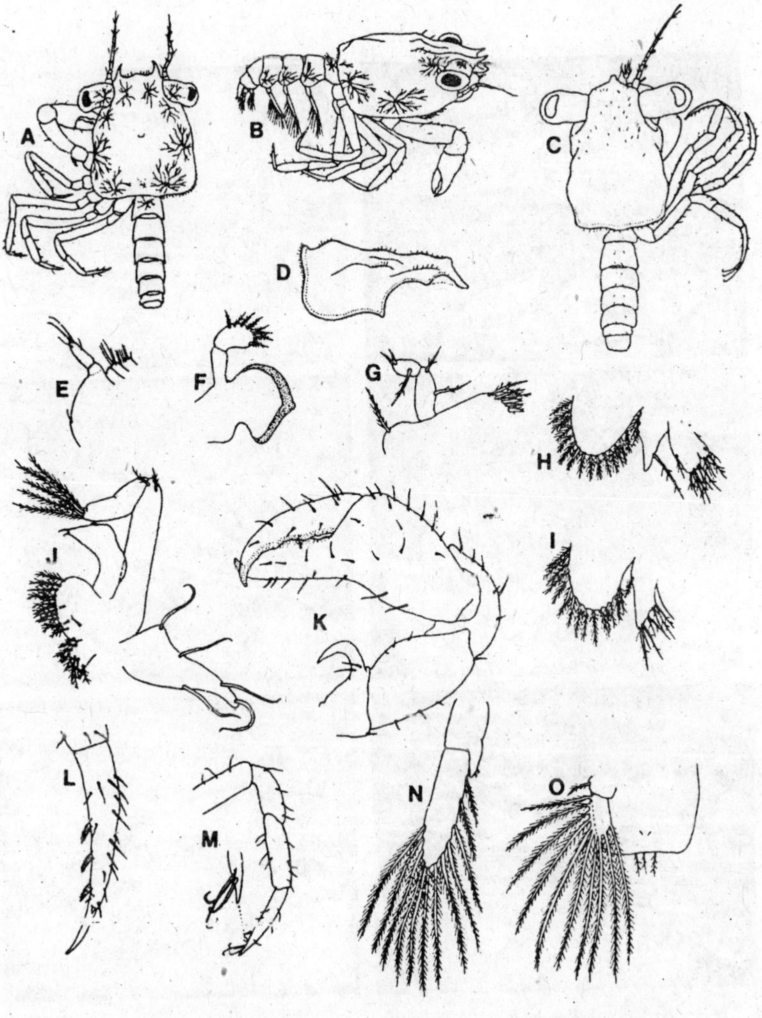

Fig. 12.64 : Selected megalopal characters of xanthid species with Group 1 zoea larvae. A, B, dorsal and lateral view, *Hexapanopeus angustifrons*; C, D, dorsal and lateral view of carapace, *Panopeus occidentalism* E, ventral ramus of antennule, *P. occidentalis*; F, mandible, *P. bermudensis*; C. endopod, *P. bermudensis*; H, endopod of maxilla (with single non-plumose seta), *P. bermudensis*; I, endopod of maxilla (with 3 + 2 + 3 setal arrangement), P. turgidus; J, first maxilliped, *P. bermudensis*; K, cheliped, *P. bermudensis*; L, dactylus of third pereiopod, *P. turgidus*; M, dactylus of fifth periopod, *P. occidentalis*; N, third pleopod, *P. turgidus*; 0, uropod and margin of telson, *P. occidentalis*. A and B from Costlow & Bookhoui (1966); C, D, E, M and 0 from Ingle (1985); all other figures from Martin, Felder *et al.* (1984). Not drawn to scale.

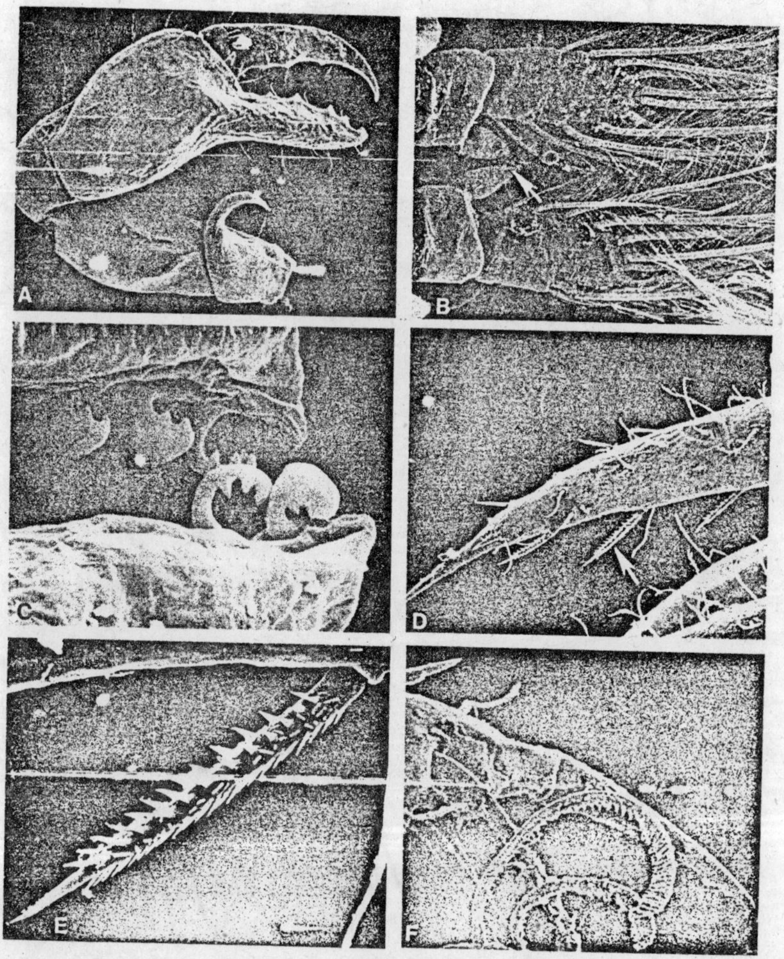

Fig. 12.65 : Scanning electron microscopy of selected xanthid megalopal characters. A, cheliped with strong ischial spine (× 85); B, fourth pair of pleopods showing endopods and natatory setae (× 225); C, hook-like setae of endopods indicated by arrow in B (× 3250); D, dactylus of third pereiopod showing three stout serrate spines on ventral border (× 200); E, higher magnification of seta indicated by arrow in D (X1150); F, tip of daciylus of fifth pereiopod showing three long serrate setac (x900). All figures are of species of *Panopeus*.

Pilumnus sayi) setae on exopod, and 2 to 3 hooks on endopod (4 in *P. lumpinus*). Uropods with 5 to 8 setae (4 in *Pilumnopeus eucratoides*) with telson variable.

Group III Megalopas

Carapace without sharp anteriolateral projections (except *Monodaeus couchii*), rostrum never acute and produced. Range of antennule setation and antenna segmentation as for Group II. Proximal segment of mandibular palp always devoid of setae, as in Groups I and II, but distal segment with 8 to 15 (usually >10) spines or setae. Range of setation of maxillule as for Group II. Maxilla with no non-plumose setae on endopod (except *Pilumnoides perlatus*, with 3 + 3 + 3 setae), with 6 to 11 plus 9 to 12 setae on proximal and distal lobes of basal endite, respectively. Scaphognathite with from 'about 50' (*Epixanthus dentatus, Monodaeus couchii*) to over 90 (*Baptozius vinosus*) fringing plumose setae. Setation of maxillipedal epipods variable; first maxilliped with from 1 or 2 (*Monodaeus couchii*) to 19 (*Eriphia verrucosa*) long smooth setae, second maxilliped with 0 to 16 setae (with the same two species having extreme values), third maxilliped with 5 (*Monodaeus couchii*) to 29 (*Baptozius vinosus*) setae. Basal protrusion of epipod of first maxilliped with 1 to 5 long setae. Ischial cheliped spines described for only two species, *Monodaeus couchii* and *Ozius truncatus*. Dactylus of pereiopods 2 to 4 with 2 to 4 spines. Fifth pereiopod with 0 to 2 spines and 2 to 3 long serrate setae (apparently 4 in Lumare & Gozzo's (1972) figure of *Eriphia verrucosa*). First pleopod with 15 to 26 (usually 21 or more) setae on distal segment of exopod and 3 to 6 hooks on endopod. Third pleopod with 15 to 24 (usually 19 or more) setae on exopod and 4 to 5 hooks on endopod. Uropods with 10 to 13 setae, rarely 7 (*Eriphia varrucosa*) or 8 (*Pilumnoides perlatus*). Telson variable.

Group IV Megalopas

No megalopas described for *Sphaerozius*; all information derived from species of *Menippe*. Carapace without anteriolateral projections, acute rostrum, or any spination. Ventral ramus of antennule with 7 or 8 setae, of which 3 or 5 terminal. Antenna with 8 to 12 roial segments. Mandibular palp with 9 to 14 setae on distal segment. Endopod of maxillule with 1 seta on proximal segment and 4 setae, in two groups of 2, on distal segment. Maxillary endopod with no long non-plumose setae on proximal border. Basal endite of maxilla with 8 to 13 setae on proximal lobe and 9 to 15 setae on distal lobe. Scaphognathite fringed with 65 to 78 setae. (Although only 38 setae are illustrated for

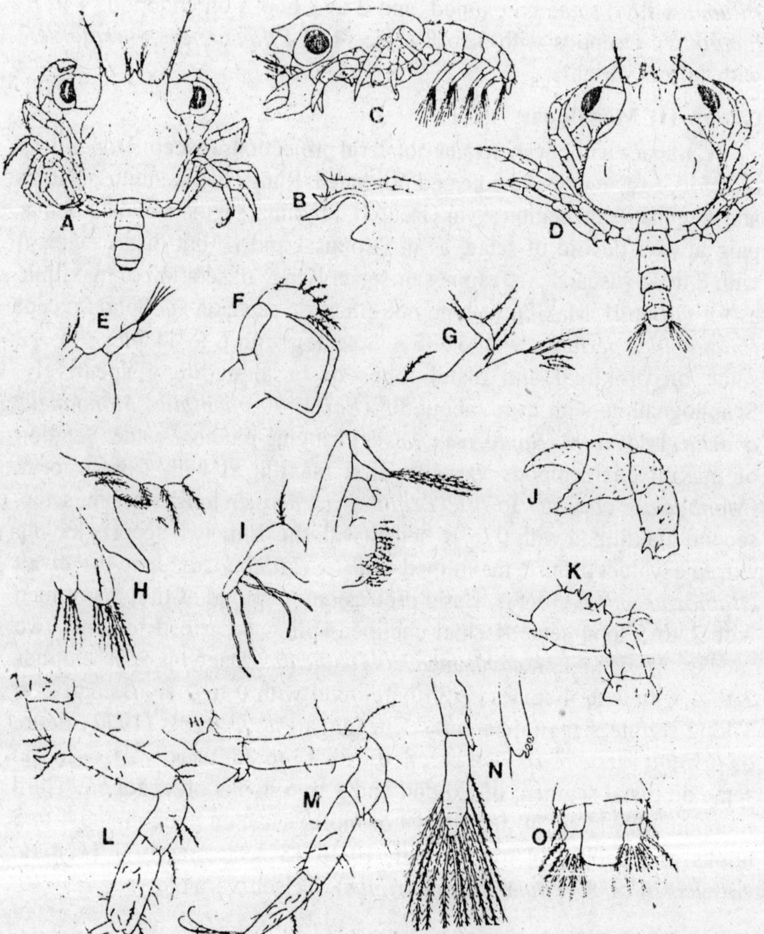

Fig. 12.66 : Selected megalopal characters of xanthid species with Group II zoea larvae. A, B, dorsal and lateral view of carapace, *Parapilumnus trispinosus*; C, D, lateral and dorsal view, *Heteropanope glabra*; E, ventral ramus of antennule, F, mandible, G, endopod of maxillule, H, endopod of maxilla lacking non-plumose setae, I, first maxilliped, *P. trispinosus*; J, cheliped of *H. glabra*; K, proxim 1 segments of cheliped of *P. trispinosus*; L, third pereiopod of *P. trispinosus*; M, fifth pereiopod of *P. trispinosus*; N, first pleopod (with enlarged endopod), *P. trispinosus*; O, uropods and telson, *P. trispinosus*. C, D and J from Greenwood & Fielder (1984b); all others from Quintana (1986c). Not drawn to scale.

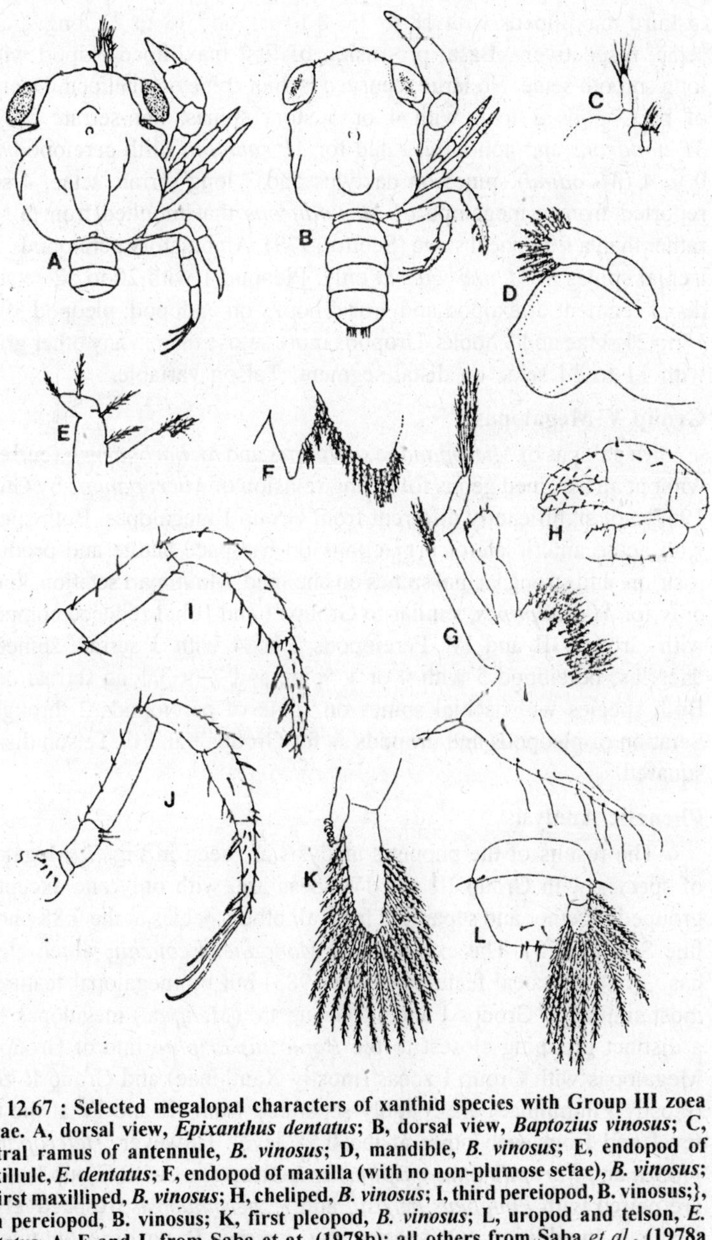

Fig. 12.67 : Selected megalopal characters of xanthid species with Group III zoea larvae. A, dorsal view, *Epixanthus dentatus*; B, dorsal view, *Baptozius vinosus*; C, ventral ramus of antennule, *B. vinosus*; D, mandible, *B. vinosus*; E, endopod of maxillule, *E. dentatus*; F, endopod of maxilla (with no non-plumose setae), *B. vinosus*; G, first maxilliped, *B. vinosus*; H, cheliped, *B. vinosus*; I, third pereiopod, B. vinosus;}, fifth pereiopod, B. vinosus; K, first pleopod, *B. vinosus*; L, uropod and telson, *E. dentatus*. A, E and L from Saba et at. (1978b); all others from Saba *et al.*, (1978a reprinted with permission).

M. rumphii, Kakari (1977) staced 'about 65' in text.) Epipods of first to third maxillipeds with 18 to 23, 8 to 10, and 18 to 22 long smooth setae, respectively. Basal protrusion of first maxilliped epipod with 3 long smooth setae. No large recurved ischial spine on cheliped. Dactylus of pereiopods 2 to 4 with 4 or 5 stout spines, non-serrate only in *M. nodifrons* and none illustrated for *M. rumphii.* Fifth pereiopod with 0 to 4 (*M. adind*) spines on dactylus and 3 long serrate setae; 4 setae reported from a megalopa of *M. nodifrons* that moulted from a sixth rather than a fifth zoeal stage (Scotto 1979). Ambulatory pereiopods with ischial spines in *M. mercenaria* only. Pleopod 1 with 20 to 22 setae on distal segment of exopod and 3 or 4 hooks on endopod; pleopod 3 with 18 to 21 setae and 3 hooks. Uropods more setose than in any other group, with 11 to 14 setae on distal segment. Telson variable.

Group V Megalopas

Megalopas of *Micropanope scitlptipes* and *M. barbadensis* (currently without an assigned genus following revision of *Micropanope* by Guinot 1967) not significantly different from Group I megalopas. Both species with acute anteriolateral projections on carapace actute and produced rostrum, and strong ischial spines on cheliped. Mouthpart setation, known only for *M. sculptipes*, similar to Groups I and II but reduced compared with Groups III and IV. Pereiopods 2 to 4 with 3 serrate spines on dactylus; pereiopod 5 with 0 or 1 ^pine and 2 or 3 long serrate setae. Both species with ischial spines on some of pereiopods 2 through 5. Setation of pleopods and uropods as for Groups I and II. Telson disrally squared.

Phenetic Analysis

The results of the phonetic analysis are seen in Fig. 7. Megalopas of species with Group III and IV zoeas are, with only one exception, grouped together and separated from all other species at the 0.88 phenon line Sokal 1973). The exception is *Monodaeus couchii*, which clearly has Group III zoeal features (Ingle 1983) but by megalopal features is most similar to Groups I and II. Group IV (*Menippe*) megalopas form a distinct grouping closest to the *Baptozius/Eriphia* line of Group III. Megalopas with Group I zoeas (mostly Xanthinae) and Group II zoeas (mostly Pilumninae) are roughly separated from III and IV at the 0.68 level and from each other at the 0.58 level. However, *Heteropanope glabra* and *Parapilumnus trispinosus*, both species with Group II zoeas, are paired with *Panopeus herbstit* and *P. bermndensis*, respectively, of Group I. In addition, two species of *Pilumnus, P. novaezealandiae* and

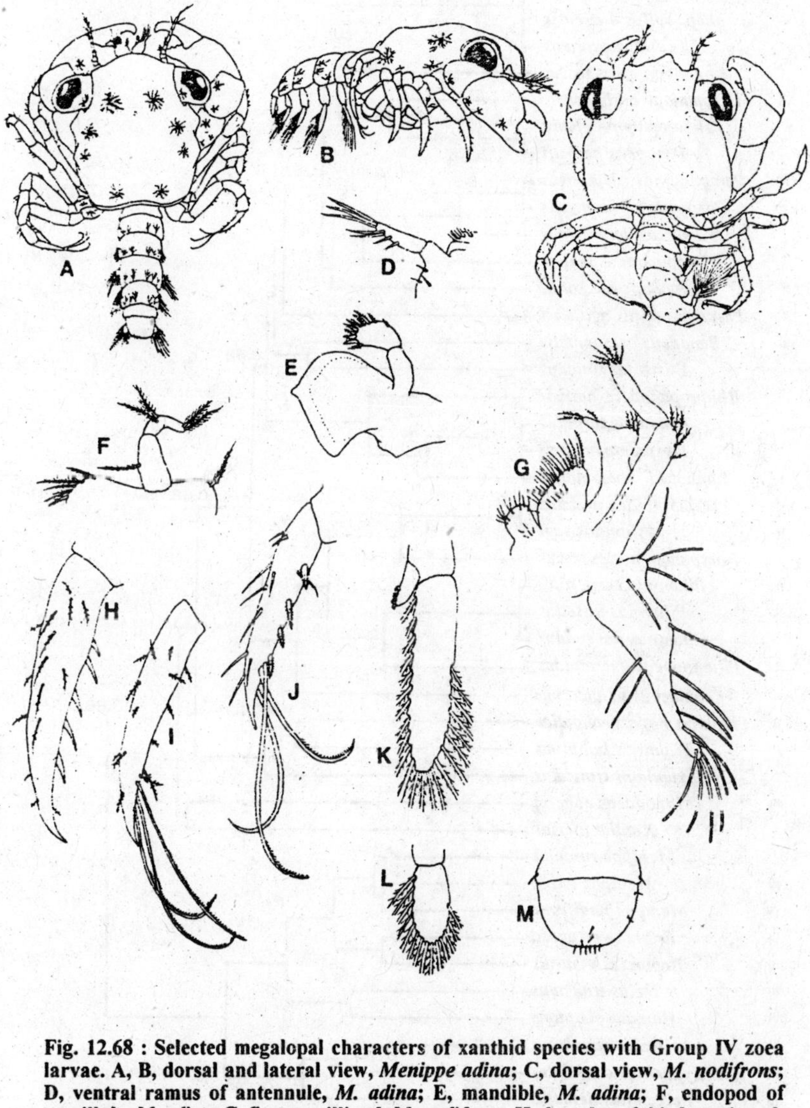

Fig. 12.68 : Selected megalopal characters of xanthid species with Group IV zoea larvae. A, B, dorsal and lateral view, *Menippe adina*; C, dorsal view, *M. nodifrons*; D, ventral ramus of antennule, *M. adina*; E, mandible, *M. adina*; F, endopod of maxillule, *M. adina*; G, first maxilliped, *M. nodifrons*; H, dactylus of third pereiopod, *M. nodifrons*; I, dactylus of fifth pereiopod, *M. nodifrons*; J, dactylus of fifth pereiopod, *M. adina*; K, first pleopod, M. adina; L, uropod, *M. adina*; M, telson, *M. adina*. C, G, H and 1 from Scotto (1979); all others are original.

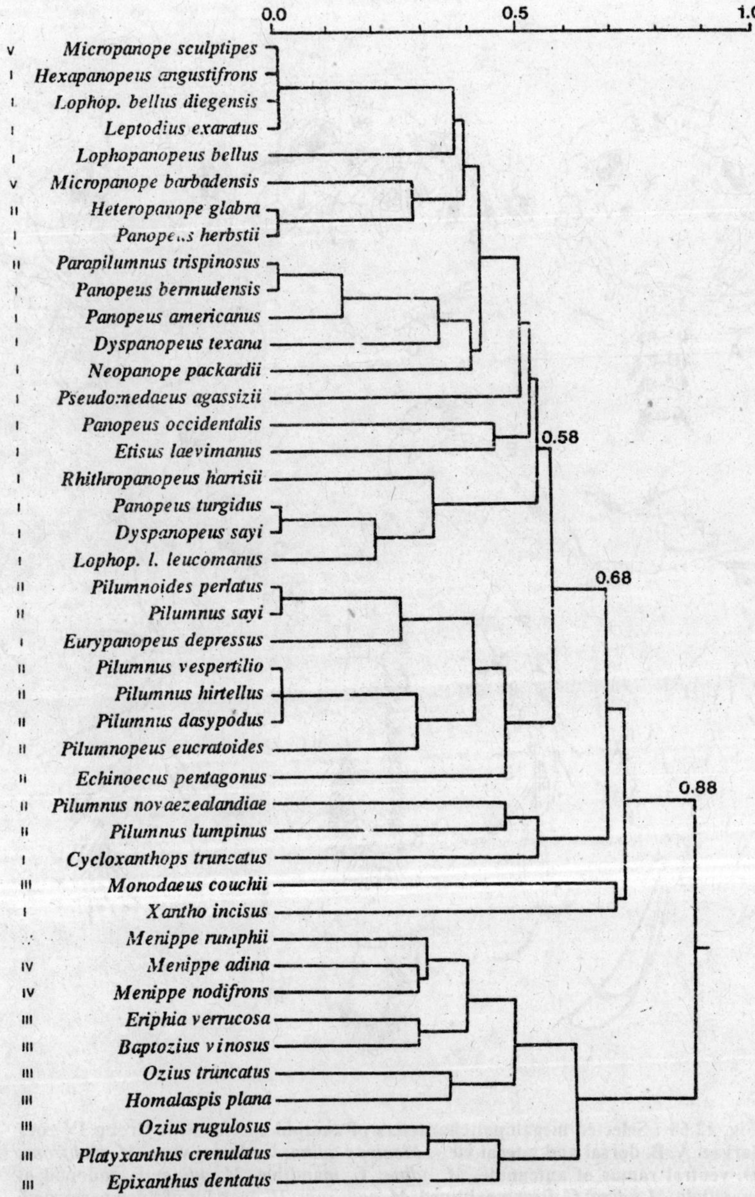

Fig. 12.69 : Phenetic relationships of 43 xanthid species based on 14 characters of the megalopa. Roman numerals to the left of species names indicate groupings of the zoea larvae (see text).

P. lumpinus, are paired with *Cycloxanthops truncatus* (which has Group 1 zoeas). All other Group I megalopas are clustered together below the 0.58 level, with the exception of *Xantho incisus*, which is most similar te *Monodaeus couchii* of Group III. Both of the latter species appear very different not only from other groups but from each other. Megalopas with Group V zoeas (*i.e. Micropanope*) cluster with Group 1 species.

General Morphology

From the characters examined, it is not possible to formulare a general description of all xanthid megalopas. There also is no single character or group of characters that serves to separate xanthid megalopas from those of many other brachyuran families. Many characters 'typical' of xanthid megalopas, such as the sti'ong recurved ischial spine on the cheliped, acute anteriolateral projections on the carapace, lack of spines on the carapace, and number of serrate setae on the fifth pereiopod, are known for members of other families and vary within the Xanthidae. For example, of the four unidentified xanthid megalopas described by Muraoka &C Shibata (1980), 'Xanthid A' and 'Xanthid B' belong to Group I (Xanthinae) species. Although anteriolateral projections occur in other groups, the single recurved ischial spine on the cheliped is diagnostic of Group I. The number of setae on the-distal segment of the uropod (10 in 'Xanthid A', 9 in 'Xanthid B') is within the range of Group I. Both unidentified megalopas may belong to Group III, as two members have ischial cheliped spines as well, but only one Group III species (*Monodaeus couchii*) has acute anteriolateral projections. (*Monodaeus couchii* differs in many characters from other Group 111 megalopas, and may be an exception.) Muraoka & Shibata's 'Xanthid C' most likely has Group III or IV zoeas. The smoothly rounded frontal border of the carapace, number of setae on the ventral ramus of the antennule (7), and number of setae on the distal segment of the uropod (17) are all typical for megalopas of Groups III and IV. No megalopas of Group I have so many setae on either the antennule or the uropod, and of Group II only the platyxanthid *Homalaspis plana* has a similar antennule. However, even Group IV megalopas, which have a relatively setose uropod, do not have as many setae as 'Xanthid C'. Setation of the uropod would also indicate that 'Xanthid D' is a Group III or IV species. Another unidentified xanthid megalopa, described by Muraoka (1986), has an ischial spine on the cheliped, as do Group I megalopas. But it also has 15 setae on the mandibular palp, 15 setae on the epipod of the first maxilliped, and 60 setae bordering the scaphognarhite. This setarion is most like Group IV megalopas. Thus, this megalopa is difficult to assign to any of the groupings here described, but it bears striking sternal spines typical of the Portuninae and may not belong to the Xanthidae.

Phonetic Analysis and Classifications

It is clear from Tables 1 and 2, that groupings based on megalopal characters do not agree with any proposed classification of the xanrhids based on adults. A weak case could be made for recognition of three of the subfamilies of Baiss (1957). The Menippinae of Baiss (1957) contained Baptozius, Eriphia and Ozius, as well as Menippe. These genera are clustered together in the phenogram and are clearly separated from all other species, except Epixanthus dentatus, Homalaspis plana, and Platyxanthus crenulatus. The Pilumninae of Baiss are represented by a large cluster of species that lies between the 'Menippinae' (the lower 10 taxa in the phenogram) and a phenon line at 0.58. This grouping includes all species of *piluminie genera*, but it also includes Cycloxanthops, Xantho, and Eurypanopeus of Balss's Xanthinae, and Monodaeus couchii. The remaining species collectively correspond to Balss's Xanthinae, except that Heteropanope glabra and Parapilumnus trispinosus, both pilumnines, are also included. Unfortunately, no megalopas are known for any genera included in Balss's Trapeziinae, except for a brief description of Quadrella nitida Smith by Garth (1961). That description is indeed unusual. A megalopa with very long chelipeds and a strongly serrate dactylus on all ambulatory pereiopods, which are much longer in relation to the carapace than are the pereiopods of any other described xanthid megalopa. If other megalopas in genera of Balss's Trapeziinae prove similar to *Q. nitida*, it would strongly support the separation of the trapeziids from all other xanthids.

The classifications of Takeda (1976), Guinot (1978), and Serene (1984) cannot be fully evaluated because so few megalopas have been described for any given family or subfamily. For Guinot's Xanthoidea, megalopas have not been described for the Carpiliidae and Trapeziidae (with the exception of Garth (1961), as noted above), and the geryonids have not been included in the present analysis because they are probably closer to ponunids than to xanthids. Megalopas of two species from Guinot's (1977) Platyxanrhidae, *Platyxanthus crenulatus* and *Homalaspis plana*, are less similar to each other than each is to different species of Ozius. The groupings already discussed in support of the subfamilies of Baiss (1957) would also support the Menippidae and possibly the Pilumnidae of Guinot, but there is no clear distinction between species in the Panopeidae and restricted Xanthidae.

There is slightly more support for recognized groupings based on zoeal morphology. Because the species were organized according to zoeal groups and then examined for characters that would serve to distinguish them, this perhaps is not surprising. All megalopas of the 'Menippinae' cluster are those that have Group III zoeas, except Menippe itself, which has Group IV zoeas. The groupings of xanthine and pilumnine megalopas support the Group I and Group II zoeal assemblages, respectively, just as they support the subfamilies Xanthinae and Pilumninae of Baiss (1957); there is close agreement between these two subfamilies of Baiss and the two zoeal assemblages (Martin, Truesdale *et al.* 1985). *Echinoecus pentagonus* was suggested to be a pilumnine xanthid, rather than a parthenopid, on the basis of zoeal morphology (Van Dover *et al.* 1986) and its megalopa groups with *Pilumnus* and its allies. A review of the description of the megalopa of *Heteroaus rotundifrons* by Wear (1968), a former xanthid listed in Martin (1984), showed it to be strikingly different from all other megalopas examined. The antenna is greatly reduced, the maxilla is devoid of setae on the endopod. the tirst and third pleopods have only eight setae, and the uropods are absent. The unusual morphology of the megalopa therefore supports the exclusion of this species from the Xanthidae on the basis of zoeal characters and adult morphology. If increased sention and spination are indicative of a plesiomorphic condition in zoeal larvae, then the Pilumninae would indeed be a cerived assemblage and not, as suggested by Martin, Truesdale *et al.* (1985) for the zoeas, a primitive group. However, Group IV (*Menippe*) megalopas also have very setose appendages, and this group is thought to be derived.

At lower taxonomic levels, there is often little agreement between megalopal groupings and those of zoeas or adults. As an example, megalopas of *Dyspanopeus sayi* and *D. texana*, nearly indistinguishable as adults and as zoeas, are very different and widely separated in the phonogram. Numerous other examples will be obvious to practising carcinoiogists.

The patten, that seems to be emerging from attempts to employ the megalopa as an aid to systematics and phylogeny is a disappointing one. Megalopal characters have occasionally clarified systematic problems at or below the level of genus, but at most they only reinforce pre-existing ideas of brachyuran relationships at higher taxonomic levels. This should not be taken as a dismissal of the importance of the megalopa stage in studies of systematics, or as a reason to cease examination of this interesting stage. Much of the problem probably lies with erroneous or incomplete descriptions or with misidentification of parental material. The fact that only 14 characters (three of the original 16 were combined; an undesirably small number, were used in the phenetic analysis points to a need for more detailed comparative studies. Indeed, it is time to echo the plea of Rice (1979) for improved standards of descriptions for this often neglected stage. But another problem is that the megalopa is subjected to selective pressures different from those acting on the planktonic zoeas and benthic adults. It is a highly specialized stage that is adapted for a specific transitional role. It should not be surprising, therefore, to see phylogenetically important information obscured by homoplasty, since the role of the megalopas of all species is essentially the same, *i.e.* leaving the plankton and locating suitable habitat for adult life. Numerous examples of non-congruence of larval and adult characters are found in the insect literature (*e.g.* Rohlf 1963).

A phenetic classification is not a reflection of evolutionary history, rather a set of hierarchical groupings based on overall similarity. The phenogram is also based on a very small number of characters, so that the addition of more characters will probably yield different results. Ideally, classifications should reflect phytogeny, *i.e.* they should be cladistic rather than phenetic. But previous groupings of the Xanthidae (adults and zoeas) are based on overall similarity, and character polarity is difficult to determine for megalopa stages. For example, we cannot assume that increased setation is a plesiomorphic condition. Some xanthids, such as *Baptozius vinosus* and *Eriphia verrucosa*, have more setose appendages than does the megalopa of *Raninoides benedicti*, although raninids are generally thought to be more primitive than any of the Eubrachyura (Rice 1980, 1983). For these reasons, and until a clearer picture of character polarity emerges for the megalopa stage, a phenetic approach is justified.

(13)

ECONOMIC IMPORTANCE

Crustaceans can be divided according to size into two main groups. The larger group, which is divided into the classes Branchiopoda, Copepoda, Ostracoda, and Cirripedia, includes the familiar *shrimp, crayfish, lobsters and crabs*. The smaller group includes species that are either microscopic or range up to a few inches (about 5 cm) in size. Most of the smaller marine forms can be found in plankton and thereby occupy an important position in the marine food chain. Other copepods supply food for small fish and still others exist as parasites on the skin and gills of fish. Best known of the smaller freshwater crustaceans are members of the genus *Daphnia* (water fleas), the fairy shrimp (a phyllopod that swims inverted), and *Cyclops* (a copepod). The order Isopoda includes the only large group of truly terrestrial crustaceans. Known as wood lice, sow bugs, or pillbugs, these small animals can be found under the bark of trees, beneath stones and rocks, and in other damp places. When disturbed they curl up armout like, withdrawing into the exoskeleton.

Introduction of Crustacea

Four obvious features with few exceptions:

- They are all aquatic arthropods.
- Respiration occurs through gifts or though general body surface.
- They all have two pairs of antennae.
- Most of the body segments or at least the more anterior ones bear paired, jointed appendages that are fundamentally birramous.
- They are almost as influential in the sea, as insect are on land.
- They are abundant in fresh-water as plankton.
- Their disappearance would greatly disturb the natural balance of lakes and ponds.
- They are found only in humid environment and have not been particularly successful.
- This class is represented in the soil by members of the ostracoda,copepoda and malacostracan, amphipoda,decapoda and isopoda.

GENERAL CLASSIFICATION-OF CRUSTACEA

Ist Sub-Class-Branchiopoda

- Most primitive to modern Crustacea.
- Oldest fossil branchiopods appeared in the middle of the Paleozoic.
- The most primitive features of the Branch-isopods are the many body segments.
- An unusual features of the branchiapods is the soft, flattened type of trunk, appendages.
- Covered with a delicate cuticle, they are flexible enough to move freely are flexible enough to move freely without segmentation.
- This type of appendages called as phyllopodium
- The branchiopods include four distinct in type of organism, arranged in three orders.

1-Order-Anostraca (Fairy Shrimps)

Branchipods without a carapace, body elongated with at least 20 somites and 11-19 pairs of trunk appendages abdomen without appendages and ending in a short furca, stalked compound eyes.

Example-*Artemia*—Cultured for food of fishes.

2-Order-Notostraca (Tadpole Shrimps)

Law, oval carapace, notched at the rear, and covering the head thorax. Abdoman uncovered, ending in a long furca up to 60 pairs of appendages, sessile compound eyes.

Example-*Apus*

3-Order-Diplostraca

Laterally compressed carapace typically enclosing the trunk and appendages.

Large second antennae used in swimming. Trunk appendages often considerably specialized sessile compound eye. Abdomen variable in length and usually ending in a pair of clam.

Sub-Order 1-Conchostraca (Clam shrimps)

Example-*Cyzicum*

Sub-Order 2-Cladocera (water fleas)

Example-*Daphnia*

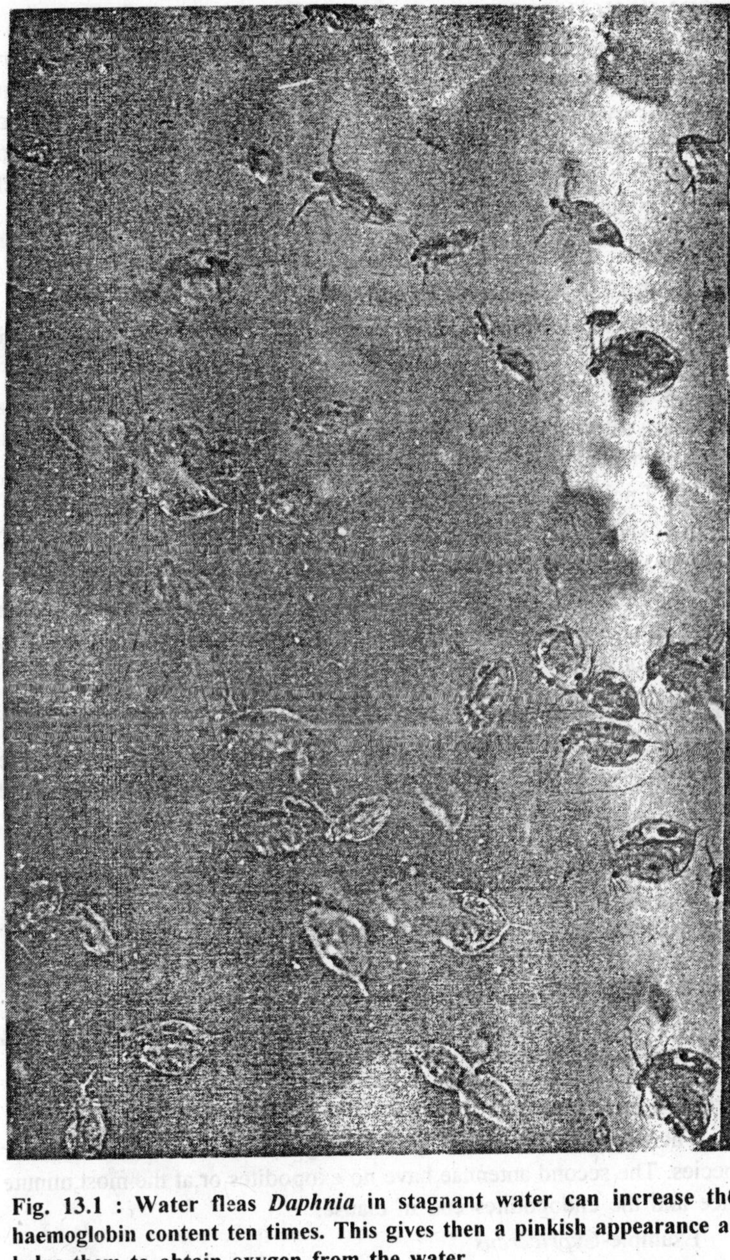

Fig. 13.1 : Water fleas *Daphnia* in stagnant water can increase their haemoglobin content ten times. This gives then a pinkish appearance and helps them to obtain oxygen from the water.

ECONOMIC IMPORTANCE-(SRANCHIOPODA)

Many years ago, the Indians in the vicinity of great salt lake are said to have used dried *Artemia* extensively as food. *Triops* is at present said to be used as food by a few natives of the federal district in Mexico. Occasionally, *Triops* becomes a part in rice field, notabaly in California. These are several report of clam sbrimps becoming a nuisance in fish rearing ponds. Especially where the fish do not feed on the shrimp (gold fish and pike)shrimp become so numerous as to clog outlet screens.

IInd Sub-Class-Ostracoda-(seed shrimp)

- About 200spp of ostracod are known of which about 2/3 are marine.
- Ostracods are small. animals, with most of the spp. not far from a milliimeter in length, but there are some fairly large spp. and the largest ostracods attain a length of over 10mm.
- Enclosed in bivalved carpace.
- The trunk of the body is very short and has lost its external segmentation, some ostracods are exellent swimmers.
- They are widely distributed in fresh water and the seas, although most spp. prefer shallow water s where vegetation is present some spp. have been taken at depth of 2000m.

Order 1-Myodocopa

Marine ostracods with second antenna having small, immobile exopadites and long endopodities with at least seven segments and with swimming setae.

Order 2-Cladocopa

Marine ostracods with both the exopdites of the second antennae well developed and important in swimming.

Order 3-Platycopa

Marine ostracods with flattened, leaf-like second antennae having well developed exopodite and endopodite.

Order 4-Podocopa

The largest order-ostracoda including both marine and fresh water species. The second antennae have no exopodites or,at the most,minute once and the endopodites end in clause.

Example-*Cypricercus*

Economic Importance-Ostacoda

With occasional exception, ostracods are only a minor element in the diet of young and adult fish. A few are intermediate hosts of *Acanthocephala parasiting* fish. Hoff (1942) estimates that 1.0 to 1.5% of the ostracods in permanent water are inflected. Ostracods are also the intermediate host of some tapeworms.

IIIrd Sub-Class-Copepoda

- Are abundant in marine and in fresh water lentic habitats and many are successful parasites.
- Free-living spp are from about I-few mm.
- About 4500 modern spp. are known.
- Are quite typical looking Crustacea.
- Head bearing five paired appendages an each somite.
- Abdomen without appendages.
- Seven orders of Copepods are recognized.
- In 4-adutts are free living or predominantly free living.
- In 3-parasitic adult stages but in some cases only the females are parasitic.

1-Order-Calanoida

Example-*Calanus*

2-Order-Cyclopoid

Example-*Cyclops*

3-Order-Harpacticoida

Example-*Parastenocaris*

4-Order-Monstrilloida

Example-*Haemocera*

5-Order-Notodelphyoidia

Example-*Entercola*

6-Order-Caligoida

Example-*Parapandarus*

7-Orer-lernaeopofioida

Example-*Lernanthropus*

ECONOMIC IMPORTANCE-COPEPODA

Under natural conditions parasitic copepods are rarely present in sufficient numbers to cause serious injury to the host. In hatchery pond,heavy and serious infection may break out, where fish are crowded in a limited area by parasites. Copepod parasites are difficult to control chiefly between the scleratized exoskeleton of the adult is resistant to chemical solution. Newly attached larval parasites are relatively delicate however,and sometimes if is feasible to combat infection by giving the fish frequent salt or acid baths. Free-living copepods constitute an essential link in the aquatic food chain. They are on an intermediate trophic level between bacteria, algae and protozoans. Fresh water copepods are not as important an element in the fish diet as cladocera. In Africa, Southern Ashia, the Western India, and part of Guianas and Brazil. Cyclops is of great important in the trammission of the nematode parasite of man. *Dracunculus inedinensid* (Guinea worm) and some spp. of *Cyclops* and *Diaptomus* are intermediate host of brood fish tapworm of man and numerous carnivorous.

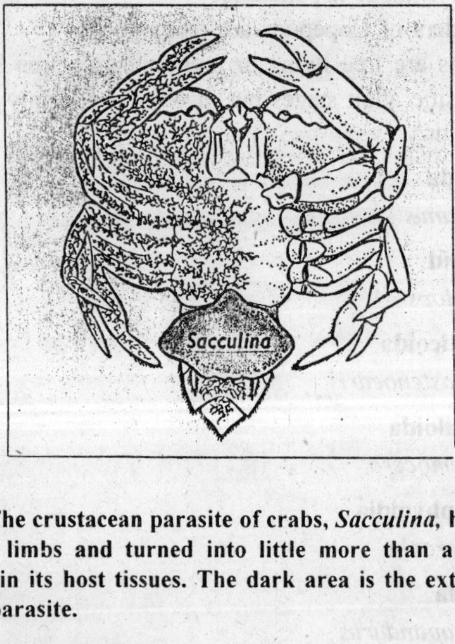

Fig. 13.2 : The crustacean parasite of crabs, *Sacculina*, has host its nervous system and limbs and turned into little more than a branching rootlet system within its host tissues. The dark area is the external reproductive sac of the parasite.

11th Sub-Class-Cirripedia

Order-1 Thoracia

Example-*Balanus*—spoils the ships or wood.

Order-2 Acrothoracica

Example-*Alcippe*

Order-3-Ascothorasica

Example-*Petrurea*

Order-4-Apoda

Example-*Proteolapas*

Order-5-Rhizocephala

Example-*Sacculina*—spoils the crabs in aquaculture.

Fig. 13.3 : The large sac-like structure under the abdomen of this crab is the reproductive part of a parasite, *Sacculina*. The rest of the parasite is not visible for it forms an absorptive branching network inside the crab's tissues. Distantly related to the barnacles this parasite has become extremely modified in adaption to its way of life.

Vth-Sub-Class-Malacotraca

- Mostly large crustaceans found both in fresh and sea-water.
- Body is always distinctly segmented.
- Thorax comprises eight segment.
- Compound eyes are usually stalked.

Order-l-Nebsiliacea

Example-*Nebalia*

Order-2-Mysidacea

Example-*Mysis*

Order-3-lsopada

Example-*Borpyrus*

Order-4-Amphipoda

Example-*Gammarus*

Order-5-Stomatopoda

Example-*Squilla*—edible and used for academics has special importance in ecology.

Order-6-Decapoda

Example-*Palaemon, Hippa, Cancer*—All have food value.

Economic Importance-Decapoda

In poorly drained clay agricultural land of west Virginia, Mississippi, Alabama and a few other southern states burrowing crayfish become so abundant (30,000 burrows per acre) that crops cannot be rised between of the feat that the cray fish denube the land of the young grain sugarcane and cotton seeding. Also their chimneys clog farm machinary. If the land can not be drained the best control is the application of some poison to the burrows. *Pyrethrum, cresote*, cyanids, carbon-dioxide unstalked lime are some of the more useful poison. They are most effective when the water table is high. Occasionally, cray fish become a nuisance in small reservoirs and irrigation ditches when their lateral burrows through earfhern dams and dikes drain the reservoir. Europeans use cray fish for the table to a much greater extent then do Americans. Most of the edible crayfish in the country come from the Great lake, certain parts of the Mississippi river system, Louisiana and Oregon, small spp. of cray fish have been raised experimentally at hatcheries for use as fish food, but the practice has not attained any importance.

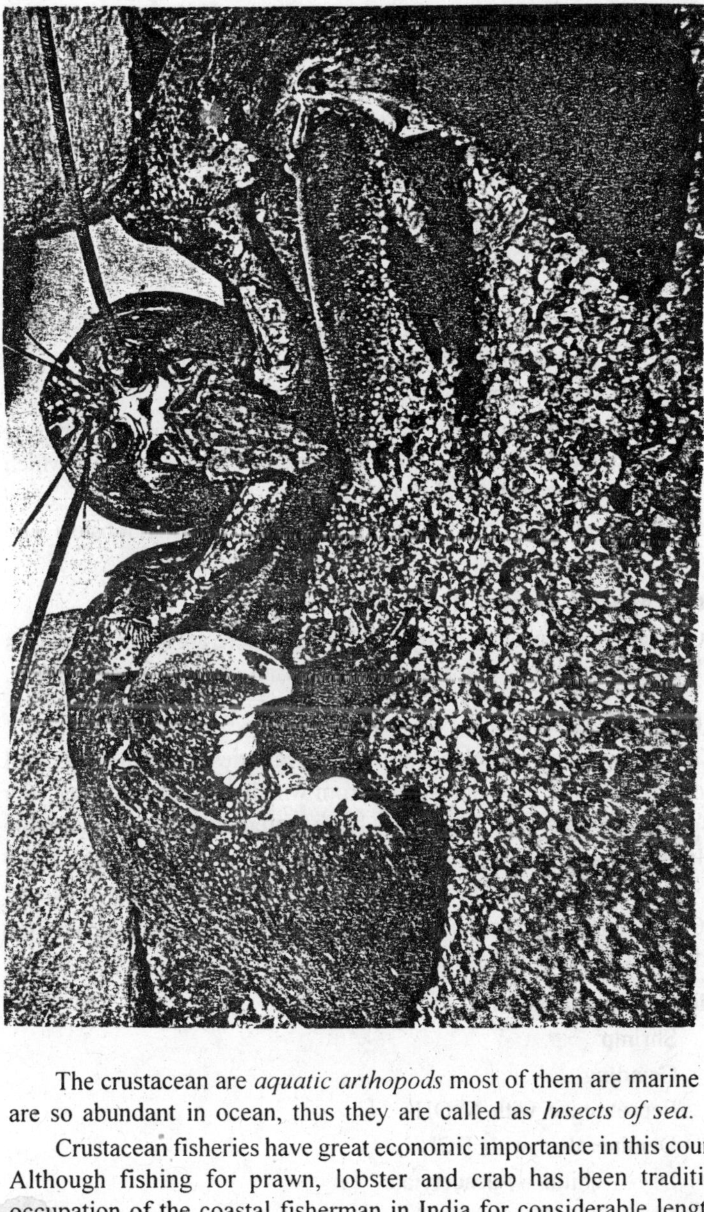

Fig. 13.4 : The European lobster. This and other crustaceans have long been prized as food.

The crustacean are *aquatic arthopods* most of them are marine they are so abundant in ocean, thus they are called as *Insects of sea.*

Crustacean fisheries have great economic importance in this country. Although fishing for prawn, lobster and crab has been traditional occupation of the coastal fisherman in India for considerable length of time reliable information, about them was not available till recently similarly the export trade of prawn pulp from west of India to Burma and Ceylon was an old as the rice trade of the later countries with India. All present among the prawn and shrimp producing countries of the world, India ranks 4th day by day production is increasing.

The commercial crustacean landing of the country are classified into three distinct group.

(i) The penaeid prawn

(ii) The nonpenaeid prawn

(iii) The other crustacean

(i) The group penaeid prawn include several species of large sized marine prawn and it totally supported our prawn export industry

(ii) Several small of small sized shrimp belonging to different genera are classified under the non penaeid prawn which have commercial values.

(iii) The other crustacean group include mostly lobsters and crab.

Positive Aspects Related To Crustaceans

Crustacean are economically very useful. There are several products which obtain from the crustacean like Chitasan etc., are given below :

1. CHITOSAN

The chitonous exoskeleton of certain varieties of shell fishes and Arthopodes provide us a rich source from which a unique polymer with wide application chitin-Chitasan can be prepared on a commercial scale

It is a biopolymer cellulose. It is natural non toxic biodegradable and of high molecular weight It is second most abundant natural polymer.

Chitosan has been sold commercially as a solution, flapped and Fine powder and in lead and fibre forms.

Method of preparation:

Shrimp

Grinding

Peproteinize with NaOH

Demineralise with HCl

Dry----chitin-----deacetylate

Chitosan----chitosan----dry

Fine mesh flabeel
Chitosan (and bead)

USES OF CHITOSAN

(1) *Medical field :* used in wound dressing, homeostatic agent in drug delivery system and as a hypocholestelolemic agent.

It is due to ability to improve *Uranus* healing and clot blood and absolve liquid and to form protective layer film and coating

(ii) In agriculture

This used on seed costing to improve the production, also added to crop and to stimulate natural microbes for intending product ion.

(iii) In Biotechnology It in commercially useful to immobilize Enzyme and cells.

It is also used in hair treatment and skin case this a cheaper moisturizer). B. chitosan is used informing contact lense from pens (chitosan from crustacean)

2. DIFFERENT CRUSTACEAN AS A FOOD

Eating use out of the joy of life. Eating sea food is not only for joy but is nutritional source too. Any food which we eat must supply protein, vitamins, minerals and fiber etc. sea food is an excellent food because it is nutrient rich. It has a best portion which supplies essential amino acid. For standard food following quality are must.

(i) Nature of product

(ii) Health and safty hazard.

(iii) Legal packaging, labelling and requirement. Most important class of value added marin product is Bread and battered products. It is included in value added products because the process of coating is done by bread and butter. Prawns economic importance is pretty high on they prefered food item among the different food item obtain from marine resource. Prawn can be battered and breaded undifferent from which are following.

1. Battered and breaded peeled prawn.
2. Battered and breaded peeled prawn.
3. Salted and flavoured butter sand wiched between two sufficiently big and identeical size peeled and deveined and fantail form prawn.

The product is wrapped in polythene bag and frozen in or transfrom. in price or the prawn catched in markted in fresh condition simple semidrying and drying condition.

There are over six hundred crab species occuring in the Indian water but only few of them are being used for food purpose therefore no sauge fishery for crab in any part of the chuntey. In the Indian market crabs fetch low price which account for the absence of any largy scale fishery the edible.

Saylla, Scylla serrata—largest most important crab for aquaculture. This is export commodity.

Sand lobster or slipper lobster one also edible but some year ago they was not considend for edible purpose because they have very hand shell and less melt constant but now a days it is export commodity and it is also exported to westen country

Part of crustacean used for consumption.

Shrimp = whole

-peeled and deveined

-cooled

-head less shell on butterfly

-fantail

Lobster-cooked lobster

Lobster tails

Lobster meat.

3. USED IN THE PURIFICATION OF BLOOD

The sea food obtain from crustacean are helpful in building better blood It helps in formation on of R.B.Cs.

4. TO SOLVE THE EMPLOYMENT PROBLEM

Now a days the emplouyment problum is a very big problem .It helps in incresing the economical statuts of people.

Crab support mostly a sustanance fishery being collected by poor section of people for their own use.

There are some other uses like following :

(1) Many of the crustaceans are used in the ornamental purpose

(2) More than 50% zooplankten are belonging to the class Crustacea and it is useful for fish food. Eg. *Artemia*—used as live for fish.

(3) *Squilla* us not useful for edible purpose but useful in industrial work.

(4) Blood of crab is used in the treatment of cancer.

(5) Many crustacean are used information of fish meal.

Negative Aspects of Crustacean and Their Farming

These are not only positive impacts of crustaceans. There are some negative impacts of crustaceans :

1. Destruction of mangrove forest. Man grove are highly fertile due to the presences of of high organic matter. They also support rich diversting of fauna and flora. Much of the mangrove forest were already cleared for shrimp culture (in AP). The degradation of the mangrove result in coastal cause soil erosion, lack of protection from cyclones and tidal waves.
2. Arising of social conflict costal region between fisherman, farmer, police, forest officials and some against the culture to save nature and some favouring the culture

In Tallarevuneer Kakinada conflict arised between the farmers and local fisherman community regarding prawn culture.

Many crustacean are individually not use full such as *Argulus*. *Argulus* in Economically, important in negative sense heavy infection of these parasite on the fisheries is result immortality and morality of fishes reducing the production of fish.

These are some crustacean Lepas produce roughness on the surface of a ship these by *ingering turn* balance average roughniess on a power vessels can rest in 3 to 10% increase in consumption . for large power vessels fuel cost can amount to 50% of the total operating cost.

The fresh water crab has a masonid which is cosidercd as menace because of its burrow making habit at the bottom and in dykes of pond weak and also causing greater seepage of water therefore perodical removal of these crab is necessary for a better pond management crabs are also removed from nursesry pond because it eat and kill the fry therefore production becomes very less.

Production, consumption and export of crustacean and their role in the economic status of country Among the various marine product shrimp command the leading position with large share total sea food market.

India topped among the list of shrimp producing countries for over a decade form 1973 to 1983 excepting 198 land these after it has been releated to a place after China. In 1980 India produced more than 15% of total shrimp production and 10% in 1985.

1990—3rd place

1991—4th place

Particularly, the prawn production is 1991-92 was 1.55 lakh tones in case of penaeid and 1.50 lakh tons in case of non penaeid prawn. Due to owing large size of carapace fast growth crab is beconing more and more popular about 10,000 t/yr eatch form the bay of Bengal in 1990 crustacean production reached 4% of the total production in Ashia pacific shrimp is very widely used for food. It has been reported that the USA consurmed the highest quatum of shrimp in 1992. it is estimated that in 1992 about 758 million ponds of shrimp were cosnurmed in USA 758 million pounds of shrimp were consumed in USA while in Japan consumed 672 million pound of shrimp.

The commercial scientific farming of *Macrobrachium* became popular in several parts of the woeld and the Global production of this spp., through cultured is in the order of 27,000 tons. Meat crab *Scylla* srrrata is more popular and total catch of it has been estimated to about 10,000 tons/yr.

It has been reported that the USA consumed the highest quantum of shrimp in 1992 in 1992, 758 million pounds of shrimp were consumed in USA while in Japan 672 millions pounds shrimp were consumed.

The export trade of marine production form India is on the increase year after year. During 1994-95 the country exported 273240 tons of sea food valued at Rs. 3270 crores which is bout 4% of India total export coming, in the 50 items of marine products shrimp contributes the major role value 73.82% and quantity 37.25%

In 1992-93 export data

For shrimp	Q-74393 tonns	value	1180.26 crores
Lobster	Q-1613 tonns	value	43.34 crores

From this productions consumption and export data we can see that the crustacean are playing great role in the economy of India.

(14)

FEW CRUSTACEANS

ARTEMIA

Introduction

Artemia, the brine shrimp, is not a true shrimp, but belongs to a quite different and primitive order of crustàceans. It is able to line in a notably wide range of salinities, from a salt concentration not much above that of sea water. Normally, however, *Artemia* is found only in highly saline lakes and ponds, probably because it falls an easy prey to the predators present in other environment. The distribution is world wide mostly is the warme and more arid regions. Dried eggs from which laboratary cultures can be started, are on sale by biological supply companies.

Body Plan

The body appears divided into three main region, but because of difficulty is defining thorax and abdomen in a manner permiting comparisons with other Branchiopoda, it seem best to speak only of a head and trunk.

The trunk has eleven, limb–bearing somites,

The telson usually is fused to the last somite but some time it distinct. Attached to the telson is a caudal furca, which is used for steering and for making sudden spurts through the water.

Exoskeleton

The while body is covered by an extemely this skeleton, which never the less is remarkably impermeable to salts, water, and many other substances. This exoskeleton is molted about every 4-6 days through out the adult life of several mouth.

Paired Appendages

The head bears the fine pairs of appendages typical for crustacea. The antennules are slender and tipped with a few special setae, believed to be chemosensory. The antennae are much thicker do not bear setae, and are modified in adult males as large claspers.

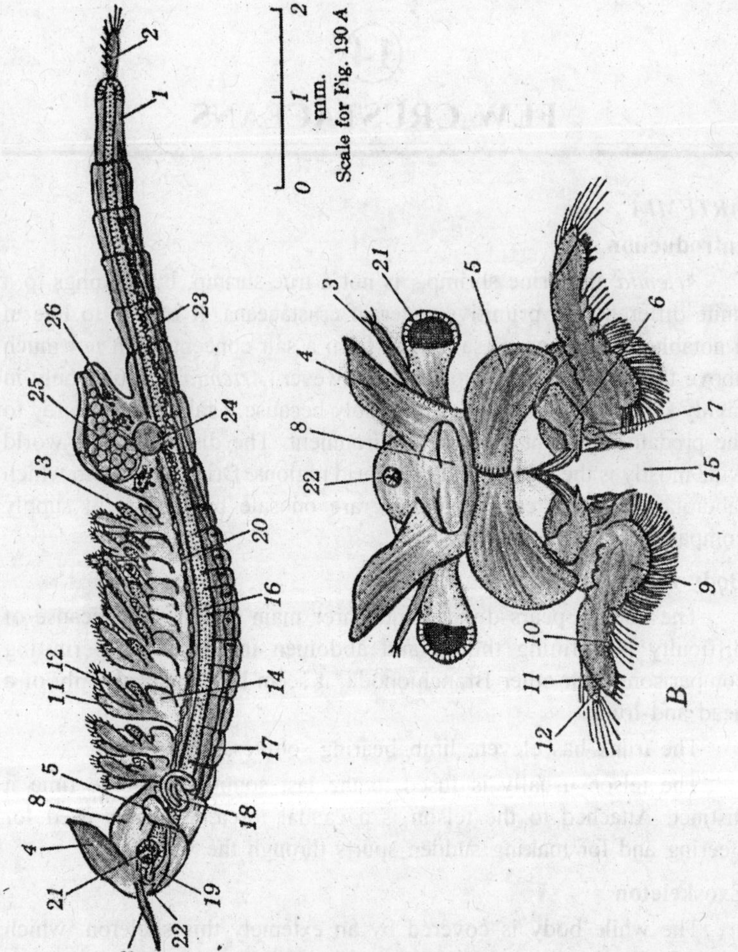

Fig. 14.1 : A. Diagram of a female *Artemia*, shown in lateral view. (Redrawn from Lochhead.) B. Ventral view of the head and first trunk somite of a female *Artemia;* the labrum turned forwards and the first pair of trunk appendages spread laterally. 1. Last trunk somite and the telson, fused together, 2. Caudual fureca, 3. Antennule, 4. Antenna, 5. Mandid!e, 6. Maxillule, 7. Maxilla, 8. Labrum, 9. First endite of a phyllopodium, 10. Fifth endite, 11. Telopodite, 12. Exopodite, 13. Metepipodite, 14. Protepipodite, 15. Food groove, 16. Midgut, 17. Heart, 18. Maxillary gland 19. Supraesophageal ganglion, 20. Ventralnerve cord, 21. Compound eye, 22. Median eye, 23. Ovary, 24. Lateral pouch, 25. Ovisac, 26. Posterior shell gland

The maxillae are greatly reduced. The trunk bears eleven pairs of leaf like appendages, or phyllopodia. A typical one of these, from near the middle of the series, should be dissected off as close to the body as possible. Along the median border there is a series of give *enditer* of which especially the first two bear a long regular fringe of filter setae. Serving as important role in the process of filter feeding. Distally, the *enditer* are succeeded by a large lobe termed by Eriksson the telopodite. This bears strong, scarping setae used is feeding on bottom detritus.

Digestive System

A short, cervical, foregut or esophagus, leads up from the mouth to open at right angles into the long midgut or stomach intestine. In front of the point of entrance of the esophagus, the midgut expands into a pair of pouches. Secretion of digestive enzymes and absorprion of food products probably occur both here and along the length of the midgut; which also plays an important part in osmoregulation. Pasteriorly the midgut joins the short hindgut or rectum, provided with dilater muscles present on all parts of the gut. The rectum opens at the anus, between thc furcal rami. Foregut and hindgut both are provided with a thin cuticular lining, continuous with the exoskeleton.

Circulatory System

The heart is a long tube, dorsal to the gut, extending the length of the trunk. Blood enter it posteriorly and through valved lateral slits or ostia, of which there is one pair in each of the posterior 14-15 somites. At the anterior and blood leaves the heart, entering the hemocoelic spaces of the head, and than circulating through the hemocoel of the trunk and appendages. Though there are no definite blood vessels, the blood never the less follows a more or less constant route. Movement of the trunk appendages undoubredly aid in the cirulation. Frequently the blood is coloured red by hemoglobin. A single type of blood cell is present, these cells behave as amoebocytes when adhering to tissue surfaces. These blood cells originate from blood-cell forming organs located near the base of each trunk limb.

Excretory System

Exceation is probably performed by the pair of maxillary glands, which form rounded prominences on the sides of the body just behind the mandibles.

Each consists of a blind, coelomic end sac, followed by a coiled tubule which opens on the maxilla. The corresponding antennary glands of the nauplius larva are claimed by some investigaters to be entirely absent is the adult, but a rudiment of the end–sac does seem to persist, as cell selectively taking up certain pigments. Uptake of pigment grannules and other foreign particles also is performed by larva phagocytic storage cells, located in many parts of the body, and perhaps most easily seen in the antennae. In well-fed individuals these cells also store fat and glycogen.

Nervous System

The supraesophageal ganglion, circumesophageal connectives, and two parallel chains of ventral ganglia. Connected in ladder like fashion by comissuces, can best be studied is frontal sections. The supraesophageal ganglion receives nerves from the eyes and antennules. Antennal nerves join circumesophageal connectives.

Antennal nerves join the circumesophageal connectives. A pair of ventral ganglia occurs in each somite, from the mandibular to the second genital, posterior to which there is only a slender pair of ventral nerves.

Sense Organs

The stalked compound eyes, each composed of numerous ommatidia, are conspicuous. The smaller found. It is formed of three hemispherical cups and is located just in front of the supraesophageal ganglion. It sections are available, the paired frontal organs can be found each consisting small group of what appear to be receptor cells, located under the hypodermis just ventrolareral to the median eye. Other presumed sense organs, including organs of taste on the mouth parts, have been described but are difficult to find.

Reproduction

In the male the testes are long and tubular, located usually in about the anterior four somites of the limbless part of the trunk. Sperm are produced in the more posterior parts, while the anterior part secretes a mucilaginous fluid. From the anterior end of each testis a coiled glandular duct leads ventroposteriorly to about the posterior border of the second genital somite. Here the duct turns and runs forwards to near the anterior border of the first genital somite, where it turns again and runs back to the tip of one of the two penes which project posteriorly from the second genital somite. In copulation the penes are everted or extended, evidently by blood pressure, to about four times their resting length, and the duct is then no longer *kinked*. Eversion of one or both penis often can be observed in a dying specimen confined under a coverglass. A retractor penis muscle runs to the tip of each penis from the midventral line.

The antennae in the male are modified as enormous claspers, ornamented and shaped so as to fit tightly around the female just in front of the ovisac. Clasping persists a number of days, the animals swimming around while thus united as a pair. Copulation occurs after each molt of the female. In some stocks of *Artemia* males are absent, reproduction then being parthenogenetic.

In the female the position of the ovaries is similar to that of the testes in the male. In the second genital somite each ovary is connected by a short duct to a lateral pouch, situated in the dorsal part of the prominent ovisac. Here the ova undergo maturation. When filled with ova, the lateral pouches are round and easily seen, but at other times they may be collapsed and more difficult, to distinguish. Each opens into the median central part of the ovisac. Here the eggs are fertilized, acquire a shell, and undergo some embryonic development. Groups of shell glands, appearing like clusters of grapes, open into the median ovisac, and muscles are present by which the eggs can be kept moving. Shortly before a molt thick-shelled eggs are distrituted.

ARGULUS

There are about 120 species of Branchyura described, of which the japanese *Argulus scutiformis,* up to 3cm long is the largest.

Branchiura, fish lice, are small csustaceans with a wide flat carapace and on unsegmented abdomen. There are compund eyes and four pairs of biramus thorocic limbs.

Anatomy

The animal is dorsoventrally flattened, and its body consests of two parts. The anterior part is modified for parasitic life, the posterrior part for locomotion. At the base of the notch is the anus and behind it a minute furca. The anterior part is a cephalothorax containing the head some of which the first thoracic somite is more or less fused. A flat fold, the carapace, extends anteriorly and to the sides, posteriorly it differs considerably from species to species. The lateral extensions can be move up and down slightly, winglike. The pereon has three limb-like metameres and is partiolly or completely covered by the carapace dorsally. The hypodermis contain mainly individual gland cells and glands and produces a delicate exoskeleton.The head appendages, adopted to parasitic life, differ considerably from those of other crustaceans. Both pairs of antennae if present are very short and may bear hooks in *Argulus.* Here both are attachment organs and have only scattered innervated sense organs.*Chonopeltis* lacks first antennac and the second antennae have no. Neither antenna pairs have hooks in *Dipteropeltis* Behind the small antennae the much larger two paires of maxillae can be seen. They are uniramous without endites and have lost their function as mouthparts.The firs maxillae of *Dolops* have a long clasping hook at the tip. The base of the first maxilla in *Argulus* has changed into an attachment organ, a deep tubelike sucker; young animals still have vestiges of the distal articles. The mandibles can be found only by careful examination. They are hidden between labrum and labium and form a pair of which face anteriorly. In *Dolops* they lie directly behind the mouth between the papillate raised lips. However, in *Argulus* distal from the mouth there is a posteriorly dircted proboscis on the tip of which, under the borders out of the lip folds. The proboscis lies flat posteriorly and can be exte nded ventrally. In *Argulus*, but not in other genera there is in the midline of the body an anteriorly directed hollow spine, the so called sting or poison, in some species reaching to the anterior border of the head. The distal tip of the spine bears the openings of large cephalothoracic glands and cab be withdrawn into the base. Since the spine is innervated by the labral ganglion (sympathetic nerve) it can be concluded that the spine is derived from the labral primordium in the embryo.

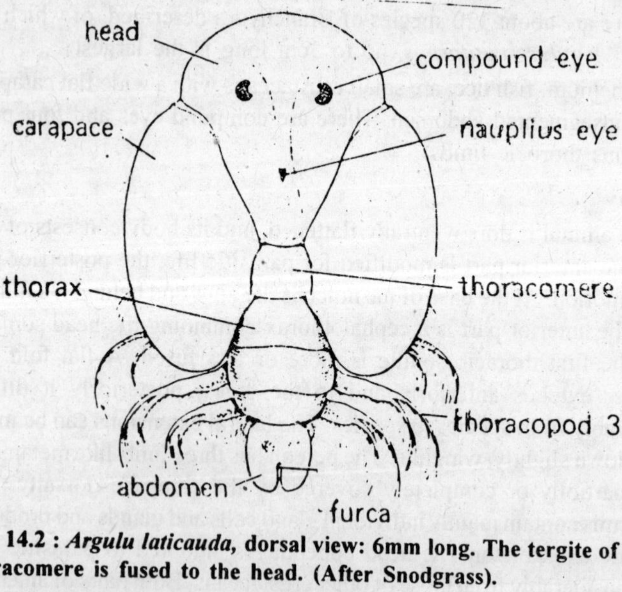

Fig. 14.2 : *Argulu laticauda*, dorsal view: 6mm long. The tergite of the first thoracomere is fused to the head. (After Snodgrass).

The morphological significance of the head appendages was not recognized for a long time. It can be studied by comparing *Argulus* with larval instars and with less specialised *Dolops*. The metanauplius, which is and embryonatic stage and does not leave the egg, has all the five pairs of anterior appendage primordia typical in form and location of the rsnt of the crustaeea. There is no evidence of the large poison spine of *Argulus*, suggesting that this spine is a secondary acquisition, and not the vertige of an appendage. The newly hatched stage has the total complement of appendages, in form and position as in other crustacean subclasses. The manible has a large endopodite, a palp, which has an obvious relationship to the hookshaped endite. This endite, unlike those of other crustaceans, is movable. At the same time, it can be seen that the main part between endopodite and endite will contribute to the lateral wall of the future proboscies. Also on the anterior of the metastomal area one can see two small swellings , out of which arise the labial spines of the adult, two lobes each with a hollow spine. These spines have been interpreted as maxillary vestiges by some authors. However, if they were maxillary vestiges, one would expect them to be large in a larval stage. As this not the case they are probably of secondary origing and

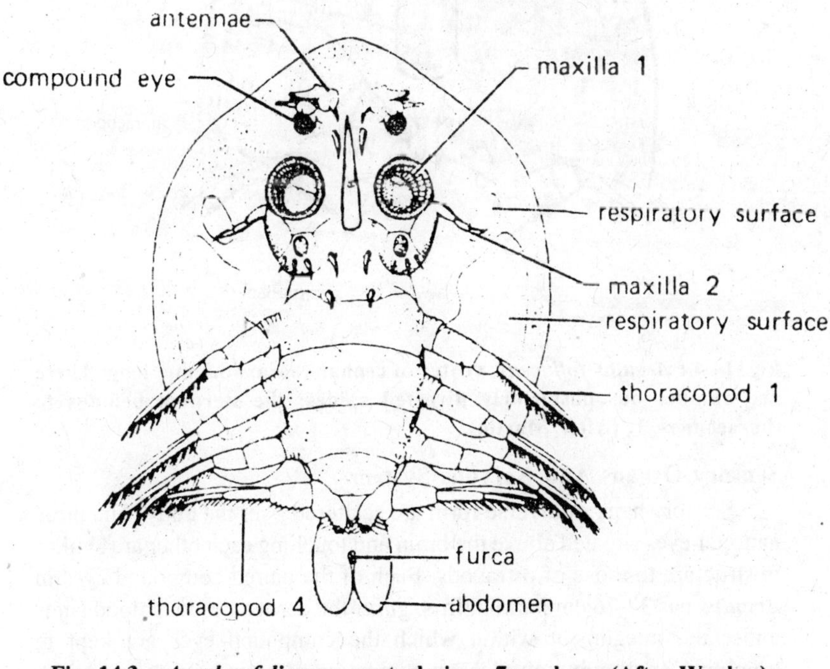

Fig. 14.3 : *Argulus foliaceus*, ventral view; 7mm long. (After Wagler.)

the two large head appendages of the adult correspond to the first and second maxillae, despite their lack of endites. further evidence is that the maxillary glands open on the second maxillary.

In marked contrast to the dodified cephalic appendages, the four pairs of thoracopods (the first one belonging to the cephalothorx)are biramus swimming limbs,having their surfaces increased by setae. The thoracic limbs are modified for claspers in males. The abdomen lacks appendages.

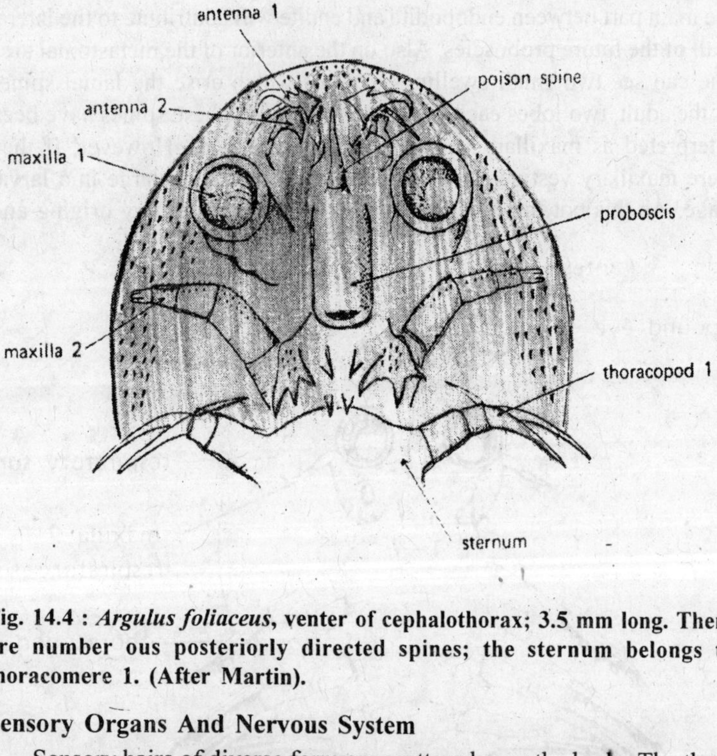

Fig. 14.4 : *Argulus foliaceus*, venter of cephalothorax; 3.5 mm long. There are number ous posteriorly directed spines; the sternum belongs to thoracomere 1. (After Martin).

Sensory Organs And Nervous System

Sensory hairs of diverse form are scattered over the body. The three naupliar eyes situated above the brain and touching each other are similar in structure to those of ostracods. Each of the paired compound eyes in *Argulus* has 30-76 emmatidia. In *Argulus* the cornea lies in a blood sinus under the integument within which the compound eyes are kept in trembling motion by muscles, an adoptation that can be compared with the ectodermal eyesac of *Triops*.

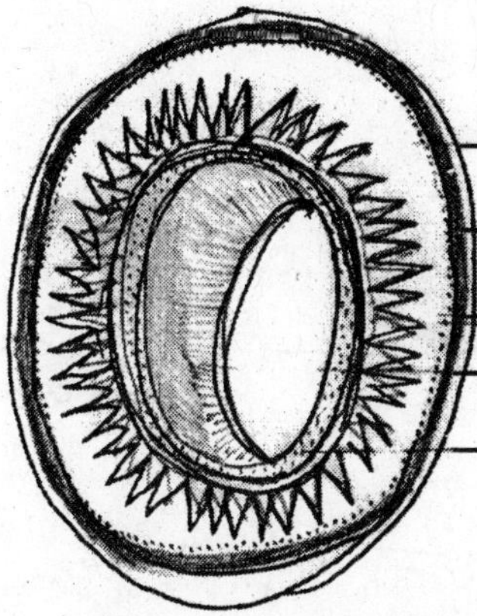

Fig. 14.5 : *Argulus foliaceus*,maxilla of a young male showing vestigial distal articles; 0.2mm in diamester (After Claus)

The nervous system consists of a supraesophgeal brain and a mass of ganglia which borely extend to the first lesgs. The mass has constrictions marking a subesophageal ganglion and five additional ganglia. The subesophageal ganglion innovates the mandibles and the first maxillae, the other ganglia innovates the following appendages, the last innervating also in the abdomen.

Digestive System

The ectodesrmal foregut continues as a wide midgut open two ducts which come from the brached caeca. The caeca lie within the carapace and grow with it, permitting the intake of large quantities of food at one time, an adoptation found in other temporary priorities. The anus opens at the base of note on the abdomen.

Excretory And Circulatory System

One pair of large maxillary glands, which open at the base of the second maxillac, constitutes the excretory system.

The heart lies in the last thorocic somite and through one pair of ostia takes in blood that comes form the platelike abdomen. The abdomen makes pumping movements of importance to circulation.The blood floues through a long cephalic aorta opens at ends but has no branches. *Dolops* has haemoglobin in the blood.

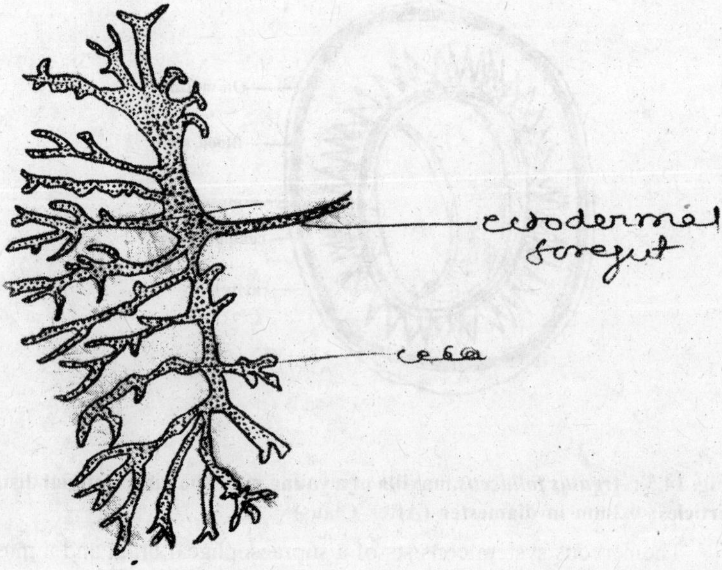

Fig. 14.6 : *Argulus folioceus*, right gut seem of a femle; 4.2mm long (After Claus)

Reproductive System

The large median ovary in the thorase produces numesrous eggs. The ovary opens through paired oviducts, only one of which is functional at time. The oviducts opens into an ectodermal pouch behind the last swimming limbs. Each of the two seminal receptacles present has its own opening on the tip of a minute sclerotized spine. These spines are located near the opening of the oviduct.

The testis lie in the abdomen. Their afferent duct extend anteriorly to fuse in the thorax into a large seminal vescicle, from the antesrior of which a pair of sperm ducts originate, pass posteriorly, and fuse into a median duct before opening between the last limb coxae.

Reproduction

During the mating season the male *Argulus* swims actively. He settles on the back of a female while she is feeding on a fish, and surrounds her last limbs with him. After grasping the female from above, the male tourists its posterior end beneath the female, first on one side, then the other, presumably filing each seminal rcceptable in turn.

The males of some species have a simple genital papilla but it is not known how it fits over the spine. *Dolops ranarum* unlike *Argulus*,employs spermatophores which form as material flows out from the male sperm. The single spermatopohre is *impaled* on the two spines bearing the opening of the seminal receptacle of the female permitting the sperm to pass into the seminal receptactes. As each egg is laid it is pricked and injected with sperm.

After 3 weeks, the time probably depending on the temperature, the female of *Argulus foliaceous* leaves the host and seeks aquatic plants or stones on which, with the secretion of oviduct, she attaches a double row of thick shelled eggs. This is a departure from the brood care practiced by other crustaceans, most of which carry their eggs in brood chambers or on appendages. About 20 to 480 eggs are laid at a time; altogether 800 to 1200 are deposited. In an aquarium the female parasitizes fishes between clutches.

Devlopment

Naupliar and metanaupliar stages of *Argulus foliaceous* are passed within the egg, in 3 weeks at 19° C, in 5 weeks at 14°C. The first free stage has a carapace and immediately attaches itself to a fish. Eight more juvenile instars follow. All nine larval stages are passed in 5 weeks at 14° C, in 4 weeks at 16° C.

The first free stages of *Argulus* differ from te adults mainly by their small size (0.6mm), the shortness of the carapace (which does cover the three percomeres), the shortness of the abdomen (which does not extend beyond furca), therudimentary condition of the three most posterior thoracic limbs, the large branches of the second antennae and mandibular palps, and the absence of a sucker on the first maxillac.

There are two types of newly hatched individuals in branchiura even in the genus *Argulus.* In one the larvae swim jerkily with the second antennae, the mandibles, and the first limbs. The other swim ith thoracopods 1–4. The first becomes like th e second after one molt. Some species of *Argulus and Dolops nanarum* remain in the egg until all thoracie limbs are functional.

The larvae swimming with antennae attach themselves to the host with the distal hooks of the first maxillae and first molt all four pairs of swimming limbs are funditional, and the second antennae and mandibular palps are shortened in the fifth stage the sucker begins to form , but become functional only after the seventh molt at intervals of 3 to 4 days, *Argulus* becomes sexually mature and the carapace and abdomen achieve their final from. The adults continuous to molt.

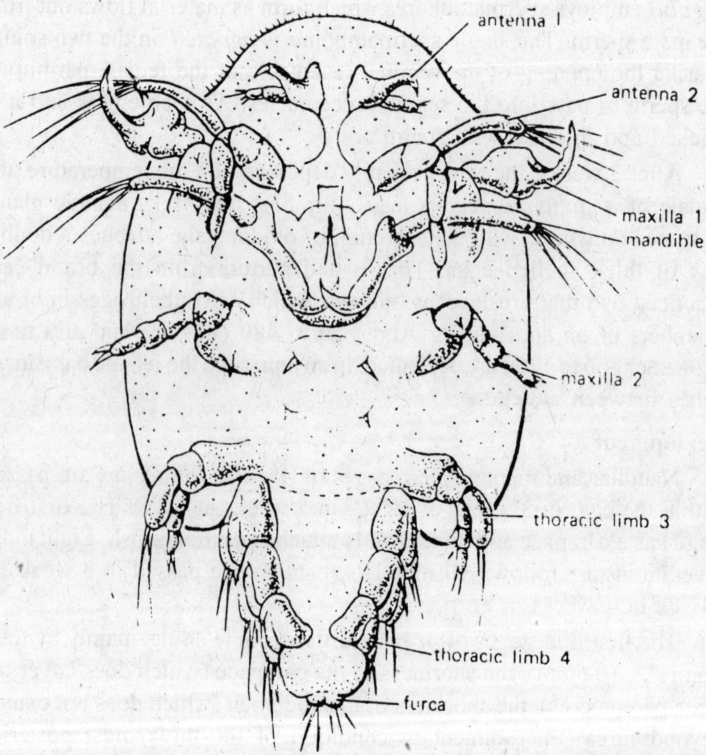

Fig. 14.7 : *Argulus foliaceus*, second larval stage, ventral view: 0.7 mm long. The hooked endites of the mandible are covered by the transparent labrum whose convex posterior border is distinct. Anteriorly is the poison spine. (After Stammer.)

The earliest stages of *Chonopeltis bresis* differ from the adult mainly in having smaller size and in lacking suckers. Instead of suckers they use segmented appendages for attachment. Although later instars have increasing number of setae on the limbs, neither larvae nor adults can swim. All stages are parasitic. The adult changes hosts (bottom-dwelling fishes) by an unknown method.

Relationships

The presence of carapace and compound eyes are the characters that distinguish Branchiura from Copepoda. The differentiation of the limbs distinguishes Branchyura from the Branchiopoda and Ostracoda. The similarity with *Caligus* is probably due to convergent evolution.

Habits

Branchiura are temprony fish parasites that feed on blood or mucus. Many are active swimmers, found in the ocean as well as in freshwater. *Chonopeltis* and some species of *Dipteropeltis* cannot swim, judging from their anatomy. *Argulus foliaceus* swims rapidly, gliding smoothly through the water propelled by the thoracic limbs. Usually the dorsum is turned towards light, a reflex that depends on the compound eyes. But *Argulus foliaceus* can swim up and down at any time. To rest they attach to objects, even glass, with their suckers. On glass, they can glide forward by pushing the suckers forward. Finding of the host by *Argulus foliaceus* does not depend on chemical stimutor passing shadowess. But an *Argulus* toches by a fish shoots in a curved path through the water and then settles, after reaching the fish that disturbd it the animal remains attached depending on tactile and chemical stimuli. Receptors of such stimuli probably are sensory hairs of which various kinds have been described without knowledge of the functions.

If clean filter paper moved through water touches *Argulus,* which is not attached to a host, the animal will shoot off but immediately it attaches to the paper for a second before swimming off again, if the experiment is repeated with filter parer rubbed over the scales of a fish, the fish louse will move about on it for a while.

A sense of taste has not been demostrated after attaching to a host, (the animal will shoot off but immediately it attached to the) rheotaxis is of importance. *Argulus* turns soon until its longitudnal axis is parallel to the host and then crawls toward the head of the fish. It settles where there is little friction with the water, usually behind gill operculi, behind fins, or in other thin *skinned* places. When swimming in the water the animal moves against the current, the seat of the rheotactic sense may be hairs that have thin filamentous tips. This cover the head and dorsum of the carapace and are innervated by a larg nerve form the tritocerebrum. Although fish lice are considered to be temperature sensitive the sense organs are not known.

Hosts are diverse fish species. It is certain that during the corse of their life they not only change host, but also host species. It is puzzling how the non-swimmers change hosts. Some *Argulus* species have been found on tadpoles. *Argulus areassonensis* has even been found on dead cuttel fish (*sepia*) *Argulus* attaches to the host mainly with the large sucker of the first maxillae, but may increase adhesion by oppressing itself tightly. It the host swims rapidly, the parasite is pressed backward toward the substrate, forcing the many thorns to hook and anchor in *Argulus and Dolops* the antennae help attach parasite to host.

Larger carp, tench and pike do not react to *Argulus foliaceus,* but carp 4.5 cm long and tadpoles shoot through the water, when this fish louse attaches. It is assumed that the sting of poison spine and the metastomal spines is irritating.

The parasite remains on the skin, cutting through with its mandibles, and drills into blood vessels, in a relativley short time the digestive caeca are filled with blood.

Several fish lice on a tadpole or small fish may be fatal to the host. But carp 4 cm long and mackeral 10 cm long do not show obvious damage from the presence of 20 *Argulus* for a few days. However, heavy infestations may occur, and 420 *A. foliaceus* have been counted on the skin of a 1kg. *Esox. sp*. Especially fish in weirs or fish ponds may be heavily attacked. A tench 28 cm long had 4250 parasites besides the fungi an infestation may make the host so restless that it feeds little. Trout may succumb to the parasites. central European species of fish lice transmit a fish disease, *Pseudomona punctata.*

"Flagella" structures, of dubious homologus, occur on some of the legs of some species of *Argulus and Dolops* and are used to wipe the ventral surface of the carapace.

Caeca and gut absorb the food and a well fed *Argulus foliaceus* feeds again after 2 weeks at 3°C, or 1 week at 14°C . The attached *Argulus* beat with the swimming limbs, constantly supplying fresh respiratory water to their venter. If *Argulus* is placed in a 1% silver nitrate solution. The Ag^- ions are immediately reduced and darken some areas of the carapoce that are presumed to be of respriatory function. The anteriour respiratory areas of the carapace are smaller than the posterior ones. Each area is framed by a sclerotized ridge, but the cuticle of the area itself is very thin, its hypodermis differentiated as are amphipod gills. Below it is a large blood sinus. The failure of other areas to reduce the salt indicates that are of less importance.

Both abdominal plates have musculature that pumps the blood from the venter of the tunk. posterioul y and them anteriorly and up, towards the heart. These "tails" thus function mainly as accessory hearts, less in respiration.

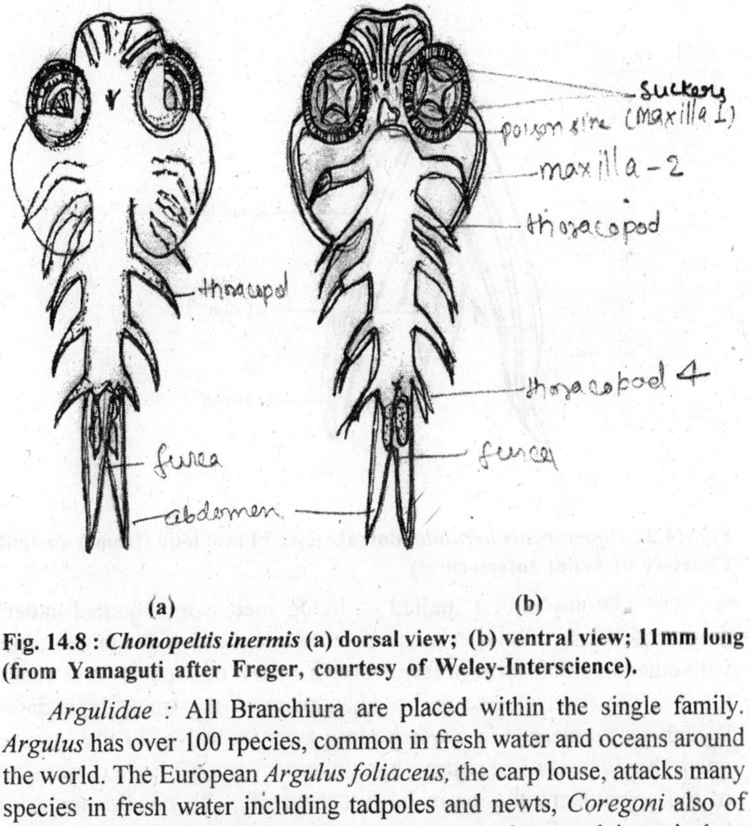

Fig. 14.8 : *Chonopeltis inermis* (a) dorsal view; (b) ventral view; 11mm long (from Yamaguti after Freger, courtesy of Weley-Interscience).

Argulidae : All Branchiura are placed within the single family. *Argulus* has over 100 rpecies, common in fresh water and oceans around the world. The European *Argulus foliaceus,* the carp louse, attacks many species in fresh water including tadpoles and newts, *Coregoni* also of Europe has a shorter carapace, up to 13 mm long and is mainly a parasite of trout. There are about 15 species of *Argulus* in North America. The African *Chonopelti* lacks a poison spine but has suckers. The largely South American *Doops* lacks both the poison spine and suckers and its preoral cavity consests only of babrum and metastonal area. Other *Dolops* species are known from Africa and Tasmania. *Dipteropeltis* is South American.

Various species of *Daphnia* are known, in fresh-water habitats ranging from puddles to open lakes. *Daphnia magna* occurs in the northern hemisphere, in small ponds or ditches having a good growth of vegetation. It tends to be rather local in distribution, but. when found may be very abundant. It has been chosen for study here because of its large size and because it is the species for which we have the best anatomical descriptions. Other species resemble it in most of the essentials.

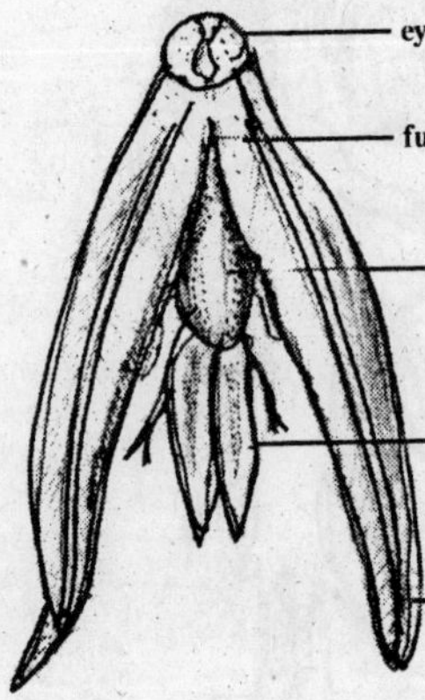

Fig. 14.9 : *Dipteropeltis herundo,* dorsal view; 14 mm long (from Yamaguti, Courtesy of Wiley-Intersecince).

The anatomy is best studied on living specimens mounted under a raised coverglass, or on specimens mounted after being killed or narcotized with ether or chloroform (a few drops to 10 ml. of water). For a ventral view of the food groove and median parts of the trunk appendages, specimens submerged in a watch glass, with the valves of the carapace open, should be studied under a binocular microscope. For a closer study of the appendages these can be dissected off with needles under the binocular, after first removing one valve of the carapace.

The body is divisible into head and trunk, terminating in a large clawed structure termed the *abreptor*. The first five trunk somites bear appendages, and perhaps three more somites are present. A bivalved carapace entirely encloses the trunk, fused dorsally with its first two somites.

Most features of the adult female can be made :

1. Exopodite of antenna; it is by means of the antennae that the animal swims in its characteristic flea-like jumps.

2. Abductor muscle of antenna; a levator muscle is shown just posterior to this, and a second abductor some distance farther forward.
3. Ventral border of left valve of carapace.
4. Mandible; from the portion indicated by the label line the part into which muscles insert curves laterally and then dorsally; its extent is indicated on the diagram by broken lines; the medially directed chewing surface is seen just dorsal to the label line; just behind this, the maxillule, with three forwardly directed. There are about 420 species of Cladocera known, the largest being *Leptodora kindtii* of north temperate regions, which may reach 1.8 cm in length.

Cladocera have only a few metameres in the trunk and have five or six pairs of limbs. The furca is clawlike. The bivalved carapace never covers the head but usually covers the trunk and limbs only. Exceptions are the few predacious species having slender limbs; in these the carapace only forms a dorsal brood chamber.

The Cladocera arc dwarf forms, usually about 1 mm long. Most appear in huge numbers of individuals and are of great importance in the economy of lakes and in food webs, making available a wealth of food to larger animals.

They differ from the Conchostraca by the reduction in the number of limb pairs, at most six, and in never having the head covered by the carapace.

Anatomy

From their appearance and habits one can easily separate the filter feeders from the predators. Filter feeders, like the Conchostraca, have leaflike limbs enclosed by the carapace. The carapace is folded into two valves; it is not hinged. Within the carapace is also the abdomen, which lacks appendages and is flexed ventrally (Fig. 14.10); the recurved section is called the postabdomen. This very successful group was probably derived from the Conchostraca. The few predators have cylindrical, raptorial legs, with their freedom of movement enabled by the reduction of the carapace, In these the minute carapace is carried on the dorsum like a *facksack*, forming the brood chamber (Fig. 14.10). The predators appear very specialized. Filter feeders have the head bent ventrally. The trunk is divided into an appendage-bearing part with 6 metameres at most, and a posterior part lacking appendages. The posterior part still has four metameres in some predators and the same number can also be seen in embryos of Daphniidae (Fig. 14.10).

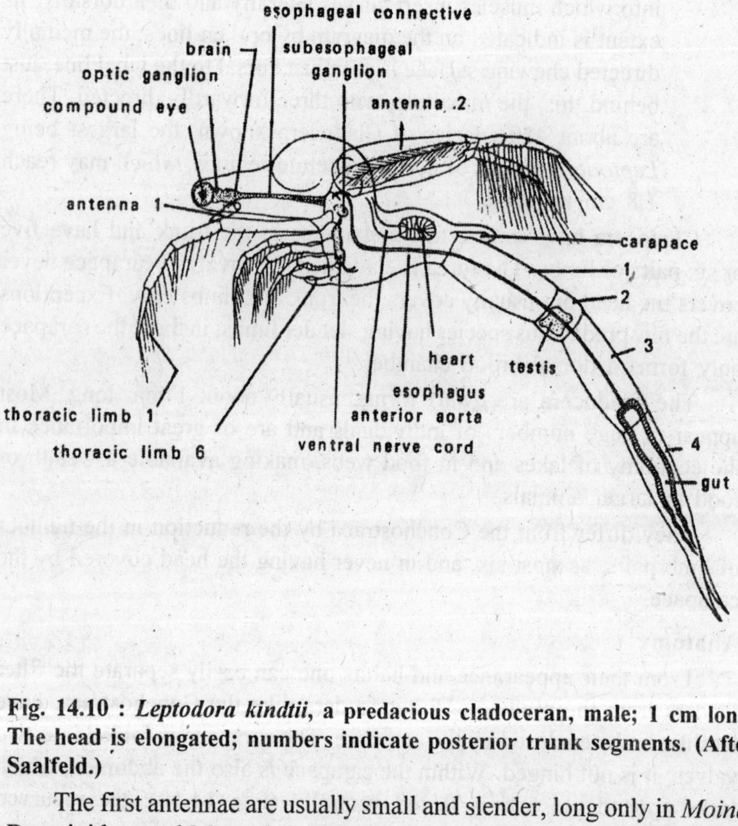

Fig. 14.10 : *Leptodora kindtii*, a predacious cladoceran, male; 1 cm long. The head is elongated; numbers indicate posterior trunk segments. (After Saalfeld.)

The first antennae are usually small and slender, long only in *Moina,* Bosminidae, and Macrothricidae. The second antennae are biramus and well developed. In the filter feeders each mandible has a grinding ridge (Fig. 14.11), while in predators they bear long teeth. Two different types of limbs are found among the filter feeders: the superfamily Sidoidea has six similar (except for the last) pairs of leaf-like limbs easily derived from the Conchostraca, and the Chydoroidea, has limbs that differ from each other (Fig. 14.15). The cylindrical, jointed limbs of the predators are completely different. *Polyphemus*, Polyphemoidea, still has exopodites, but in the related marine genera *Evadne* and *Bythotrephes*, the exopodites are reduced. These are completely absent in *Leptodora.*

The cuticle of the carapace frequently has a polygonal pattern reflecting the borders of the hypodermal cells below. Most Cladocera are transparent; any coloration present is due to the hemoglobin of the blood and to oil droplets in the mixocoel of the trunk and carapace, in which red carotenes and blue reserves have accumulated.

Sense Organs

While the three to four nauplius eyes are small, may even lack pigment or be absent, the compound eyes are relatively large. These are fused in the midline of the body and are moved by a number of muscles within a completely enclosed eye chamber. There are all together 22 ommatidia in the eyes of *Daphnia*, but many more in predators (Evadne 80, *Leptodora ca.* 500), indicating that they might be used to find prey.

Digestive System

The labrum contains gland cells and extends posteriorly far beyond the mouth, forming a preoral cavity within which the mandibles work. The midgut is usually straight, coiled in most chydorids and macrothricids; often it has an anterior pair of tubelike caeca, sometimes a posterior median cecum (some Chydoridae) or no caeca *at all* (*Sida, Bosmina*).

Circulatory System

The heart, except in *Sida*, is very short, barrel-shaped, and has only one pair of ostia. Delicate membranes within the body and limbs guide the circulation.

Reproductive System

The paired gonads are in the appendage-bearing part of the trunk and extend into the abdomen (Fig. 14.10). They are shorter in *Polyphemus* and *Leptodora*. The ovaries open laterally, through paired short oviducts, into the space between dorsum and the carapace, the brood chamber. The testes may be connected to a pair of penis.

Reproduction

Males are much smaller than females : Mating has only rarely been carefully observed. During mating the furca with the male gonopores is extended between the valves of the female to the brood chamber. All species produce eggs which initially enter the brood chamber.

In the Sidoidea the brood chamber is closed toward the sides and posteriorly by a horseshoe-shaped ridge of the inner wall of the shell, in Daphniidae it is closed laterally by a paired swelling of the dorsum and posteriorly by fingerlike projections. In other families it is limited

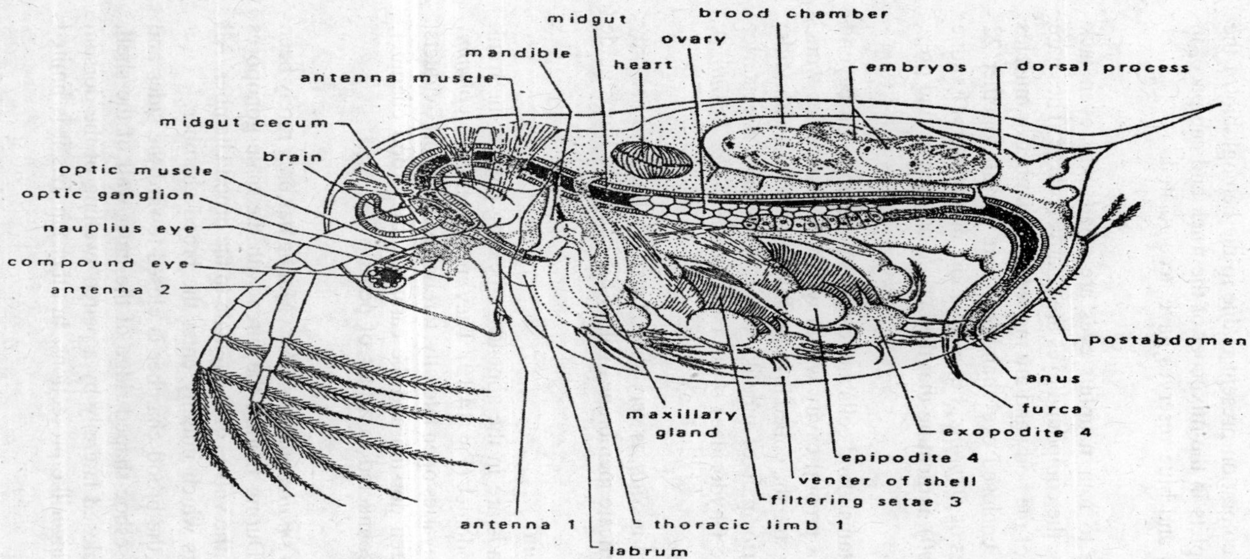

Fig. 14.11 : ***Daphnia pulex*, a filter feeder, lateral view in morphological orientation; 3.3 mm long. The first antennae hang into and chemically test the water current produced by the limbs. The second are used for swimming. The exopodite marked is that of leg 4, which closes the interlimb chamber ventrally by placing its posterior border against the 5th limb. The filter setae of leg 3 lie in the plane of this page while the main part of the limb is transverse to it and only the edge shows. The brain consists of proto- and deutocerebrum; behind it is the esophageal connective, horizontal due to bending of the head. The dorsal process closes the brood pouch posteriorly. (After Matthes.)**

by a bend of the posterior of the trunk. In predators the brood pouch is tightly closed and filled with a nutrient fluid that nourishes the eggs, which have little yolk. According to old observations a food organ is present in the predatory Cladocera, and in *Moina* and *Moina-Daphnia*, a swelling of the hypodermis, which may secrete nutrients into the fluid. The young, which are completely formed within the brood chamber, escape usually through an opening produced by lowering and bending the posterior part of the trunk.

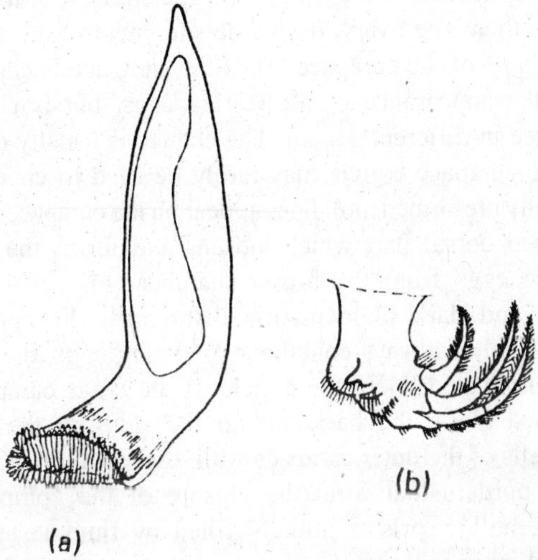

Fig. 14.12 : Mouthparts. (a) The vertically oriented mandible of *Daphnia pulex*, inner surface; 0.4 mm long. One can see the large area of insertion and the chewing surface with rows of teeth, (b) Maxilla. (After Lilljeborg.)

In reproduction, parthenogenetic and bisexual generations alternate. Within the brood sac the parthenogenetically produced substaneous eggs (summer eggs) develop without reduction division directly into young cladocerans. The haploid eggs are fertilized; they are winter (resting) eggs, out of which hatch, sometimes after a few days instead of over winter, parthenogenetic, usually *amictic*, females.

In parthenogenetic reproduction, the offspring are not necessarily all alike. In *Daphnia pulex*, before the only maturation (equatorial) division of the summer eggs there is a kind of meiosis within the nuclear membrane. Bivalents conjugate without separating during anaphase. There may be gene exchange before, which may produce breaks or new linkages. In this case the chromosomes of the eggs have a combination of genes different from that of the mother.

The large resting eggs, rich in yolk, develop in the brood chamber of the mother up to the appearance of the mesoderm. Then there is a diapause lasting days or months, during which time the eggs are resistant to drying and freezing. In Sidoidea, Polyphemoidea, and Haplopoda the double-shelled eggs are released into the water while the mother molts. Chydoroidea, however, do not release the thin-shelled eggs from the brood chamber; instead, the eggs are packed into the *exuvia* during a molt of the mother. The ovary, during development of the resting eggs, changes the shape of the carapace. Thus the carapace is changed in the molt before the appearance of the resting eggs, but is changed to a different degree in different genera. The change is usually considerable and the whole carapace exuvia may rarely be used to encase the eggs (Alona). Usually preformed molt lines appear on the carapace surrounding a saddle-shaped dorsal part which loosens and forms the ephippium, containing the eggs from the brood chamber (Fig. 14.12b). Strong sclerotization and dark pigmentation distinguish the ephiphium. In *Daphnia*, in which it always contains two resting eggs, the structure is especially complicated. Here the cuticle of the outer carapace wall is much thickened along the back and to the sides of the eggs. Also hypodennal cells of the outer carapace wall, especially in the area of the wide ventral borders that form the closure of the ephiphium, have secreted high, hollow prisms initially filled by fluid instead of solid cuticle. All ephiphia sink with their contents to the bottom, but later rise

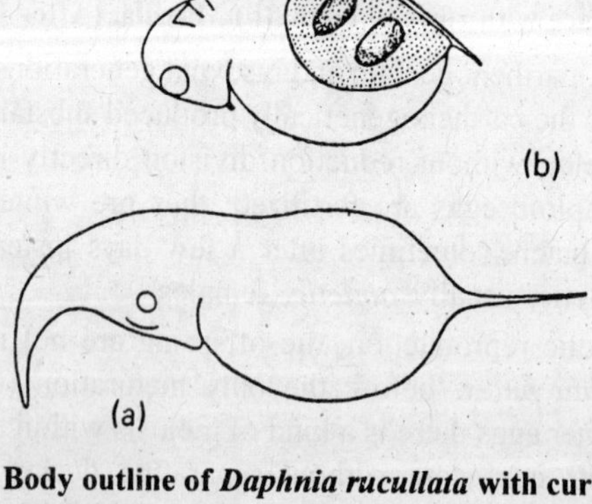

Fig. 14.13 : Body outline of *Daphnia rucullata* with curved helmet; (a) June generation; (b) fall generation carrying 2 eggs in the ephiphium; 1-2 mm long. (After Wagler.)

to the surface, especially in *Daphniidae*, due to the filling of the hollow prisms of their walls with gas (perhaps air). Such floating, unwettable ephiphia adhere to the feathers of water birds, which carry the ephiphia to other waters. The resting eggs may even survive passage through the bird's digestive tract.

Unlike rotifers the same female of many species in different succession can produce parthenogenetic diploid female eggs, parthenogenetic diploid male eggs, or haploid resting eggs. Also ephippia may be formed several times, one after another.

Experiments have been made with *Daphnia magna, D. pulex, D. cucullata; Moina brachiata*, and *Chydorus sphaericus* which demonstrated that unspecific stimuli within the cultures may cause a change in the ovary of the parthenogenetic female. The change does not occur if environmental conditions remain constant. In *Moina brachiata*, lipid materials of the food determined the eggs to be male. In overpopulated cultures fed with normal yeast, 30% of the eggs are male. Populations fed with yeast from which the lipids had been removed produced only females. After adding the fatty extracts to the cultures, 30% of the eggs were again male. The same results were obtained when the lipid extracts were replaced by a mixture of 1 part ergosterin and 1 part olive oil. Only feeding with certain materials will activate the production of males, normally the female factors are more common. The resting eggs also develop as a result of change of food. Fed with normal yeast, 66% of females produce overwintering eggs; fed with lipid-extracted yeast, 5.9% of females. If ergosterin was added to lipid-extracted yeast, 28% of females produced overwintering eggs.

With plenty of food, favorable temperature and aeration, cultures of *Daphnia magna* existed for 4 years, and *Moina brachiata* for 3 years without bisexual reproduction. In contrast to conditions in nature only parthenogenetic generations appeared, in *Moina* 180 of them. Reduction of temperature below 15°C, poor nourishment, and overpopulation, however, cause the production of males and resting eggs. Comparison may be made with similar experiments made with rotifers.

Day length and water temperature influence the determination of eggs of parthenogenetic females of *Daphnia magna*. A culture of amictic females, all offspring of a single amictic mother, reproduced parthenogenetically at 11, 18, and 27°C and a day length of 12 hr. When day length increased to 20 hr, bisexual reproduction was favoured, especially if at the same time the temperature was low. At long day lengths and 30°C, besides amictic females only males appeared, at 8°C only ephippial females.

The changes in the ovary are of great importance for the upkeep of cladoceran populations. During steady favorable conditions and parthenogenetic reproduction, in which each individual reproduces, a large population appears rapidly. Each individual of *Daphnia pulex* and *D. magna* can have more than 40 eggs in the brood chamber, and each individual can produce such broods several times. During adverse conditions (*e.g.,* drying out), a new sexual generation makes its appearance and with it the production of resting eggs.

Water fleas that live in the center of a large pond or lake where temperature only slowly follows air temperatures have usually only one sexual generation in the fall when the temperature falls. Thus they are monocyclic (*e.g., Diaphanosoma, Holopedium, Leptodora, Bythotrephes*, and some *Bosmina*). They over-winter as resting eggs. In very large lakes, *Daphnia longispina, D. cucullata*, and *Bosmina longirostris* may leave out the fall sexual period (acyclic) so that one finds no or very few ephippial females and only a small number of overwintering, brood-carrying females under the ice.

In small ponds the spring and autumn environments may be very different from those during the rest of the year and their cladoceran inhabitants may reproduce sexually twice during the year (dicyclic). But in such waters there may also be changes during other seasons: changes in temperature and dissolved chemicals, etc., and other sexual periods may be present, especially in small shallow vernal pools. *Moina brachiata* may, in the first parthenogenetic generation, hatch out some mictic females from resting eggs. The second generation can already produce resting eggs which, should the water dry up, will prevent the extermination of the population. Each succeeding' generation has a larger percentage of males than the preceding, so that the number of resting eggs steadily increases. In the Aretic, there may not be parthenogenetic reproduction at all. However, *Daphnia middendorffiana*, a species almost totally restricted to the Aretic, may form resting eggs in the absence of males.

Development

All Eucladocera develop directly, completely within the egg (Fig. 2-5). Only the resting eggs of Haplopoda (*Leptodora*) have a metanauplius that undergoes a postembryonal anamorphic development. In Europe *Daphnia pulex* has a life span of about 62-85 days during spring, late summer and fall, 51-63 days in late spring and early summer, with abundant food. All cladocerans molt, even females that have had eggs. The young molt frequently, the adults less frequently. In summer, adult *Daphnia* may molt every 2 days.

In molting the cuticle of the posterior border of the head breaks transversely, and the head exuvia folds anteriorly and down. The head emerges first through the opening; valves and trunk follow. During the molt, which takes only 1-3 min, the individual grows in size and changes in proportions (*e.g.,* the brood chambers).

Cyclomorphosis

In addition to changes of shape resulting from postembryonic growth, some permanent swimming, limnetic species *e.g., Daphnia longispina, D. cucullata, Bosmina coregoni* and *B. longirostris*) have parthenogenetic generations that differ in appearance. There are differences in average length and body shape, as well as in the number of parthenogenetic eggs in the brood chamber. The changes consist mainly of increased length of the anterior margin of the head or the tail spine in *Daphnia,* or the height of the dorsal shell hump or length of the first antenna in *Bosmia.* In Bosmina one can see the differences already in the young within the mother's brood chamber. These changes begin in early summer, increase in the following generations, and disappear completely in fall. All individuals of a population undergo the same changes. Changes in size are mainly due to increased cell size or sometimes to increased cell number. Populations of the same species in neighboring takes may also have a cyclomorphosis but the resultant appearance of each population might be very different. There may be a straight or a curved helmet spine, or a wide head protection. These morphological differences between races are very distinct in the summer populations, while in winter all local races look the same. These changes, as well as those of post-embryology, make the study of cladoceran taxonomy very difficult. Similar changes occur in rotifers.

The dependence of cyclomorphosis on the environment and especially on water temperature, has been studied in field experiments with races of *Daphnia* and with *Bosmina coregoni.* With each favorable increase in temperature there was an increase in the length of the extensions (*e.g.,* the hump of *Bosmina*), and the number of eggs within the brood chamber. If the temperature optimum was passed, the length of extensions and number of eggs decrease, just as if the temperature fell below the optimum, flow the environment affects the organ system is not known.

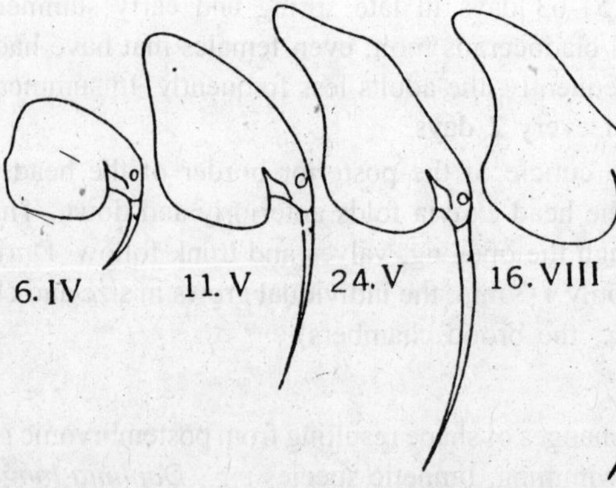

Fig. 14.14 : Cyclomorphosis of *Bosmina coregoni thersites*; body outline in swimming position; up to 1 mm long. Below, the collecting dates; Roman numerals indicate month collected. (After Licder).

A laboratory study of parthenogenetic females of *Daphnia galeata mendotae* in individual cultures indicated that in addition to temperature, other environmental factors, such as light intensity and photoperiod which increase during the summer, and the amount of food and turbulence of water all affect cyclomorphosis. The various external conditions cause increased growth rate of the upper head or helmet, anterior to the eyes, as compared to the carapace (from the mouth to the tail spine).

The influence of temperature and turbulence is expressed as early as the second embryonal period. The clones from Bantam Lake, Connecticut, proved to be genetically polymorphic in their cyclomorphic reaction. Also the cultures indicated that water temperature produced changes even during the lifespan of an individual in the relative head growth. The offspring of mothers kept at 24.5°C have relatively long helmets. If their postnatal development takes place in water at 7.5°C the positive allometric growth rate of the heads declines so much that as adults they resemble members of clones that passed also their embryonal stages at the lower temperature. The reverse is true for the newborn, short-headed offspring of mothers kept at 7.5°C if they are reared at 24.5°C. In both groups the body proportions approach those of individuals that spent their entire lives at these temperatures. This change supports the view that cyclomorphosis is adaptive to the surroundings, but the function of the adaptation is not known.

Perhaps in predation by fishes, long-helmeted *D. galeata* raised at 25°C, which swim and sink more horizontally, have a selective advantage over the short-headed ones raised at 5°C. Five *Daphnia galeata* at a time and three guppies (*Lebistes reticulatus*) were placed in 2-liter glass cylinders, 15 cm in diameter. In trials lasting 2-15 min, *Daphnia* raised at 25°C were less preyed upon than those having the spine amputated or those raised at 5°C (Table 4-2). However, it cannot be assumed that the long spines compensate for the lower viscosity of warmer water and facilitate swimming in upper water layers, because cyclomorphosis occurs in tropical waters at constant temperatures, and in small temperature changes of only 2-3°C accompanied by negligible viscosity changes. Also the extensions may be longer in cold water than in warm in *D. cucullata* and *Bosmina coregoni*.

Habits

Cladocera usually live in standing freshwater. Very few are marine. Though they may inhabit quiet streams, typically they live in shallow ponds, especially among plants close to shore, and swim only short distances, occasionally resting on plants. *Sida* can attach itself to leaf surfaces with an attachment disc of the anterior, dorsum; *Simocephalus* attaches, with a hooked seta of the second antennae, *Chydorus* with limb spines.

Some Macrothricidae and Chydoridae live on the bottom sediment or within its surface layer. *Daphnia magna and D. pulex* live in bays of lakes and ponds but not among plants. *Scapholeberis* lives below the water surface, a habit for which it is adapted by having straight ventral valve borders armed with laterally directed unwettable hairs. The animal penetrates the surface and hangs itself up on the surface film by the hairs. The second antennae then move the animal along the underside of the surface while it feeds on microorganisms and detritus. Scapholeberis can also swim down away from the surface. *Notodromus* has similar habits.

Table 14.1 : Results of fish predation experiments on differently shaped Daphnia galeata.

No. of trials	Temp., °C	Survival of long-helmeted forms, %	Survival of short-helmeted forms, %	Survival of long-helmeted forms whose helmet was cut off, %
28	25	71.4	42.1	
40	15	65	46.5	
	25	48.7		34.4

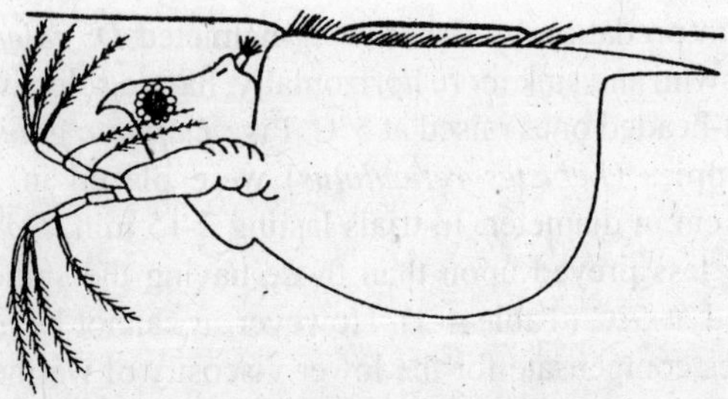

Fig. 14.15 : *Srapholeberis mucronata*, moving along water surface; 1 mm long. (After Wagler.)

Other species, transparent like many other planktonic organisms, are found far out in the center of lakes where they cannot rest, but must keep permanently in motion; this is probably a secondary habitat for Cladocera (*e.g., Daphnia longispina, D. cucullata, Diaphanosoma, Holopedium,* some *Bosmina,* and *Leptodora*). One rarely finds these species near shore. If at night they drift toward shore, they swim out to open water before sunrise, together with the copepods *Eudiaptomus gracilis* and *Cyclops tatricus*. Experiments with these copepods and *Daphnia longispina* in the Lake of Lunz, Austria, indicated that orientation in these migrations depended on landmarks above the horizon, such as mountains and forest borders. By shifting an artificial horizon, a plate 1.2 m high, the direction of migration could be changed. The migration of the large females of *Acanthodiaptomus denticornis* and *Daphnia longispina*, carrying ephippia in the fall, as in the reverse direction, toward shore and water only 0.1-0.5 m deep.

Vertical migration of *Daphnia longispina* resembles that of the copepods *Eudiaptomus*.

Locomotion

Most species swim with their second antennae. If the second antennae are large (*Daphnia*), their propelling movements are separated by pauses during which the cladoceran sinks with antennae spread, as if attached to a parachute. If the second antennae are small (*Bosmina*), they are moved very rapidly. Some Chydoridae (*Alona*) also propel themselves with the long posterior end of the body, proceeding by jerky movements through the water.

The swimming position of the trunk differs among genera. Some limnetic species can float, so that they have to paddle only once in a while. The specific gravity of *Diaphanosoma* is extremely low, and in

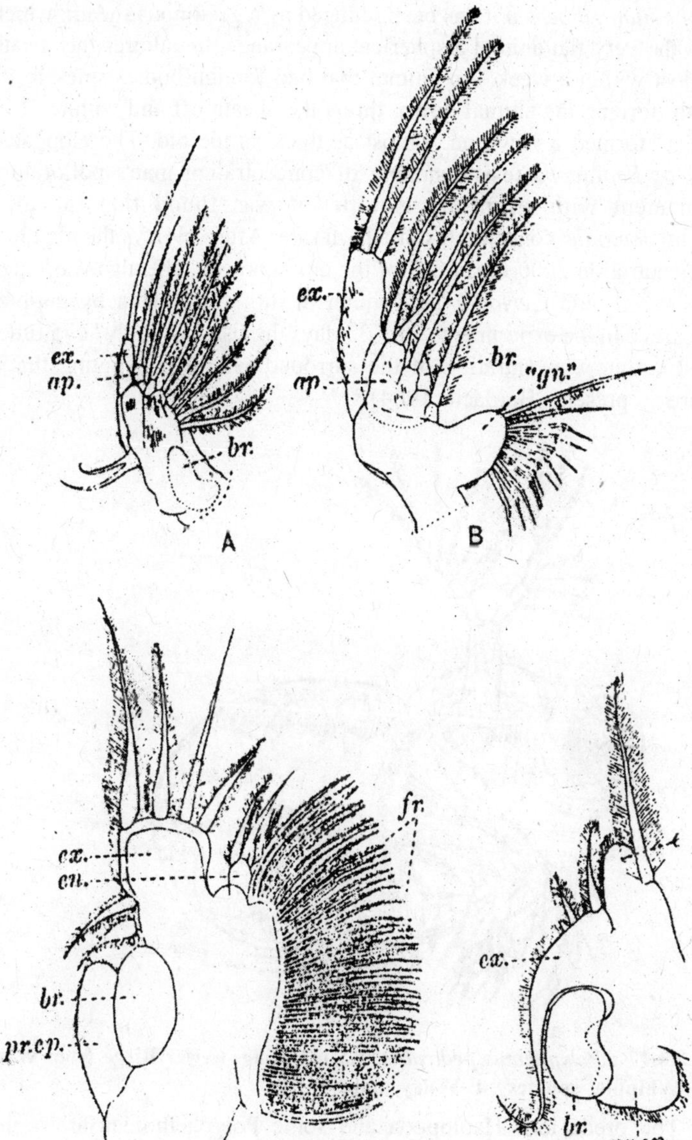

Fig. 14.16

Holopedium gibberum it has been reduced by a gelatinous sheath, which gives the very flat animal a spherical appearance. In cultures the sheath was lost within a week. An animal that had foreign bodies stuck to its sheath, forcing the animal down, threw the sheath off and within 13 hr at 15°C formed a new one, almost as thick as the old. The elongated body of *Leptodora* resists sinking of concentrating man's polloratory experiment with strontium isotopes it was found that the alga *Chlamydomonas* concentrates at a high rate. After 35 days the alga had a concentration 2000 times that of the environment. The alga was eaten by *Daphnia* and *Ceriodaphnia* which in turn were eaten by guppies (*Lebistes*). In the experiments after 35 days the fish had "only" 175 times the strontium concentration of the surroundings, demonstrating that a hazard is present (Bardach, 1964).

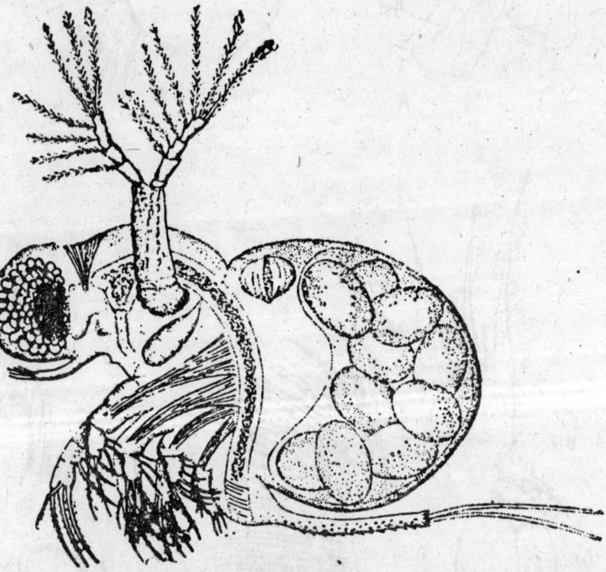

Fig. 14.17 : *Polyphemus pediculus*, 1.3 mm long. (After Birge from Ward and Whipple, courtesy J. Wiley and Sons.)

The predacious Haplopoda and some Polyphemoidea are armed with slender legs. But predation has been only little studied. *Leptodora* has the venter of the limb-bearing trunk almost at right angles to the head, facing anteriorly and with the grasping limbs covering an anterior area as a trapping basket. Between the limbs of preserved *Bythotrephes* have been found Cladocera and, within the gut, remains of *Diaptomus*. Among the legs of *Polyphemus* ostracods have been found, but it is uncertain whether this genus is exclusively predacious. Their slender legs have exopodites that enclose a space bordered anteriorly by the head; into this chamber water enters when the head is lifted, and with the water come ciliates and other small animals.

Circulation

The hemoglobin content of the hemolymph is inversely proportional to the oxygen content of the surrounding water, independent of its iron content. Also an increase of metabolism may increase hemoglobin content. Synthesis and destruction of hemoglobin takes place in fat cells and probably in the maxillary gland. In mature females considerable hemoglobin enters the eggs. The different clones of *Daphnia* differ in their ability to synthesize hemoglobin. Various surfaces permit exchange of gases.

KELP CRAB *(PUGELLA PRODUCIA)*

Xiphosura polyphemne. long known to zoologists a *Limulus* is commonly celled the king crab or horse-shoe crab. It however, not really a crab at all, but is a member of an isolated of the order of the Chelicerata, bearing the same name. *Xiphosura*, as has been given to the genus. It is abundant on the eastern coast of North America, from Maine to Mexico. Attempts to establish it in Europe and on the California coast have succeeded only temporarily. Two closely related genera, embracing four species occur in Asiatic waters, from Japan to India.

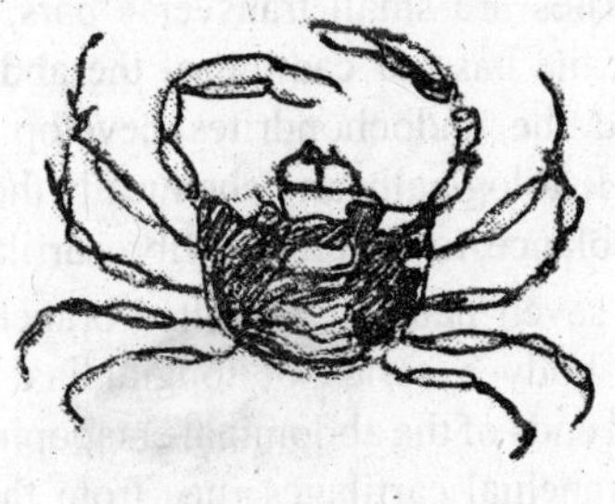

Fig. 14.18. Kelp Crab.

The bizarre appearance, large size, and isolated taxonomic position of *Xiphosura* have long excited the interest of zoologists. The closest relatives of the order appear to be found in the extinct Eurypterida (= Gigantostomata), with which the *Xiphosura* often are allied in the

Class Merostomata. This is an exclusively aquatie class, though a theory was once proposed that it had its origin in terrestrial Arachnida. Anatomically the closest living relatives of the *Xiphosura* probably are the scorpions. At one time the view was vigorously upheld that *Xiphosura* lies close to the ancestral tree of the vertebrates, this view being advanced in two quite conflicting theories by Gaskell and by Patten. Patten always used the terms haemal and neural in place of dorasl and ventral. because of his belief that the functionally ventral surface in *Xiphosura* is represented by the functionally dorsal surface in a vertebrate. The larva of *Xiphosura* has been said to resemble the extinct trilobites, but this resemblance would appear to be purely superficial. However, certain features of the embryos do seem to suggest a trilobite relationship.

Three distinct classes of structures make up the endoskeleton, differing in their chemical composition, histology, and embryology.

First of all there are a number of infoldings of the exoskeleton. Most noticeable of these are the six pairs of dorsal entapophyses in the opisthosoma, plus the single large pair at the posterior margin of the prosoma, and ventrally the six pairs of tendious apodemes, which project inwards from the stigmata visible on the proximal posterior surfaces of the operculur and gill flaps. The cuticular linings of the fore and hindgut also develop as infoldings of the exoskeleton.

Second, there are the six median endochondrites and the endocranium. The endocranium is a structure of complex shape, mainly in the form of a horizontal shelf, located below the anterior third of the stomach-intestine, dorsal to the esophagus and the vascular ring. A posterior portion of the endocranium encircles the ventral artery near its anterior end. The endochondrites are small transverse bars, located below the ventral artery, one at the base of cadium of the abdominal flaps. Both the endocranium and the endochondrites develop from the ends of embryonic muscles. Histologically and chemically they have been found to show some resemblance to vertebrate fibrocartilage.

Lastly there are seven pairs of so-called branchial cartilages, and on each side of the body a series of longitudinal cartilaginous bars connecting the ventral ends of the abdominal entapophyses. Each member of the first pair of branchial cartilages runs from the posterior margin of the endocranium into one of the *chelaria*. The remaining six pairs run from the ventral ends of the abdominal entapophyses into the operculum and the gill flaps. Those that enter the gill flaps can be found just medial to the gills. Embryologically these cartilages develop from mesoderm of the somite walls. Histologically and perhaps chemically, they resemble the hyaline cartilage of some of the lower vertebrates.

The entapophyses, tendinous apodemes, cndochondrites, endocranium, and branchial cartilages all serve for the attachment of muscles, some of which were mentioned earlier.

Nervous System

When dissecting for the esophagus, a median strip probably was removed from the endocranium. If necessary this strip can be widened so as to expose more fully the nerve ring and associated nerves, Care should be taken not to proceed too far laterally, since parts of the excretory organs lie just lateral to the endocranium. If an uninjected specimen is used for the dissection of the nervous system. application of Schaudinn's fluid will be found to whiten and harden the nerves. The time usually allowed for the dissection probably will not permit more than the identification of the nerve ring. ventral cord. and the proximal parts of the larger nerves. Tracing out the more distal parts of the larger nerves requires much more time and eare. In any event, after the nervous system has been studied in situ from the dorsal surface, as much as possible of the more detailed study, more especially of the nerve roots and of the ventral surface.

Those nerves which run inside arteries need not be described here, since the student can refer to the descriptions already given for the corresponding arteries. This is true for the the neural ring, with its three or four postoral commissures and anterior cerebral swelling the nerves to each of the prosomal appendages, the stomodeal nerves, the olfactory nerves, and thee nerves to the median and lateral eyes. The nerve ring is made up of fused ganglia. but details of its structure are not easily made out in an ordinary dissection.

Some of the nerves not adequately covered in the description of the circulatory system are the following. Emerging through the dorsal surface of the vascular ring there are eight dorsal nerves on each side, which do not run within arteries. The first of thses is a very small nerves, which arises just lateroposterior to the root of the lateral eye nerve. Although, small, this nerve to which it at first is lateral, becoming medial as the two curve posteriorly. Passing medioventrally to the lateral eye, it continues back in the ventral epidermis, curves posteriomidially around the base of the sixth prosomal appendage, then curves slightly laterally again near the adductor muscle for the first gill flap, and finally ends in the anterior part of the metasoma. It akppears to supply ventral sense organs in the opisthosoma. The largest dorsal nerve is the stomodeal nerve, thin passis forward just lateral to the first dorsal nerve. At the point where this and the lateral eye nerve curve laterally it divides into three main branches, which continue forwards, one to the dorsal and two to the ventral epidermis. Each of the third, fourth, fifth, and sixth dorsal nerves arises dorsal to the nerves for the corresponding appendage and runs out in front of that appendage, to supply the lateral parts of the epidermis, usually with three main branches, one dorsal and two ventral. The fifth and sixth of these nerves at first run back above the nerves to their corresponding appendage, then turn sharply laterally. The seventh and eighth dorsal nerves arise close together, between the origin of the sixth nerve and the start of the ventral nerve cord. They branch like the more anterior dorsal nerves, to supply the lateral epidermis in the posterior part of the prosoina and anterior part of the opisthosoma respectively.

On the ventral surface of the nerve ring, a preoral transverse commissure will be found, at the posterior border of the cerebral ganglia. From it three tiny nerves pass to the anterior lip of the mouth.

The ventral nerve cord is actually double, as in all arthroposs, though this can be seen only after removal of the arterial sheath. Opposite each of the first four gill gaps it is enlarged as a ganglion, from each side of which an anterior dorsal nerve passes outwards to the. lateral epidermis and into one of ten marginal spine and a posterior ventral nerve passes into the corresponding gill flap. In addition two small nerves arise on either side from each ganglion close to the dorsal nerve, and pass dorsolaterally to supply, the ventral abdominal muscles and the stomach-intestine. Intestinal branches also arise from the sixth, seventh, and eighth dorsal nerves, near the point where these curve laterally. A network of anastomoses connects together the various intestinal nerves and also the nerves to the ventral abdominal muscle.

Behind the fourth abdominal ganglion, the ventral nerve cord appears to fray out on each side into about eight nerves, which supply the last gill flap, the stomach-intestine, the rectum, the more posterior abdominal musculature, the epidermis and the last two marginal spines, the telson, and the posterior parts of the heart and pericardium.

On the dorsal surface of the heart three longitudinal nerves probably have already been noticed. There is a median ganglionated cord, from which a number of minute nerves pass out to a plexus, with some of the strands on each side forming a more definite lateral nerve. Both the two lateral nerves and the median ganglionated cord run nearly the full length of the heart.

Dissection of the remaining cardiac nerves and of certain associated nerves require more time than is usually available. However, a description will be given here to complete the functional picture and to aid those who may wish to attempt the dissection.

A series of segmental cardiac nerves arises from the sixth to the thirteenth dorsal nerves, inclusive, just lateral to the ventral abdominal flexor. From the base of each of these, excepting the first two, or from the adjacent part. of the corresponding dorsal nerve, connection is made with a lateral longitudinal nerve, which runs forward onto the ventral surfaces of the two dorsal abdominal flexors. This latter nerve receives a further connection from the fourteenth dorsal nerve. In the opisthosoma the lateral longitudinal nerve lies just lateral to the tendinous apodemes and medial to the branchial cartilages.

The segmental cardiac nerve arising from the sixth dorsal nerve passes up to the wall of the pericardial sinus, on which it branches at a level midway between the first two pairs of ostia. No connection has been found from this nerve to the heart, so that, it may not be a functional cardiac nerve, although it is so named because it is in series with the others. The next two of the segmental cardiac nerves join together to form a somewhat larger nerve, which reaches the pericardial sinus at a level between the second and third pairs of ostia. This nerve and each of the succeeding five segmental cardiac nerves send branches to the dorsal epidermis, connect with a longitudinal nerve running in the wall of the pericardial sinus, and also connect with the median ganglionated cord on the surface of the heart.

The beat of the heart is maintained by a rhythmic discharge of impulses from the ganglionated cord. Unless shifted by experimental means, the pacemaker for the beat occurs in this cord at the level of the fifth and sixth pairs of ostia, which is where the greatest number of nerve cell bodies are located. The beat is subject to regulation by impulses arriving through the segmental cardiac nerves. Of these, the one which takes its origin from the seventh and eighth dorsal nerves is an inhibitor nerve. Those which originate from the ninth, tenth, and eleventh dorsal nerves, and which join the cardiac ganglion at about the level of the fourth, fifth, and sixth pairs of ostia, are augmenter nerves. Presumably the last two segmental cardiac nerves also are augmenters, but this point appears not to have been determined.

After removal from the body of the animal, the *Xiphosura* heart will continue to beat for hours, if merely moistened with sea water, provided that some nervous connection is left between the dorsal ganglionated cord and the heart muscle.

Sense Organs

The most complex sense organs are the large lateral eyes. The exoskelctal covering over each of them is divided into numerous, relatively large facets, corresponding with an internal organization into separate receptor units. Each unit has about, ten to fifteen receptor cells, and a conical lens formed by an ingrowth of the exoskeleton. There is no curtain of surrounding pigment cells, and no crystalline cone, and there are no *tapetal* cells. The structure of these receptor units is so different from that of the ommatidia in the compound eyes of insects and Crustacea that their homology has been doubted. The differences, however, may well be the result of degeneration, associated with the burrowing and nocturnal habits of *Xiphosura.*

The median eyes are located just in front of the most anterior of a row of small spines on the dorsal surface of the prosoma. Normally there are two, close together on either side of the median line. Each is a hemispherical cup. with a nearly spherical lens formed from an ingrowth of the exoskeleton, beneath which are a columnar epidermis and a layer of receptor cells which tend to be arranged in groups. The two nerves from these cups shortly unite, and in the angle thus formed there is a mass of tissue, which when examined histologically proves to be a vestige of two medianly fused eyes. Occasionally adult specimens are found with this posterior double eye in a functional condition, provided with a lens like the pair of eyes just in front.

On the ventral surface, a median, *wart*-like elevation of the exoskeleton, about 5-8 mm. wide, can be seen just in front of the "elbows" of the thoracic. A distinctive oval area on each side of the *wart* will be noted. This is the structure supplied by the two so-called olfactory nerves. Its internal structure changes remarkably during development. In the larva there is a median, pigmented "frontal organ," of doubtful significance, on either side of which is a fully functional ventral eye, supplied by one of the so-called olfactory nerves. Actually the fibers of these nerves originate, even in the adult, from the same ganglionic centers within the cerebral ganglia as do the fibers of the lateral eye nerves. Each of the ventral eyes has an exoskeletal lens, a cup of receptor cells, and a backing of pigment cells around this cup. During further development the frontal organ largely breaks down, the pigment cells become arranged in scattered groups, and apparently the photoreceptor cells become transformed into bipolar and multipolar sensory neurons, the distal fibers of which extend into tiny sensory setae on the surface of the organ! There are also numerous vertical pores through the exoskele-stomach-intestine is first, lifted out uf the way after cutting it across near its anterior end.

If the digestive tract is slit open longitudinally, all of it except the stomach-intestine will be seen to have a cuticular lining. The lining of the esophagus is longitudinally folded. That of the crop is wrinkled, permitting wide detention of this organ. The gizzard is much less distensible, and its lining is thrown into longitudinal ridges studded with soinc'what flattened teeth. From the gizzard a conical valve projects into the stomach-intestine. The cuticular lining of this valve is tin-own into about thirteen longitudinal ridges which extend freely at their posterior ends. The inner wall of the stomach-intestine is considerably folded, and in the part overlapped by the conical valve, bears numerous circle set papillae.

When studied histologically the lining of the stomach-intestine is found to consist of columnar, glandular, epithelial cells, having a striated border. In the hepatopancreas there are absorptive cells with a striated border, and glandular cells which secrete the digestive enzymes. All parts of the gut, including the tubules of the hepatopancreas, are provided with both circular and longitudinal-muscle fibers.

Food is ground up in the gizzard and admitted to the stomach-intestine through the conical valve. Coarse fragments are ejected through the mouth. Digestion occurs in the stomach-intestine, by means of enzyme from the hepatopancreas, except for the splitting of dipeptides, which occurs intracellularly in the hepatopancreas after absorption as occurred. A second set of enzymes, acting in acid medium in the foregut, has been reported, but where these enzymes would come from is not clear. The stomach-intestine secretes a mucilaginous coating for the fecal pellets, which may be up to 10 cm. long. Fat, glycogen, and protein are stored in the connective-tissue cells between the tubules of the hepatopancreas, rather than in the cells of the hepatopancreas itself. The absorptive cells also have an excretory function, as is illustrated by the fact that at times they pass considerable quantities of Calcium phosphate into the lumen of the hepatopancreas, whence crystals of this material are conveyed to join the faeces in the stomach-intestine.

Endoskeleton

A full dissection of the endoskelcton is not possible without damage to other structures yet to be described. Only the endocranium and portions of the entapophyses should be exposed to view at tins stage of the dissection. The other parts can be located by feeling with a probe ton, smaller than those leading to the setae and quite different from the pores found in other parts of the exoskeleton. Possibly these proes serve for the escape of a glandular seeretion but a recent worker has been unable to find gland cells. Whether this organ in the adult condition is actually olfactory as has been summerised, has not been determined. The possession of ventral eyes by the larva probably can be correlated with the larval habit of swimming on the baek, contrasting with the more usual burrowing habits of the adult.

The movable spines on the gnathobases of the second through fifth prosomal appendages appear to contain taste receptors. When they are touched with appropriate food, reflex chewing movements of the limb, which is thus stimulatd, and bending and snapping movements of the chelieerate are initiated. Below each of the spines is a group of bipolar sensory neuron the distal tibers of which extend into the cavity of the spine. Taste receptors of some sort may also be present in the chelae of the first five pairs of appendages, which have been reported to open when touched with food.

The external spatualte process of the sixth prosomal appendage receives an unusually large nerve the fibers of which have been reported to be connected with an enormous number of receptor cells located underneath the anterior surface of the process. Since water going to the gfills passes over this anterior surface, it has been surmised that the function of these sense receptors is to test some quality of the water in the respiratory current.

Relatively large nerves supply the telson and the six movable spines on each side of the mesosoma, suggesting that these structures, too are sensory.

On the jonted distal processes on either side of the midline of the gill flaps a small number of blister-like elevations are present. The exoskeleton of these has been reported to be perforated by peculiar pores, with underlying receptor cells, but what sense, if any, is here involved has not been determined. On other parts of the gill flaps sensory setae have been reported likewise of unknown function.

Xiphosura has been claimed to be remarkably sensitive to slight increases of temperature. The margins of the gill chamber, the margins of the carapace and the gill flaps are said to be the most sensitive areas, with other areas of the body only slightly less sensitive. The presumed temperature receptors have not been found.

Excretory System

In the ventrolateral parts of the prosoma a pair of complicated coxal glands in present, presumed from their structure to be exeretory. The glandular portion of each of these organs consists of four brick-red lobes, located near the bases of the second through fifth prosomal appendages, and each containing a maze of fine, partly intracellular ducts. Each elongated lobe runs laterally from a point near the nerve ring into the basal part of the corresponding appendage, just dorsal to the smaller nerves supplying that part of the appendage, and between anterior and posterior muscles running from the endocranium to the appendage. The first lobe often is considerably smaller than the others. All four lobes receive a copious arterial blood supply. The ducts of the first three lobes open into a labyrinth of slightly larger ducts in a longitudinal stolen. which connects the inner ends of the lobes. The stolen runs at right angles to the lobes, at a level dorsal to the nerves of the appendages and ventral to the dorsal nerves which pass outwards between the appendages. Posteriorly the stolen widens into a coelomosac, into the cavity of which open the ducts of the stolen and of the. fourth lobe. From the coelomosac a relatively wide, convoluted tubule leads to the outside, followhig a torturous course. First it passes above the fourth lobe to a point posterodorsal to the outer third of this lobe. Here the tubule coils a few times, then runs forward with many coils, dorsal to the outer thirds of the third and second lobes. Anterior to the second lobe, or dorsal to the first lobe, the tubule widens considerably, passes medial, and then runs posteriorly above the middle portions of the second and third lobes, close to the lateral border of the endocranium. Above the fourth lobe it turns ventromedial and passes through part of the lobe on its way to the external aperture. This aperture is located on the articular membrane of the fifth prosomal appendage, proximal and posterior to the basal segment. When the folds in this membrane are smoothed out by stretching, the position of the opening will be seen to lie indicated by a small liglit. area surrounded by a ring of darker gray, located in the membrane near the midpoint of the posterior proximal horder of the appendage. In young specimens the position of the opening may lie very hard to see.

Of the internal parts of the coxal gland only the lobes and the *colon* are at all likely to be seen in an ordinary dissection. The tubule is very transparent, and probably will be seen only if a successful injection is achieved through the external aperture. Except for a short length of the excretory tube near the external aperture, all parts of the coxal gland are mesodermal in origin.

The excretory activity of the hepatopancreas has already been mentioned.

Passes forward along the ventrolateral surface of the esophagus, supplying the esophagus, crop, and gizzard and then anastomoses with the aorta where this starts to bend upwards behind the crop. From the anterior border of the cerebral part of the vascular ring, three small arteries can be seen running forwards, medial to the lateral eye arteries. Two of these arise from the ring slightly ventrally and supply the so-called olfactory organ, located midventrally in front of the *chelicerate*. The median one of these three arteries arises slightly dorsally, a little of center, runs forward below the esophagus. curves upwards in front, of and slightly to the right of the crop, and filially reaclie the median eyes on the dorsal surface.

The large ventral artery gives off branches to the five gill flaps the abdominal muscles, and the stomach-intestine. Near the posterior end of the opisthosoma a branch passes dorsally on each side to anastomose with the superior abdominal artery. Nerves for the telson emerge from branches of the ventral artery just behind the anus, the arteries not continuing any farther.

All the arteries finally discharge into hemocoelic spaces, as in other arthropods. But in two different features confirms; these spaces *Xiphosura* is perhaps unique. In the first place the hemocoel develops by a coalesence of the primary body cavity with several coelomic pouches. The cavity so formed has been termed a myxoeoel. Second, there are present a number of minutely branched collecting channels which drain the blood from the lateral parts of the prosoma. The system of collecting channels is not likely to be seen unless specially injected, but will be described here to complete the functional picture. Ventrally one main collecting channel runs nearly the full length of the body on each side. In the. prosoma it hes medial to the outer row of large limb muscles, and ror part of its course is close, to the stolen of the coxal gland (described under the excretory system). In the opisthosoma it the just lateral to the ventral abdominal flexor. Three large branches enter it laterally in the prosoma, each in turn greatly branched, so that the drainage of the lateral parts of the prosomia is perhaps as complete as that provided by the lymphatics in a vertebrate. Drainage of other parts of the body and of the appendages is through wider and less clearly defined hemocoelic spaces, which gain access to the main collecting channel on each side through ventral and lateral openings. Opposite each of the gill flaps, an cffeient branchial channel leaves each of the main channels ventrally and passes to the inner border of the gills.

CYCLOPS

The taxonomy of genus *Cyclops* is complex, in that a considerable number of sub-genera are distinguished, while some of the species are split into varieties or environment. The extent of the genus is in dispute , sine certain sub- genera of some authors are elevated to distinct genera by other authors. From the morphological point of view no one species has been adequately described. The present account is based on publication dealing with several species, more specially *Cyclops viridis.* Except in some small details the account should be found applicable to any species of the genus.

Among the Copepoda, the Calanoids rather than the cyclopoids probably are the most primitive. But Cyclops is described here, chiefly because it is so easily available. Almost any small pond or lake will provide specimens, and even is mid-winter some species continue reproduction. Many spp. readily become established in fresh water aquaria. For experimental purposes individual specimens can be reared through all stages of the life history, in two or three drops of culture India on a cavity slide, kept in a moist chamber. A medium based on manure, such as is commonly used for *Daphnia,* has been found suitable. The supply to the animals should be renewed daily or more often.

Living specimens can be mounted under a raised coverglass, or in a drop of water suspended in a wire loop as described by Williams. Dorsal and ventral views are easily obtained, but for a lateral view it side the body. To kill or narcotize the specimen other or chloroform can be used (a few drops to 10ml of water). If described, the appendages can be dissected of with mounted needles, under a binocular microscope, preferably using a simple micro-dissector such as is referred to by canon. Care should be taken not to confuse copepodiets larval stages with the adults. The copepodiets are of adult form, but have fewer thoracic somatics and are not sexually matured.

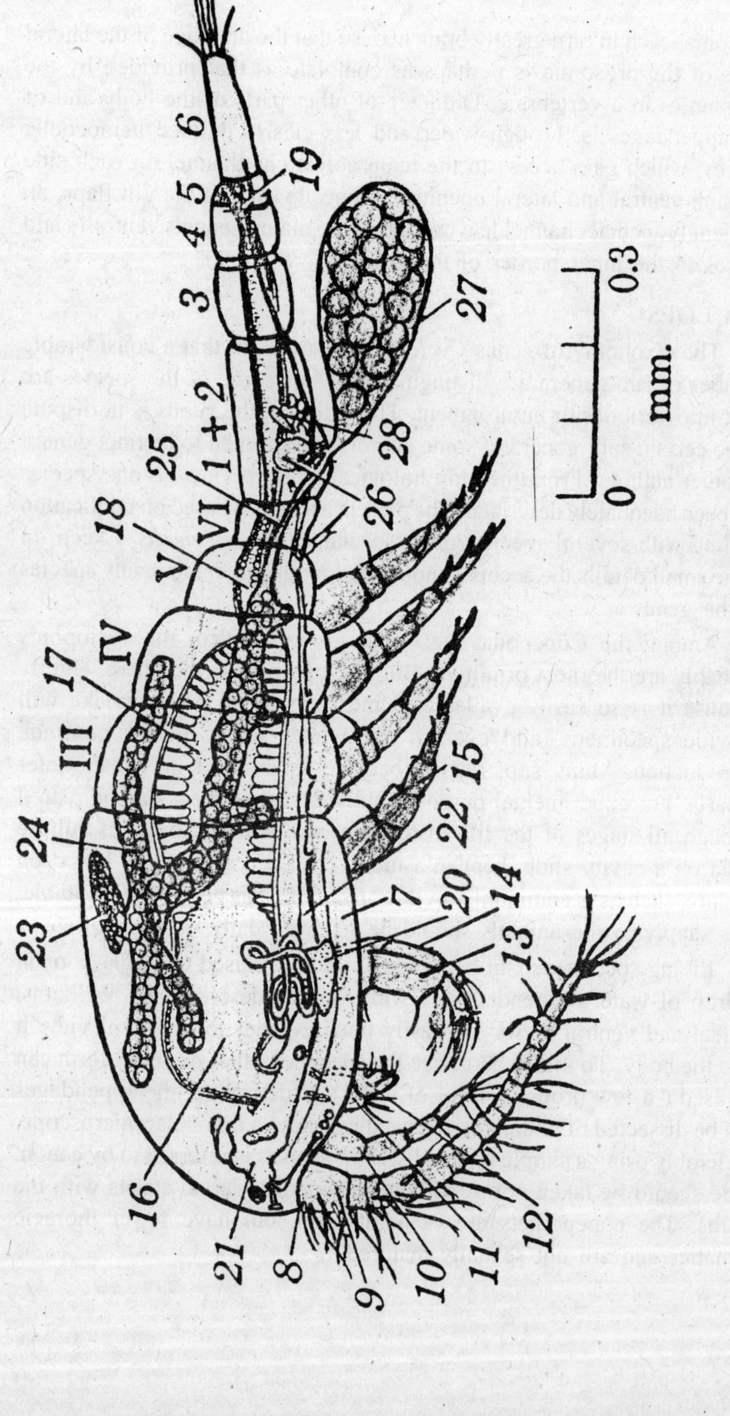

Fig. 14.19 : Diagram of a typical female cyclops.

The anatomy of an adult female *Cyclops* is diagrammed in Fig. 13.20-III-IV. Third, to sixth somites of the thorax; the first and second thoracic somited are fused with the head; a jointed waste occurs between somites five and six, rather than, as in the calanoids, between the thorax and abdomen. 1-4, The four abdominal somites; the first two of these are distinct in the male but are fused in the female. 5, Telson. 6, left ramus of caudal furca. 7, Pleural fold, which extends downward from the terga of the head and thorax. 8, Rostrum. 9, Antennule. 10, Antenna, Consisting of a protopodite of two segments and an indopodite of two segments; the exopodite, present in calanoids, is here absent. 11, Mandible, this appendage has a spiny kitting surface and a small palp which bears a few setae, this is in contrast to the condition in calanoids, where both the indopodite and the exopodite are well developed. 12, Maxillule. 13, Maxilla. 14, Maxilliped; this is the first thoracic appendage, the position of its attachment, medial to maxilla, is a pecular feature the homologies of the different lobes and segments on the maxillules, maxillae, and maxillipeds are not entirely clear. All three pair of appendages are armed with spine like setae used in feeding. Their appearance suggest a predacious feeding habit, and in deed cyclops has been reported to feed on smaller crustacea, ratifiers and sometimes even larger animals such as mosquito larvae and young fish. However, three is some evidence that the food more usually consist of diatoms, ciliates, flagellates and possibly even fine diteritus and bacteria. No filter apparatus is present, yet in some way not understood the animals do appear to be able to abstract surprisingly small food particles from the swimming current. 15, Second thoracic appendage, the first of four biramus swimming feet, each foot consists of a two segmented protopodite, a three segmented exopodite, and a three segmented endopodite (not shown in the diagram), a median transverse plate unites the basal segment of each pair of swimming feet, so that the pair beats as a unit; the four pairs are ordinarily held pointing upwards, and during swimming are suddenly thrust backward together the sudden jumps which Cyclops makes are so rapid that it is difficult to observe the mechanism by which they are produced. It seems, however, that at least on occasion they are the result of simultaneous action by the antennules, antenna and swimming feet, perhaps together with a flick of the abdomen and caudal furea.

The sixth thoracic appendages are much reduced, each consisting of only two small segments. A pair of first abdominal appendage perhaps is reduced by small valvular plates, which partly cover the openings of the genital duct. In the female, each valve is a very small plate or papilla, with one or two tiny setae.

Structures bordening the mouth are not shown in the diagram. Anteriorly three is a quite large labrum, packed with labral gland calls which appear to open dorsoposteriorly to the exterior by a common, median duct. Behind the mouth are a postoral bar and a pair of small, spiny paragnaths. And 16, Anterior part of the stomach; the short, verticle oesophagus and this part of the stomach together form the foregut, having a cuticular living continuous with the exoskeleton. 17, posterior part of the stomach, the large cells of this region probably secrete the digestive enzymes and apparently also have an secretory function; they contain yellowish, refractile granules and at intervals are pinched off into the lumess of the gut, other cells growing to take their place, in a starved specimen the cost-off cells have been observed to dump together and then to pass posteriorly through the intestine, rectum and anus. 18, Intestine, belonging with the posterior part of the stomach to the midgut. 19, Rectum or hindgut, provided with a cuticular living and with special dilator muscles.

Water is rhythmically taken into and expelled from the rectum through the anus, perhaps for respiratory purposes.

Muscles attached to the front of the stomach alternately pull the stomach and intestine forwards and permit them to retract backwards, these movements being in time with those of the rectum. When retracted, the anterior part of the intestine may be coiled into a verticle sigmoid loop. These movements of the gut aid the circulation of the blood, apparently replacing the action of heart, which is absent

Each time that the gut moves forwards, blood in the hemocoel is squeezed forward dorsally. Each time that the gut moves backwards, blood is forced back ventrally.

The blood contains a single type of amoeboid blood cells. Often, in a ventral view of the animal, some of these cells can be seen creeping over the surface of the nerves cord and of adjacent nerves in the in the last two thoracic somites.

Except perhaps for the rectum, no distinct morphological structure is present for respiration. Perhaps a respiratory exchange of gases occurs through most parts of the body surface. However, it may be, that dorsal part of the three thoracic somites is especially permeable. The exoskeleton and hypodermis in this region can be stained selectively in the living animals by a variety of oxidizable salts and leuco dyes. This may indicate respiratory activity, or it may indicate a region of active salt uptake 20, the coelomosac of the maxillary gland. A curled tubule bads from the coelomosac to the external aperture on the medial surface of the basal segment of the maxilla; not shown is a bladder like enlargement of the tubule, is easily seen, near the external aperture. On living specimens the tubule is easily seen, but the coelomosac usually cannot be seen unless stained (as for example by keeping the animal for a week in a 0.5 percent solution of trypan pad). It is presumed that the maxillary gland functions in excretion, osmoregulation, or both.

SUBCLASS-CEPHALOCARIDS

Only four species of cephalocarida is known, the largest of which is upto 3.6mm long.

The trunk of cephalocaridans is composed of head, thorax and abdomen. Limbs are present on all interior but not the additional somites. The telson bears a caudal furca. Their are neither eyes nose carapace. The thoracic appendages are similar to each other and to the second maxilla and in both external appearance and musculature. They are turgor appendages with the proto-and exo-and pseudepipodites flattened and leaf like, the many articles endopodites, however are tubular.

The subclass was discovered in 1955. Its members are common where they occur, and are remarkable for their primitive appearance.

Anatomy

The fairly large head and each of the eight thoracomeres bear a pair of lateral folds of the tergites (epimeres). The abdomen consist of 11 ringlike somites and a telson with a furca, each branch of which has a setae sometimes half as long as the animal. Both pairs of antennae are multiarticulate, the second antennae are similar except for the first maxillae which have especially long basal endites and lack pseudopipodites. The eighth thoracic limbs have lost their endopodites. There are also limb rudiments on the first pleomere, used for carrying eggs.

The second maxillae and the following seven pairs of thoracic limbs have the same general plan. They are turgor appendages whose anteroposteriorly flattened protopodite bears a leaf-like exopodite and pseudopipodite laterally, a many tubular endopodite distally, and a row of setose endites medially. The expodite consists of two articles.

The labrum is ventral and extends posteriorly to cover the mandibular endites and posteriorly directed mouth. A pair of cephalic digestive glands extends posteriorly to open into the anterior end of the midgut. An antennae gland is active in the larva. It persists in the adult, but here the maxillary gland is the dominant excretory organ.

The cephalocarida are hermaphroditic. Each individual bears paired testes and ovaries. The common gonoduct exists on the sixth thoracic limb. The egg sacs, carried on the first abdominal somite each contains only one embryo.

Development

The first free-living stage is a metanauplius with the usual complement of uniramous first antennae, biramus second antennae, mandibles, and with rudimentary maxillae. The second antennae are armed with the typical basal endite called the naupliar process. All stages are benthic.

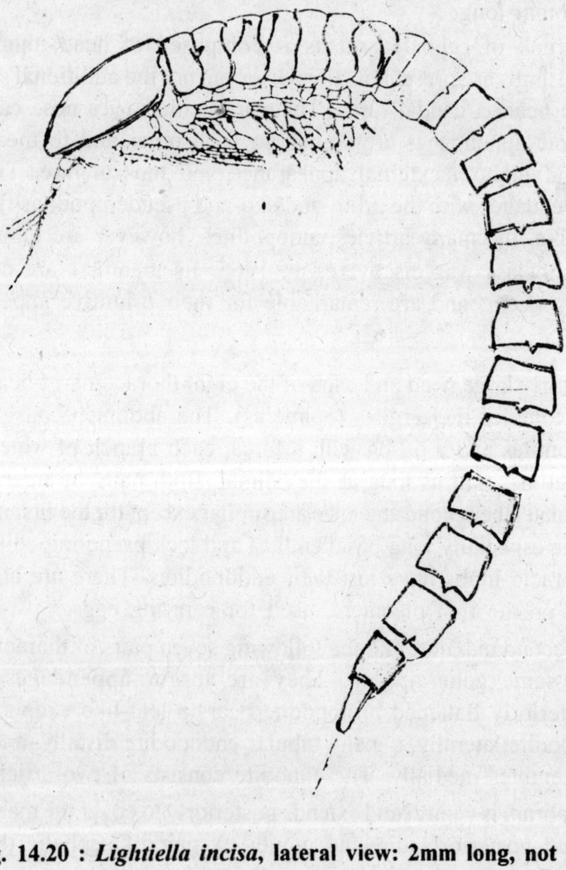

Fig. 14.20 : *Lightiella incisa*, lateral view: 2mm long, not including.

In the naupliar stages the animal uses the expodites of the second antennae and mandible to bring detritus to the mouth, and endites of the second antennae and mandibles push the particles under thc labrum.

Lepidocaris (Suborder *Lipostraca*), a minute, blind, freshwater form from the Middle Devonian, was closely related to the Anostraca which survive (*Euanostraca*), hut differed from them in the following, among other respects. It had biramus antennae which recall those of the Cladocera; a clasping organ on the maxillule of the male, instead of on the antenna; and the trunk limbs without branchiae and differentiated into two sets—the first three pairs adapted for gathering food, with gnathobase and with the last endite directed inwards and the exopodite lateral, and the remaining pairs adapted for swimming, with the last endite and the exopodite directed distally side by side at the end of the limb.

ORDER NOTOSTRACA

Branchiopoda with a carapace in the form of a broad shield above the trunk; the compound eyes sessile and close together; the antennules and antennae much reduced; the trunk limbs numerous, the first two pairs of them differing considerably from the rest; and slender, multiarticulate caudal rami.

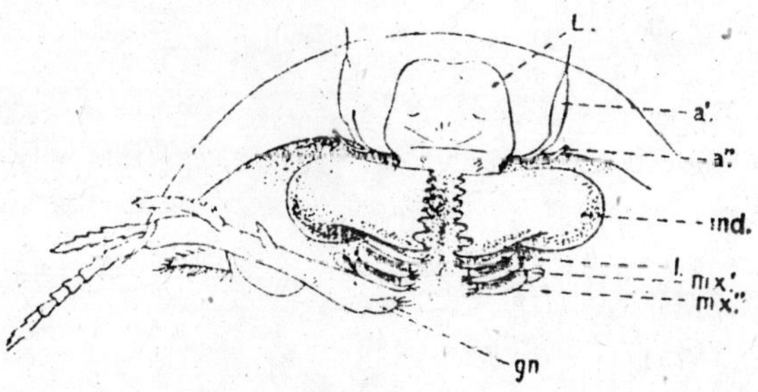

Fig. 14.21 : A ventral view of the head region of *Lepidurus glacialis*. From Calman, *a.*' antennule; *a.*" antenna; gn. gnathobase; *L.* labrum (turned forwards); *l.* paragnathum; *md.* mandible; *mx.*' maxillule; *mx.*" maxilla.

This order contains only the genera *Apus* and *Lepidurus*, which differ in but minor features. *Apus cancriformis* (Fig. 14.22) is British, but is now very rarely found in these islands. The head is broad and depressed, flat below and arched above, and forms with the carapace a horseshoe-shaped structure, which bears the eyes above and the small

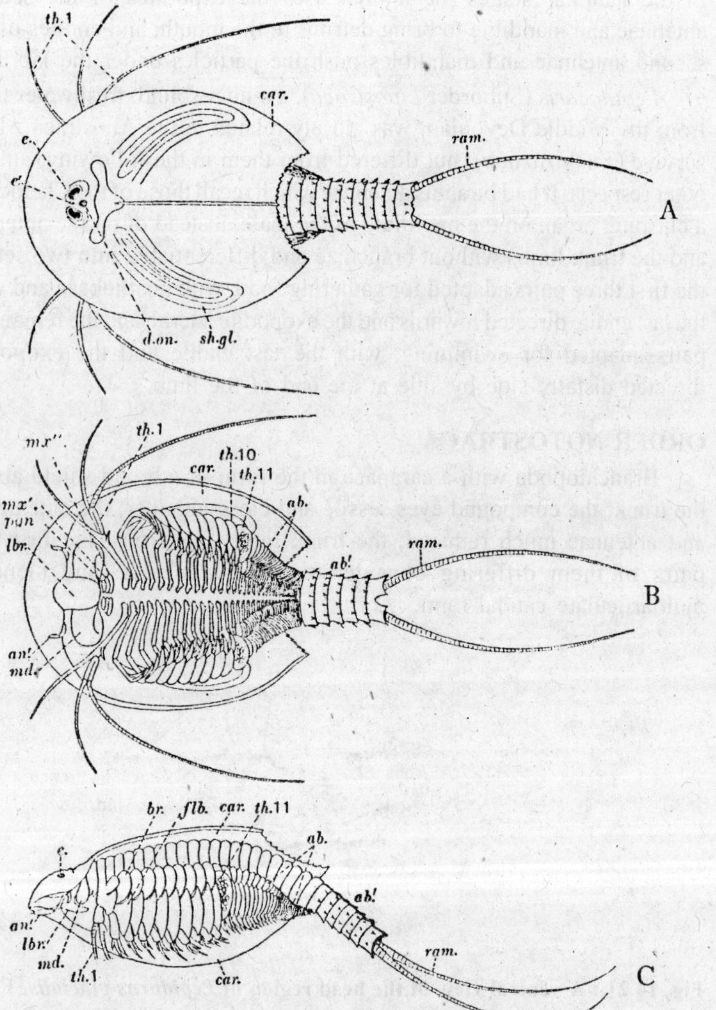

Fig. 14.22 : *Apus cancriformis*. A, Dorsal view. *car.* carapace; *d.on.* dorsal organ; *e.* compound eye; *e.*' median eye; ram. ramus of caudal fork; *sh.gl.* shell gland (maxillary gland) seen through the carapace; th. l, processes of the first thoracic limb. B, Ventral view. *ab.* abdominal limbs; *ab.*' limbless somites of the abdomen; *an.*' antennule; *car.* carapace; *lbr.* labrum; *md.* mandible; *mx.*' maxillule; *mx.*" maxilla; *pgn.* paragnathum; ram. ramus of caudal fork; *th. I*, first thoracic limb; *th.*10, tenth thoracic limb; *th.* 11, egg pouch on eleventh thoracic limb. C, Side view with the left half of the carapace cut away. *br.* branchia; *flb.* flagellum. Other letters as above.

antennules and antennae beneath, at some distance from the sharp front edge. There is a dorsal organ, which is not used for fixation, but no nuchal sense organ. From under the carapace the hinder part of the trunk projects backwards, ending in two long, jointed caudal rami. The genital opening is on the 11th of the trunk somites, Each of these bears a pair of limbs until the 13th (second of the abdomen) is reached, after which there are two to five pairs to a somite

REFERENCES

1. Invertebrate Zoology by Alferd Kertner.
2. The Invertebrates, Hyman.
3. Neurobiology of Invertebrates, Plenum.
4. Ecology of Fresh Water Invertebrates by Hart Fuller.